Lung Biology and Pathophysiology

The lungs are the organ for gas exchange between the body and the external environment. Dysfunction of upper airway epithelium and smooth muscle cells leads to pathogenesis of asthma, chronic obstructive pulmonary disease (COPD), cystic fibrosis, and other conditions, resulting in airway inflammation and narrowing. Injury to alveolar epithelium and endothelium causes influx of neutrophil and protein-rich fluid from circulation, resulting in edema and disruption of gas exchange. In addition to lung structural cells, immune cells including alveolar macrophages and lymphocytes play critical roles in the maintenance of lung function. This book contributes to the understanding homeostasis of lung cells in the physiological and pathological conditions critical to the development of novel therapeutics.

KEY FEATURES

- Highlights the role the lungs play as an interface between the body and the environment.
- Describes the underlying mechanism of lung diseases.
- Emphasizes the ways nutrition contributes to lung health and the ways pollution adversely affects lung function.
- Includes contributions from leading researchers.

Yutong Zhao is Professor of Physiology and Cell Biology at The Ohio State University. He received his clinical medical degree from Dalian Medical University in China and PhD degree from Gifu University School of Medicine in Japan. He was trained in biochemistry, molecular biology, and active lipid signaling during his PhD student period. He joined a pulmonary postdoctoral training program at Johns Hopkins University School of Medicine and received his first National Institutes of Health R01 funding at the University of Chicago as a Research Assistant Professor. Dr. Zhao is an expert in research fields of lung biology, bioactive lysophospholipid, and protein ubiquitination. His research focuses on investigating the pathogenesis of lung diseases including acute respiratory distress syndrome, pulmonary fibrosis, pulmonary arterial hypertension, asthma, lung cancer, and COPD. The goal of Dr. Zhao's laboratory research is to develop new therapeutic targets for treating lung diseases. Dr. Zhao is active in academic professional services as a reviewer, associate editor, and member of editorial boards of scientific journals and grant review study sections.

Methods in Signal Transduction

Series Editors
Michael X. Zhu and Harpreet Singh

The overall theme of this series continues to be the presentation of the wealth of up-to-date research methods applied to the many facets of signal transduction. Each volume is assembled by one or more editors who are pre-eminent in their specialty. In turn, the guiding principle for editors is to recruit chapter authors who will describe procedures and protocols with which they are intimately familiar in a reader-friendly format. The intent is to assure that each volume will be of maximum practical value to a broad audience, including students and researchers just entering an area, as well as seasoned investigators.

Lipid-Mediated Signaling Transduction, Second Edition
Eric Murphy, Thad Rosenberger, and Mikhail Golovko

Calcium Entry Channels in Non-Excitable Cells
Juliusz Ashot Kozak and James W. Putney, Jr.

Autophagy and Signaling
Esther Wong

Signal Transduction and Smooth Muscle
Mohamed Trebak and Scott Earley

Polycystic Kidney Disease
Jinghua Hu and Yong Yu

New Techniques for Studying Biomembranes
Qiu-Xing Jiang

Ion and Molecule Transport in Lysosomes
Bruno Gasnier and Michael X. Zhu

Neuron Signaling in Metabolic Regulation
Qingchu Tong

Non-Classical Ion Channels in the Nervous System
Tian-Le Xu, Long-Ju Wu

Immune Cells, Inflammation and Cardiovascular Diseases
Shyam S. Bansal

Opsin-Free Optogenetics: Technology and Applications
Yubin Zhou and Kai Zhang

Lung Biology and Pathophysiology
Yutong Zhao

For more information about this series, please visit: https://www.routledge.com/Methods-in-Signal-Transduction-Series/book-series/CRCMETSIGTRA

Lung Biology and Pathophysiology

Edited by Yutong Zhao

CRC Press
Taylor & Francis Group
Boca Raton London New York

CRC Press is an imprint of the
Taylor & Francis Group, an **informa** business

Designed cover image: Shutterstock Contributor: VectorMine

First edition published 2024
by CRC Press
2385 NW Executive Center Drive, Suite 320, Boca Raton FL 33431

and by CRC Press
4 Park Square, Milton Park, Abingdon, Oxon, OX14 4RN

CRC Press is an imprint of Taylor & Francis Group, LLC

ISBN: 978-1-032-40902-3 (hbk)
ISBN: 978-1-032-40906-1 (pbk)
ISBN: 978-1-003-35524-3 (ebk)

DOI: 10.1201/9781003355243

Typeset in Times
by Apex CoVantage, LLC

The Open Access version of chapters 8 and 13 was funded by The Ohio State University.

Contents

PART IV Traditional Chinese Medicine in Lung Diseases

Preface

The lungs are the organs responsible for gas exchange between the body and the external environment. Respiration provides oxygen for the body while eliminating carbon dioxide. Multiple cell types participate in gas exchange and maintenance of lung homeostasis. Dysfunction of upper airway epithelium can lead to several diseases including asthma, chronic obstructive pulmonary disease, and cystic fibrosis, resulting in airway inflammation and narrowing. Injury to alveoli causes an influx of neutrophil and protein-rich fluid from circulation, resulting in edema and disruption of gas exchange. Acute respiratory distress syndrome (ARDS) can be caused by direct pathogen infections, such as SARS-CoV-2–induced COVID-19, or by systemic infection, such as sepsis. Lungs have the ability to self-repair and remodel after injury; however, excessive lung repair and remodeling may cause differentiation of lung fibroblasts into myofibroblasts and accumulation of extracellular matrix, eventually resulting in lung fibrosis. Pulmonary vascular dysfunction contributes to the development of ARDS, pulmonary fibrosis, and pulmonary arterial hypertension. In addition to lung structural cells, immune cells, including alveolar macrophages and lymphocytes, play critical roles in the maintenance of lung function. Understanding homeostasis of lung cells in physiological and pathological conditions is critical to the development of novel therapeutics to treat lung diseases.

This book, consisting of four parts, is the result of the combined efforts of a distinguished group of investigators in the fields of lung biology and pulmonary diseases. Part I describes the most recent discoveries in antimicrobial peptides, lung epithelium, environmental factors, epigenetic modification, and hypoxia-induced factors in lung health and pulmonary diseases. Part II introduces methods and research strategies in lung organoids, animal models of pulmonary fibrosis, protein stability assays, and lipid oxidation. Part III focuses on pathogenesis and therapies of COVID-19. Part IV reviews the use of traditional Chinese medicine in pulmonary diseases. Each chapter in this book describes the lung biology and pathogenesis of pulmonary diseases and raises potential research questions that need to be addressed in the future.

Yutong Zhao

Acknowledgments

I appreciate all the contributors for their considerable enthusiasm, time, and effort to complete their chapters. I would like to thank Sarah Taleb and Patrick J. Oliver for assisting with the preparation of the book.

Contributors

Joseph S. Bednash
The Ohio State University
Columbus, Ohio

Bill B. Chen
University of Pittsburgh
Pittsburgh, Pennsylvania

Mo Chen
Nankai University
Tianjin, China

Samuel J. Cochran
The Ohio State University
Wexner Medical Center
Columbus, Ohio

Elliott D. Crouser
The Ohio State University
Wexner Medical Center
Columbus, Ohio

Zhiyu Dai
University of Arizona
Tucson, Arizona

Yuanpu Peter Di
University of Pittsburgh
Pittsburgh, Pennsylvania

Austin C. Elliott
The Ohio State University
Wexner Medical Center
Columbus, Ohio

Moemen Eltobgy
The Ohio State University
Columbus, Ohio

Kymberly M. Gowdy
The Ohio State University
Wexner Medical Center
Columbus, Ohio

Mingxia Gu
University of Cincinnati
Cincinnati, Ohio

SeungHye Han
Northwestern University
Chicago, Illinois

Thomas J. Hund
The Ohio State University
Wexner Medical Center
Columbus, Ohio

Hongpeng Jia
Johns Hopkins University
School of Medicine
Baltimore, Maryland

Finny Johns
The Ohio State University
Columbus, Ohio

Lasya Kopuri
The Ohio State University
Wexner Medical Center
Columbus, Ohio

Jenna M. Kuhn
University of Pittsburgh
Pittsburgh, Pennsylvania

Travis B. Lear
University of Pittsburgh
Pittsburgh, Pennsylvania

Lian Li
Nankai University
Tianjin, China

Qing Lin
Johns Hopkins University
School of Medicine
Baltimore, Maryland

Yuan Liu
University of Pittsburgh
Pittsburgh, Pennsylvania

Krishna Rao Maddipati
Wayne State University
Detroit, Michigan

Surya Narla
The Ohio State University
Wexner Medical Center
Columbus, Ohio

Patrick J. Oliver
The Ohio State University
Wexner Medical Center
Columbus, Ohio

Narasimham L. Parinandi
The Ohio State University
Wexner Medical Center
Columbus, Ohio

Nicole Pek
University of Cincinnati
Cincinnati, Ohio

Yan Y. Sanders
Eastern Virginia Medical School
Norfolk, Virginia

Brett Saunders
The Ohio State University
Wexner Medical Center
Columbus, Ohio

Evangeline Schott
The Ohio State University
Wexner Medical Center
Columbus, Ohio

Xiaohong Shu
Dalian Medical University
Liaoning, China

Maolin Yao
Nankai University
Tianjin, China

Anita Yenigalla
The Ohio State University
Wexner Medical Center
Columbus, Ohio

Hong Zhang
Dalian Medical University
Liaoning, China

Baochang Zhao
Dalian Medical University
Liaoning, China

Yutong Zhao
The Ohio State University
Columbus, Ohio

Part I

Lung Health and Pulmonary Diseases

1 Antimicrobial Peptides in Lung Health and Pulmonary Diseases

Jenna M. Kuhn and Yuanpu Peter Di

1.1 INTRODUCTION

Pulmonary disease remains a critical threat to human health globally. In the United States in 2020 and 2021, lung cancer, severe acute respiratory syndrome coronavirus 2 (SARS-CoV-2), and chronic lower respiratory diseases were the second, third, and sixth leading causes of death, respectively (1). In 2020, lung cancer was attributed as the leading source of cancer mortality in the United States, responsible for over 136,000 deaths (2). These threats to pulmonary health have fueled a vast expansion of research dedicated to a group of innate antimicrobial peptides (AMPs) and their roles in immune defense against respiratory pathogens and immunomodulation of pulmonary disease. AMPs, also known as host defense peptides (HDPs), are oligopeptides produced and released by immune and epithelial cells in response to pathogenic infection but also have roles in regulating disease progression and even tumor formation (3–5). In this chapter, we will discuss the activities of AMPs in the pathogenesis of respiratory infection, the roles of these agents in immunomodulation, and the promising potential of novel AMP designs for treating and preventing pulmonary diseases, including lung cancer.

1.2 OVERVIEW OF INFECTIOUS PULMONARY DISEASE

Pulmonary diseases are pathological conditions that impair respiratory function and overall lung health. These diseases or disorders may develop through infection by pathogens, including bacteria, viruses, and fungi, or exposure to environmental chemicals. The pathogenesis and overall health impact of various pulmonary diseases or disorders are complex based on the diverse lung anatomy. The US Centers for Disease Control and Prevention has recently reported that chronic respiratory diseases, including chronic obstructive pulmonary disease (COPD), emphysema, and chronic bronchitis, were the sixth-ranked leading cause of death in the United States in 2021 (1). The scope of this chapter will focus on pathogenic lung infection, the innate immune response to infection by lung AMPs, and the inflammatory response associated with AMP activity.

Until recently, the lung has been considered a sterile environment (6). It is now agreed that many microbial organisms, including bacteria, fungi, and viruses, exist in the lung environment, collectively known as the lung microbiome. The healthy lung microbiome includes upwards of 100 different taxa, and the airways are vulnerable to exposure to diverse microorganisms in the environment simply through daily inhalation of approximately 10,000 L of air each day for the average sedentary adult (7–9). In addition, the diversity and abundance of various taxa of the lung microbiota differ in healthy and diseased individuals, suggesting that these microorganisms may influence the pathogenesis of lung infections and health outcomes (10–12). Due to the significant morbidity and mortality of respiratory infections globally, research progression has focused on the mechanisms of innate immune defense against pathogens as well as advancements in novel therapeutic options to eradicate these infections.

DOI: 10.1201/9781003355243-2

1.2.1 Burden of Respiratory Infections

Infections of the upper respiratory tract are some of the most prevalent diseases globally, with a total of over 17 billion incident cases occurring in 2019 (13). Upper respiratory infections (URIs) include laryngitis, sinusitis, pharyngitis, nasopharyngitis (common cold), and rhinitis, which often present with coughing, headache, sore throat, and nasal congestion (14–16). Global incident cases of URIs were on the rise from 1990 to 2019, with the highest age-standardized incidence and disability-adjusted life year rates occurring in high sociodemographic index regions (13). URIs are often caused by viral agents, most often from the families of adenovirus, influenza A/B/C, coronavirus, parainfluenza, respiratory syncytial virus (RSV), and rhinovirus (15). Due to the colonization of bacteria in the nasopharyngeal region of the upper respiratory tract, there is potential for these host microbes to invade and induce infection within the lower respiratory tract (17). Respiratory bacterial-viral coinfections are especially problematic with amplified disease burden. *Streptococcus pneumoniae* is a major causal agent for pneumonia in young children, and coinfection with viruses such as RSV or influenza may lead to both synergistic and antagonistic immune responses, directly impacting disease severity (18). For instance, coinfection of mice with pneumococcal-murine pneumonia virus (the murine equivalent to RSV) led to elevated inflammatory cytokines, necrosis of olfactory sensory neurons and airway epithelium, and enhanced pneumococcal nasopharyngeal density (18). Therefore, stimulated inflammation and modified immune response triggered by an initial viral infection likely contribute to bacteria proliferation and improved bacteria adherence (18).

Lower respiratory tract infections (LRIs) are associated with high mortality rates, especially in children, the elderly, and immunocompromised populations. The global incidence of LRIs has been estimated to be almost 500 million cases, contributing to over 2 million deaths (19). However, there has been a decline in global incidence and mortality rates of LRIs since 1990, regardless of age or sex, mainly due to targeted interventions, including vaccinations (19). Acute bronchiolitis and community-acquired pneumonia are dominant sources of LRI burden and a leading cause of morbidity and mortality in children under the age of 5. Bronchiolitis is presented with tachypnea and wheezing, whereas symptoms of severe pneumonia include fever, cough, difficulty breathing, and tachypnoea (20, 21). The predominant causal agent for bronchiolitis is RSV, among other viruses that may occur as coinfections such as human metapneumovirus, adenovirus, influenza, rhinovirus, and human coronaviruses (20). About one-third of all cases of community-acquired pneumonia involve a dual infection of virus and bacteria, with *S. pneumoniae* as the major causal agent, followed by *Haemophilus influenzae* type b, *Mycoplasma pneumoniae*, and *Staphylococcus aureus* (21). Increases in antibiotic resistance as a global public health threat, influencing the rate and severity of these respiratory infections, calls for further investigation of the innate defense mechanisms and alternative therapeutics strategies necessary to fight these pathogens.

1.2.2 Pathogenesis of Lung Infections

1.2.2.1 Bacterial Infection

The pathogenesis of acute respiratory infections by bacteria involves stages of initiation of inflammation, amplification of inflammatory cells, and (potential) resolution of the infection. Initiation consists of the activation of the host inflammatory response, arising from sufficient microbial exposure necessary to overcome clearance by the host immunity (22). At this stage, the epithelial cells are first stimulated to release chemokines and specific proinflammatory cytokines to initiate immune responses, including the recruitment of leukocytes. Next, amplification involves neutrophil recruitment from the bloodstream to the lung apical surface through neutrophil transepithelial migration and heightened immune response activation, leading to elevated chemoattractant concentration in the airways, including interleukin (IL)-8, leukotriene B4, and tumor necrosis factor α (TNF-α) (23, 24). The increase in neutrophil activation leads to further phagocytosis of bacteria and the release of additional proinflammatory cytokines. Finally, in the resolution stage, adequate phagocytosis of microbes occurs, which leads to a reduction of inflammatory markers and inactivation of elastase produced from activated neutrophils (22). The decrease of elastase prevents further damage to the lung epithelium and host defense factors (22). Bacteria may colonize the lung environment when complete resolution cannot occur as a result of inadequate or dysfunctional host defenses. For instance, a physiologically acidic environment in cystic fibrosis (CF) airways may promote chronic *Pseudomonas aeruginosa* colonization and biofilm formation (25). In chronic lung infection, there is a continuous and heightened immune response and, in many cases, the production of bacteria biofilms. Bacterial biofilms are complex clusters of bacteria attached to a surface or each other by polymer matrixes they produce. Chronic infections involving biofilms reduce antimicrobial efficacy, even for the most updated antibiotics, and an increase in evasion of host defenses (26). Those with weakened innate immunity, such as CF patients, are more susceptible to chronic infections. Another pathological condition may be the interference of bacteria on mucociliary clearance (MCC), potentially causing further damage to the epithelium and excess mucus secretion (22). In CF, there is reduced MCC, increased mucous production, and persistent infection, causing damage to the lung and other organ tissues (26).

1.2.2.2 Viral Infections

Respiratory viral infections, a significant cause of mortality, such as influenza A and B, contribute to millions of clinical infectious diseases and almost 500,000 deaths yearly (27). Various viruses have a tropism for a particular tissue of the host. For instance, seasonal influenza viruses have a dominant tropism and bind to ciliated epithelial cells of the upper respiratory tract (URT), trachea, and pulmonary bronchi. In contrast, the avian virus H5N1 shows high tropism for the lower respiratory tract (LRT), especially the bronchioles and alveoli (27). The pandemic H1N1 virus has shown strong tropism for both the URT and LRT, primarily infecting the alveoli's type I and II pneumocytes, contributing to significant diffuse alveolar damage (28). COPD is one of the leading causes of morbidity and mortality globally. The inhalation of harmful gases from occupational exposure, air pollution, and primarily cigarette smoke are the main contributing factors to COPD. Those with COPD show heightened inflammatory response in the airway epithelium, with increased neutrophil production and disruption of adaptive immunity in chronic inflammation (29, 30). Airway obstruction occurs through the accumulation of leukocytes such as neutrophils, macrophages, and lymphocytes, as well as inflammatory exudates in the small airways and thickening of the airway walls due to impaired host defense and the tissue repair/remodeling process (29). When infected with the virus responsible for causing the 2019 coronavirus disease, SARS-CoV-2, those patients with COPD experience significantly worse clinical outcomes, including severe pneumonia (31). The expression of the angiotensin-converting enzyme 2 receptor, the primary target of the SARS-CoV-2 envelope spike protein necessary for fusion and intracellular entry of the virus, is significantly increased in COPD patients (32). Therefore, healthy airways and nonsmoking behaviors may influence the incidence and outcomes of respiratory viruses.

1.2.2.3 Coinfection

The coinfection of a primary viral infection with bacteria in the lung largely contributes to the complexity and severity of the illness. The virulence factors expressed by viruses may facilitate bacterial infection in the host. Some of these virulence factors lead to modifications in lung physiology, increases in respiratory epithelial sloughing, improved availability for bacterial adherence, modifications in viral tropism from the URT and LRT, and increased inflammation (33). The co-pathogenesis of secondary bacterial infection is strongly linked to the proinflammatory response of viruses, allowing for the invasion of highly pathogenic bacteria, leading to deadly pneumonia in many cases (33). For example, the effects of type I interferon on type 17 immune suppression during influenza A virus (IAV) infection contribute to increased host susceptibility to secondary Gram-positive and

Gram-negative bacterial infection (34, 35). More specifically, IAV-bacterial coinfection led to a significant increase in inflammation and enhanced lung damage compared to bacterial infection alone, likely due to delayed bacterial clearance achieved through impaired host immunity (35). Importantly, IAV-associated inhibition of type 17 immunity and reduced expression of AMPs bacterial permeability family member A1, otherwise known as short palate, lung, and nasal epithelium clone 1 (SPLUNC1) and lipocalin-2, during coinfection of IAV and *P. aeruginosa* were determined to contribute to a reduction in bacterial clearance, without a deficiency in inflammatory cell recruitment in response to the lung infection (35). The airway epithelial secretory cells secrete SPLUNC1 as a defense mechanism against Gram-negative bacteria infections, whereas lipocalin-2 has roles in host mucosal defense against both Gram-positive and Gram-negative bacterial species (36–39).

1.2.3 Roles of Lung Infections on Exacerbations

Respiratory infections by both bacteria and viruses contribute to acute exacerbations in patients with chronic respiratory conditions, especially COPD and asthma. Acute exacerbations are characterized by a sudden increase in disease severity, reduced lung function, and decreased quality of life. Exacerbations are the primary cause of morbidity and mortality in chronic lung disease patients. It has been shown that the URT microflora differs in healthy individuals compared to those with COPD (40). The composition of the pulmonary microbiota also changes throughout COPD progression (40). *H. influenzae, S. pneumonia*, and *Moraxella catarrhalis* are the predominant microbes cultured from COPD patients (40). However, with increased disease progression leading to more severe airway obstruction and during exacerbations, *P. aeruginosa* and other Gram-negative bacterial species are more frequently identified (41, 42). Additionally, compared with stable disease, the frequency of exacerbations in COPD is likely associated with infection and isolation of a new bacterial strain (40). Recent studies have shown that respiratory viral infections, predominantly rhinoviruses, have been detected in over half of all COPD exacerbations (40).

1.3 INTRODUCTION TO AMPs

AMPs, also known as HDPs, are a group of evolutionarily conserved molecules that are part of the innate host defense mechanism of virtually all kingdoms of life. These peptides have integral roles in the host's response to infection and inflammation. Sources of AMPs range from bacteriophages, bacteria (both Gram-positive and Gram-negative), fungi, plants, invertebrates, fish, amphibians, reptiles, birds, and mammals (3). Gramicidin is considered to be the first AMP discovered in 1939, derived from the soil aerobic sporulating bacillus species *Bacillus brevis*, with bactericidal activity against some Gram-positive bacteria (43). Since this first discovery, the number of AMPs has risen exponentially. Naturally derived AMPs are usually short (10–100 amino acid residues), cationic amphipathic peptides that bind to the membrane of negatively charged bacteria through electrostatic interactions (3, 4). The mechanism of action of AMPs is highly complex due to their diverse structural and physiochemical properties. Ultimately, these peptides can be characterized based on their target organism or region, which may include viruses, fungi, the bacterial cell wall and/or cell membrane, and intracellular regions or cellular functions of bacteria (3). This chapter will focus on the function and activity of lung-associated AMPs.

1.4 AMPs IN THE LUNG ENVIRONMENT

During normal respiration, the lung serves as a critical environment for immune defense through MCC of microbes, chemokine and cytokine production, and the release of numerous AMPs (44). Antimicrobial molecules of the lung (e.g., defensins, cathelicidins, PLUNC [palate, lung, and nasal epithelium clone] family proteins, lysozyme, secretory leukocyte protease inhibitor, lactoferrin, collectins, elafin, and mucins) are predominantly produced by airway epithelial cells and released into the airway lumen (45). In addition, myeloid cells, including neutrophils and macrophages, are essential in AMP production in the lung environment (45). The most abundant AMPs observed in lung tissue and airway secretions are neutrophil α-defensins, or human neutrophil peptides; human β-defensins, and the cathelicidin hCAP-18, or LL-37 (46). Defensin families include the α- and β-defensins, based on the structural pairing of six cysteine residues within the formation of three disulfide bridges. α-Defensin peptides are highly cationic, about 29–35 amino acid residues in length, and have shown broad-spectrum antimicrobial activity against Gram-negative and Gram-positive bacteria, viruses, fungi, and mycobacteria (47–51). β-Defensins are also highly cationic, about 36–42 amino acid residues in length, and have shown broad-spectrum antimicrobial activity against both Gram-positive and Gram-negative bacteria, fungi, and mycobacteria (49). Human β-defensins are the most well-studied of the β-defensins due to their antimicrobial activity, immunomodulation, and modulation of tissue repair (52).

Cathelicidins are α-helical cationic peptides synthesized by proteolytic processing of the precursor hCAP18, which possesses a cathelin-like domain attached to a carboxy-terminal active peptide (52). The only identified cathelicidin in humans is LL-37, which is released after proteolytic processing by proteinase 3 (53). Human LL-37 has a high range of antimicrobial activity against both Gram-positive and Gram-negative bacteria, may aid in neutralizing proinflammatory response, increases proinflammatory gene expression, and is important in epithelial integrity, angiogenesis, and immunomodulation (52, 54).

Several innate host defense antimicrobial proteins exist in the mammalian airways, including the bactericidal/permeability increasing protein (BPI) and the lipopolysaccharide-binding protein (LBP). Both LBP and BPI bind to the lipopolysaccharide (LPS) of bacterial cell membranes and elicit either a stimulatory cellular response or a neutralizing, inhibitory effect, respectively (55). BPI has also shown bactericidal activity against Gram-negative bacteria (56). Therefore, because of their antagonistic response, a balance of BPI and LBP activity is essential to innate host defense against pathogens. In the upper respiratory tract, PLUNC family proteins are abundantly secreted by the airway epithelia and share sequence homology to BPI and LBP (57). As an airway surfactant, SPLUNC1 modulates the surface tension of secretions of the airway epithelium and has been shown to inhibit the biofilm formation of *Klebsiella pneumoniae* and *P. aeruginosa* without needing a significant bactericidal effect (39, 58). In addition, generated transgenic mice carrying human-derived PLUNC (hPLUNC) directed to the airway epithelium showed significantly improved survival and resistance to one of the most critical respiratory pathogens of CF infection, *P. aeruginosa* (59). Following *P. aeruginosa* challenge, transgenic mice carrying human-derived PLUNC showed increased bacteria clearance and reduction in proinflammatory cytokines IL-1β, TNF-α, and IL-6 (59). The surfactant activity of SPLUNC1 may be related to the inhibition of early stages of biofilm formation, including attachment and micro-colonization (58). SPLUNC1 is a critical factor in mucosal immunity through establishing and maintaining a reduction in airway surface liquid tension and, by this mechanism, may enhance MCC upon exposure to environmental microbes (39).

1.5 AMPs AND HOST DEFENSE

1.5.1 AMP Modes of Action

When the lung epithelium is exposed to different stimuli, such as microbes and their products, the host will secret a panel of factors, including Pattern Recognition Receptors (PRRs), cytokines, or growth factors, which modulate the expression and release of natural AMPs such as defensins and cathelicidins (60). In general, AMPs have two main bioactivities: antimicrobial activity and modulation of the adaptive and innate immune response (61). There is a series of different modes of action by which AMPs defend against pathogens: membrane permeabilization, membrane pore formation, inhibition of cell wall synthesis, inhibition of DNA and protein synthesis, and impairment of intracellular signaling cascades (61). AMPs may selectively bind to essential lipids of peptidoglycan, leading to the inhibition of cell wall synthesis and destruction of cell wall integrity (62). There are two main ways that AMPs target the cell membrane. The transmembrane pore model involves the formation of oligomers with peptide monomers, allowing AMPs to be inserted within the hydrophobic core of the cell membrane based on a threshold concentration (62). In addition, AMPs may be inserted vertically into the membrane bilayer, altering the directionality of membrane phospholipids to form pores (62). The nonmembrane pore model includes two models where AMPs may form a membrane "carpet," or show detergent-like activity. AMPs are attracted to the negatively charged, acidic lipids of the outer membrane of bacteria and form a "carpet" over its surface (63, 64). At high concentrations above a threshold, AMPs alter the bending of lipids and increase membrane fluidity, leading to the formation of micelles and cell lysis (63, 64).

Contrary to previous thought that AMPs required membrane pores or peptide channels to reach the cytoplasm, the mode of entry for AMPs may include direct penetration or endocytosis of the cell membrane (65). Some peptides may directly permeate the membrane through instantaneous pore formation, including cell-penetrating peptides. Another mode of intracellular entry for AMPs is endocytosis through a membrane receptor-mediate (62). Following cellular entry, AMPs may target critical physiological processes, including cellular replication, disrupt the cell cycle, alter DNA transcription and translation, or degrade organelles (62, 66–68).

1.5.2 AMPs against Bacterial Pathogens

The cationic charge and amphipathic properties of most AMPs facilitate their interaction with the negatively charged membrane of prokaryotes selectively, in contrast to the neutrally charged host cells (Figure 1.1) (69, 70). Many studies have described the in vitro activity of AMPs in chronic lung infections. Human β-defensin 3 and LL-37 showed antibacterial activity against clinical isolates frequently associated with CF, including *P. aeruginosa, S. aureus, Streptococcus* spp., *Achromobacter* spp., and *Stenotrophomonas maltophilia*, even when these isolates were resistant to the standard CF antibiotics ciprofloxacin, tobramycin, and meropenem (71). It has been shown that human β-defensin 2 activity decreases *Lactobacillus* ssp. bacterial biofilm survival at micromolar concentrations (72). Furthermore, a more recent study showed that human β-defensin 2 administered at nanomolar levels successfully reduced the biofilm production of both *P. aeruginosa* and *Acinetobacter baumannii* without impairing the metabolic activity of these strains (73).

Based on the broad-spectrum antibacterial activity of innate AMPs, novel peptides are under development for therapeutic potential. Two synthetic peptides, referred to as P5 and P6.2, showed in vitro antibiofilm activity against both *P. aeruginosa* and *S. aureus* and reduced the production of proinflammatory cytokines and bacteria load in the lungs of neutropenic mice (74). More novel synthetic antimicrobial and antibiofilm peptides (SAAPs) have shown efficient bactericidal activity, even for drug-resistant, established infections (75). SAAP-148 treatment was effective in eliminating methicillin-resistant *S. aureus* as well as multidrug-resistant (MDR) *A. baumannii* biofilm-associated infections and persisting cells without enhanced drug resistance (75). In addition, a novel AMP α4-short, sequentially optimized from respiratory host defense SPLUNC1 protein, showed significant efficacy in an in vivo murine model of *P. aeruginosa* respiratory infection, with minimal hemolysis and leukocyte toxicity (76). α4-Short was also found to have bactericidal and antibiofilm activity against three *P. aeruginosa* clinical MDR strains, even at a low concentration of 2 μM (76). These research findings support further structural advancement of SPLUNC1-derived AMPs for eradicating MDR, complicated biofilm-associated respiratory infections with lowered host cell toxicity.

Although AMPs serve as significant elements of lung immune defense, AMP activity may be impaired for some bacteria species in patients with COPD, cystic fibrosis, and asthma due to the increased expression of serine proteases, matrix metalloproteases, and cysteine cathepsins associated with these chronic respiratory diseases (44, 60). Advances in novel peptide engineering have improved the therapeutic efficacy and stability of AMPs. For instance, a lead of engineered cationic AMP (eCAP), WLBU2, was modified by generating a D-Val for L-Val substitution, which led to decreased protease degradation, improved efficacy against biofilm growth, and lowered erythrocyte

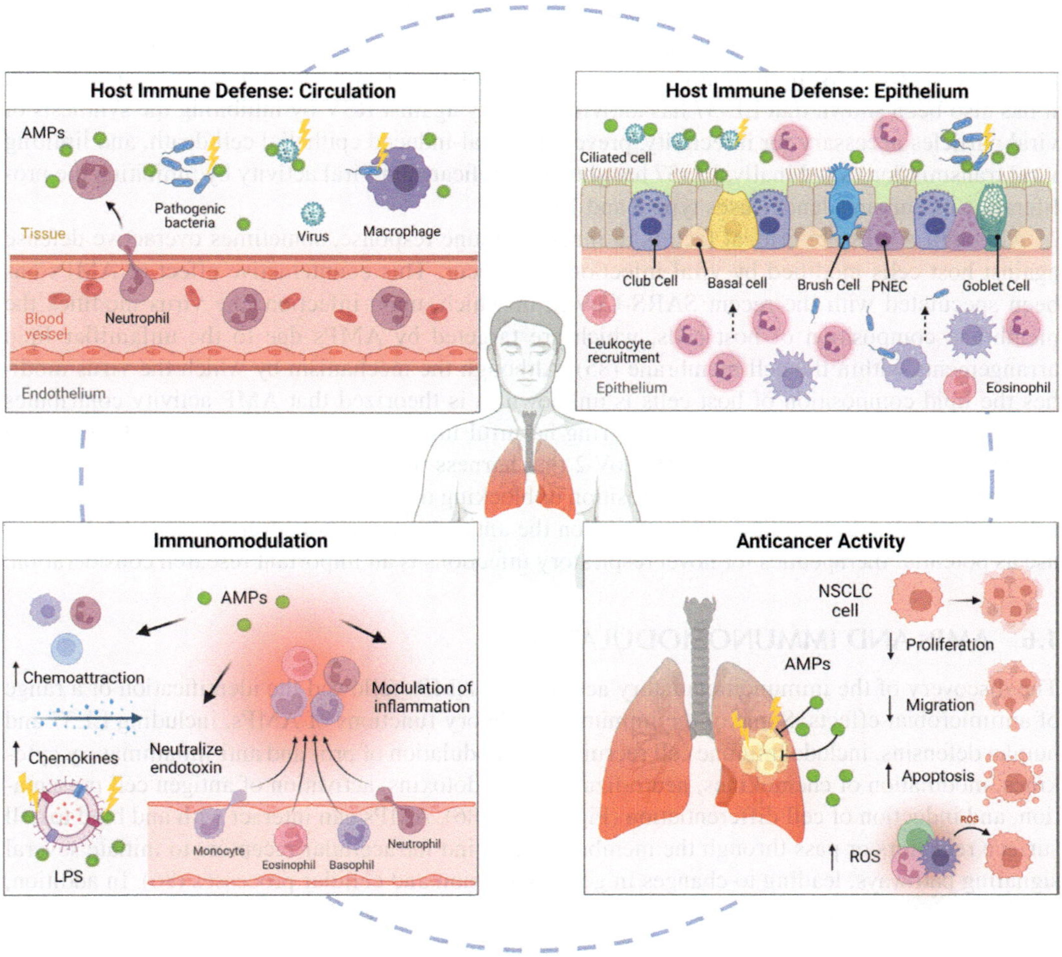

FIGURE 1.1 Interconnected roles of AMPs in host defense. AMPs directly enact antimicrobial activity against environmental pathogens within the circulation and respiratory endothelium and epithelium. Through immune cell recruitment, modulation of cytokines and chemokine signaling molecules, and modulation of inflammation, AMPs have critical roles in immunomodulation. Future directions of AMPs aim to utilize anticancer properties against NSCLC cells by inhibiting cell proliferation and migration while increasing apoptosis and production of ROS. LPS, lipopolysaccharide; PNEC, pulmonary neuroendocrine cells. Created with BioRender.com.

and leukocyte toxicity in mice infected with MDR pathogens *P. aeruginosa* and *A. baumannii* (77). Thus, D-enantiomerization of synthetic peptides is one example of how structural improvements in eCAP design can enhance drug stability and also increase the therapeutic window and bioavailability of these treatments for use in serious MDR bacterial respiratory infections (77).

1.5.3 AMPs against Viral Pathogens

Both human defensins and LL37 have shown activity against both enveloped and non-enveloped viruses (78, 79). Some of the targets of defensins are viral envelopes, glycoproteins, and capsids (78). Defensins enact antiviral activity by disrupting viral fusion with host cells, preventing neutralization after intracellular entry, and inhibiting viral replication (78). Through their ability to mobilize leukocytes as chemokines, defensins may also alter the immune response to viral infection (78). A group of α-defensins known as human neutrophil peptides may stimulate viral aggregation and inhibit the infectivity of IAV (80). β-defensins may also indirectly inhibit IAV through immunomodulatory activity (80). Similarly, it is suggested that LL-37 may interact and bind with IAV through neutrophil extracellular traps in response to infection (81). LL-37 mediates innate immunity against IAV by inhibiting viral replication and activating inflammatory response (82). It has also been shown that LL-37 has antiviral activity against RSV by inhibiting the synthesis of viral particles necessary for infectivity, preventing viral-induced epithelial cell death, and limiting viral transmission (83). Finally, LL-37 has shown significant antiviral activity by inhibiting the proliferation of human adenoviruses type 8 and 19 (84).

Although AMPs are critical factors in innate immune response, sometimes overactive defense against host cells modified by viral infection may occur. This counteractive effect of AMPs has been speculated with the recent SARS-CoV-2, in which, upon infection, the virus modifies the membrane composition of host cells, which are targeted by AMPs due to the unfamiliar lipid arrangements within the cell membrane (85). Although the mechanism by which the virus modifies the lipid composition of host cells is unknown, it is theorized that AMP activity contributes to an overactive immune response, triggering harmful inflammation (85). It is possible to design alternative therapeutics against SARS-CoV-2 that harness the beneficial activity of AMPs against pathogens through restoring lipid composition or blocking the effect of AMPs from destroying host cells with modified membranes (85). Based on the antiviral activity of several innate AMPs, their use as potential therapeutics for novel respiratory infections is an important research consideration.

1.6 AMPs AND IMMUNOMODULATION

The discovery of the immunomodulatory activities of AMPs followed the identification of a range of antimicrobial effects. Some of the immunomodulatory functions of AMPs, including LL37 and human defensins, include immune cell recruitment, modulation of pro- and anti-inflammatory cytokines, modulation of chemokines, neutralization of endotoxins, activation of antigen cell presentation, and induction of cell differentiation (Figure 1.1) (86). AMPs can interact with and bind to cell surface receptors or pass through the membrane and bind intracellular receptors to initiate several signaling pathways, leading to changes in gene expression and cellular processes (86). In addition, AMPs are integral in both pro- and anti-inflammatory responses. Chemoattraction of leukocytes and activation of cytokine expression are two proinflammatory activities regulated by AMPs that aid in infection clearance (86). However, AMPs may suppress proinflammatory cytokines or stimulate the production of anti-inflammatory cytokines, demonstrating their complex role in inflammation regulation (87). Both defensins and LL-37 may recruit leukocytes through interaction with chemokine and formyl-peptide receptors (88). Furthermore, AMPs can regulate the function of neutrophils. For instance, LL-37 may stimulate the release of neutrophil extracellular traps and induce the production of various α-defensins to enhance neutrophil-associated antimicrobial activity (87). LL-37 has also been shown to modify dendritic cell differentiation, suggesting its potential therapeutic use for modulating T helper cell polarization (89). Furthermore, AMPs are involved in the activation of antigen-presenting cells, recruitment and polarization of lymphocytes, and induction of phagocytosis or a pro-inflammatory response to DNA/RNA, and have been shown to accelerate angiogenesis and wound healing (90). The balance of immunomodulatory effects by AMPs is critical for the activation, modulation, and clearance of pathogenic infection in the host.

1.7 BACTERIAL RESISTANCE MECHANISMS TO AMPs

Despite targeted mechanisms of bacteria killing, AMPs are not immune to the natural evolution of bacteria under exposure to numerous AMPs, which enhances the opportunity for adaptive resistance to these peptides (91). However, the push toward advances in AMP design offers promise as alternative therapeutics to antibiotics, for which bacteria have evolved resistance to almost all available treatments at a much faster and alarming pace (92). The following resistance mechanisms have been identified for some AMPs in certain Gram-positive and Gram-negative species, but the identified genetic alterations conferring resistance should not be assumed as standard for all AMPs or in all bacterial strains. Because of their importance as a last resort clinical therapy for the most severe MDR infections, polymyxins such as colistin (polymyxin E) have been broadly studied. Therefore, the timing and main mechanisms of colistin resistance have been identified, and the development of enhanced AMPs with reduced potential for resistance while overcoming multidrug (including colistin) resistance is underway (93).

1.7.1 Production of Proteases and "Trapping" Molecules

Bacteria have multiple strategies of defense against AMPs. First, bacteria may release extracellular molecules that bind, form complexes with, and inhibit the function of AMPs (94). For example, *S. aureus* can release the protein staphylokinase, which can bind to α-defensins to inhibit peptide bacteriostatic activity (95). In addition, the secretion of streptococcal inhibitors of complement protein and M serotype proteins by *Staphylococcus pyogenes* facilitates inhibition of LL-37 activity, promoting bacterial survival persistence in infection (96). Also, bacteria can secrete proteases that prevent AMPs from interacting with the cell membrane, thereby preventing any AMP effect on the pathogen (97). The metalloprotease aureolysin, synthesized by *S. aureus*, inhibits LL-37 through the hydrolysis of peptide bonds at the C-terminal domain (98). Additionally, *P. aeruginosa* inactivates LL-37 through the production of the protease elastase, which hydrolyzes Asn-Leu and Asp-Phe peptide bonds (99). It has also been shown that *P. aeruginosa* exposure to LL-37 stimulated an elevation in quorum sensing and virulence-associated compounds, including elastase, hydrogen cyanide, rhamnolipids, and phenazine, along with adaptive resistance to fluoroquinolone and aminoglycoside antibiotics (100). However, the structure and composition of AMPs facilitate improved host defense against bacterial resistance compared to standard-of-care antibiotic agents. For example, the structural diversity, amphipathic and predominant cationic properties, and cyclic structure of AMPs support enhanced efficacy against bacteria and stability as clinically significant therapeutics (92). AMPs primarily target the bacterial cell membrane and do not rely on specific metabolic pathways like many conventional antibiotics, allowing for the effective killing of all stages of bacteria life, even slow or non-growing forms (92). Also, under cationic AMP pressure, bacteria have a propensity to randomly acquire genetic alterations that reduce the electronegative charge at the cell membrane, which may, in turn, lead to decreased fitness and membrane integrity, lessening the opportunity for AMP resistance (92). Still, the ability of bacteria to readily adapt and regulate according to environmental stress greatly confounds methods of innate host defense. Further advancement of AMPs for the treatment of severe respiratory infections proves to be essential, especially with a substantial rise in both community and nosocomial MDR pathogens.

1.7.2 Capsule Production

The production of extracellular polysaccharide matrices, the main component of the bacteria capsule, provides potential defense against AMPs (94). Ultimately, the capsule prevents direct interaction of AMPs with the bacterial membrane, thereby facilitating resistance. Negatively charged molecules, such as the teichoic acids, peptidoglycan, and LPS, lead to a cumulative net negative charge at the cytoplasmic membrane of bacteria. This charge creates an affinity of cationic molecules, such as AMPs, to bind or interact with the cell membrane. Therefore, a decrease in the synthesis or modifications in the structure of negatively charged molecules at the cell envelope facilitates another mechanism of resistance to AMPs (94).

1.7.3 Cell Membrane Charge Neutralization

The mechanism of bacteria resistance to AMPs differs in Gram-negative versus Gram-positive bacteria due to the structural components of the cell membrane. The Gram-negative membrane contains an inner (cytoplasmic) and outer membrane composed of LPS surrounding a middle layer of peptidoglycan (cell wall). The Gram-positive membrane is composed of a cytoplasmic membrane surrounded by a complex layer of peptidoglycan containing glycolipid polymers. These polymers include teichoic acid and lipoteichoic acid that bind to peptidoglycan or the cytoplasmic membrane, respectively (101). LPS, lipids, peptidoglycan and teichoic acids contribute to the net negative charge of the bacterial membrane, which facilitates the electrostatic connections between cationic AMPs and the membrane (94). In Gram-negative species, modifications in the lipid A moiety of LPS constitute rapid resistance to many AMPs and the polymyxin class of antibiotics, which target and perturb the membrane of Gram-negative bacteria (102). Resistance to polymyxins has been observed in clinically important respiratory and intestinal pathogens such as *P. aeruginosa, A. baumannii, K. pneumoniae*, and *Escherichia coli* through modifications of LPS and LPS loss (103, 104). Modifications in lipid A include additions of aminoarabinose or phosphoethanol amine, which offset the negative charge of LPS, reducing the binding of AMPs and polymyxins. The pmrAB two-component system regulates the expression of genes involved in the synthesis of these molecules. In addition, the PhoPQ two-component system regulates the transcription of the pmrAB system (105). Therefore, mutations associated with these regulatory systems are important in generating resistance to AMPs (102, 105). Another mechanism of resistance to AMPs is facilitated through the loss of lipid A synthesis, regulated by genes *lpxA, lpxC*, and *lpxD*. Because negatively charged lipid A is a primary target for interactions with AMPs, including polymyxins, loss of this target has been shown to generate rapid polymyxin resistance in *A. baumannii* and potentially *K. pneumoniae* (93, 106).

For Gram-positive bacteria species, substitutions of teichoic acids by the positively charged amino acid D-alanine have been linked to increased AMP resistance (107, 108). D-Alanylation involves the operon *dltABCD*, and the inactivation of *dltA-D* genes facilitates the removal of D-alanine esters from teichoic acids. Inactivation of *dltA* in *S. aureus* and other Gram-positive species, including *S. pneumoniae* and *Streptococcus gordonii*, leads to increased susceptibility to defensins and polymyxins (108–110). Similar to the loss in negative charge due to D-alanine substitutions of teichoic acids, modifications in phosphatidylglycerol with positively charged lysine through MprF activity have been linked to increased AMP resistance in *S. aureus* clinical isolates (111).

1.7.4 Cell Membrane Fluidity

Fluctuations in cell membrane fluidity are associated with increased bacterial resistance to AMPs (94). PhoPQ-mediated regulation of *pagP* and *pagL* expression is responsible for increased acylation and diacylation of lipid A, respectively. The acylation of lipid A is directly related to membrane integrity, particularly membrane fluidity, and the ability of AMPs to permeabilize. Therefore, modifications of lipid A acylation lead to increased bacterial resistance to AMPs, as a result of enhanced membrane permeability (111–113). The length of unsaturated fatty acids in the cell membrane also influences membrane fluidity. Resistance to small cationic peptides, known as platelet microbicidal proteins, by *S. aureus* has been observed because of elevated levels of longer-chain unsaturated fatty acids and significantly enhanced membrane fluidity (114). Cell membrane charge neutralization and changes in membrane fluidity are critical mechanisms of bacterial resistance to AMPs.

1.7.5 Efflux Pumps

Mutations in genes associated with efflux pumps have been shown to contribute to AMP resistance in Gram-negative bacteria (115). For example, mutations in the AcrAB-TolC multidrug efflux pump were found to contribute to enhanced resistance to the AMP 4DK4L7 in *Salmonella enterica* serovar *typhimurium* under experimental evolution (116). Similarly, reduced susceptibility to LL-37 is modulated by the multiple transferable resistance efflux system in *Neisseria gonorrhoeae* (115). The efflux pump MexXY/OprM is linked to the pathway of lipid A modifications of LPS in *P. aeruginosa*, which is associated with acquired resistance to colistin (117). Furthermore, a novel plasmid co-harboring genes *tmexCD1-toprJ1* (encoding a resistance-nodulation-cell division efflux pump) and *mcr* identified in *K. pneumoniae* was found to confer resistance to both tigecycline and colistin (118). The expulsion of AMPs by various drug efflux systems harbored by some bacterial species continues to be an important area of research regarding the potential mechanisms of resistance to novel AMPs.

1.7.6 Biofilm Formation

An organized exopolysaccharide matrix produced by an aggregate environment, known as a biofilm, shields bacteria from environmental stressors. The components of the biofilm, especially secretions such as cellulose, nucleic acids, polysaccharides, and teichoic acids, may bind to and inhibit AMP activity by inducing structural modifications (96, 98). For instance, biofilms of *P. aeruginosa* secrete the polysaccharide alginate that aids in surface attachment, but may also bind with and introduce an α-helical conformation in cationic AMPs (119). Furthermore, polysaccharides synthesized by clinically important respiratory pathogens *P. aeruginosa, K. pneumoniae*, and *Burkholderia cepacian* species inhibited the activity of AMPs LL-37 and human β-defensin 3, which supports chronic CF infection persistence (120). Finally, access to negatively charged nucleic acids within the biofilm facilitates chelation with nearby cations and triggers the potential adaptive resistance to AMPs through PhoPQ and PmrAB-mediated LPS modifications in *P. aeruginosa* (121). Therefore, the materials that facilitate and maintain biofilm formation in some bacterial species also protect against invasion by AMPs.

1.8 FUTURE DIRECTIONS FOR AMP DEVELOPMENT

1.8.1 AMPs as Novel Drug Therapies for Pulmonary Diseases

There is vast potential for novel synthetic AMPs to be used as new therapies for infectious and non-infectious lung diseases, with improved efficacy and lowered propensity for resistance and host toxicity. Although there has been a substantial rise in the number of clinical trials for synthetic AMPs, very few have been approved as therapeutic candidates, such as polymyxins. Novel peptides Esc(1–21), or Esc peptides, have shown to be promising agents in combating CF pathogens, especially *P. aeruginosa*, without damaging the lung epithelia or inducing inflammatory cell recruitment in lung alveoli of treated mice, when compared to vehicle controls (122). In addition, peptide Esc(1–21)-1c showed notably high tolerability with a high therapeutic index of 70 or greater without inducing damage to the liver, spleen, or kidneys and remained bioavailable in the bronchoalveolar lavage 48 hours post-instillation (122). Esc peptides have also been shown to interact with the cystic fibrosis transmembrane conductance regulator (CFTR) protein, acting as potentiators of its mutated form and altering ion currents regulated by the protein, with a promising role in blocking the persistent lung infection predominantly established by *P. aeruginosa* (123). There have also been advancements in AMP design to treat and prevent *Mycobacterium tuberculosis*, the pathogen responsible for airborne tuberculosis infection, one of the leading causes of death globally (124). Human AMPs, including defensins, cathelicidin, and lactoferrin, are active components of the host immune defense response against mycobacteria. At the same time, there are many synthetic peptides with antimicrobial properties such as bacteria pore formation, cell wall disruption, cell membrane permeability, inhibition of cell wall synthesis, ATPase inhibition, immunomodulation, and anti-inflammatory activity (124). As novel synthetic AMPs continue to be designed, critical limitations regarding their use in preventing and treating pulmonary disease are mainly peptide stability, bioavailability, toxicity, and proteolytic degradation (124). One successful example is demonstrated by a newly developed AMP named D8 with a designed substitution of D-Val for L-Val of eCAP WLBU2, which resulted in increased resistance to protease enzymatic degradation, higher activity against bacterial biofilm, noticeably lower cytotoxicity to erythrocytes and white blood cells, and increased safety in mice compared to the original WLBU2 (77). Significantly, direct airway delivery of D8 revealed a therapeutic index of greater than 140 compared to less than 35 for WLBU2 in treating *P. aeruginosa*–induced respiratory infections (77).

1.8.2 Anticancer Activity of AMPs

Along with their use as prospective therapeutics against pulmonary disease, AMPs are under investigation for their use in anticancer therapy. Traditional chemotherapeutic treatment is challenged by acquired drug resistance, so AMPs are valuable alternatives as anticancer agents based on their selective targeting and cytotoxicity toward cancer cells versus healthy cells (5). AMPs derived from the Anura frog species *Rana temporaria* (temporin A and temporin L) and *Bombina variegata* (bombinin H2 and bombinin H4) were investigated for their efficacy as anticancer agents, with bombinin H4 and temporin A showing selective cytotoxicity against non–small cell lung carcinoma (NSCLC) cells (125). However, the cell membrane composition of the NSCLC cell lines tested affected the ability of the peptides to enact cytotoxicity (125). It is suggested that enhanced electrostatic connections between AMPs and cancer cells are due to the predominance of negatively charged phospholipids within the outer membrane leaflets of cancerous cells (125).

A novel AMP derivative of LL-37 identified as 17BIPHE2 was found to significantly raise the rate of apoptosis of the human lung adenocarcinoma cell line A549 and reactive oxygen species (ROS), as well as to inhibit the extracellular signal–regulated kinase (ERK) pathway (126). The serine/threonine protein kinase ERK pathway regulates cellular proliferation, differentiation, apoptosis, invasion, and migration (127). Through ERK pathway interference, reduction in mitochondrial membrane potential, and migration activity of A549 cells, 17BIPHE2 effectively reduced tumor formation (126). Importantly, 17BIPHE2 showed antimicrobial effects by preventing biofilm formation and cell proliferation of methicillin-resistant *S. aureus* (126).

Additionally, the AMP nisin ZP, derived from the Gram-positive species *Lactococcus lactis*, showed selective toxicity against NSCLC cells A549 and H1299 in vitro, through triggering apoptosis, cell cycle arrest, mitochondrial membrane depolarization, and a rise in ROS release (128). Nisin ZP also reduced the proliferation and migration of cancer cells, which supports its potential utility in more serious metastatic NSCLC (128). Another novel AMP derived from the skin secretion of amphibian species *Phyllomedusa camba*, PSN-PC, has broad-spectrum activity against several notable clinical pathogens, including *S. aureus*, *E. coli*, *P. aeruginosa*, and *Candida albicans*, while inhibiting the proliferation of NSCLC line NCI-H157 with minimal toxicity on the human microvascular endothelial cell line HMEC-1 (129). With an upsurge in research dedicated to novel AMP development, the clinical value of these agents has expanded, showing high effectiveness against drug-resistant bacteria and tumor generation.

1.8.3 AMPs and Cancer Immunotherapy

The systemic administration of AMPs, along with clinically approved cancer immunotherapies, is another critical avenue for cancer prevention and treatment advancements. For example, a systemic administration of oncolytic peptide LTX-315, derived from bovine lactoferrin, along with CpG adjuvant and anti-PD-1 treatment, showed increased immunotherapy against malignant B16F10 melanoma in an in vivo murine model (130). The systemic delivery led to a long-term protective effect due to increased splenic memory T cells and effector memory T cells, generating a heightened immune response and resulting in a cure rate of 30% (130). Therefore, oncolytic peptides show great clinical promise in systemic delivery with cancer immunotherapies to boost antitumor effects and to support long-term immune defense against malignancy. Another study investigated the co-administration of the hydroxymethylglutaryl coenzyme A reductase inhibitor fluvastatin along with cationic peptide melittin to treat colorectal cancer (131). This study administered a combination agent consisting of fluvastatin and melittin along with phospholipid and alpha lipoic acid nanoparticles, which was found to show enhanced toxicity against human colon cancer Caco2 cells, when compared to conventional fluvastatin treatment alone (131). Further research is necessary to investigate the heightened anticancer potential of combination cancer immunotherapy agents and novel AMPs for lung cancer treatment.

1.9 CONCLUSIONS

In summary, AMPs are essential factors in both innate and adaptive immunity. These peptides protect against pathogenic microbial infection and impact respiratory disease progression by regulating inflammation and mediating lung injury. The primary roles of AMPs are interconnected with antimicrobial activities, triggering the immune system, and regulating the inflammatory response and immunomodulatory activities, facilitating anticancer effects in the host. As respiratory diseases and lung cancer continue to be primary sources of all-cause mortality and burden, research in advancing AMP design is critical for improving lung health by developing enhanced therapeutic alternatives that overcome the deficiencies of traditional antibiotics and cancer immunotherapies.

ABBREVIATIONS

AMP	antimicrobial peptide
BPI	bactericidal/permeability-increasing protein
CF	cystic fibrosis
CFTR	cystic fibrosis transmembrane conductance regulator
COPD	chronic obstructive pulmonary disease
eCAP	engineered cationic antimicrobial peptide
ERK	extracellular signal–regulated kinase
HDP	host defense peptide
IAV	influenza A virus
LBP	lipopolysaccharide-binding protein
LPS	lipopolysaccharide
LRI	lower respiratory tract infection
LRT	lower respiratory tract
MCC	mucociliary clearance
MDR	multidrug resistant
NSCLC	non–small cell lung carcinoma
PLUNC	palate, lung, and nasal epithelium clone
ROS	reactive oxygen species
RSV	respiratory syncytial virus
SAAP	synthetic antimicrobial and antibiofilm peptide
SARS-CoV-2	severe acute respiratory syndrome coronavirus 2
SPLUNC1	short palate, lung, and nasal epithelium clone 1
URI	upper respiratory infection
URT	upper respiratory tract

REFERENCES

1. Xu, J., Murphy, S., Kochanek, K., and Arias, E. (2022) *Mortality in the United States, 2021*, National Center for Health Statistics (U.S.). https://doi.org/10.15620/cdc:122516
2. An Update on Cancer Deaths in the United States (2022) *Centers for Disease Control and Prevention* [online]. www.cdc.gov/cancer/dcpc/research/update-on-cancer-deaths/index.htm (Accessed May 24, 2023)
3. Bin Hafeez, A., Jiang, X., Bergen, P. J., and Zhu, Y. (2021) Antimicrobial peptides: An update on classifications and databases. *Int J Mol Sci.* **22**, 11691

4. Hancock, R. E. W., Haney, E. F., and Gill, E. E. (2016) The immunology of host defence peptides: Beyond antimicrobial activity. *Nat Rev Immunol.* **16**, 321–334
5. Zhang, Y., Wang, C., Zhang, W., and Li, X. (2023) Bioactive peptides for anticancer therapies. *Biomater Transl.* **4**, 5–17
6. Dickson, R. P., Erb-Downward, J. R., and Huffnagle, G. B. (2013) The role of the bacterial microbiome in lung disease. *Expert Rev Respir Med.* **7**, 245–257
7. Cookson, W. O. C. M., Cox, M. J., and Moffatt, M. F. (2018) New opportunities for managing acute and chronic lung infections. *Nat Rev Microbiol.* **16**, 111–120
8. Pettigrew, M. M., Tanner, W., and Harris, A. D. (2021) The lung microbiome and pneumonia. *J Infect Dis.* **223**, S241–S245
9. Adams, W. C. (1993) *Measurement of Breathing Rate and Volume in Routinely Performed Daily Activities: Final Report, Contract No. A033–205*, The Division, Sacramento, CA
10. Bassis, C. M., Erb-Downward, J. R., Dickson, R. P., Freeman, C. M., Schmidt, T. M., Young, V. B., Beck, J. M., Curtis, J. L., and Huffnagle, G. B. (2015) Analysis of the upper respiratory tract microbiotas as the source of the lung and gastric microbiotas in healthy individuals. *mBio.* **6**, e00037
11. Hilty, M., Burke, C., Pedro, H., Cardenas, P., Bush, A., Bossley, C., Davies, J., Ervine, A., Poulter, L., Pachter, L., Moffatt, M. F., and Cookson, W. O. C. (2010) Disordered microbial communities in asthmatic airways. *PLoS ONE.* **5**, e8578
12. Dickson, R. P., Martinez, F. J., and Huffnagle, G. B. (2014) The role of the microbiome in exacerbations of chronic lung diseases. *Lancet.* **384**, 691–702
13. Jin, X., Ren, J., Li, R., Gao, Y., Zhang, H., Li, J., Zhang, J., Wang, X., and Wang, G. (2021) Global burden of upper respiratory infections in 204 countries and territories, from 1990 to 2019. *EClinicalMedicine.* https://doi.org/10.1016/j.eclinm.2021.100986
14. Incze, M., Grady, D., and Gupta, A. (2018) I have a cold—What do I need to know? *JAMA Intern Med.* **178**, 1288
15. Grief, S. N. (2013) Upper respiratory infections. *Prim Care Clin Off Pract.* **40**, 757–770
16. Wald, E. R., Applegate, K. E., Bordley, C., Darrow, D. H., Glode, M. P., Marcy, S. M., Nelson, C. E., Rosenfeld, R. M., Shaikh, N., Smith, M. J., Williams, P. V., Weinberg, S. T., and American Academy of Pediatrics (2013) Clinical practice guideline for the diagnosis and management of acute bacterial sinusitis in children aged 1 to 18 years. *Pediatrics.* **132**, e262–e280
17. Claassen-Weitz, S., Lim, K. Y. L., Mullally, C., Zar, H. J., and Nicol, M. P. (2021) The association between bacteria colonizing the upper respiratory tract and lower respiratory tract infection in young children: A systematic review and meta-analysis. *Clin Microbiol Infect.* **27**, 1262–1270
18. Manna, S., McAuley, J., Jacobson, J., Nguyen, C. D., Ullah, M. A., Sebina, I., Williamson, V., Mulholland, E. K., Wijburg, O., Phipps, S., and Satzke, C. (2022) Synergism and antagonism of bacterial-viral coinfection in the upper respiratory tract. *mSphere.* **7**, e0098421
19. Li, Y., and Nair, H. (2022) Trends in the global burden of lower respiratory infections: The knowns and the unknowns. *Lancet Infect Dis.* **22**, 1523–1525
20. Dalziel, S. R., Haskell, L., O'Brien, S., Borland, M. L., Plint, A. C., Babl, F. E., and Oakley, E. (2022) Bronchiolitis. *Lancet.* **400**, 392–406
21. Mathur, S., Fuchs, A., Bielicki, J., Van Den Anker, J., and Sharland, M. (2018) Antibiotic use for community-acquired pneumonia in neonates and children: WHO evidence review. *Paediatr Int Child Health.* **38**, S66–S75
22. Stockley, R. A. (1998) Lung infections. 1. Role of bacteria in the pathogenesis and progression of acute and chronic lung infection. *Thorax.* **53**, 58–62
23. Khair, O. A., Devalia, J. L., Abdelaziz, M. M., Sapsford, R. J., Tarraf, H., and Davies, R. J. (1994) Effect of *Haemophilus influenzae* endotoxin on the synthesis of IL-6, IL-8, TNF-alpha and expression of ICAM-1 in cultured human bronchial epithelial cells. *Eur Respir J.* **7**, 2109–2116
24. Adams, W., Espicha, T., and Estipona, J. (2021) Getting your neutrophil: Neutrophil transepithelial migration in the lung. *Infect Immun.* https://doi.org/10.1128/IAI.00659-20
25. Lin, Q., Pilewski, J. M., and Di, Y. P. (2021) Acidic microenvironment determines antibiotic susceptibility and biofilm formation of *Pseudomonas aeruginosa. Front Microbiol.* **12**, 747834
26. Bjarnsholt, T. (2013) The role of bacterial biofilms in chronic infections. *APMIS.* **121**, 1–58
27. Kuiken, T., Riteau, B., Fouchier, R., and Rimmelzwaan, G. (2012) Pathogenesis of influenza virus infections: The good, the bad and the ugly. *Curr Opin Virol.* **2**, 276–286

28. Shieh, W.-J., Blau, D. M., Denison, A. M., Deleon-Carnes, M., Adem, P., Bhatnagar, J., Sumner, J., Liu, L., Patel, M., Batten, B., Greer, P., Jones, T., Smith, C., Bartlett, J., Montague, J., White, E., Rollin, D., Gao, R., Seales, C., Jost, H., Metcalfe, M., Goldsmith, C. S., Humphrey, C., Schmitz, A., Drew, C., Paddock, C., Uyeki, T. M., and Zaki, S. R. (2010) 2009 pandemic influenza A (H1N1): Pathology and pathogenesis of 100 fatal cases in the United States. *Am J Pathol.* **177**, 166–175
29. Hogg, J. C., Chu, F., Utokaparch, S., Woods, R., Elliott, W. M., Buzatu, L., Cherniack, R. M., Rogers, R. M., Sciurba, F. C., Coxson, H. O., and Paré, P. D. (2004) The nature of small-airway obstruction in chronic obstructive pulmonary disease. *N Engl J Med.* **350**, 2645–2653
30. Linden, D., Guo-Parke, H., Coyle, P. V., Fairley, D., McAuley, D. F., Taggart, C. C., and Kidney, J. (2019) Respiratory viral infection: A potential "missing link" in the pathogenesis of COPD. *Eur Respir Rev.* **28**, 180063
31. Leung, J. M., Niikura, M., Yang, C. W. T., and Sin, D. D. (2020) COVID-19 and COPD. *Eur Respir J.* **56**, 2002108
32. Leung, J. M., Yang, C. X., Tam, A., Shaipanich, T., Hackett, T.-L., Singhera, G. K., Dorscheid, D. R., and Sin, D. D. (2020) ACE-2 expression in the small airway epithelia of smokers and COPD patients: Implications for COVID-19. *Eur Respir J.* **55**, 2000688
33. McCullers, J. A. (2014) The co-pathogenesis of influenza viruses with bacteria in the lung. *Nat Rev Microbiol.* **12**, 252–262
34. Kudva, A., Scheller, E. V., Robinson, K. M., Crowe, C. R., Choi, S. M., Slight, S. R., Khader, S. A., Dubin, P. J., Enelow, R. I., Kolls, J. K., and Alcorn, J. F. (2011) Influenza a inhibits Th17-mediated host defense against bacterial pneumonia in mice. *J Immunol.* **186**, 1666–1674
35. Lee, B., Robinson, K. M., McHugh, K. J., Scheller, E. V., Mandalapu, S., Chen, C., Di, Y. P., Clay, M. E., Enelow, R. I., Dubin, P. J., and Alcorn, J. F. (2015) Influenza-induced type I interferon enhances susceptibility to Gram-negative and Gram-positive bacterial pneumonia in mice. *Am J Physiol Lung Cell Mol Physiol.* **309**, L158–L167
36. Chan, Y. R., Liu, J. S., Pociask, D. A., Zheng, M., Mietzner, T. A., Berger, T., Mak, T. W., Clifton, M. C., Strong, R. K., Ray, P., and Kolls, J. K. (2009) Lipocalin 2 is required for pulmonary host defense against Klebsiella infection. *J Immunol.* **182**, 4947–4956
37. Wu, H., Santoni-Rugiu, E., Ralfkiaer, E., Porse, B. T., Moser, C., Høiby, N., Borregaard, N., and Cowland, J. B. (2010) Lipocalin 2 is protective against *E. coli* pneumonia. *Respir Res.* **11**, 96
38. Di, Y. P. (2011) Functional roles of SPLUNC1 in the innate immune response against Gram-negative bacteria. *Biochem Soc Trans.* **39**, 1051–1055
39. Liu, Y., Bartlett, J. A., Di, M. E., Bomberger, J. M., Chan, Y. R., Gakhar, L., Mallampalli, R. K., McCray, P. B., and Di, Y. P. (2013) SPLUNC1/BPIFA1 contributes to pulmonary host defense against *Klebsiella pneumoniae* respiratory infection. *Am J Pathol.* **182**, 1519–1531
40. Beasley, V., Joshi, P. V., Singanayagam, A., Molyneaux, P. L., Johnston, S. L., and Mallia, P. (2012) Lung microbiology and exacerbations in COPD. *Int J Chron Obstruct Pulmon Dis.* **7**, 555–569
41. Murphy, T. F., Brauer, A. L., Eschberger, K., Lobbins, P., Grove, L., Cai, X., and Sethi, S. (2008) *Pseudomonas aeruginosa* in chronic obstructive pulmonary disease. *Am J Respir Crit Care Med.* **177**, 853–860
42. Miravitlles, M., Espinosa, C., Fernández-Laso, E., Martos, J. A., Maldonado, J. A., and Gallego, M. (1999) Relationship between bacterial flora in sputum and functional impairment in patients with acute exacerbations of COPD. Study Group of Bacterial Infection in COPD. *Chest.* **116**, 40–46
43. Dubos, R. J. (1939) Studies on a bactericidal agent extracted from a soil bacillus. *J Exp Med.* **70**, 1–10
44. Lecaille, F., Lalmanach, G., and Andrault, P.-M. (2016) Antimicrobial proteins and peptides in human lung diseases: A friend and foe partnership with host proteases. *Biochimie.* **122**, 151–168
45. Hiemstra, P. S., Amatngalim, G. D., van der Does, A. M., and Taube, C. (2016) Antimicrobial peptides and innate lung defenses: Role in infectious and noninfectious lung diseases and therapeutic applications. *Chest.* **149**, 545–551
46. Hiemstra, P. S. (2006) Defensins and cathelicidins in inflammatory lung disease: Beyond antimicrobial activity. *Biochem Soc Trans.* **34**, 276–278
47. Diamond, G., Beckloff, N., and Ryan, L. K. (2008) Host defense peptides in the oral cavity and the lung: Similarities and differences. *J Dent Res.* **87**, 915–927
48. Ganz, T., and Lehrer, R. I. (1995) Defensins. *Pharmacol Ther.* **66**, 191–205
49. Lehrer, R. I., and Ganz, T. (2002) Defensins of vertebrate animals. *Curr Opin Immunol.* **14**, 96–102

50. Ganz, T. (2003) Defensins: Antimicrobial peptides of innate immunity. *Nat Rev Immunol.* **3**, 710–720
51. Yang, D., Biragyn, A., Hoover, D. M., Lubkowski, J., and Oppenheim, J. J. (2004) Multiple roles of antimicrobial defensins, cathelicidins, and eosinophil-derived neurotoxin in host defense. *Annu Rev Immunol.* **22**, 181–215
52. Hiemstra, P. S. (2007) The role of epithelial β-defensins and cathelicidins in host defense of the lung. *Exp Lung Res.* **33**, 537–542
53. Sørensen, O. E., Follin, P., Johnsen, A. H., Calafat, J., Tjabringa, G. S., Hiemstra, P. S., and Borregaard, N. (2001) Human cathelicidin, hCAP-18, is processed to the antimicrobial peptide LL-37 by extracellular cleavage with proteinase 3. *Blood.* **97**, 3951–3959
54. Larrick, J. W., Hirata, M., Balint, R. F., Lee, J., Zhong, J., and Wright, S. C. (1995) Human CAP18: A novel antimicrobial lipopolysaccharide-binding protein. *Infect Immun.* **63**, 1291–1297
55. Tobias, P. S., Soldau, K., Iovine, N. M., Elsbach, P., and Weiss, J. (1997) Lipopolysaccharide (LPS)-binding proteins BPI and LBP form different types of complexes with LPS. *J Biol Chem.* **272**, 18682–18685
56. Levy, O. (2000) A neutrophil-derived anti-infective molecule: Bactericidal/permeability-increasing protein. *Antimicrob Agents Chemother.* **44**, 2925–2931
57. Bartlett, J. A., Gakhar, L., Penterman, J., Singh, P. K., Mallampalli, R. K., Porter, E., and McCray, P. B. (2011) PLUNC: A multifunctional surfactant of the airways. *Biochem Soc Trans.* **39**, 1012–1016
58. Gakhar, L., Bartlett, J. A., Penterman, J., Mizrachi, D., Singh, P. K., Mallampalli, R. K., Ramaswamy, S., and McCray, Jr., P. B. (2010) PLUNC is a novel airway surfactant protein with anti-biofilm activity. *PLoS ONE.* **5**, e9098
59. Lukinskiene, L., Liu, Y., Reynolds, S. D., Steele, C., Stripp, B. R., Leikauf, G. D., Kolls, J. K., and Di, Y. P. (2011) Antimicrobial activity of PLUNC protects against *Pseudomonas aeruginosa* infection. *J Immunol.* **187**, 382–390
60. Hiemstra, P. S. (2007) The role of epithelial beta-defensins and cathelicidins in host defense of the lung. *Exp Lung Res.* **33**, 537–542
61. Rahnamaeian, M. (2011) Antimicrobial peptides. *Plant Signal Behav.* **6**, 1325–1332
62. Luo, Y., and Song, Y. (2021) Mechanism of antimicrobial peptides: Antimicrobial, anti-inflammatory and antibiofilm activities. *Int J Mol Sci.* **22**, 11401
63. Ciumac, D., Gong, H., Hu, X., and Lu, J. R. (2019) Membrane targeting cationic antimicrobial peptides. *J Colloid Interface Sci.* **537**, 163–185
64. Shai, Y. (1999) Mechanism of the binding, insertion and destabilization of phospholipid bilayer membranes by α-helical antimicrobial and cell non-selective membrane-lytic peptides. *Biochim Biophys Acta (BBA)—Biomembr.* **1462**, 55–70
65. Ulmschneider, J. P. (2017) Charged antimicrobial peptides can translocate across membranes without forming channel-like pores. *Biophys J.* **113**, 73–81
66. Graf, M., and Wilson, D. N. (2019) Intracellular antimicrobial peptides targeting the protein synthesis machinery. In *Antimicrobial Peptides: Basics for Clinical Application* (Matsuzaki, K., ed.), pp. 73–89, Advances in Experimental Medicine and Biology, Springer International Publishing, Singapore. https://doi.org/10.1007/978-981-13-3588-4_6
67. Su, B.-C., Liu, Y.-C., Ting, C.-H., Lyu, P.-C., and Chen, J.-Y. (2020) Antimicrobial peptide TP4 targets mitochondrial adenine nucleotide translocator 2. *Mar Drugs.* **18**, 417
68. Kamli, M. R., Sabir, J. S. M., Malik, M. A., and Ahmad, A. (2022) Human β defensins-1, an antimicrobial peptide, kills *Candida glabrata* by generating oxidative stress and arresting the cell cycle in G0/G1 phase. *Biomed Pharmacother.* **154**, 113569
69. Bobone, S., and Stella, L. (2019) Selectivity of antimicrobial peptides: A complex interplay of multiple equilibria. *Adv Exp Med Biol.* **1117**, 175–214
70. Matsuzaki, K. (2009) Control of cell selectivity of antimicrobial peptides. *Biochim Biophys Acta.* **1788**, 1687–1692
71. Payne, J. E., Dubois, A. V., Ingram, R. J., Weldon, S., Taggart, C. C., Elborn, J. S., and Tunney, M. M. (2017) Activity of innate antimicrobial peptides and ivacaftor against clinical cystic fibrosis respiratory pathogens. *Int J Antimicrob Agents.* **50**, 427–435
72. Goeke, J. E., Kist, S., Schubert, S., Hickel, R., Huth, K. C., and Kollmuss, M. (2018) Sensitivity of caries pathogens to antimicrobial peptides related to caries risk. *Clin Oral Investig.* **22**, 2519–2525

73. Parducho, K. R., Beadell, B., Ybarra, T. K., Bush, M., Escalera, E., Trejos, A. T., Chieng, A., Mendez, M., Anderson, C., Park, H., Wang, Y., Lu, W., and Porter, E. (2020) The antimicrobial peptide human beta-defensin 2 inhibits biofilm production of *Pseudomonas aeruginosa* without compromising metabolic activity. *Front Immunol.* **11**, 805
74. Martínez, M., Polizzotto, A., Flores, N., Semorile, L., and Maffía, P. C. (2020) Antibacterial, anti-biofilm and in vivo activities of the antimicrobial peptides P5 and P6.2. *Microb Pathog.* **139**, 103886
75. de Breij, A., Riool, M., Cordfunke, R. A., Malanovic, N., de Boer, L., Koning, R. I., Ravensbergen, E., Franken, M., van der Heijde, T., Boekema, B. K., Kwakman, P. H. S., Kamp, N., El Ghalbzouri, A., Lohner, K., Zaat, S. A. J., Drijfhout, J. W., and Nibbering, P. H. (2018) The antimicrobial peptide SAAP-148 combats drug-resistant bacteria and biofilms. *Sci Transl Med.* **10**, eaan4044
76. Jiang, S., Deslouches, B., Chen, C., Di, M. E., and Di, Y. P. (2019) Antibacterial properties and efficacy of a novel SPLUNC1-derived antimicrobial peptide, α4-short, in a murine model of respiratory infection. *mBio.* **10**. https://doi.org/10.1128/mBio.e00226-19
77. Di, Y. P., Lin, Q., Chen, C., Montelaro, R. C., Doi, Y., and Deslouches, B. (2020) Enhanced therapeutic index of an antimicrobial peptide in mice by increasing safety and activity against multidrug-resistant bacteria. *Sci Adv.* **6**, eaay6817
78. Wilson, S. S., Wiens, M. E., and Smith, J. G. (2013) Antiviral mechanisms of human defensins. *J Mol Biol.* **425**, 4965–4980
79. Gabriel, M., and Holtappels, R. (2021) Inactivation of an enveloped virus by immobilized antimicrobial peptides. *Bioconjug Chem.* **32**, 2480–2484
80. Hsieh, I.-N., and Hartshorn, K. L. (2016) The role of antimicrobial peptides in influenza virus infection and their potential as antiviral and immunomodulatory therapy. *Pharmaceuticals.* **9**, 53
81. Tripathi, S., Wang, G., White, M., Rynkiewicz, M., Seaton, B., and Hartshorn, K. (2015) Identifying the critical domain of LL-37 involved in mediating neutrophil activation in the presence of influenza virus: Functional and structural analysis. *PLoS ONE.* **10**, e0133454
82. Barlow, P. G., Svoboda, P., Mackellar, A., Nash, A. A., York, I. A., Pohl, J., Davidson, D. J., and Donis, R. O. (2011) Antiviral activity and increased host defense against influenza infection elicited by the human cathelicidin LL-37. *PLoS ONE.* **6**, e25333
83. Currie, S. M., Findlay, E. G., McHugh, B. J., Mackellar, A., Man, T., Macmillan, D., Wang, H., Fitch, P. M., Schwarze, J., and Davidson, D. J. (2013) The human cathelicidin LL-37 has antiviral activity against respiratory syncytial virus. *PLoS ONE.* **8**, e73659
84. Uchio, E., Inoue, H., and Kadonosono, K. (2013) Anti-adenoviral effects of human cationic antimicrobial protein-18/LL-37, an antimicrobial peptide, by quantitative polymerase chain reaction. *Korean J Ophthalmol.* **27**, 199–203
85. Garcia-Fandino, R., and Piñeiro, Á. (2020) Delving into the origin of destructive inflammation in COVID-19: A betrayal of natural host defense peptides? *Front Immunol.* **11**, 610024
86. Martell, E. M., González-Garcia, M., Ständker, L., and Otero-González, A. J. (2021) Host defense peptides as immunomodulators: The other side of the coin. *Peptides.* **146**, 170644
87. Drayton, M., Deisinger, J. P., Ludwig, K. C., Raheem, N., Müller, A., Schneider, T., and Straus, S. K. (2021) Host defense peptides: Dual antimicrobial and immunomodulatory action. *Int J Mol Sci.* **22**, 11172
88. Yang, D., Chen, Q., Schmidt, A. P., Anderson, G. M., Wang, J. M., Wooters, J., Oppenheim, J. J., and Chertov, O. (2000) Ll-37, the neutrophil granule- and epithelial cell-derived cathelicidin, utilizes formyl peptide receptor-like 1 (Fprl1) as a receptor to chemoattract human peripheral blood neutrophils, monocytes, and T cells. *J Exp Med.* **192**, 1069–1074
89. Davidson, D. J., Currie, A. J., Reid, G. S. D., Bowdish, D. M. E., MacDonald, K. L., Ma, R. C., Hancock, R. E. W., and Speert, D. P. (2004) The cationic antimicrobial peptide LL-37 modulates dendritic cell differentiation and dendritic cell-induced T cell polarization1. *J Immunol.* **172**, 1146–1156
90. Mookherjee, N., Anderson, M. A., Haagsman, H. P., and Davidson, D. J. (2020) Antimicrobial host defence peptides: Functions and clinical potential. *Nat Rev Drug Discov.* **19**, 311–332
91. Weidenmaier, C., Kristian, S. A., and Peschel, A. (2003) Bacterial resistance to antimicrobial host defense—An emerging target for novel antiinfective strategies? *Curr Drug Targets.* **4**, 643–649
92. Deslouches, B., Montelaro, R. C., Urish, K. L., and Di, Y. P. (2020) Engineered cationic antimicrobial peptides (eCAPs) to combat multidrug-resistant bacteria. *Pharmaceutics.* **12**, 501

93. Kuhn, J. M., and Di, Y. P. (2023) Determination of mutational timing of colistin-resistance genes through *Klebsiella pneumoniae* evolution. *Pharmaceutics.* **15**, 270
94. Koprivnjak, T., and Peschel, A. (2011) Bacterial resistance mechanisms against host defense peptides. *Cell Mol Life Sci.* **68**, 2243–2254
95. Nguyen, L. T., and Vogel, H. J. (2016) Staphylokinase has distinct modes of interaction with antimicrobial peptides, modulating its plasminogen-activation properties. *Sci Rep.* **6**, 31817
96. Band, V. I., and Weiss, D. S. (2015) Mechanisms of antimicrobial peptide resistance in Gram-negative bacteria. *Antibiotics (Basel).* **4**, 18–41
97. Schmidtchen, A., Frick, I.-M., Andersson, E., Tapper, H., and Björck, L. (2002) Proteinases of common pathogenic bacteria degrade and inactivate the antibacterial peptide LL-37. *Mol Microbiol.* **46**, 157–168
98. Moravej, H., Moravej, Z., Yazdanparast, M., Heiat, M., Mirhosseini, A., Moosazadeh Moghaddam, M., and Mirnejad, R. (2018) Antimicrobial peptides: Features, action, and their resistance mechanisms in bacteria. *Microb Drug Resist.* **24**, 747–767
99. Nizet, V. (2006) Antimicrobial peptide resistance mechanisms of human bacterial pathogens. *Curr Issues Mol Biol.* **8**, 11–26
100. Strempel, N., Neidig, A., Nusser, M., Geffers, R., Vieillard, J., Lesouhaitier, O., Brenner-Weiss, G., and Overhage, J. (2013) Human host defense peptide LL-37 stimulates virulence factor production and adaptive resistance in *Pseudomonas aeruginosa. PLoS ONE.* **8**, e82240
101. Weidenmaier, C., and Peschel, A. (2008) Teichoic acids and related cell-wall glycopolymers in Gram-positive physiology and host interactions. *Nat Rev Microbiol.* **6**, 276–287
102. Moffatt, J. H., Harper, M., and Boyce, J. D. (2019) Mechanisms of polymyxin resistance. In *Polymyxin Antibiotics: From Laboratory Bench to Bedside* (Li, J., Nation, R. L., and Kaye, K. S., eds.), pp. 55–71, Advances in Experimental Medicine and Biology, Springer International Publishing, Cham. https://doi/10.1007/978-3-030-16373-0_5
103. Raetz, C. R. H., Reynolds, C. M., Trent, M. S., and Bishop, R. E. (2007) Lipid A modification systems in Gram-negative bacteria. *Annu Rev Biochem.* **76**, 295–329
104. Vila-Farrés, X., Ferrer-Navarro, M., Callarisa, A. E., Martí, S., Espinal, P., Gupta, S., Rolain, J.-M., Giralt, E., and Vila, J. (2015) Loss of LPS is involved in the virulence and resistance to colistin of colistin-resistant *Acinetobacter nosocomialis* mutants selected in vitro. *J Antimicrob Chemother.* **70**, 2981–2986
105. Gunn, J. S., and Miller, S. I. (1996) PhoP-PhoQ activates transcription of pmrAB, encoding a two-component regulatory system involved in *Salmonella typhimurium* antimicrobial peptide resistance. *J Bacteriol.* **178**, 6857–6864
106. Moffatt, J. H., Harper, M., Harrison, P., Hale, J. D. F., Vinogradov, E., Seemann, T., Henry, R., Crane, B., St. Michael, F., Cox, A. D., Adler, B., Nation, R. L., Li, J., and Boyce, J. D. (2010) Colistin resistance in *Acinetobacter baumannii* is mediated by complete loss of lipopolysaccharide production. *Antimicrob Agents Chemother.* **54**, 4971–4977
107. Xia, G., Kohler, T., and Peschel, A. (2010) The wall teichoic acid and lipoteichoic acid polymers of *Staphylococcus aureus. Int J Med Microbiol.* **300**, 148–154
108. Peschel, A., Otto, M., Jack, R. W., Kalbacher, H., Jung, G., and Götz, F. (1999) Inactivation of the *dlt* operon in *Staphylococcus aureus* confers sensitivity to defensins, protegrins, and other antimicrobial peptides. *J Biol Chem.* **274**, 8405–8410
109. Chan, K. G., Mayer, M., Davis, E. M., Halperin, S. A., Lin, T.-J., and Lee, S. F. (2007) Role of D-alanylation of *Streptococcus gordonii* lipoteichoic acid in innate and adaptive immunity. *Infect Immun.* **75**, 3033–3042
110. Kovács, M., Halfmann, A., Fedtke, I., Heintz, M., Peschel, A., Vollmer, W., Hakenbeck, R., and Brückner, R. (2006) A functional *dlt* operon, encoding proteins required for incorporation of D-alanine in teichoic acids in Gram-positive bacteria, confers resistance to cationic antimicrobial peptides in *Streptococcus pneumoniae. J Bacteriol.* **188**, 5797–5805
111. Bayer, A. S., Mishra, N. N., Chen, L., Kreiswirth, B. N., Rubio, A., and Yang, S.-J. (2015) Frequency and distribution of single-nucleotide polymorphisms within mprF in methicillin-resistant *Staphylococcus aureus* clinical isolates and their role in cross-resistance to daptomycin and host defense antimicrobial peptides. *Antimicrob Agents Chemother.* **59**, 4930–4937
112. Guina, T., Yi, E. C., Wang, H., Hackett, M., and Miller, S. I. (2000) A PhoP-regulated outer membrane protease of *Salmonella enterica* serovar *typhimurium* promotes resistance to alpha-helical antimicrobial peptides. *J Bacteriol.* **182**, 4077–4086

113. Boll, J. M., Tucker, A. T., Klein, D. R., Beltran, A. M., Brodbelt, J. S., Davies, B. W., and Trent, M. S. (2015) Reinforcing lipid a acylation on the cell surface of *Acinetobacter baumannii* promotes cationic antimicrobial peptide resistance and desiccation survival. *mBio*. **6**. https://doi.org/10.1128/mBio.00478-15
114. Bayer, A. S., Prasad, R., Chandra, J., Koul, A., Smriti, M., Varma, A., Skurray, R. A., Firth, N., Brown, M. H., Koo, S. P., and Yeaman, M. R. (2000) In vitro resistance of *Staphylococcus aureus* to thrombin-induced platelet microbicidal protein is associated with alterations in cytoplasmic membrane fluidity. *Infect Immun*. **68**, 3548–3553
115. Shafer, W. M., Qu, X.-D., Waring, A. J., and Lehrer, R. I. (1998) Modulation of *Neisseria gonorrhoeae* susceptibility to vertebrate antibacterial peptides due to a member of the resistance/nodulation/division efflux pump family. *Proc Natl Acad Sci*. **95**, 1829–1833
116. Kapach, G., Nuri, R., Schmidt, C., Danin, A., Ferrera, S., Savidor, A., Gerlach, R. G., and Shai, Y. (2020) Loss of the periplasmic chaperone Skp and mutations in the efflux pump AcrAB-TolC play a role in acquired resistance to antimicrobial peptides in *Salmonella typhimurium*. *Front Microbiol*. **11**, 189
117. Puja, H., Bolard, A., Noguès, A., Plésiat, P., and Jeannot, K. (2020) The efflux pump MexXY/OprM contributes to the tolerance and acquired resistance of *Pseudomonas aeruginosa* to colistin. *Antimicrob Agents Chemother*. **64**. https://doi.org/10.1128/aac.02033-19
118. Sun, S., Gao, H., Liu, Y., Jin, L., Wang, R., Wang, X., Wang, Q., Yin, Y., Zhang, Y., and Wang, H. (2020) Co-existence of a novel plasmid-mediated efflux pump with colistin resistance gene mcr in one plasmid confers transferable multidrug resistance in *Klebsiella pneumoniae*. *Emerg Microbes Infect*. **9**, 1102–1113
119. Chan, C., Burrows, L. L., and Deber, C. M. (2004) Helix induction in antimicrobial peptides by alginate in biofilms. *J Biol Chem*. **279**, 38749–38754
120. Benincasa, M., Mattiuzzo, M., Herasimenka, Y., Cescutti, P., Rizzo, R., and Gennaro, R. (2009) Activity of antimicrobial peptides in the presence of polysaccharides produced by pulmonary pathogens. *J Pept Sci*. **15**, 595–600
121. Mulcahy, H., Charron-Mazenod, L., and Lewenza, S. (2008) Extracellular DNA chelates cations and induces antibiotic resistance in *Pseudomonas aeruginosa* biofilms. *PLoS Pathog*. **4**, e1000213
122. Cappiello, F., Casciaro, B., Loffredo, M. R., Puglisi, E., Lin, Q., Yang, D., Conte, G., d'Angelo, I., Ungaro, F., Ferrera, L., Barbieri, R., Cresti, L., Pini, A., Di, Y. P., and Mangoni, M. L. (2022) Pulmonary safety profile of Esc peptides and Esc-peptide-loaded poly(lactide-co-glycolide) nanoparticles: A promising therapeutic approach for local treatment of lung infectious diseases. *Pharmaceutics*. **14**, 2297
123. Ferrera, L., Cappiello, F., Loffredo, M. R., Puglisi, E., Casciaro, B., Botta, B., Galietta, L. J. V., Mori, M., and Mangoni, M. L. (2021) Esc peptides as novel potentiators of defective cystic fibrosis transmembrane conductance regulator: An unprecedented property of antimicrobial peptides. *Cell Mol Life Sci*. **79**, 67
124. Mehta, K., Sharma, P., Mujawar, S., and Vyas, A. (2022) Role of antimicrobial peptides in treatment and prevention of *Mycobacterium tuberculosis:* A review. *Int J Pept Res Ther*. **28**, 132
125. Swithenbank, L., Cox, P., Harris, L. G., Dudley, E., Sinclair, K., Lewis, P., Cappiello, F., and Morgan, C. (2020) Temporin a and bombinin H2 antimicrobial peptides exhibit selective cytotoxicity to lung cancer cells. *Scientifica (Cairo)*. **2020**, 3526286
126. Yang, T., Li, J., Jia, Q., Zhan, S., Zhang, Q., Wang, Y., and Wang, X. (2021) Antimicrobial peptide 17BIPHE2 inhibits the proliferation of lung cancer cells in vitro and in vivo by regulating the ERK signaling pathway. *Oncol Lett*. **22**, 501
127. Lavoie, H., Gagnon, J., and Therrien, M. (2020) ERK signalling: A master regulator of cell behaviour, life and fate. *Nat Rev Mol Cell Biol*. **21**, 607–632
128. Patil, S. M., and Kunda, N. K. (2022) Nisin ZP, an antimicrobial peptide, induces cell death and inhibits non-small cell lung cancer (NSCLC) progression in vitro in 2D and 3D cell culture. *Pharm Res*. **39**, 2859–2870
129. Wu, X., Pan, J., Wu, Y., Xi, X., Ma, C., Wang, L., Zhou, M., and Chen, T. (2017) PSN-PC: A novel antimicrobial and anti-biofilm peptide from the skin secretion of phyllomedusa-camba with cytotoxicity on human lung cancer cell. *Molecules*. **22**, 1896

130. Xia, Y., Wei, J., Zhao, S., Guo, B., Meng, F., Klumperman, B., and Zhong, Z. (2021) Systemic administration of polymersomal oncolytic peptide LTX-315 combining with CpG adjuvant and anti-PD-1 antibody boosts immunotherapy of melanoma. *J Control Release*. **336**, 262–273
131. Alfaleh, M. A., Fahmy, O., Al-Rabia, M. W., Abourehab, M. A. S., Ahmed, O. A. A., Fahmy, U. A., Alsulimani, H. H., Badr-Eldin, S. M., Aldawsari, H. M., Aldhabi, B. M., Alharbi, A. S., and Alhakamy, N. A. (2022) Hybrid nanoparticulate system of Fluvastatin loaded phospholipid, alpha lipoic acid and melittin for the management of colon cancer. *Sci Rep*. **12**, 19446

2 Alveolar Epithelium in Lung Injury and Repair

SeungHye Han

2.1 INTRODUCTION

The lung comprises multiple cell types including epithelial cells, endothelial cells, mesenchymal cells, and immune cells (1), with different functions and interactions. Specifically, lung epithelial cells play important roles in gas exchange, mucociliary clearance, host defense, and tissue homeostasis. Anatomically, the lung is divided into three functionally distinct but continuously connected regions: the proximal conducting airways (trachea and bronchi), the distal conducting airways (bronchioles), and the terminal gas exchange region (alveoli), each with a distinct epithelial cell population (1, 2). The proximal conducting airways are lined by pseudostratified columnar epithelium with basal cells, goblet cells, ciliated cells, and rare neuroendocrine cells, tuft cells, and ionocytes. The distal conducting airways are lined by simple cuboidal epithelium with club cells and neuroendocrine cells. The terminal alveolar region comprises alveolar epithelial type I (AT1) cells, alveolar epithelial type II (AT2) cells, and capillary endothelial cells. These anatomically distinct epithelial cell populations undergo temporally and/or spatially separated developmental processes and repair processes after injury. Previously it was believed that the human lung was a terminally differentiated organ in adults, mainly due to the very long turnover times of lung epithelial cell populations in homeostasis (3, 4). However, recent advances in experimental techniques such as single-cell RNA sequencing analysis, in vivo lineage tracing technology, and 3D organoid culture systems allow us to confirm that multiple pools of region-specific epithelial stem/progenitor cells exist throughout the airways and alveoli, ensuring homeostasis and timely repair of damaged lung tissue (5–7). In fact, patients with severe viral pneumonia often recover lung function close to pre-infection baseline (8). Here, we focus on novel epithelial cell populations that were recently found, the emerging mechanisms of how these epithelial stem/progenitor cells are regulated during repair upon injury, and their therapeutic potentials.

2.2 ALVEOLAR EPITHELIAL CELLS DURING DEVELOPMENT AND REPAIR AFTER INJURY

The alveolus is a terminal respiratory unit and is responsible for the main physiologic function of the lung: gas exchange between circulating blood and air. The alveolar epithelium consists of AT1 and AT2 cells. AT1 cells are large, thin, membranous cells, covering greater than 90% of the alveolar surface (9). Recent studies suggest that an AT1 cell covers multiple alveolar sacs by folding its cytoplasm (10, 11). AT1 cells are located next to the alveolar capillary endothelium, forming a blood gas barrier (12). The AT1 cell's extremely thin shape allows passive diffusion of oxygen and carbon dioxide between the blood stream and air in the alveolar sac. On the other hand, AT2 cells are small cuboidal cells and cover about 5% of the alveolar surface. AT2 cells mainly synthesize and secrete surfactant to reduce lung surface tension (13). AT2 cells are also facultative stem cells to self-renew and give rise to AT1 cells (14, 15). Additionally, AT2 cells transport sodium and fluid to keep the alveoli relatively dry, and modulate the innate immune system (16). Recent evidence indicates, other than AT2 and AT1 cells, intermediate alveolar

DOI: 10.1201/9781003355243-3

epithelial cell types exist during normal postnatal development and repair after injury. Furthermore, evidence suggests that AT1 cells, previously considered a terminally differentiated cell type, may have plasticity in certain conditions.

2.2.1 Embryonic Lung Development—Branching Morphogenesis

The development of the mammalian lung is a complex process that involves the interaction of two germ layers: the endoderm and the mesoderm (17, 18). The endoderm gives rise to the epithelial lining of the respiratory tract, whereas the mesoderm contributes to the surrounding mesenchyme and blood vessels. The lung originates from the ventral foregut endoderm, which forms two lateral outgrowths called the lung buds at around embryonic day 9.5 in mice and postconception week 4 in humans (17, 18). The embryonic lung develops through different morphological stages—pseudoglandular, canalicular, saccular, and alveolar—that shape the respiratory tree with more cell diversification (19). The pseudoglandular stage is defined by extensive branching of the lung buds into the surrounding mesenchyme to generate a highly branched network of conducting airways that reach the peripheral regions of the lung. The canalicular stage involves further branching and differentiation of epithelial cells into proximal and distal types. The saccular stage is characterized by the formation of terminal sacs or saccules, which are the precursors of the alveoli. The final stage of lung development is the alveolar stage, which begins after birth and continues postnatally. During this stage, the alveolar spaces expand (alveolarization) by further alveolar septation accompanied by full differentiation of AT1 and AT2 cells. The alveolar stage lasts for about 4–5 weeks in mice and at least 3 years in humans.

2.2.2 Postnatal Alveolar Development

At mouse embryonic day 17.5, bipotent progenitor cells appear at the tips of distal tubules, expressing both early AT1 (Ager) and AT2 (SftpC, Mucl) markers (15). By the time of birth, bipotent progenitor cells differentiate to either AT1 cells or AT2 cells by turning off inappropriate cell markers and by turning on mature cell markers (Aqp5 for AT1 cells; SftpB, SftpD, and Abca3 for AT2 cells). Bipotent progenitor cells are exhausted by postnatal day 4, when sacculation has completed throughout the lung. A recently published single-cell atlas of mouse lung development (20) also reported a similar cell population that emerges at mouse embryonic day 15 and remains present through birth. A similar cell population has been observed in human fetal lungs as well. At around 20–22 weeks postconception, the late-stalk cell state appears expressing both AT1 (*AGER*) and AT2 (*SFTPC*) markers, and they are predicted to differentiate to either AT1 or AT2 per RNA velocity analysis (21).

According to the single-cell atlas of mouse lung development (20), transitional cells expressing both AT1 and AT2 markers also normally exist during murine postnatal development. Time series single-cell RNA sequencing of CD45/Ter119 depleted mouse lung cells showed that transitional cells expand from postnatal day 7 and persist through postnatal day 14. The relative abundance of postnatal transitional cells by postnatal day 14 was 7.1 ± 4.8% by sequencing and 1%–2% by quantitative RNA in situ hybridization. Postnatal transitional cells express both mature AT1 (*Hopx*, *Aqp5*) and AT2 (*Sftpa1*, *Sftpc*) cell markers, as well as cytokeratin genes and *Cdkn1a*. They share similar transcriptomic features with an intermediate cell type that was observed during adult repair in lung injury models (see Section 2.2.3.2). Postnatal transitional cells are cuboidal, similar to AT2 cells. RNA velocity analysis suggests that these transitional cells arise mainly from AT2 cells with contributions from the prenatal epithelium and are predicted to become AT1 cells with a 90.6% median probability. This finding, however, needs to be confirmed with an in vivo lineage tracing study. Although postnatal transitional cells highly express *Cdkn1a*, a gene that encodes a protein that inhibits cell cycle progression, there was no evidence that they are arrested in the G0 or G1 stage of the cell cycle (20, 22).

2.2.3 Alveolar Repair after Injury

2.2.3.1 AT2 Cells as Alveolar Stem Cells

AT2 cells have been long recognized as potential stem cells because isolated AT2 cells spontaneously changed to AT1-resembling cells in vitro (23, 24). With the advent of in vivo lineage tracing techniques, several studies confirmed that AT2 cells self-renew and give rise to AT1 in adult homeostasis and during repair after injury in vivo (14, 15). It appears that only a subset of AT2 cells serve as stem cells in homeostasis, whereas the majority of AT2 cells perform their standard function only: biosynthesis and secretion of surfactant (25, 26). A careful mouse genetic study showed that Axin2$^+$ AT2 cells self-renew and differentiate into AT1 cells in adult homeostasis (25). Using in vivo lineage tracing technique, investigators showed that Axin2$^+$ AT2 cells are maintained as a stable population at baseline and increased sixfold a year later by its clonal expansion with daughter AT1 and daughter AT2 cells (25). Another group also observed a stable population of Axin2$^+$ AT2 cells in uninjured adult murine lungs using an independent *Axin2* inducible *Cre* mouse strain (26). Furthermore, Axin2$^+$ AT2 cells act as a major stem cell population to regenerate AT1 and AT2 cells after mild to moderate influenza-induced lung injury. Importantly, both studies elegantly showed that Wnt signaling controls the cell fate of AT2 stem cells (26). When 40% of alveolar epithelial cells were ablated with diphtheria toxin, Wnt signaling was induced and 73% of AT2 cells became Axin2$^+$ AT2 cells serving as stem cells (25). Similarly, in vitro, both mouse (25) and human (26) AT2 cell proliferation was promoted by activating Wnt, whereas AT2-to-AT1 differentiation was promoted by inhibiting Wnt. Interestingly, hypercapnia frequently observed in patients with lung disease limits AT2 cell proliferation by inhibiting β-catenin signaling (27). A recent study reported hypertrophic binucleated AT2 cells observed after bleomycin-induced lung injury where AT2-to-AT1 differentiation occurs (28). Polyploidy occurs when cells increase their DNA content without mitosis, and this is frequently observed in epithelium after injury in vertebrates (29). Further studies are necessary to understand the functional role of AT2 polyploidization during repair after injury.

2.2.3.2 Intermediate Transitional Cells

Recent advances in single-cell RNA sequencing technology enable us to characterize heterogeneous cell types in the tissue at a single-cell level. Using single-cell RNA sequencing analysis, several groups have reported an intermediate cell type during the AT2-to-AT1 differentiation (hereinafter *transitional cells*) in various murine models of lung injury (30–32), normal postnatal murine lung development (20), organoid culture of murine and human AT2 cells (33), and in human lung explants from patients with pulmonary fibrosis, including patients with COVID-19 pneumonia (34–38). Although these intermediate transitional cells were initially named differently by the various reporting groups—Krt8+ alveolar differentiation intermediate (ADI) (31), damage-associated transition progenitors (DATPs) (32), and pre-alveolar type-1 transitional cell state (PATS) (33), they all share similar transcriptional features. Transitional cells co-express AT2 and AT1 marker genes at lower levels than AT2 and AT1 cells, respectively, and highly express cytokeratin genes (early basal cell markers *Krt8*, *Krt18*, *Krt7*, and *Krt19* in mice, and *KRT17* in humans) and genes inhibiting cell-cycle progression such as *Cdkn1a*. Lineage tracing studies and RNA velocity analysis indicate these transitional cells originate from AT2 cells and differentiate into AT1 cells. These intermediate transitional cell types are also present, but very rarely, in adult murine homeostasis lungs. One study reported that 0.3 ± 0.2% of murine adult alveolar epithelial cells were found to be in an intermediate state (Muc1$^+$ Pdpn$^+$) by flow cytometry (15). Krt8+ ADI cells existed very rarely during the homeostasis of adult murine lungs (31) and 6- to 10-week-old female mouse lungs (39). Collectively, transitional cells are in an intermediate state during AT2-to-AT1 differentiation, and their transient accumulation probably represents a common transition in cellular development, homeostasis, and repair after injury. The functional role and significance of these transitional cells in lung physiology and pathology are still unclear and warrant further investigation.

2.2.3.3 AT1 to AT2 Transdifferentiation

AT1 cells have been believed to be a quiescent cell type that was terminally differentiated. However, recent evidence suggests that AT1 cells may have the ability to exit their quiescent terminally differentiated state in certain conditions. During murine postnatal development, ectopic SOX2 expression induced AT1 cell proliferation and reprogramming of cell fate (10). Using *Hopx-CreER*^{T2} mice, investigators observed that *Hopx*-lineage positive cells express SFTPC, an AT2 cell marker, when YAP/TAZ were deleted (40) or actin polymerization was disturbed (41). Murine *Hopx*-lineage positive cells also gave rise to SFTPC$^+$ AT2 cells after pneumonectomy (42), but it appears to be only limited to Hopx$^+$Igfbp2$^-$ AT1 cells, and Hopx$^+$Igfbp2$^+$ AT1 cells (>85% of AT1 cells) were not able to transdifferentiate into AT2 cells during post-pneumonectomy alveolar growth (43). Future studies should focus on investigating whether AT1-to-AT2 transdifferentiation occurs in humans and the role of AT1-to-AT2 transdifferentiation in physiological and pathological conditions.

2.3 AIRWAY EPITHELIAL CELLS DURING ALVEOLAR REPAIR AFTER INJURY

During embryonic development, the ventral foregut begins as an epithelial tube that becomes the future trachea. Primary lung buds appear from the tube-like structure around embryonic day 9.5 in mice (44). Throughout embryonic development, distal lung buds elongate and bifurcate to build tree-like tubular airway structures as epithelial cells are differentiating appropriately to line proximal and distal conducting airways, respectively (19). The trachea is lined with basal cells, rare neuroendocrine cells, and rare tuft cells (1). The luminal layers of the bronchi and bronchioles are lined with goblet cells, ciliated cells, and secretory cells (club cells). Using single-cell RNA sequencing analysis, a rare columnar epithelial cell population, ionocytes, has been recently identified as related to cystic fibrosis (45, 46). Subgroups of airway epithelial cells are found to have regenerative potential and serve as stem/progenitor cells to maintain and repair airway epithelial cells; they also have a role in alveolar epithelial repair. We will briefly overview airway epithelial cells that have stem cell properties participating in airway epithelial maintenance and alveolar epithelial repair, and discuss newly found airway epithelial cells in humans.

2.3.1 Airway Epithelial Cells Engaging in Airway Homeostasis and Repair

Basal cells are multipotent stem cells that reside in the trachea and bronchi, and have the capacity to self-renew and give rise to other airway cells (47). Several in vivo lineage tracing studies and 3D organoid assays showed that murine basal cells self-renew and give rise to ciliated cells and club cells during postnatal development and during adult homeostasis and repair after tracheal and bronchial injury (48–50). Similarly, club cells are the major stem cells in bronchioles; they self-renew and give rise to ciliated cells during postnatal development, adult homeostasis, and repair after injury in bronchioles (51). Club cells in the trachea appeared to dedifferentiate into basal cells when basal cells in the trachea were ablated, according to in vivo lineage tracing and 3D organoid assay (52).

2.3.2 Airway Epithelial Cells Engaging in Alveolar Repair

Severe alveolar damage can impair the normal repair process by AT2 cells and activate alternative sources of stem cells for alveolar epithelial repair. Several groups have reported that metaplastic alveolar Krt5$^+$ cells emerge and form Krt5$^+$ pods after severe influenza-induced alveolar injury (26, 53–57). This is likely an immediate adaptive response to severe alveolar injury in order to rapidly mitigate the loss of the epithelial barrier. However, this can lead to pathologic repair because Krt5$^+$ cells have limited capacity to differentiate into functional AT2 cells (55) unless HIF1α is deleted or Wnt is activated (56). In mice, Krt5$^+$ cells are not derived from alveolar epithelial cells, but rather from SOX2$^+$Krt5$^-$ progenitor cells in distal airways (26, 57, 58). On the other hand, human AT2 cells appear to be able to transdifferentiate into Krt5$^+$ cells (59). Another stem cell population that can contribute to alveolar epithelial repair is a bronchioalveolar stem cell (BASC) (60). BASCs are located at the junction of the mouse bronchioalveolar duct and co-express AT2 and club cell markers (61). The current evidence indicates that BASCs minimally contribute to adult alveolar homeostasis (51), and their contribution to alveolar epithelial regeneration varies depending on the injury model (62–64).

2.3.3 Novel Distal Airway Cell Populations in Human Lungs

The human and mouse lungs share similarities in their anatomy and cellular composition. Thus, many murine models of lung injury and disease have been widely used providing important biological insights that can be applied to humans. However, there are also distinct anatomic features between murine and human lungs (65), which may limit the applicability of murine findings to human lung disease pathology. Likely from the size differences, airway branching in mice gives rise to about 13–17 generations of branching from the trachea to the gas exchange region of the lung, alveoli (66), whereas airway branching in humans gives rise to about 16–23 generations of branching (67). In mice, the distal conducting airways (terminal bronchioles) abruptly transition to alveolar ducts at the bronchoalveolar duct junction. On the contrary, human terminal bronchioles make a more gradual transition by branching into two to three generations of respiratory bronchioles before branching into alveolar ducts. It has been believed that some of the human lung diseases could not be studied in mouse models due to this anatomical difference: the lack of respiratory bronchioles in mice.

Using single-cell transcriptomic profiling techniques, several novel distal airway cell populations were recently found in the human lungs that were not observed in the mouse lungs (68, 69). Investigators specifically focused on the human terminal and respiratory bronchioles (TRBs) using two parallel approaches: spatial transcriptomic profiling and single-cell transcriptome profiling of microdissected distal airways (<1 mm in diameter and contiguous with alveoli) from surrounding tissue (68). Through these approaches, they identified a cell population characterized by the expression of *SFTPB, SCGB3A2*, and *SFTPC* and spatial association with alveolar septae contiguous with respiratory bronchioles. These cells share transcriptional markers with both AT2 and AT1 cells (*SFTPC* and *AGER*), as well as airway epithelial cells (*SCGB3A2* and *SOX2*) and thus were termed TRB-specific alveolar type-0 (AT0) cells. They also found a population that expresses *SFTPB*$^+$*SCGB3A2*$^+$ but lacks *SCGB1A1, FOXJ1*, and *SFTPC* expression. This population is localized to TRBs, and thus they are termed TRB secretory cells (TRB-SCs). With RNA velocity analysis and organoid culture systems, they showed that AT0 cells originate from AT2 cells and can differentiate into either AT1 cells or TRB-SC cells. These AT0 cells and TRB-SCs were also observed in primate lungs and during human lung development. Another study also observed a novel secretory cell population in the human distal lungs (69). Investigators prepared single-cell suspension from five healthy nonsmoker donor lungs by mechanically mincing distal lung tissue that was absent of pleura and visible airways/blood vessels. By single-cell RNA sequencing analysis, they identified a novel secretory cell population, *SCGB3A2*$^+$*SCGB1A1*$^+$ cells, in the human distal lung that they call respiratory airway secretory (RAS) cells. RAS cells were not found in single-cell RNA sequencing analysis of the first segmental airway branch, suggesting they reside mainly in respiratory bronchioles. Isolated RAS cells were able to give rise to SFTPC$^+$ organoids ex vivo. The transcriptional profile of RAS cells was altered in lungs with chronic obstructive pulmonary disease (COPD), suggesting RAS cells may be associated with the pathology of human airway disease. Further studies are needed to consolidate the findings from the two studies and to understand how these novel human cells differentiate and affect lung disease development and progression.

2.4 EMERGING MECHANISMS OF ALVEOLAR EPITHELIAL DIFFERENTIATION AND THERAPEUTIC IMPLICATIONS

The alveolar epithelium is a critical component of the lung that maintains gas exchange and barrier function. However, it is also susceptible to various injuries that can compromise its integrity and function. Alveolar epithelial repair is a complex process that involves the proliferation and differentiation of epithelial progenitor cells, as well as the interaction with other cell types including mesenchymal cells and immune cells (Figure 2.1). Several molecular mechanisms have been investigated to explain how damaged alveolar epithelium is repaired upon injury, and these have been reviewed recently (5, 70, 71). Here, we focus on recent work with therapeutic potential that has improved our understanding of alveolar epithelial cell differentiation during physiologic and pathologic repair.

2.4.1 The Integrated Stress Response and Mitochondrial Dysfunction

The lungs are constantly exposed to environmental stress and infectious pathogens. Cellular stress is associated with chronic lung diseases and has been implicated in the pathogenesis of lung diseases such as COPD and pulmonary fibrosis. The integrated stress response (ISR) is an intracellular defense system that is evolutionally conserved. It restores cellular homeostasis in response to diverse stress stimuli such as infection, perturbed proteostasis, nutrient deprivation, and redox imbalances (72, 73). Genetic mutations in components of the ISR have been linked with cognitive dysfunction, intellectual disability, and diabetes in humans (74–78). ISR activation requires activation of at least one of four specialized kinases (PERK, PKR, GCN2, and HRI) that lead to the phosphorylation of eukaryotic translation initiation factor 2 subunit alpha (eIF2α). eIF2α phosphorylation blocks the action of eIF2's guanine nucleotide exchange factor termed eIF2B, resulting in global inhibition of protein synthesis while inducing the translation of selective mRNAs, including the transcription factors ATF3, ATF4, and ATF5. These transcription factors, especially ATF4, increase the expression of a host of genes including those involved in stress responses, one-carbon metabolism, and *Ddit3* (encodes CHOP). Consequently, ISR activation drives a transcriptional program that can maintain cellular homeostasis or activate apoptosis. Although the ISR has been implicated in lung disease (79), the underlying mechanisms have been poorly understood.

ISRIB is a small molecule ISR inhibitor, and the precise structural mechanisms of ISRIB function have been extensively studied (72, 80–82). ISRIB binds to eIF2B and enhances its activity, improving global protein translation and inhibiting ATF4 induction. ISRIB has been shown to be beneficial in a host of disease models as well as normal aging (83–93). The inhibitory effects of ISRIB are overcome during conditions of high ISR stress (e.g., viral infections), which might explain the lack of toxicity when ISRIB is administered in vivo. Furthermore, ISRIB attenuates bleomycin- and asbestos-induced lung fibrosis in young and old mice (90, 91). Lineage tracing study suggests ISRIB targets transitional epithelial cells to reduce ISR signaling and promote their differentiation into AT1 cells after bleomycin-induced lung injury (90). Notably, binucleated AT2 cells seen after bleomycin-induced lung injury were also reduced by ISRIB (28), implying the high persistent ISR may stall the AT2-to-AT1 cell differentiation (Figure 2.1).

Mitochondrial dysfunction, specifically in the electron transport chain (ETC), has been identified as a trigger to activate the ISR in vitro and in vivo (94–105). Previously, two independent groups simultaneously reported that ETC inhibition activates the mitochondrial protease OMA1, which cleaves a mitochondrial protein DELE1 (100, 101). Cleaved DELE1 accumulates in the cytosol to activate HRI (100, 101). Mitochondrial dysfunction and altered cellular metabolism have been consistently reported in fibrotic lungs in both humans and animal models of pulmonary fibrosis and, thus, have been implicated in the pathogenesis of pulmonary fibrosis (106–111). However, the exact causal mechanisms are unknown. Interestingly, the analysis of genetic rare variants revealed that genes associated with mitochondrial ETC complex I assembly was enriched in patients with familial pulmonary fibrosis (112). Mitochondrial complex I is a large 45-subunit complex that catalyzes the transfer of electrons from NADH to ubiquinone in the ETC, in the process of regenerating NAD^+ (contributing to tricarboxylic acid cycle metabolites) and pumping protons across the mitochondrial inner membrane (contributing to ATP synthesis). Mitochondrial complex I also generates reactive oxygen species (ROS) as by-products of electron leakage, which can act as signaling molecules that modulate various cellular processes, such as cell proliferation, differentiation, and death. Recently, we found that loss of mitochondrial ETC complex I function in distal lung epithelium during development results in a dramatic expansion of postnatal transitional epithelial cells through an increase in $NADH/NAD^+$ ratio, which leads to subsequent death of the mice from respiratory failure in postnatal weeks 5–9 (22). Transitional cells with no mitochondrial complex I function displayed very high levels of ISR activation compared to other epithelial cell types or transitional cells in other settings (20, 31–33). Administration of ISRIB in mice with no complex I function in their lung epithelium lowered the ISR activation and rescued mortality of these animals. Furthermore, pathologic ISR activation and impaired epithelial differentiation were rescued by genetic and pharmacologic strategies that normalized the $NADH/NAD^+$ ratio. Our findings causally link mitochondrial dysfunction, alterations of the $NADH/NAD^+$ ratio, and pathologic ISR activation to failed epithelial differentiation during postnatal alveolar development. Traditionally, the ISR has been viewed as an adaptive process in response to stressors and can induce cellular apoptosis. However, our study uncovered an unappreciated role of mitochondrial ISR whereby they can dictate lung epithelial cell fate in vivo, independent of their role in generating ATP. Targeting mitochondrial metabolism, specifically NAD^+ regeneration, and pathologic ISR activation may be attractive strategies to promote alveolar epithelial differentiation during repair and prevent or treat pulmonary fibrosis by reducing aberrant repair (Figure 2.1).

2.4.2 Wnt Signaling

Wnt/beta-catenin signaling is a highly conserved pathway that regulates cell proliferation and differentiation during development and tissue homeostasis. Wnt signaling is essential for lung development and repair upon injury (113), and plays an important role in the pathobiology of various lung diseases (114). Wnt signaling is required to maintain a stable population of $Axin2^+$ AT2 stem cells during adult homeostasis, and the neighboring niche *Pdgfrα*$^+$ fibroblasts provide Wnt signals to these AT2 stem cells (25). Similarly, investigators identified mesenchymal cells expressing Lgr5 and Lgr6 that participate in alveolar epithelial differentiation (115). However, modulating Wnt signaling in lung disease is challenging because the beneficial effects of Wnt activation are cell type specific (114) and it can stimulate fibroblast activation leading to pulmonary fibrosis (116–118). Recently, investigators mapped the gene expression of Wnt receptors in human and mouse alveoli and found that different Wnt receptors are expressed by alveolar epithelial cells (Fzd5/6) and stromal cells (Fzd1) (119). By using Fzd-specific agonists that target only Fzd5 or Fzd6, they were able to enhance lung epithelial stem cell activity in organoid culture. Also, mice tolerated 2-week dosing of Fzd5 or Fzd6 agonists and did not show overt toxicity in other organs, suggesting this may be a promising strategy to promote lung epithelial stem cell activity without inducing pulmonary fibrosis. Further studies are required to evaluate the off-target effects of these Fzd-specific agonists and toxicity in vivo.

2.4.3 Cell Senescence

Cellular senescence is defined as the cell state in mostly irreversible cell-cycle arrest with secretory phenotype and altered metabolism (120). Cellular senescence was originally described from an observation that normal human diploid cells stop proliferating after certain numbers of passage in vitro cultures (121). It has been since recognized as a response to stress including DNA damage, nutrient deprivation, perturbed proteostasis, and mitochondrial dysfunction. The number of cells expressing senescence markers generally increases with age, and it was reported that eliminating those cells was beneficial, thus cellular senescence has been generally perceived as detrimental to health (122–125). However, recent evidence suggests that senescence marker-positive cells are present in young healthy animals although they are rare, and eliminating those cells is not always beneficial (126, 127). In senescent cells, $p21^{WAF1/Cip1}$, encoded by the cyclin-dependent kinase 1A (*CDKN1A*), and $p16^{INK4A}$, encoded by *CDKN2A*, accumulate resulting in cell-cycle arrest as they slow the cell cycle progression from the G1 to the S phase. Hence, they are considered senescence markers. A recently published study carefully dissected the function and role of senescence marker ($p16^{INK4a}$)-positive cells in lung repair after injury (128). Investigators generated a sensitive reporter mouse for $p16^{INK4a}$, and showed that $p16^{INK4a+}$ cells present in young adult murine lungs, mainly in immune and mesenchymal cells. Focusing on $p16^{INK4a+}$ mesenchymal cells, they showed that $p16^{INK4a+}$ fibroblasts are a stable cell population located at the epithelial-mesenchymal interface during postnatal development and adult homeostasis, where $p16^{INK4a}$ expression increased with age. Naphthalene-induced airway epithelial injury increased $p16^{INK4a}$ expression only in immune cells, not in fibroblasts. Although $p16^{INK4a+}$ fibroblasts themselves were less proliferative than $p16^{INK4a-}$ fibroblasts in vivo after naphthalene injury, $p16^{INK4a+}$ fibroblasts promoted epithelial cell growth compared to $p16^{INK4a-}$ fibroblasts when co-cultured with $Scgb1a1^{+}$ airway stem cells in 3D organoid assay, likely through senescent-associated secretory phenotype (SASP) sensing inflammatory signaling. Using mice with mesenchymal-specific deletion of $p16^{INK4a}$, they showed that mesenchymal $p16^{INK4a}$ expression is required for airway epithelial repair in vivo. This study highlights that the senescence response can be adaptive, rather than a maladaptive deleterious response, depending on the biological context. Increased senescent fibroblasts in aged animals may be an effort to overcome dysfunctional epithelial stem cells or disrupted stem cell niches in the lung as $p16^{INK4a+}$ fibroblasts promote epithelial repair. Continued exploration of in vivo senescent cells during tissue repair upon injury will further elucidate the mechanisms of age-related lung diseases and potential therapeutic targets.

2.4.4 Changes in the Extracellular Matrix and Biophysical Force: YAP/TAZ Signaling

Extracellular matrix (ECM) alterations in the lungs are associated with chronic lung disease and have emerged as a potentially important player during lung development, homeostasis, and repair after injury (129). Recent studies suggest that changes in lung mechanical tension may be a cue for AT2 cells to proliferate and differentiate in order to repair damaged AT1 cells or damaged whole alveoli (130, 131). YAP is a transcription coactivator and acts as a stiffness sensor, regulating mechanotransduction. Using conditional knockout mice and pharmacologic tools, investigators demonstrated that YAP in AT2 cells is necessary for post-pneumonectomy alveolar growth, which was regulated by JNK and p38 MAPK signaling upstream (130). They showed that Cdc42 was required for the activation of JNK, p38, and subsequently YAP in post-pneumonectomy alveolar growth. Furthermore, the loss of Cdc42 in AT2 cells induced spontaneous progressive lung fibrosis in aged mice and led to impaired alveolar growth and fibrosis after pneumonectomy (131). In contrast to its beneficial role in alveolar epithelial regeneration, YAP/TAZ promote fibroblast activation and exacerbate pulmonary fibrosis (132, 133). Therefore, modulating YAP/TAZ activity in a cell-specific manner is a major challenge for therapeutic intervention. A recent study identified a $G\alpha_s$-coupled dopamine receptor D1 (DRD1), as a potential target to selectively inhibit YAP/TAZ signaling in lung mesenchymal cells (134). DRD1 is preferentially expressed in lung mesenchymal cells, and DRD1 agonist inhibits YAP/TAZ signaling selectively in mesenchymal cells and ameliorated fibrosis in a murine model of bleomycin-induced lung injury (134). However, DRD1 agonists may also have side effects on the brain. Thus, developing more selective DRD1 agonists with higher potency in the lung is an important direction for future research.

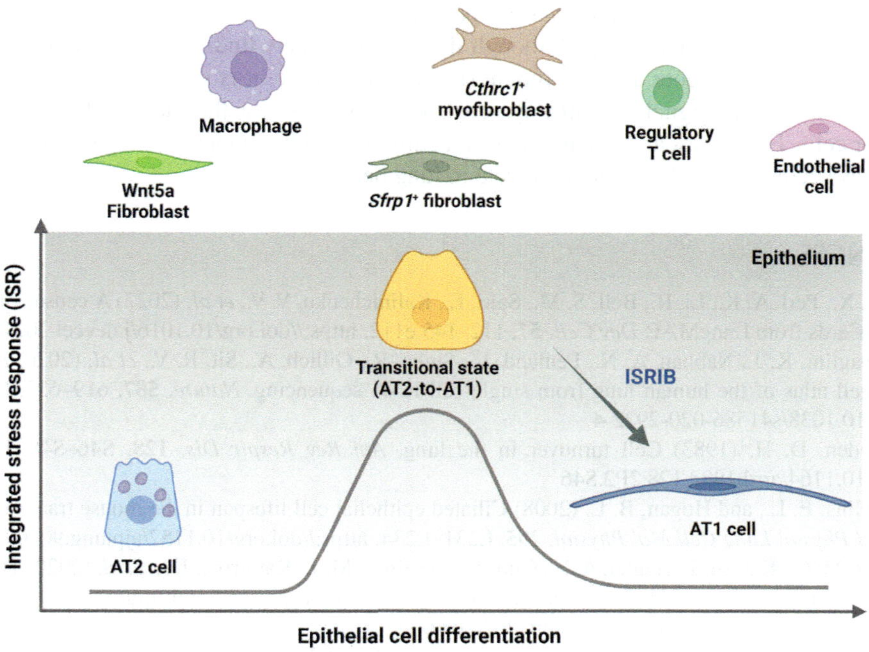

FIGURE 2.1 The alveolar epithelium comprises alveolar epithelial type 1 (AT1) cells and type 2 (AT2) cells. AT1 cells are responsible for gas exchange, but susceptible to injury. AT2 cells are facultative stem cells that can give rise to AT1 cells. When AT1 cells are damaged or die, AT2 cells give rise to transitional cells that have an intermediate phenotype and eventually become AT1 cells. During this process, the integrated stress response (ISR) is transiently activated and subsequently decreases as transitional cells differentiate into AT1 cells. However, if the ISR remains high, it inhibits the differentiation of transitional cells into AT1 cells. A small molecule ISR inhibitor can enhance the differentiation of transitional cells into AT1 cells by lowering the ISR level. The differentiation of AT2 cells into AT1 cells also involves interactions with other cell types in the niche, such as mesenchymal cells, immune cells, and endothelial cells. The mechanisms of these interactions need to be further investigated.

2.4.5 Interactions with Mesenchymal Cells and Immune Cells

The alveolar stem cell niche includes mesenchymal cells and immune cells as well as AT2 and AT1 cells (Figure 2.1). Previous studies showed that AT2 stem cells receive Wnt signals from adjacent fibroblasts to maintain their stem cell activity (25, 115). A recent single-cell RNA sequencing study revealed that a fibroblast population that highly expresses *Sfrp1* is expanded after bleomycin-induced lung injury (135). Per trajectory analysis, *Sfrp1*⁺ fibroblasts appear to be the transitional precursor to *Cthrc1*⁺ myofibroblasts that are observed in fibrotic lungs (136). Interestingly, when we disrupted mitochondrial complex I function in lung epithelium during development, we observed the expansion of *Sfrp1*⁺ fibroblasts (22), suggesting the crosstalk between epithelial cells and mesenchymal cells. There is emerging evidence suggesting that immune cells such as monocytes, alveolar macrophages, and regulatory T cells are involved in the alveolar repair process and development of pulmonary fibrosis (137–141), but the exact mechanisms remain elusive. Investigating how the cell-cell interaction affects epithelial repair and contributes to the development of lung disease will be an important area of future research.

2.5 CONCLUSIONS

Recent advances in experimental tools and methods enabled us to identify novel cell populations during lung development, homeostasis, and repair after injury. Interacting with other types of cell populations, region-specific epithelial stem/progenitor cells self-renew and differentiate during homeostasis and repair after injury. Although these exciting new findings open the door to identifying target cells to promote alveolar repair and reduce disease, it is still largely unknown which metabolic and cellular signaling pathways control the behavior of the identified epithelial stem/progenitor cells. Future research should focus on gaining a deeper understanding of the mechanism of how these cells are regulated and their role in lung health and disease.

REFERENCES

1. Sun, X., Perl, A. K., Li, R., Bell, S. M., Sajti, E., Kalinichenko, V. V., *et al.* (2022) A census of the lung: CellCards from LungMAP. *Dev Cell.* **57**, 112–145.e112. https://doi.org/10.1016/j.devcel.2021.11.007
2. Travaglini, K. J., Nabhan, A. N., Penland, L., Sinha, R., Gillich, A., Sit, R. V., *et al.* (2020) A molecular cell atlas of the human lung from single-cell RNA sequencing. *Nature.* **587**, 619–625. https://doi.org/10.1038/s41586-020-2922-4
3. Bowden, D. H. (1983) Cell turnover in the lung. *Am Rev Respir Dis.* **128**, S46–S48. https://doi.org/10.1164/arrd.1983.128.2P2.S46
4. Rawlins, E. L., and Hogan, B. L. (2008) Ciliated epithelial cell lifespan in the mouse trachea and lung. *Am J Physiol Lung Cell Mol Physiol.* **295**, L231–L234. https://doi.org/10.1152/ajplung.90209.2008
5. Basil, M. C., Katzen, J., Engler, A. E., Guo, M., Herriges, M. J., Kathiriya, J. J., *et al.* (2020) The cellular and physiological basis for lung repair and regeneration: Past, present, and future. *Cell Stem Cell.* **26**, 482–502. https://doi.org/10.1016/j.stem.2020.03.009
6. Alysandratos, K. D., Herriges, M. J., and Kotton, D. N. (2021) Epithelial stem and progenitor cells in lung repair and regeneration. *Annu Rev Physiol.* **83**, 529–550. https://doi.org/10.1146/annurev-physiol-041520-092904
7. Hogan, B., and Tata, P. R. (2019) Cellular organization and biology of the respiratory system. *Nat Cell Biol.* https://doi.org/10.1038/s41556-019-0357-7
8. Mylvaganam, R. J., Bailey, J. I., Sznajder, J. I., Sala, M. A., and Northwestern Comprehensive COVID Center Consortium. (2021) Recovering from a pandemic: Pulmonary fibrosis after SARS-CoV-2 infection. *Eur Respir Rev.* **30**. https://doi.org/10.1183/16000617.0194-2021
9. Crapo, J. D., Barry, B. E., Gehr, P., Bachofen, M., and Weibel, E. R. (1982) Cell number and cell characteristics of the normal human lung. *Am Rev Respir Dis.* **126**, 332–337. https://doi.org/10.1164/arrd.1982.126.2.332
10. Yang, J., Hernandez, B. J., Martinez Alanis, D., Narvaez del Pilar, O., Vila-Ellis, L., Akiyama, H., *et al.* (2016) The development and plasticity of alveolar type 1 cells. *Development.* **143**, 54–65. https://doi.org/10.1242/dev.130005
11. Little, D. R., Gerner-Mauro, K. N., Flodby, P., Crandall, E. D., Borok, Z., Akiyama, H., *et al.* (2019) Transcriptional control of lung alveolar type 1 cell development and maintenance by NK homeobox 2–1. *Proc Natl Acad Sci U S A.* **116**, 20545–20555. https://doi.org/10.1073/pnas.1906663116
12. Gillich, A., Zhang, F., Farmer, C. G., Travaglini, K. J., Tan, S. Y., Gu, M., *et al.* (2020) Capillary cell-type specialization in the alveolus. *Nature.* **586**, 785–789. https://doi.org/10.1038/s41586-020-2822-7
13. Han, S., and Mallampalli, R. K. (2015) The role of surfactant in lung disease and host defense against pulmonary infections. *Ann Am Thorac Soc.* **12**, 765–774. https://doi.org/10.1513/AnnalsATS.201411-507FR
14. Barkauskas, C. E., Cronce, M. J., Rackley, C. R., Bowie, E. J., Keene, D. R., Stripp, B. R., *et al.* (2013) Type 2 alveolar cells are stem cells in adult lung. *J Clin Invest.* **123**, 3025–3036. https://doi.org/10.1172/JCI68782
15. Desai, T. J., Brownfield, D. G., and Krasnow, M. A. (2014) Alveolar progenitor and stem cells in lung development, renewal and cancer. *Nature.* **507**, 190–194. https://doi.org/10.1038/nature12930
16. Mason, R. J. (2006) Biology of alveolar type II cells. *Respirology.* **11 Suppl**, S12–S15. https://doi.org/10.1111/j.1440-1843.2006.00800.x
17. Herriges, M., and Morrisey, E. E. (2014) Lung development: Orchestrating the generation and regeneration of a complex organ. *Development.* **141**, 502–513. https://doi.org/10.1242/dev.098186
18. Nikolić, M. Z., Sun, D., and Rawlins, E. L. (2018) Human lung development: Recent progress and new challenges. *Development.* **145**. https://doi.org/10.1242/dev.163485

19. Chen, J. (2015) The regulation of branching morphogenesis in the developing lung. In *Stem Cells in the Lung: Development, Repair and Regeneration* (Bertoncello, I., ed.), pp. 3–16, Springer International Publishing, Cham, Chapter 1
20. Negretti, N. M., Plosa, E. J., Benjamin, J. T., Schuler, B. A., Habermann, A. C., Jetter, C. S., *et al.* (2021) A single-cell atlas of mouse lung development. *Development.* **148**, https://doi.org/10.1242/dev.199512
21. He, P., Lim, K., Sun, D., Pett, J. P., Jeng, Q., Polanski, K., *et al.* (2022) A human fetal lung cell atlas uncovers proximal-distal gradients of differentiation and key regulators of epithelial fates. *Cell.* **185**, 4841–4860.e4825. https://doi.org/10.1016/j.cell.2022.11.005
22. Han, S., Lee, M., Shin, Y., Giovanni, R., Chakrabarty, R. P., Herrerias, M. M., *et al.* (2023) Mitochondrial integrated stress response controls lung epithelial cell fate. *Nature.* **620**, 890–897. https://doi.org/10.1038/s41586-023-06423-8
23. Danto, S. I., Shannon, J. M., Borok, Z., Zabski, S. M., and Crandall, E. D. (1995) Reversible transdifferentiation of alveolar epithelial cells. *Am J Respir Cell Mol Biol.* **12**, 497–502. https://doi.org/10.1165/ajrcmb.12.5.7742013
24. Demaio, L., Tseng, W., Balverde, Z., Alvarez, J. R., Kim, K. J., Kelley, D. G., *et al.* (2009) Characterization of mouse alveolar epithelial cell monolayers. *Am J Physiol Lung Cell Mol Physiol.* **296**, L1051–L1058. https://doi.org/10.1152/ajplung.00021.2009
25. Nabhan, A. N., Brownfield, D. G., Harbury, P. B., Krasnow, M. A., and Desai, T. J. (2018) Single-cell Wnt signaling niches maintain stemness of alveolar type 2 cells. *Science.* **359**, 1118–1123. https://doi.org/10.1126/science.aam6603
26. Zacharias, W. J., Frank, D. B., Zepp, J. A., Morley, M. P., Alkhaleel, F. A., Kong, J., *et al.* (2018) Regeneration of the lung alveolus by an evolutionarily conserved epithelial progenitor. *Nature.* **555**, 251–255. https://doi.org/10.1038/nature25786
27. Dada, L. A., Welch, L. C., Magnani, N. D., Ren, Z., Han, H., Brazee, P. L., *et al.* (2023) Hypercapnia alters stromal-derived Wnt production limiting β-catenin signaling and proliferation in alveolar type 2 cells. *JCI Insight.* **8**. https://doi.org/10.1172/jci.insight.159331
28. Weng, A., Maciel Herrerias, M., Watanabe, S., Welch, L. C., Flozak, A. S., Grant, R. A., *et al.* (2022) Lung injury induces alveolar type 2 cell hypertrophy and polyploidy with implications for repair and regeneration. *Am J Respir Cell Mol Biol.* **66**, 564–576. https://doi.org/10.1165/rcmb.2021-0356OC
29. Schoenfelder, K. P., and Fox, D. T. (2015) The expanding implications of polyploidy. *J Cell Biol.* **209**, 485–491. https://doi.org/10.1083/jcb.201502016
30. Jiang, P., Gil de Rubio, R., Hrycaj, S. M., Gurczynski, S. J., Riemondy, K. A., Moore, B. B., *et al.* (2020) Ineffectual type 2-to-type 1 alveolar epithelial cell differentiation in idiopathic pulmonary fibrosis: Persistence of the Krt8(hi) transitional state. *Am J Respir Crit Care Med.* **201**, 1443–1447. https://doi.org/10.1164/rccm.201909-1726LE
31. Strunz, M., Simon, L. M., Ansari, M., Kathiriya, J. J., Angelidis, I., Mayr, C. H., *et al.* (2020) Alveolar regeneration through a Krt8+ transitional stem cell state that persists in human lung fibrosis. *Nat Commun.* **11**, 3559. https://doi.org/10.1038/s41467-020-17358-3
32. Choi, J., Park, J.-E., Tsagkogeorga, G., Yanagita, M., Koo, B.-K., Han, N., *et al.* (2020) Inflammatory signals induce AT2 cell-derived damage-associated transient progenitors that mediate alveolar regeneration. *Cell Stem Cell.* **27**, 366–382.e367. https://doi.org/10.1016/j.stem.2020.06.020
33. Kobayashi, Y., Tata, A., Konkimalla, A., Katsura, H., Lee, R. F., Ou, J., *et al.* (2020) Persistence of a regeneration-associated, transitional alveolar epithelial cell state in pulmonary fibrosis. *Nat Cell Biol.* **22**, 934–946. https://doi.org/10.1038/s41556-020-0542-8
34. Habermann, A. C., Gutierrez, A. J., Bui, L. T., Yahn, S. L., Winters, N. I., Calvi, C. L., *et al.* (2020) Single-cell RNA sequencing reveals profibrotic roles of distinct epithelial and mesenchymal lineages in pulmonary fibrosis. *Sci Adv.* **6**, eaba1972. https://doi.org/10.1126/sciadv.aba1972
35. Adams, T. S., Schupp, J. C., Poli, S., Ayaub, E. A., Neumark, N., Ahangari, F., *et al.* (2020) Single-cell RNA-seq reveals ectopic and aberrant lung-resident cell populations in idiopathic pulmonary fibrosis. *Sci Adv.* **6**, eaba1983. https://doi.org/10.1126/sciadv.aba1983
36. Delorey, T. M., Ziegler, C. G. K., Heimberg, G., Normand, R., Yang, Y., Segerstolpe, Å., *et al.* (2021) COVID-19 tissue atlases reveal SARS-CoV-2 pathology and cellular targets. *Nature.* **595**, 107–113. https://doi.org/10.1038/s41586-021-03570-8

37. Bharat, A., Querrey, M., Markov, N. S., Kim, S., Kurihara, C., Garza-Castillon, R., *et al.* (2020) Lung transplantation for patients with severe COVID-19. *Sci Transl Med.* **12**. https://doi.org/10.1126/scitranslmed.abe4282
38. Chen, J., Wu, H., Yu, Y., and Tang, N. (2020) Pulmonary alveolar regeneration in adult COVID-19 patients. *Cell Res.* **30**, 708–710. https://doi.org/10.1038/s41422-020-0369-7
39. Han, X., Wang, R., Zhou, Y., Fei, L., Sun, H., Lai, S., *et al.* (2018) Mapping the mouse cell atlas by microwell-seq. *Cell.* **172**, 1091–1107.e1017. https://doi.org/10.1016/j.cell.2018.02.001
40. Penkala, I. J., Liberti, D. C., Pankin, J., Sivakumar, A., Kremp, M. M., Jayachandran, S., *et al.* (2021) Age-dependent alveolar epithelial plasticity orchestrates lung homeostasis and regeneration. *Cell Stem Cell.* **28**, 1775–1789.e1775. https://doi.org/10.1016/j.stem.2021.04.026
41. Shiraishi, K., Shah, P. P., Morley, M. P., Loebel, C., Santini, G. T., Katzen, J., *et al.* (2023) Biophysical forces mediated by respiration maintain lung alveolar epithelial cell fate. *Cell.* https://doi.org/10.1016/j.cell.2023.02.010
42. Jain, R., Barkauskas, C. E., Takeda, N., Bowie, E. J., Aghajanian, H., Wang, Q., *et al.* (2015) Plasticity of Hopx(+) type I alveolar cells to regenerate type II cells in the lung. *Nat Commun.* **6**, 6727. https://doi.org/10.1038/ncomms7727
43. Wang, Y., Tang, Z., Huang, H., Li, J., Wang, Z., Yu, Y., *et al.* (2018) Pulmonary alveolar type I cell population consists of two distinct subtypes that differ in cell fate. *Proc Natl Acad Sci U S A.* **115**, 2407–2412. https://doi.org/10.1073/pnas.1719474115
44. Morrisey, E. E., and Hogan, B. L. (2010) Preparing for the first breath: Genetic and cellular mechanisms in lung development. *Dev Cell.* **18**, 8–23. https://doi.org/10.1016/j.devcel.2009.12.010
45. Montoro, D. T., Haber, A. L., Biton, M., Vinarsky, V., Lin, B., Birket, S. E., *et al.* (2018) A revised airway epithelial hierarchy includes CFTR-expressing ionocytes. *Nature.* **560**, 319–324. https://doi.org/10.1038/s41586-018-0393-7
46. Plasschaert, L. W., Žilionis, R., Choo-Wing, R., Savova, V., Knehr, J., Roma, G., *et al.* (2018) A single-cell atlas of the airway epithelium reveals the CFTR-rich pulmonary ionocyte. *Nature.* **560**, 377–381. https://doi.org/10.1038/s41586-018-0394-6
47. Rock, J. R., Randell, S. H., and Hogan, B. L. M. (2010) Airway basal stem cells: A perspective on their roles in epithelial homeostasis and remodeling. *Dis Model Mech.* **3**, 545–556. https://doi.org/10.1242/dmm.006031
48. Hong, K. U., Reynolds, S. D., Watkins, S., Fuchs, E., and Stripp, B. R. (2004) In vivo differentiation potential of tracheal basal cells: Evidence for multipotent and unipotent subpopulations. *Am J Physiol Lung Cell Mol Physiol.* **286**, L643–L649. https://doi.org/10.1152/ajplung.00155.2003
49. Rock, J. R., Onaitis, M. W., Rawlins, E. L., Lu, Y., Clark, C. P., Xue, Y., *et al.* (2009) Basal cells as stem cells of the mouse trachea and human airway epithelium. *Proc Natl Acad Sci U S A.* **106**, 12771–12775. https://doi.org/10.1073/pnas.0906850106
50. Hong, K. U., Reynolds, S. D., Watkins, S., Fuchs, E., and Stripp, B. R. (2004) Basal cells are a multipotent progenitor capable of renewing the bronchial epithelium. *Am J Pathol.* **164**, 577–588. https://doi.org/10.1016/S0002-9440(10)63147-1
51. Rawlins, E. L., Okubo, T., Xue, Y., Brass, D. M., Auten, R. L., Hasegawa, H., *et al.* (2009) The role of Scgb1a1+ Clara cells in the long-term maintenance and repair of lung airway, but not alveolar, epithelium. *Cell Stem Cell.* **4**, 525–534. https://doi.org/10.1016/j.stem.2009.04.002
52. Tata, P. R., Mou, H., Pardo-Saganta, A., Zhao, R., Prabhu, M., Law, B. M., *et al.* (2013) Dedifferentiation of committed epithelial cells into stem cells in vivo. *Nature.* **503**, 218–223. https://doi.org/10.1038/nature12777
53. Kumar, P. A., Hu, Y., Yamamoto, Y., Hoe, N. B., Wei, T. S., Mu, D., *et al.* (2011) Distal airway stem cells yield alveoli in vitro and during lung regeneration following H1N1 influenza infection. *Cell.* **147**, 525–538. https://doi.org/10.1016/j.cell.2011.10.001
54. Zuo, W., Zhang, T., Wu, D. Z., Guan, S. P., Liew, A. A., Yamamoto, Y., *et al.* (2015) p63(+)Krt5(+) distal airway stem cells are essential for lung regeneration. *Nature.* **517**, 616–620. https://doi.org/10.1038/nature13903
55. Vaughan, A. E., Brumwell, A. N., Xi, Y., Gotts, J. E., Brownfield, D. G., Treutlein, B., *et al.* (2015) Lineage-negative progenitors mobilize to regenerate lung epithelium after major injury. *Nature.* **517**, 621–625. https://doi.org/10.1038/nature14112
56. Xi, Y., Kim, T., Brumwell, A. N., Driver, I. H., Wei, Y., Tan, V., *et al.* (2017) Local lung hypoxia determines epithelial fate decisions during alveolar regeneration. *Nat Cell Biol.* **19**, 904–914. https://doi.org/10.1038/ncb3580

57. Ray, S., Chiba, N., Yao, C., Guan, X., McConnell, A. M., Brockway, B., *et al.* (2016) Rare SOX2(+) airway progenitor cells generate KRT5(+) cells that repopulate damaged alveolar parenchyma following influenza virus infection. *Stem Cell Rep.* **7**, 817–825. https://doi.org/10.1016/j.stemcr.2016.09.010
58. Yang, Y., Riccio, P., Schotsaert, M., Mori, M., Lu, J., Lee, D. K., *et al.* (2018) Spatial-temporal lineage restrictions of embryonic p63(+) progenitors establish distinct stem cell pools in adult airways. *Dev Cell.* **44**, 752–761.e754. https://doi.org/10.1016/j.devcel.2018.03.001
59. Kathiriya, J. J., Wang, C., Zhou, M., Brumwell, A., Cassandras, M., Le Saux, C. J., *et al.* (2022) Human alveolar type 2 epithelium transdifferentiates into metaplastic KRT5+ basal cells. *Nat Cell Biol.* **24**, 10–23. https://doi.org/10.1038/s41556-021-00809-4
60. Jones-Freeman, B., and Starkey, M. R. (2020) Bronchioalveolar stem cells in lung repair, regeneration and disease. *J Pathol.* **252**, 219–226. https://doi.org/10.1002/path.5527
61. Kim, C. F. B., Jackson, E. L., Woolfenden, A. E., Lawrence, S., Babar, I., Vogel, S., *et al.* (2005) Identification of bronchioalveolar stem cells in normal lung and lung cancer. *Cell.* **121**, 823–835. https://doi.org/10.1016/j.cell.2005.03.032
62. Salwig, I., Spitznagel, B., Vazquez-Armendariz, A. I., Khalooghi, K., Guenther, S., Herold, S., *et al.* (2019) Bronchioalveolar stem cells are a main source for regeneration of distal lung epithelia in vivo. *EMBO J.* **38**. https://doi.org/10.15252/embj.2019102099
63. Liu, K., Tang, M., Liu, Q., Han, X., Jin, H., Zhu, H., *et al.* (2020) Bi-directional differentiation of single bronchioalveolar stem cells during lung repair. *Cell Discov.* **6**, 1. https://doi.org/10.1038/s41421-019-0132-8
64. Liu, Q., Liu, K., Cui, G., Huang, X., Yao, S., Guo, W., *et al.* (2019) Lung regeneration by multipotent stem cells residing at the bronchioalveolar-duct junction. *Nat Genet.* **51**, 728–738. https://doi.org/10.1038/s41588-019-0346-6
65. Pan, H., Deutsch, G. H., Wert, S. E., Ontology, S., and Consortium, N. M. A. o. L. D. P. (2019) Comprehensive anatomic ontologies for lung development: A comparison of alveolar formation and maturation within mouse and human lung. *J Biomed Semant.* **10**, 18. https://doi.org/10.1186/s13326-019-0209-1
66. Irvin, C. G., and Bates, J. H. (2003) Measuring the lung function in the mouse: The challenge of size. *Respir Res.* **4**, 4. https://doi.org/10.1186/rr199
67. Weibel, E. R., and Gomez, D. M. (1962) Architecture of the human lung. Use of quantitative methods establishes fundamental relations between size and number of lung structures. *Science.* **137**, 577–585. https://doi.org/10.1126/science.137.3530.577
68. Kadur Lakshminarasimha Murthy, P., Sontake, V., Tata, A., Kobayashi, Y., Macadlo, L., Okuda, K., *et al.* (2022) Human distal lung maps and lineage hierarchies reveal a bipotent progenitor. *Nature.* **604**, 111–119. https://doi.org/10.1038/s41586-022-04541-3
69. Basil, M. C., Cardenas-Diaz, F. L., Kathiriya, J. J., Morley, M. P., Carl, J., Brumwell, A. N., *et al.* (2022) Human distal airways contain a multipotent secretory cell that can regenerate alveoli. *Nature.* **604**, 120–126. https://doi.org/10.1038/s41586-022-04552-0
70. Aspal, M., and Zemans, R. L. (2020) Mechanisms of ATII-to-ATI cell differentiation during lung regeneration. *Int J Mol Sci.* **21**. https://doi.org/10.3390/ijms21093188
71. Hogan, B. L. M. (2020) *The Alveolar Stem Cell Niche of the Mammalian Lung*, pp. 7–12, Springer International Publishing, Singapore
72. Costa-Mattioli, M., and Walter, P. (2020) The integrated stress response: From mechanism to disease. *Science.* **368**. https://doi.org/10.1126/science.aat5314
73. Bilen, M., Benhammouda, S., Slack, R. S., and Germain, M. (2022) The integrated stress response as a key pathway downstream of mitochondrial dysfunction. *Curr Opin Physiol.* **27**, 100555. https://doi.org/10.1016/j.cophys.2022.100555
74. Borck, G., Shin, B. S., Stiller, B., Mimouni-Bloch, A., Thiele, H., Kim, J. R., *et al.* (2012) eIF2γ mutation that disrupts eIF2 complex integrity links intellectual disability to impaired translation initiation. *Mol Cell.* **48**, 641–646. https://doi.org/10.1016/j.molcel.2012.09.005
75. Gregory, L. C., Ferreira, C. B., Young-Baird, S. K., Williams, H. J., Harakalova, M., van Haaften, G., *et al.* (2019) Impaired EIF2S3 function associated with a novel phenotype of X-linked hypopituitarism with glucose dysregulation. *EBioMedicine.* **42**, 470–480. https://doi.org/10.1016/j.ebiom.2019.03.013

76. Moortgat, S., Désir, J., Benoit, V., Boulanger, S., Pendeville, H., Nassogne, M. C., *et al.* (2016) Two novel EIF2S3 mutations associated with syndromic intellectual disability with severe microcephaly, growth retardation, and epilepsy. *Am J Med Genet A*. **170**, 2927–2933. https://doi.org/10.1002/ajmg.a.37792
77. Skopkova, M., Hennig, F., Shin, B.-S., Turner, C. E., Stanikova, D., Brennerova, K., *et al.* (2017) EIF2S3 mutations associated with severe X-linked intellectual disability syndrome MEHMO. *Hum Mutat*. **38**, 409–425. https://doi.org/10.1002/humu.23170
78. Kernohan, K. D., Tétreault, M., Liwak-Muir, U., Geraghty, M. T., Qin, W., Venkateswaran, S., *et al.* (2015) Homozygous mutation in the eukaryotic translation initiation factor 2alpha phosphatase gene, PPP1R15B, is associated with severe microcephaly, short stature and intellectual disability. *Hum Mol Genet*. **24**, 6293–6300. https://doi.org/10.1093/hmg/ddv337
79. Emanuelli, G., Nassehzadeh-Tabriz, N., Morrell, N. W., and Marciniak, S. J. (2020) The integrated stress response in pulmonary disease. *Eur Respir Rev*. **29**. https://doi.org/10.1183/16000617.0184-2020
80. Kashiwagi, K., Yokoyama, T., Nishimoto, M., Takahashi, M., Sakamoto, A., Yonemochi, M., *et al.* (2019) Structural basis for eIF2B inhibition in integrated stress response. *Science*. **364**, 495–499. https://doi.org/10.1126/science.aaw4104
81. Sidrauski, C., Tsai, J. C., Kampmann, M., Hearn, B. R., Vedantham, P., Jaishankar, P., *et al.* (2015) Pharmacological dimerization and activation of the exchange factor eIF2B antagonizes the integrated stress response. *eLife*. **4**, e07314. https://doi.org/10.7554/eLife.07314
82. Zyryanova, A. F., Weis, F., Faille, A., Alard, A. A., Crespillo-Casado, A., Sekine, Y., *et al.* (2018) Binding of ISRIB reveals a regulatory site in the nucleotide exchange factor eIF2B. *Science*. **359**, 1533–1536. https://doi.org/10.1126/science.aar5129
83. Wong, Y. L., LeBon, L., Basso, A. M., Kohlhaas, K. L., Nikkel, A. L., Robb, H. M., *et al.* (2019) eIF2B activator prevents neurological defects caused by a chronic integrated stress response. *Elife*. **8**. https://doi.org/10.7554/eLife.42940
84. Wong, Y. L., LeBon, L., Edalji, R., Lim, H. B., Sun, C., and Sidrauski, C. (2018) The small molecule ISRIB rescues the stability and activity of Vanishing White Matter Disease eIF2B mutant complexes. *Elife*. **7**. https://doi.org/10.7554/eLife.32733
85. Krukowski, K., Nolan, A., Frias, E. S., Boone, M., Ureta, G., Grue, K., *et al.* (2020) Small molecule cognitive enhancer reverses age-related memory decline in mice. *Elife*. **9**. https://doi.org/10.7554/eLife.62048
86. Chou, A., Krukowski, K., Jopson, T., Zhu, P. J., Costa-Mattioli, M., Walter, P., *et al.* (2017) Inhibition of the integrated stress response reverses cognitive deficits after traumatic brain injury. *Proc Natl Acad Sci U S A*. **114**, E6420–E6426. https://doi.org/10.1073/pnas.1707661114
87. Halliday, M., Radford, H., Sekine, Y., Moreno, J., Verity, N., le Quesne, J., *et al.* (2015) Partial restoration of protein synthesis rates by the small molecule ISRIB prevents neurodegeneration without pancreatic toxicity. *Cell Death Dis*. **6**, e1672. https://doi.org/10.1038/cddis.2015.49
88. Kabir, Z. D., Che, A., Fischer, D. K., Rice, R. C., Rizzo, B. K., Byrne, M., *et al.* (2017) Rescue of impaired sociability and anxiety-like behavior in adult cacna1c-deficient mice by pharmacologically targeting eIF2α. *Mol Psychiatry*. **22**, 1096–1109. https://doi.org/10.1038/mp.2017.124
89. Sidrauski, C., Acosta-Alvear, D., Khoutorsky, A., Vedantham, P., Hearn, B. R., Li, H., *et al.* (2013) Pharmacological brake-release of mRNA translation enhances cognitive memory. *eLife*. **2**, e00498. https://doi.org/10.7554/eLife.00498
90. Watanabe, S., Markov, N. S., Lu, Z., Piseaux Aillon, R., Soberanes, S., Runyan, C. E., *et al.* (2021) Resetting proteostasis with ISRIB promotes epithelial differentiation to attenuate pulmonary fibrosis. *Proc Natl Acad Sci U S A*. **118**. https://doi.org/10.1073/pnas.2101100118
91. Dobrinskikh, E., Hennessy, C. E., Kurche, J. S., Kim, E., Estrella, A. M., Cardwell, J., *et al.* (2023) Epithelial endoplasmic reticulum stress enhances the risk of Muc5b-associated lung fibrosis. *Am J Respir Cell Mol Biol*. **68**, 62–74 https://doi.org/10.1165/rcmb.2022-0252OC
92. Onat, U. I., Yildirim, A. D., Tufanli, Ö., Çimen, I., Kocatürk, B., Veli, Z., *et al.* (2019) Intercepting the lipid-induced integrated stress response reduces atherosclerosis. *J Am Coll Cardiol*. **73**, 1149–1169. https://doi.org/10.1016/j.jacc.2018.12.055
93. Nguyen, H. G., Conn, C. S., Kye, Y., Xue, L., Forester, C. M., Cowan, J. E., *et al.* (2018) Development of a stress response therapy targeting aggressive prostate cancer. *Sci Transl Med*. **10**. https://doi.org/10.1126/scitranslmed.aar2036
94. Mick, E., Titov, D. V., Skinner, O. S., Sharma, R., Jourdain, A. A., and Mootha, V. K. (2020) Distinct mitochondrial defects trigger the integrated stress response depending on the metabolic state of the cell. *Elife*. **9**. https://doi.org/10.7554/eLife.49178

95. Ryan, D. G., Yang, M., Prag, H. A., Blanco, G. R., Nikitopoulou, E., Segarra-Mondejar, M., *et al.* (2021) Disruption of the TCA cycle reveals an ATF4-dependent integration of redox and amino acid metabolism. *Elife*. **10**. https://doi.org/10.7554/eLife.72593
96. Krall, A. S., Mullen, P. J., Surjono, F., Momcilovic, M., Schmid, E. W., Halbrook, C. J., *et al.* (2021) Asparagine couples mitochondrial respiration to ATF4 activity and tumor growth. *Cell Metab*. **33**, 1013–1026.e1016. https://doi.org/10.1016/j.cmet.2021.02.001
97. Quiros, P. M., Prado, M. A., Zamboni, N., D'Amico, D., Williams, R. W., Finley, D., *et al.* (2017) Multi-omics analysis identifies ATF4 as a key regulator of the mitochondrial stress response in mammals. *J Cell Biol*. **216**, 2027–2045. https://doi.org/10.1083/jcb.201702058
98. Kaspar, S., Oertlin, C., Szczepanowska, K., Kukat, A., Senft, K., Lucas, C., *et al.* (2021) Adaptation to mitochondrial stress requires CHOP-directed tuning of ISR. *Sci Adv*. **7**, eabf0971. https://doi.org/10.1126/sciadv.abf0971
99. Croon, M., Szczepanowska, K., Popovic, M., Lienkamp, C., Senft, K., Brandscheid, C. P., *et al.* (2022) FGF21 modulates mitochondrial stress response in cardiomyocytes only under mild mitochondrial dysfunction. *Sci Adv*. **8**, eabn7105. https://doi.org/10.1126/sciadv.abn7105
100. Guo, X., Aviles, G., Liu, Y., Tian, R., Unger, B. A., Lin, Y.-H. T., *et al.* (2020) Mitochondrial stress is relayed to the cytosol by an OMA1–DELE1–HRI pathway. *Nature*. **579**, 427–432. https://doi.org/10.1038/s41586-020-2078-2
101. Fessler, E., Eckl, E. M., Schmitt, S., Mancilla, I. A., Meyer-Bender, M. F., Hanf, M., *et al.* (2020) A pathway coordinated by DELE1 relays mitochondrial stress to the cytosol. *Nature*. **579**, 433–437. https://doi.org/10.1038/s41586-020-2076-4
102. Bar-Ziv, R., Bolas, T., and Dillin, A. (2020) Systemic effects of mitochondrial stress. *EMBO Rep*. **21**, e50094. https://doi.org/10.15252/embr.202050094
103. Khan, N. A., Nikkanen, J., Yatsuga, S., Jackson, C., Wang, L., Pradhan, S., *et al.* (2017) mTORC1 regulates mitochondrial integrated stress response and mitochondrial myopathy progression. *Cell Metab*. **26**, 419–428.e415. https://doi.org/10.1016/j.cmet.2017.07.007
104. Forsström, S., Jackson, C. B., Carroll, C. J., Kuronen, M., Pirinen, E., Pradhan, S., *et al.* (2019) Fibroblast growth factor 21 drives dynamics of local and systemic stress responses in mitochondrial myopathy with mtDNA deletions. *Cell Metab*. **30**, 1040–1054.e1047. https://doi.org/10.1016/j.cmet.2019.08.019
105. Fiorese, C. J., Schulz, A. M., Lin, Y. F., Rosin, N., Pellegrino, M. W., and Haynes, C. M. (2016) The transcription factor ATF5 mediates a mammalian mitochondrial UPR. *Curr Biol*. **26**, 2037–2043. https://doi.org/10.1016/j.cub.2016.06.002
106. Bueno, M., Calyeca, J., Rojas, M., and Mora, A. L. (2020) Mitochondria dysfunction and metabolic reprogramming as drivers of idiopathic pulmonary fibrosis. *Redox Biol*. **33**, 101509. https://doi.org/10.1016/j.redox.2020.101509
107. Li, X., Zhang, W., Cao, Q., Wang, Z., Zhao, M., Xu, L., *et al.* (2020) Mitochondrial dysfunction in fibrotic diseases. *Cell Death Dis*. **6**, 80. https://doi.org/10.1038/s41420-020-00316-9
108. Rangarajan, S., Bernard, K., and Thannickal, V. J. (2017) Mitochondrial dysfunction in pulmonary fibrosis. *Ann Am Thorac Soc*. **14**, S383–S388. https://doi.org/10.1513/AnnalsATS.201705-370AW
109. Zank, D. C., Bueno, M., Mora, A. L., and Rojas, M. (2018) Idiopathic pulmonary fibrosis: Aging, mitochondrial dysfunction, and cellular bioenergetics. *Front Med (Lausanne)*. **5**, 10. https://doi.org/10.3389/fmed.2018.00010
110. Han, S., and Chandel, N. S. (2021) Lessons from cancer metabolism for pulmonary arterial hypertension and fibrosis. *Am J Respir Cell Mol Biol*. **65**, 134–145. https://doi.org/10.1165/rcmb.2020-0550TR
111. Cuevas-Mora, K., Roque, W., Shaghaghi, H., Gochuico, B. R., Rosas, I. O., Summer, R., *et al.* (2021) Hermansky-Pudlak syndrome-2 alters mitochondrial homeostasis in the alveolar epithelium of the lung. *Respir Res*. **22**, 49. https://doi.org/10.1186/s12931-021-01640-z
112. Liu, Q., Zhou, Y., Cogan, J. D., Mitchell, D. B., Sheng, Q., Zhao, S., *et al.* (2023) The genetic landscape of familial pulmonary fibrosis. *Am J Respir Crit Care Med*. **207**, 1345–1357. https://doi.org/10.1164/rccm.202204-0781OC
113. Aros, C. J., Pantoja, C. J., and Gomperts, B. N. (2021) Wnt signaling in lung development, regeneration, and disease progression. *Commun Biol*. **4**, 601. https://doi.org/10.1038/s42003-021-02118-w
114. Baarsma, H. A., and Konigshoff, M. (2017) "WNT-er is coming": WNT signalling in chronic lung diseases. *Thorax*. **72**, 746–759. https://doi.org/10.1136/thoraxjnl-2016-209753

115. Lee, J.-H., Tammela, T., Hofree, M., Choi, J., Marjanovic, N. D., Han, S., *et al.* Anatomically and functionally distinct lung mesenchymal populations marked by Lgr5 and Lgr6. *Cell.* **170**, 1149–1163.e1112. https://doi.org/10.1016/j.cell.2017.07.028
116. Königshoff, M., Balsara, N., Pfaff, E. M., Kramer, M., Chrobak, I., Seeger, W., *et al.* (2008) Functional Wnt signaling is increased in idiopathic pulmonary fibrosis. *PLoS ONE.* **3**, e2142. https://doi.org/10.1371/journal.pone.0002142
117. Lam, A. P., Flozak, A. S., Russell, S., Wei, J., Jain, M., Mutlu, G. M., *et al.* (2011) Nuclear β-catenin is increased in systemic sclerosis pulmonary fibrosis and promotes lung fibroblast migration and proliferation. *Am J Respir Cell Mol Biol.* **45**, 915–922. https://doi.org/10.1165/rcmb.2010-0113OC
118. Lam, A. P., Herazo-Maya, J. D., Sennello, J. A., Flozak, A. S., Russell, S., Mutlu, G. M., *et al.* (2014) Wnt coreceptor Lrp5 is a driver of idiopathic pulmonary fibrosis. *Am J Respir Crit Care Med.* **190**, 185–195. https://doi.org/10.1164/rccm.201401-0079OC
119. Nabhan, A. N., Webster, J. D., Adams, J. J., Blazer, L., Everrett, C., Eidenschenk, C., *et al.* (2023) Targeted alveolar regeneration with Frizzled-specific agonists. *Cell.* https://doi.org/10.1016/j.cell.2023.05.022
120. Gorgoulis, V., Adams, P. D., Alimonti, A., Bennett, D. C., Bischof, O., Bishop, C., *et al.* (2019) Cellular senescence: Defining a path forward. *Cell.* **179**, 813–827. https://doi.org/10.1016/j.cell.2019.10.005
121. Hayflick, L., and Moorhead, P. S. (1961) The serial cultivation of human diploid cell strains. *Exp Cell Res.* **25**, 585–621. https://doi.org/10.1016/0014-4827(61)90192-6
122. Baker, D. J., Wijshake, T., Tchkonia, T., LeBrasseur, N. K., Childs, B. G., van de Sluis, B., *et al.* (2011) Clearance of p16Ink4a-positive senescent cells delays ageing-associated disorders. *Nature.* **479**, 232–236. https://doi.org/10.1038/nature10600
123. Baker, D. J., Childs, B. G., Durik, M., Wijers, M. E., Sieben, C. J., Zhong, J., *et al.* (2016) Naturally occurring p16(Ink4a)-positive cells shorten healthy lifespan. *Nature.* **530**, 184–189. https://doi.org/10.1038/nature16932
124. Chang, J., Wang, Y., Shao, L., Laberge, R. M., Demaria, M., Campisi, J., *et al.* (2016) Clearance of senescent cells by ABT263 rejuvenates aged hematopoietic stem cells in mice. *Nat Med.* **22**, 78–83. https://doi.org/10.1038/nm.4010
125. Bussian, T. J., Aziz, A., Meyer, C. F., Swenson, B. L., van Deursen, J. M., and Baker, D. J. (2018) Clearance of senescent glial cells prevents tau-dependent pathology and cognitive decline. *Nature.* **562**, 578–582. https://doi.org/10.1038/s41586-018-0543-y
126. Grosse, L., Wagner, N., Emelyanov, A., Molina, C., Lacas-Gervais, S., Wagner, K. D., *et al.* (2020) Defined p16(High) senescent cell types are indispensable for mouse healthspan. *Cell Metab.* **32**, 87–99.e86. https://doi.org/10.1016/j.cmet.2020.05.002
127. Omori, S., Wang, T. W., Johmura, Y., Kanai, T., Nakano, Y., Kido, T., *et al.* (2020) Generation of a p16 reporter mouse and its use to characterize and target p16(high) cells in vivo. *Cell Metab.* **32**, 814–828.e816. https://doi.org/10.1016/j.cmet.2020.09.006
128. Reyes, N. S., Krasilnikov, M., Allen, N. C., Lee, J. Y., Hyams, B., Zhou, M., *et al.* (2022) Sentinel p16(INK4a+) cells in the basement membrane form a reparative niche in the lung. *Science.* **378**, 192–201. https://doi.org/10.1126/science.abf3326
129. Zhou, Y., Horowitz, J. C., Naba, A., Ambalavanan, N., Atabai, K., Balestrini, J., *et al.* (2018) Extracellular matrix in lung development, homeostasis and disease. *Matrix Biol.* **73**, 77–104. https://doi.org/10.1016/j.matbio.2018.03.005
130. Liu, Z., Wu, H., Jiang, K., Wang, Y., Zhang, W., Chu, Q., *et al.* (2016) MAPK-mediated YAP activation controls mechanical-tension-induced pulmonary alveolar regeneration. *Cell Rep.* **16**, 1810–1819. https://doi.org/10.1016/j.celrep.2016.07.020
131. Wu, H., Yu, Y., Huang, H., Hu, Y., Fu, S., Wang, Z., *et al.* (2020) Progressive pulmonary fibrosis is caused by elevated mechanical tension on alveolar stem cells. *Cell.* **180**, 107–121.e117. https://doi.org/10.1016/j.cell.2019.11.027
132. Liu, F., Lagares, D., Choi, K. M., Stopfer, L., Marinković, A., Vrbanac, V., *et al.* (2015) Mechanosignaling through YAP and TAZ drives fibroblast activation and fibrosis. *Am J Physiol Lung Cell Mol Physiol.* **308**, L344–L357. https://doi.org/10.1152/ajplung.00300.2014
133. Noguchi, S., Saito, A., Mikami, Y., Urushiyama, H., Horie, M., Matsuzaki, H., *et al.* (2017) TAZ contributes to pulmonary fibrosis by activating profibrotic functions of lung fibroblasts. *Sci Rep.* **7**, 42595. https://doi.org/10.1038/srep42595

134. Haak, A. J., Kostallari, E., Sicard, D., Ligresti, G., Choi, K. M., Caporarello, N., *et al.* (2019) Selective YAP/TAZ inhibition in fibroblasts via dopamine receptor D1 agonism reverses fibrosis. *Sci Transl Med.* **11**. https://doi.org/10.1126/scitranslmed.aau6296
135. Mayr, C. H., Sengupta, A., Ansari, M., Pestoni, J. C., Ogar, P., Angelidis, I., *et al.* (2022) Autocrine Sfrp1 inhibits lung fibroblast invasion during transition to injury induced myofibroblasts. *bioRxiv*. https://doi.org/10.1101/2022.07.11.499594
136. Tsukui, T., Sun, K.-H., Wetter, J. B., Wilson-Kanamori, J. R., Hazelwood, L. A., Henderson, N. C., *et al.* (2020) Collagen-producing lung cell atlas identifies multiple subsets with distinct localization and relevance to fibrosis. *Nat Commun.* **11**, 1920. https://doi.org/10.1038/s41467-020-15647-5
137. Lechner, A. J., Driver, I. H., Lee, J., Conroy, C. M., Nagle, A., Locksley, R. M., *et al.* (2017) Recruited monocytes and type 2 immunity promote lung regeneration following pneumonectomy. *Cell Stem Cell.* **21**, 120–134.e127. https://doi.org/10.1016/j.stem.2017.03.024
138. Misharin, A. V., Morales-Nebreda, L., Reyfman, P. A., Cuda, C. M., Walter, J. M., McQuattie-Pimentel, A. C., *et al.* (2017) Monocyte-derived alveolar macrophages drive lung fibrosis and persist in the lung over the life span. *J Exp Med.* **214**, 2387–2404. https://doi.org/10.1084/jem.20162152
139. Joshi, N., Watanabe, S., Verma, R., Jablonski, R. P., Chen, C. I., Cheresh, P., *et al.* (2020) A spatially restricted fibrotic niche in pulmonary fibrosis is sustained by M-CSF/M-CSFR signalling in monocyte-derived alveolar macrophages. *Eur Respir J.* **55**. https://doi.org/10.1183/13993003.00646-2019
140. McQuattie-Pimentel, A. C., Ren, Z., Joshi, N., Watanabe, S., Stoeger, T., Chi, M., *et al.* (2021) The lung microenvironment shapes a dysfunctional response of alveolar macrophages in aging. *J Clin Invest.* **131**. https://doi.org/10.1172/JCI140299
141. Morales-Nebreda, L., Helmin, K. A., Torres Acosta, M. A., Markov, N. S., Hu, J. Y., Joudi, A. M., *et al.* (2021) Aging imparts cell-autonomous dysfunction to regulatory T cells during recovery from influenza pneumonia. *JCI Insight.* **6**. https://doi.org/10.1172/jci.insight.141690

3 Environmental Toxicants in Lung Diseases

Samuel J. Cochran, Evangeline Schott, Brett Saunders, and Kymberly M. Gowdy

3.1 INTRODUCTION

Lung biology and lung diseases are shaped by both endogenous and exogenous factors. Generally, endogenous factors refer to unique genetic differences giving rise to altered and potentially deleterious health outcomes. Exogenous factors include any kind of physical, chemical, biological, or behavioral stimulus that induces adverse health outcomes in an individual. However, there can be interaction between both types of factors. Individuals may be genetically predisposed to developing lung cancer; they may additionally or independently encounter an environmental toxicant that induces a genetic mutation and favors lung carcinogenesis. Because of this interaction, lung diseases are numerous with multitudinous etiologies. Not only can environmental exposure alter DNA stability, but it may also alter other cellular and molecular mechanisms involved in injurious and inflammatory pathways that underlie disease. There are numerous compounds capable of inducing toxicological or carcinogenic effect, although some are more potent and more ubiquitously encountered than others.

Some environmental toxicants arise from natural sources; however, anthropogenic sources, such as air pollution, heavily contribute to environmental toxicant exposure. In fact, worldwide air pollution is associated with worsening morbidity and mortality for many diseases. These diseases are not exclusive to the pulmonary system; however, there are many associated lung diseases, as inhalation is a primary route of exposure. Moreover, changing trends in air pollution have dramatic implications for human lung health. Worldwide, air pollution is estimated to account for 7 million premature deaths annually (1). However, air pollution is increasing due in part to the effects of climate change and despite regulations from organization such as the US Environmental Protection Agency and the European Environment Agency (2–5). Thus, morbidity and mortality for pulmonary diseases are predicted to worsen. It is imperative to understand how exposure occurs, how it may change over time, and how toxicants drive lung disease to attenuate associated human health burdens.

This chapter will provide a review of environmental toxicants and their role in initiating and exacerbating lung disease. Section 3.2 will provide a synopsis of the epidemiology and broad pathology for four prevalent types of lung disease. Section 3.3 will describe mechanisms of disease pathogenesis for common toxicants associated with these lung diseases. Toxicants will be classified by physical properties (gaseous phase, volatile organic compounds, and solid particulate matter) with attention paid to anthropogenic versus natural sources. We will also discuss natural and anthropogenic sources of these toxicants. Finally, Section 3.4 will highlight the importance of regulation and climate change in the future of environmental toxicant exposure.

3.2 ENVIRONMENTAL LUNG DISEASES AND ASSOCIATED PATHOLOGIES

3.2.1 Lung Cancer

Lung cancer remains the leading cause of cancer death worldwide, accounting for approximately 1 million deaths per year. Global incidence of lung cancer has been increasing over the last two

DOI: 10.1201/9781003355243-4

decades, though distribution primarily reflects demographic shifts in tobacco use. Most cases of lung cancer are attributable to cigarette smoking; however, 15% of lung cancers in men and 53% in women arise in never smokers (6). Variability in lung cancer suggests that different types are present in smokers versus nonsmokers. There are four major types of lung cancer: (1) small cell lung cancer (SCLC), which stands alone; as well as (2) squamous cell carcinoma (SCC); (3) adenocarcinoma; and (4) large cell carcinoma, which comprises the broader category of non–small cell lung cancer (7). Both SCLC and SCC are more strongly associated with smoking than adenocarcinoma. In contrast, adenocarcinomas are the most common lung cancer presentations in nonsmokers (8). All four lung cancer types share certain common features with most cancer pathophysiology.

Carcinogenesis, or the development of cancer, is a multistep process that occurs over many years. The four major steps in carcinogenesis are (1) initiation, (2) promotion, (3) progression, and (4) sometimes, metastasis. Though the cell types and genetic mutations may differ between cancers, these four steps are always necessary for carcinogenesis. Initiation begins with a stochastic, or random, mutation in a cancer-controlling gene, in a cell type with potential to propagate that mutation. The mutation can arise as a result of exogenous influence (e.g., ultraviolet radiation causing thymine dimers) or DNA copying errors. Genes with potential for cancer development are classified as either proto-oncogenes (POGs) or tumor suppressor genes (TSPs). Functioning normally, POGs stimulate healthy cell division, whereas TSPs monitor and repair DNA mutations, or initiate programmed cell death (apoptosis) if DNA damage is irreparable. TSP are said to be "caretaker" genes if they repair damage and "gatekeeper" genes if they prevent further division of a mutated cell (9). If a mutation occurs in the promoter region of a POG, such that it is constitutively activated, then that cell and its progeny may undergo rampant cell proliferation. If a mutation occurs in a TSP such that it becomes nonfunctional, then mutated cells can divide and accumulate more mutations. In order for this to occur, mutated cells need to be of a type with potential to divide further. Generally, these cells are adult stem cells. Adult stem cells are present in the epithelia of most tissues and exist to replace aged cells or cells that slough off. Unlike embryonic stem cells, adult stem cells are not totipotent/pluripotent (meaning that they cannot differentiate into any cell type). Instead, these cells are committed to a specific cell lineage (9).

Following initiation, a stem cell with a mutation in a POG or TSG may undergo promotion. This means it will undergo multiple divisions to propagate the mutation to form an abnormal growth of cells, or neoplasm. This process is also referred to as clonal expansion. The mutated cell "clone" is expanded many times, increasing the potential for acquisition of additional mutations. This accumulation of further mutations is the hallmark of carcinogenesis progression. In contrast to the initiation stage, mutations that arise during progression can be either stochastic or deterministic, as the cell clone with the most robust mutations is favored for propagation. This favoritism is known as clonal selection. Some mutations, referred to as passenger mutations, do not contribute to the neoplasm phenotype. Other mutations, known as driver mutations, provide molecular heterogeneity and gain of function necessary for neoplasms to become malignant and even to metastasize to other tissues. These are the general stages to carcinogenesis; however, nuances to this pathophysiology give rise to the four primary types of lung cancer (9).

Epidermal growth factor receptor (EGFR) is one of three major genes involved in lung cancer pathogenesis and is overexpressed in approximately 50% of lung cancers (10). EGFR belongs to a family of surface receptor tyrosine kinases that form intrafamily dimers when bound to their ligands, peptide growth factors. Normally this promotes the proliferation of healthy cells. When overexpressed, this initiates an anti-apoptotic signaling cascade and results in cell proliferation, angiogenesis, healthy tissue invasion, and metastasis. This cascade is made possible by activation of three pathways: (1) the Akt pathway, which inhibits apoptosis; (2) the Ras-mitogen-activated protein kinase (MAPK) pathway, which increases cell proliferation; and (3) the signal transducer and activator of transcription proteins (STAT) signaling pathway, which induces expression of various cancer and inflammation mediators (10). Single nucleotide polymorphisms and amplifying mutations in EGFR may have a role in adenocarcinoma development among nonsmokers. In fact, EGFR is the first gene in which mutations have been correlated with never-smoker status (11, 12). However, it has been associated with exposure to particulate air pollution. Alveolar macrophages exposed to particulate matter less than 2.5 μm in diameter ($PM_{2.5}$) produce interleukin IL-1β, a common pro-inflammatory cytokine. This induces a progenitor state ($KRT8^+$) in EGFR mutant alveolar type II epithelial cells that is associated with lung adenocarcinoma (13). Thus, particulate air pollution may be important for the promotion of EGFR mutants in development of lung adenocarcinomas among never smokers.

Similar to mutations in EGFR, mutations in the Kirsten rat sarcoma virus (KRAS) gene are primary to adenocarcinomas in the peripheral airways (14, 15). However, unlike EGFR mutations, KRAS mutations occur more frequently in cancers arising in smokers (16). Active forms of the carcinogens nicotine-derived nitrosamine ketone and benzo-[a]-pyrene—both of which are found in tobacco smoke—produce DNA adducts in KRAS (8). Missense mutations result in a loss of GTPase activity in Ras proteins, which promote cell proliferation. This loss of function prevents active Ras proteins from returning to the inactive GDP-bound form, and results in constitute activation of Ras and oncogenic cell proliferation (17).

A third mutation common to lung carcinogenesis occurs in the TP53 gene. Under normal circumstances, this gene encodes for the tumor suppressor p53 protein, which acts as a transcription factor for many genes, and ultimately regulates responses to DNA damage, including apoptosis (18). Mutation in this gene can result in loss of DNA binding activity in the translated protein, leading to dysregulation of programmed cell death and therefore increased potential for clonal expansion. Though they occur less commonly in nonsmokers, mutations in TP53 are observed in both smokers and nonsmokers (19).

Finally, some initiators or promoters of lung carcinogenesis occur not at the genetic but instead the epigenetic level. Epigenetic alterations affect the expression of genes without impacting the genetic code itself. These alterations include increases to expression through mechanisms such as acetylation or euchromatin remodeling. They also include decreases in expression, or inactivation, through mechanisms such as methylation or heterochromatic remodeling. Increased methylation of genetic promoter regions containing many cysteine-guanine repeats (CpG islands) can lead to inactivation (gene silencing) of tumor suppressor genes, leading to carcinogenesis (20).

3.2.2 Asthma

Asthma is a chronic inflammatory disease of the airways, characterized by cough, wheeze, chest tightness, and shortness of breath. Inflammation is driven by infiltration and activation of immune cells in the airspace, as well as mucus production, airway remodeling, and reversible airway hyperresponsiveness (AHR). Airway remodeling is characterized by muscle hyperplasia, goblet cell metaplasia, and increased subepithelial collagen deposition. AHR is generally reversible with corticosteroid treatment, though this is not phenotypic for all asthma (21). Asthma phenotypes are the observable characteristics or traits for the disease state, which have historically been used to classify and treat asthma subgroups. More recently, asthma endotypes have been defined. Whereas phenotypes do not account for mechanisms of disease biology, endotypes describe distinct pathophysiology (22, 23). Differences between the two primary asthma endotypes, type 2–high and type 2–low, are outlined in this section and in Figure 3.1.

3.2.2.1 Type 2–High Asthma

Type 2–high asthma is often synonymous with T_H2-high or T_H2-mediated asthma, essentially eosinophilic, allergic, and severe asthma (21). This is due to the cell types and cellular responses typically associated with progression of a type 2–high asthma response to environmental stimuli. Pathogen-associated molecular patterns (PAMPs) and other stimulatory molecules from exogenous sources may be inhaled into the airspace where they can bind to pattern recognition receptors (PRRs) on epithelial cells. These PRRs include toll-like receptors (TLRs), nucleotide-binding

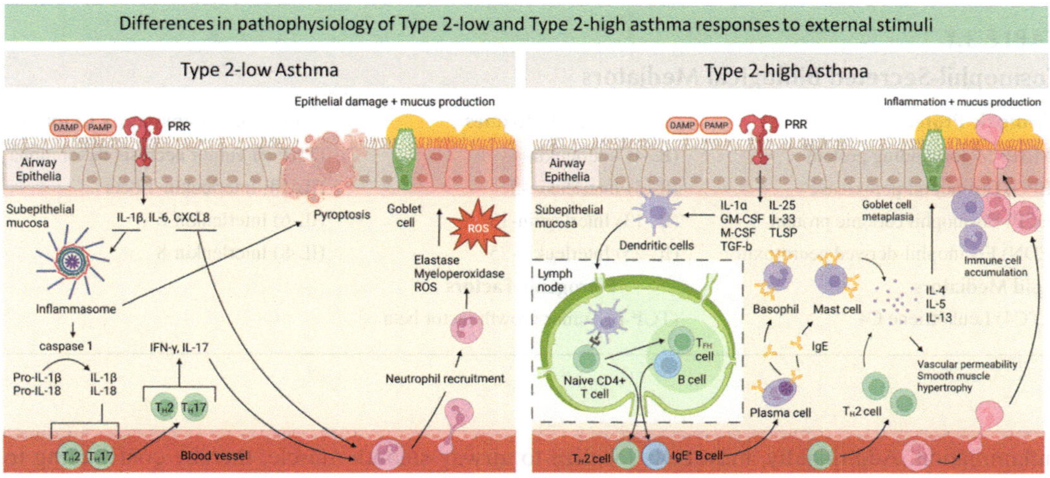

FIGURE 3.1 Differences in type 2–low and type 2–high asthma pathophysiology. Type 2–low asthma responses are triggered by PAMPs or DAMPs that bind to PRRs on the apical surface of airway epithelial cells. Receptor ligand interaction releases IL-1β, IL-6, and CXCL8. IL-1β activates inflammasomes, which release caspase 1. Caspase 1 induces pyroptosis and catalyzes transformation of pro–IL-1β and pro–IL-18 into IL-1β and IL-18. These cytokines recruit T_H2 and T_H17 lymphocytes, which secrete IFN-γ and IL-17, creating a microenvironment favorable for neutrophil recruitment. CXCL8 also drives neutrophil recruitment. Neutrophils release elastase, myeloperoxidase, and ROS, all of which induce epithelial injury and induce mucus production. In type 2–high asthma responses, cytokines and growth factors (IL-1α, IL-25, IL-33, TLSP, M-CSF, GM-CSF, TGF-b) activate resident DCs, which then migrate to draining lymph nodes. DCs induce maturation of naïve CD4+ T cells into T_{FH} and T_H2 cells. T_{FH} cells induce antibody class switching to IgE in follicular B cells. T_H2 and IgE^+ B cells migrate to the airway, where B cells differentiate into IgE producing plasma cells. IgE binds to FcεRI on mast cells and basophils, where it can cross-link with other IgE antibodies and antigens. Cross-linking stimulates release of IL-4, IL-5, and IL-13, which induce goblet cell metaplasia and mucus overproduction, as well as vascular permeability. Permeability encourages immune cells accumulation at epithelial and smooth muscle layers, causing inflammation and smooth muscle hypertrophy. DAMP, damage-associated molecular pattern; DC, dendritic cell; IFN, interferon; Ig, immunoglobulin; IL, interleukin; PAMP, pathogen-associated molecular pattern; PRR, pattern recognition receptor; ROS, reactive oxygen species.

Source: Figure adapted from Fahy (2015) with modifications from Hammad and Lambrecht (2021), created with BioRender.com.

oligomerization domain–like receptors, c-type lectin receptors, retinoic acid-inducible gene I–like receptors, protease-activated receptors, and purinergic receptors. In response to this receptor-ligand interaction, airway epithelial cells express pro-inflammatory chemokines and cytokines, including IL-1α; granulocyte-macrophage colony–stimulating factor (GM-CSF); macrophage colony–stimulating factor (M-CSF); and transforming growth factor beta (TGF-β), as well as the well-studied class of cytokines, alarmins (24–27). Alarmins include IL-25, IL-33, and thymic stromal lymphopoietin. In turn, these cytokines activate tissue-resident dendritic cells (DCs), which then migrate to draining lymph nodes. There, DCs promote development of T helper 2 (T_H2) lymphocytes, which trigger maturation of localized B cells into plasma cells (21, 28–30). Another type of T lymphocytes, follicular helper T lymphocytes (T_{FH}), release IL-4 and IL-13 cytokines, which stimulate these plasma cells to undergo class switching from primarily IgG producing to long-lived IgE producing (31, 32). IgE produced by these plasma cells can bind to FcεRI on basophils, mast cells, eosinophils, DCs, smooth muscle cells, endothelial cells, and epithelial cells (33). IgE on the surface of basophils and mast cells can cross-link in response to antigen exposure, stimulating these cells to release histamine, leukotrienes, and prostaglandins (PGD2), as well as IL-4, IL-5, and IL-9 (21, 34). This cascade rapidly induces vascular permeability and cell recruitment to the lung, increasing

TABLE 3.1
Eosinophil-Secreted Biological Mediators

Cytotoxic Proteins	T_H2 Cytokines	Acute Pro-Inflammatory Cytokines
(MBP) Major binding protein	(IL-4) Interleukin-4	(TNF-α) Tumor necrosis factor alpha
(EPO) Eosinophil peroxidase	(IL-5) Interleukin-5	(IL-1β) Interleukin-1 beta
(ECP) Eosinophil cationic protein	(IL-13) Interleukin-13	(IL-6) Interleukin-6
(EDN) Eosinophil-derived neurotoxin	(IL-25) Interleukin-25	(IL-8) Interleukin-8
Lipid Mediators	**Fibrogenic Factors**	
(LTC4) Leukotriene C4	(TGF-β) Tumor growth factor beta	

inflammation. Additionally, mast cells localize to airway smooth muscle, thereby contributing to hypertrophy, hyperplasia, and AHR (35). They also localize near airway submucosal glands and begin to secrete PGD2, IL-4, and IL-13. This stimulates goblet cells to produce mucus in excess. IgE can also bind directly to epithelial cells and DCs. In response, epithelial cells proliferate, contract, and release pro-inflammatory cytokines. IgE facilitates antigen presentation by dendritic cells to T_H2 cells, creating a positive feedback cycle of IgE production.

T_H2 cells are critically important to the type 2–high response by producing many modulating cytokines and chemokines. IL-4 signaling stimulates rampant IgE production and airway inflammation. IL-13 increases AHR, signals airway goblet cells to undergo metaplasia, and stimulates airway epithelia to produce inducible nitric oxide synthase, a key mediator of immune activation and inflammation (21, 36). IL-5 drives recruitment of eosinophils to the airway, hence why type 2–high asthma is sometimes referred to as "essentially eosinophilic asthma." Once recruited to the airways, eosinophils release cytotoxic proteins (Table 3.1) that further exacerbate AHR and goblet cell metaplasia (37). These proteins can also cause damage to structural cells in the airway during periods of chronic inflammation (38). Thus, eosinophilia is both a hallmark of type 2–high asthma and an exacerbating phenotype of this endotype.

3.2.2.2 Type 2–Low Asthma

The type 2–low asthma endotype is also synonymous with T_H2-low, and sometimes T_H1/T_H17 mediated or neutrophilic asthma. However, the extent to the role of neutrophils in type 2–low pathology is still uncertain (39). In some cases, it is a less severe endotype compared to type 2–high; however, patients with type 2–low asthma are generally corticosteroid refractory (23, 40). Despite this, there is generally a high corticosteroid use among patients with type 2–low asthma, which contributes to obesity, an associated morbidity of type 2–low asthma. In addition to higher body mass index, type 2–low asthma is associated with later disease onset, increased IL-1β, inflammasome activation through nuclear factor kappa B (NF-κB) signaling, and neutrophilic inflammation (41, 42). Upon exposure to PAMPs or damage-associated molecular patterns (DAMPs), airway epithelial cells and tissue resident macrophages secrete inflammatory cytokines (IL-1β and IL-6). They also secrete CXCL8 (IL-8), a chemokine that recruits neutrophils to the airways. Activated inflammasomes also increase pro-inflammatory cytokines in the airway by recruiting and activating caspase-1, which cleaves pro–IL-1β and pro–IL-18 into IL-1β and IL-18, respectively. Caspase-1 additionally induces pyroptosis, an inflammatory cell death (43). The unique cytokine environment recruits T_H1 and T_H17 cells, which in turn secrete interferon gamma (IFN-γ) and IL-17. These molecules are involved in the recruitment and stimulation of neutrophils to the airways, increasing the inflammatory response. Stimulated neutrophils release elastase, myeloperoxidase, and reactive oxygen species (ROS), which damage airway epithelial cells and induce mucus production (21).

3.2.3 Chronic Obstructive Pulmonary Disease

Chronic obstructive pulmonary disease (COPD) is characterized by a progressively limited airflow that is not fully reversible. Generally, this is brought on by an aggravated inflammatory response to harmful airborne particulates or gases (44, 45). Another key characteristic of COPD is emphysema, or the breakdown of alveolar septa, which permanently increases the size of the airspace and decreases elastic recoil and gas exchange (46, 47). Hyperinflation caused by alveolar wall destruction also leads to dysfunction of the diaphragm. The aberrant inflammatory response characteristic of COPD further weakens diaphragm functionality through infiltration of inflammatory cells, as well as oxidative stress, causing damage to the diaphragm itself (48–52). Oxidative stress is the primary driver of COPD pathogenesis, resulting in pro-inflammatory gene expression, cell death, extracellular tissue remodeling, and mucus secretion (47, 53, 54).

The most common source of oxidative stress in COPD pathology comes from cigarette smoke; however, airway immune cells also produce ROS in response to harmful exposures (55, 56). These exposures can include bacterial and viral infections, which significantly contribute to COPD exacerbation, morbidity, and mortality (57, 58). Patients with COPD have elevated numbers of macrophages, neutrophils, and T lymphocytes in their airspace (47). This contributes to overall inflammation. Additionally, increased nicotinamide adenine dinucleotide phosphate (NADP) signaling in phagocytes and epithelial cells, as well as myeloperoxidase generated by neutrophils, induce ROS production and tissue damage in the COPD lung (55, 56). NF-κB, activated by particulate matter, also promotes tissue damage by induction of cytokines, chemokines, and other inflammatory mediators (54).

In general, the pathophysiology of COPD can be summarized as follows. Airway epithelial cells release CXCL8 in response to air pollutants (59). This chemokine recruits and activates neutrophils to the airspace. Epithelial cells also release CXCL8 CXCL9, CXCL10, CXCL11, and CXCL12, which recruit and activate T helper 1 and T cytotoxic 1 lymphocytes (60). Cytotoxic T lymphocytes release perforins, granzyme B, and tumor necrosis factor alpha (TNF-α), which trigger unprogrammed (cytolysis) and programmed (apoptosis) death of alveolar epithelial cells (61). Loss of alveolar epithelium integrity results in emphysema and consequently reduced lung function. T lymphocytes also secrete IFN-γ, which activates alveolar macrophages (62, 63). Activated macrophages also produce the aforementioned T lymphocyte chemoattractants, exacerbating the injurious effects of cytotoxic T lymphocyte recruitment (44). Additionally, macrophages produce the chemokines CXCL1 and CXCL8, which recruit and activate neutrophils. In turn, neutrophils can secrete elastases and proteases, leading to destruction of alveolar integrity, increased mucus secretion, and ultimately declining lung function (44, 64). Both activated neutrophils and macrophages produce ROS, which drives lung injury, and in the case of macrophage-derived ROS, promote secretion of elastolytic enzymes such as matrix metalloproteinases (MMPs) (65). This family of enzymes drives tissue remodeling in the COPD lung. Fibrosis also contributes to tissue remodeling. Airway epithelial cells release fibroblast growth factor (FGF) and TGF-β, which respectively recruit fibroblasts and differentiate them into myofibroblasts. Myofibroblasts produce extracellular matrix (ECM), leading to lung fibrosis and making it difficult for gas exchange to occur (65, 66). Similarly, goblet cells in the COPD lung are stimulated by TGF-α to hypersecrete mucus, diminishing oxygen diffusion (67–69). The resulting airflow restriction, as well as inflammation, tissue remodeling, and alveolar destruction, manifest in loss of lung function in individuals with COPD.

3.2.4 Interstitial Lung Disease

Interstitial lung disease encompasses a host of diseases with varied etiologies; however, many of them share similar pathophysiology. They are all characterized by chronic inflammation that results in scarring (fibrosis) of the lung, making the lung less elastic and compliant. This loss of elasticity not only makes it more difficult to take in air (reduces the forced vital capacity [FVC]), but it also restricts the available surface area for gas exchange, making it more difficult to obtain sufficient oxygen. Symptoms include chest pain, shortness of breath, non-productive cough, fatigue, and pulmonary edema. Additional disease presentation is determined by location and cause of lung fibrosis. Inflammation can occur in the bronchioles, alveoli, or in the capillaries surrounding the alveoli (bronchiolitis, alveolitis, and vasculitis, respectively) (70). Sources of inflammation can be endogenous or exogenous. Some types of granulomatous interstitial lung disease can result from immune hyperactivity of unknown endogenous source (71). In the case of sarcoidosis, these granulomas occur throughout the body (including the lung) but are rarely life-threatening (72). Another granulomatous interstitial lung disease, hypersensitivity pneumonitis, is characterized by an allergy to certain environmental dusts. Exposure to these dusts can sometimes cause chronic inflammation leading to scarring (fibrosis) in the lung. Other exogenous dusts can cause interstitial lung disease through non-allergic mechanisms. These diseases, termed pneumoconioses, encompass coal pneumoconiosis (sometimes called "black lung disease"); byssinosis caused from inhalation of cotton, textile, or other plant-related dusts ("brown lung disease"); asbestosis from exposure to asbestos; and silicosis from exposure to silica dusts (71, 73). These diseases occur most commonly in the workplace and so are considered to be occupationally derived. Pneumoconioses can be considered "simple" if they result in limited fibrosis, whereas complicated pneumoconiosis result in significant scarring and progressive massive fibrosis. Finally, interstitial lung disease that result in pulmonary scarring and arise from unknown sources are classified as idiopathic pulmonary fibrosis. Interstitial lung diseases are discussed in later sections in greater detail and with respect to the unique exposure that gives rise to different disease etiologies.

3.3 ENVIRONMENTAL TOXICANTS: SOURCES AND MODES OF ACTION

A toxicant is any abiotic element or compound that induces harmful effects upon exposure. Toxicants differ from toxins, which are biotic poisons produced by living organisms. Toxicants are environmentally ubiquitous, occurring both indoors and outdoors. Some occur naturally, whereas others are produced from anthropogenic activities. Many are involved in pathogenesis of lung diseases, but due to their highly variable backgrounds and properties, they represent heterogenous etiologies. We will discuss some of the more common and well-researched environmental toxicants in the following sections. They may be classified based on a number of criteria. Mode of action includes allergens, irritants, and carcinogens. Exposures settings are generally indoors/outdoors or residential/occupational. We have chosen to distinguish the following toxicants based on their physical phase when encountered. This includes gaseous phase toxicants, volatile organic compounds, and respirable particulate matter.

3.3.1 Gaseous Phase

3.3.1.1 Carbon Monoxide (CO)

CO is a colorless, odorless gas regulated by the EPA as one of six criteria air pollutants. Sources include vehicle exhaust, malfunctioning gas- and oil-burning furnaces, and fire emissions (74). CO binds with high affinity to hemoglobin (Hb) in the blood, displacing oxygen and stabilizing oxygen already bound to Hb (75). These two properties reduce oxygen delivery to oxygen-depleted tissues, which can lead to CO poisoning. However, the role of CO in disease pathogenesis is complicated, demonstrating both toxic and protective potential.

In addition to fatal CO poisoning, high concentrations of exogenous CO can induce organ toxicity. Despite being an inhaled toxicant, the organ systems most affected are cardiac/circulatory and nervous. Specifically, CO induces endothelial inflammation ultimately leading to cardiac injury. CO can bind ferrous heme a_3 in the COX active site of mitochondria, disrupting normal phosphorylation, and triggering the release of superoxide and free heme (76). Free heme increases both oxidative stress and inflammation. It also induces upregulation of heme oxygenase-1 (HO-1), which degrades free heme in biliverdin, iron, and CO, thereby sustaining CO concentrations in localized tissues (77, 78). CO can also bind to platelets, displacing surface NO, which in turn forms peroxynitrite when combined with superoxide produced by inhibited mitochondria. Both CO and peroxynitrite activate platelets, which subsequently recruit neutrophils and stimulate them to degranulate and release myeloperoxidase (MPO) (79–81). MPO also stimulates neutrophils to release proteases, which induce ROS production in endothelia. ROS generated by protease activity and MPO lead to lipid peroxidation, which produces adducts in myelin, and subsequently triggers lymphocyte and microglia activity (82).

CO also exhibits anti-inflammatory potential and may be generally protective against some lung diseases. In fact, it is generated at low concentrations as part of the HO-1 antioxidative pathway. Furthermore, low concentrations of CO can upregulate MAPK pathways p38 and JNK1/2, reducing production of pro-inflammatory cytokines IL-6, TNF-α, IL-1β, and MIP1α (77). Relatively low concentrations of CO (250 ppm) can also reduce airway inflammation, fibrin deposition, and pulmonary edema after hyperoxic lung injury (83). A higher dose (500 ppm) may reduce airway neutrophilia but can also result in high carboxyhemoglobin levels, which adversely affects oxygen delivery (84). Further research is required, but several studies have found promise investigating CO as potential therapy in lung injury (76).

Low concentrations of CO demonstrate anti-inflammatory potential and protection against lung disease, whereas high concentrations can induce cardiac and neurological disease. Variability in protective effects among studies speaks to the difficulty in refining a CO therapy. Finally, despite being an inhaled toxicant, CO primarily induces adverse effects in cardiac function and may even offer some protection against lung injury. This principle is important to consider throughout this chapter as many toxicants demonstrate dual and opposing effects depending on exposure route and concentration.

3.3.1.2 Carbon Dioxide (CO_2)

CO_2 is a colorless gas with significant implications for human health, both in its role as a greenhouse gas and as a hazardous inhalant. CO_2 is the primary anthropogenic-sourced greenhouse gas, defined by the EPA as a gas that traps heat in the atmosphere (85). Changes in atmospheric CO_2 have implications for global temperatures and altered patterns in severe weather events. For humans, acute and chronic exposure to CO_2 in both outdoor and indoor environments has significant implications for inflammation. Outdoor CO_2 exposure primarily comes from anthropogenic sources (e.g., vehicle exhaust), and urban environments contribute disproportionately to CO_2 emissions (86–88). Furthermore, geographic and meteorological effects, such as inversions, can combine to contain rising concentrations of CO_2 within a region (e.g., in Phoenix, Arizona, the urban CO_2 dome reaches 650 ppm) (89, 90). Indoor exposure to CO_2 is influenced by outdoor concentrations but is primarily driven by occupancy and ventilation rates. Average indoor CO_2 concentrations range from 600 to 1000 ppm but can exceed 2000 ppm because of variability in occupancy and ventilation (91). Mean personal exposure levels to CO_2 that exceed 1000 ppm for more than 2.5 hours have been designated the exposure level of possible concern (92). Some buildings, including some schools, regularly reach levels between 1000 ppm and 2000 ppm (93). Exposure to CO_2 at the upper end of this range can induce inflammatory responses, including nucleotide-binding domain-like receptor 3 (NLRP3) inflammasome activation, and IL-1β production by neutrophils (94). Exposure to higher concentrations can result in inflammation of the bronchial epithelium and even the nasal passages (95). Chronic exposure can deoxygenate hemoglobin, resulting in mild hypoxemia and production of peroxynitrite. The subsequent oxidative damage in erythrocytes can activate macrophages, which in turn release pro-inflammatory cytokines, further contributing to systemic inflammation and tissue-specific oxidative stress (96, 97).

3.3.1.3 Ozone (O_3)

Ground-level ozone is another criteria air pollutant. It is important to distinguish ground-level ozone (ozone in the lower troposphere) from ozone in the higher atmospheres. Humans are not exposed to stratospheric ozone, which acts as a shield against ultraviolet radiation. However, humans can be exposed to ozone in the troposphere, resulting in exacerbation of cardiopulmonary diseases. Ozone is also a secondary air pollutant, meaning it forms from primary air pollutants in the atmosphere. Nitrogen oxides (NO_X) and volatile organic compounds (VOCs) combine with heat and sunlight to photochemically produce ozone at ground level. NO_X and VOC species are produced from incomplete combustion of carbon fuels, both naturally (e.g., forest fires) and from anthropogenic sources (e.g., vehicle exhaust) (98). VOCs are also found in manufactured products (especially building materials) as solvents, adhesives, and preservatives (99). Outdoor ozone concentrations generally range between 20–40 ppb but may increase with higher temperatures, more intense sunlight, and presence of primary air pollutants (100, 101). The primary source of indoor ozone is infiltrating outdoor air, and generally indoor ozone concentration ranges from 30% to 70% of outdoor levels. Other indoor sources include common commercial and residential appliances such as laser printers, photocopiers, and some air purifiers (102, 103). Ambient concentrations of ozone are predicted to increase with rising global temperatures, due to the effects of climate change (101). This has significant implications for global health burdens. Each 10 ppb increase in daily ozone exposure is associated with a 0.87% increase in total mortality. Currently, ozone exposure is estimated to account for 800 premature deaths, 4500 hospital admissions, and 900,000 school absences each year. The associated economic burden is estimated to be $5 billion annually (104).

Ozone is a highly reactive oxidizing chemical. Despite its relative insolubility, approximately 80% of inhaled ozone is deposited in the airways. However, insolubility and reactivity of ozone prevents it from directly acting upon the airway epithelium and reaching underlying tissues. Indeed, the immediate effects of ozone exposure are limited to the respiratory tract. When ozone interacts with the airway surface liquid (ASL), it readily reacts with carbon double bonds in phospholipids, cholesterol, and mucins to form primary ozonide (105, 106). Further reactions result in formation of DAMPs and oxidative metabolites, including ROS. These intermediates can then stimulate downstream inflammatory responses, including activation of pulmonary macrophages and recruitment of neutrophils to the airway. DAMPs are recognized by TLRs located on macrophage cellular membranes (107). This activates the NF-kB protein complex, which stimulates production of pro-inflammatory cytokines (TNF-α, IL-1β, IL-6, GM-CSF) (108, 109). These cytokines induce permeability as well as apoptosis and cellular damage in airway epithelial cells. They also induce, in tandem with ROS, cytokine production by airway epithelial cells. ROS also activates the NLRP3 inflammasome, which induces pyroptosis and production of IL-1β and IL-18 through caspase 1. Epithelial-derived cytokines further induce inflammation (IL-6 and GM-CSF), including recruitment of neutrophils (via IL-8 signaling) (110). In turn, neutrophils can release oxidants and proteases, which also contribute to injury and inflammation. Oxidants from neutrophils can generate disulfide bonds in mucin molecules, which increases their elasticity and reduces their mobility, making it more difficult to clear mucus accumulation (111). That said, acute exposure to ozone induces a greater restrictive effect in the lung, rather than an obstructive effect. This is due to stimulation of airway neural receptors, including C-fibers, which may respond to a host of ozone-induced nociceptive signals (e.g., prostaglandin [PGE_2]) (112, 113). This airway restriction results in lower FVC and overall lung function. Finally, ozone and its metabolites oxidize phospholipids in ASL, triggering a lipid peroxidation cascade (114, 115). This rampant peroxidation results in compromised cell membrane integrity and can lead to cell programmed death. In fact, accumulation of intracellular lipid peroxide pools is characteristic of ferroptosis, a type of cell death mediated by iron overload (116). This corresponds to an increase in iron accumulation in the airspace and increased concentrations of iron-sequestering complexes (lactoferrin and transferrin) after ozone exposure (117).

3.3.1.4 Radon (Rn)

Radon is estimated to be associated with approximately 20,000 lung cancer deaths in the United States every year. This burden is disproportionately shared by uranium miners; however, 30%

of lung cancer deaths among non-occupationally exposed never-smokers are attributable to indoor exposure including exposure in residences (118). Radon is a ubiquitous, colorless, odorless gas found outdoors and indoors. The average concentration of radon in outdoor air is 0.4 pCi/L, whereas the average indoor radon concentration for America's homes is about 1.3 pCi/L. The EPA bases its estimate of 21,000 radon-related lung cancers a year on this indoor average (119). Radon is considered a carcinogen (no exposure level is regarded as safe); therefore, the EPA recommends that Americans consider home radon remediation for levels between 2 pCi/L and 4 pCi/L.

Radon forms from the breakdown of uranium in the Earth's crust. It is chemically inert, but readily decays into active products ("radon progeny"), which can attach to airborne particulate or aerosols. Decay of radon or its progeny emits alpha particles that can damage DNA directly—causing double-stranded breaks and chromosomal aberrations (120)—or indirectly through the production of radical species (121, 122). This DNA damage can lead to mutations in tumor suppressor genes or proto-oncogenes involved in lung cancer progression. Pro-inflammatory cytokines are also produced in response to alpha particle exposure, suggesting an inflammatory role in radon-induced carcinogenesis (123).

3.3.2 Volatile Organic Compounds

Volatile organic compounds (VOCs) are a class of compounds that readily volatize, meaning that they have a high vapor pressure and may shift from a solid or liquid to a gaseous state. As such, they occupy a space between gaseous toxicants and particulates. However, some VOCs (e.g., polyacrylic aromatic hydrocarbons [PAHs]) can be considered particulates. Other than aldehydes, these compounds are generally composed of one or more aromatic rings, which increases their vapor pressure. They may be produced from combustion sources and are commonly used as solvents and preservatives in adhesive compounds, building materials, and cosmetics (124). Exposure to VOCs may directly impact human health, or VOCs can combine with other air pollutants to form additional, potentially hazardous exposure (e.g., combines with NO_X species to form O_3). There are many compounds that are classified as VOCs. We will discuss two compounds with high exogenous prevalence and important implications for human health.

3.3.2.1 Formaldehyde (H_2CO)

Formaldehyde exposure can occur from endogenous and exogenous sources. Endogenous sources are generally low concentration and result from metabolic reactions, including the deamination of creatine or adrenaline metabolic products. Breakdown of folate derivatives and demethylation of nucleic acid can also result in endogenous formaldehyde (125–127). However, endogenous formaldehyde can be short lived when formaldehyde catabolic pathways are present and functional. Glutathione (GSH) and alcohol dehydrogenase 5 react with formaldehyde and S-formylglutathione to yield formate and prevent formaldehyde toxicity (128, 129). However, deficiency in this machinery can lead to DNA adducts and carcinogenesis.

Exogenous formaldehyde is also considered to be a carcinogen. Incomplete combustion, including industrial processes, vehicle exhaust, and cigarette smoking all are possible sources of exogenous formaldehyde exposure (130–132). Formaldehyde is also commonly found in building materials and household products where it is used as a preservative and solvent due to its toxicity and activity as a VOC. Exposure to high concentrations of formaldehyde is implicated in pathogenesis of lung cancer, asthma, COPD, and lung fibrosis. Formaldehyde can react with amine groups in the nuclear bases that form DNA and RNA, resulting in base adducts and DNA cross-linkage (133). If unrepaired, this DNA damage can result in genomic instability, leading to accumulating mutations and eventually carcinogenesis. Formaldehyde can also induce global methylation changes and histone alterations in respiratory cells, leading to further instability (134). Additionally, formaldehyde promotes degradation of the tumor suppressor gene, BRCA2; however, the mechanisms are still not fully understood (135).

Formaldehyde exposure also induces pulmonary inflammation, which has implications for lung cancer as well as other lung diseases. Inhalation induces production of IgE as well as airway hyperresponsiveness, which are important factors in asthma exacerbation (134). Formaldehyde can also activate protein complexes, which upregulate expression of pro-inflammatory cytokines, furthering asthma exacerbation and acute lung injury (136). NF-κB signaling can also activate the NLRP3 inflammasome, which contributes to production of pro-inflammatory cytokines, including IL-1β (137). This pathway is involved in airway hyperresponsiveness characteristic in asthma, especially type 2–high asthma.

NF-κB signaling is also important for formaldehyde in lung fibrosis and COPD pathogenesis. Formaldehyde exposure induces hypomethylation of CpG islands in the gene encoding for the G-protein coupled receptor, 5-hydroxytryptamine receptor 4 (HTR4). Overexpression of HTR4 increases NF-κB activation and subsequent secretion of TGF-β1 (138). TGF-β1 induces fibrotic responses by activating myofibroblasts and stimulating the secretion of extracellular matrix. Formaldehyde exacerbates fibrosis by inducing inflammation in the airspace, including recruitment of granulocytes and generation of pro-inflammatory cytokines (IL-1 and IL-17) and collagenase (139). These factors increase inflammation and reduce overall lung elastance, exacerbating fibrosis and injuring cellular architecture in the lung. These conditions also reduce lung structural integrity, thereby promoting emphysema, a characteristic step in the pathogenesis of COPD (140).

3.3.2.2 Benzene (C_6H_6)

Benzene is a chemical in a special class of VOCs, termed PAHs, known for their aromatic rings (covalently bonded, resonate carbon rings). Benzene is the simplest PAH, with only one ring and no peripheral chemical structures. This structure is why benzene is commonly used in various industries. It acts as a base structure for manufacturing other chemicals, including polymers, plastics, and pesticides. As an aromatic chemical, it is commonly used in solvents and adhesive products, making it prevalent in building materials. Benzene also occurs naturally in petroleum and exposure can occur through petroleum products. However, the general population is predominately exposed to benzene through cigarette smoke. Occupational exposure usually occurs in industries handling chemical production or refinement as well as production of construction materials. Currently, the occupational exposure standard set by the US Occupational Safety and Health Administration (OSHA) is 1 ppm, although adverse health outcomes are possible at exposures below this threshold. Benzene is known to be a hematotoxicant (meaning it causes toxicity in blood cells) and a carcinogen (mainly resulting in blood and bone marrow cancers) (141). However, it may also induce lung injury, lung fibrosis, and acute symptoms such as dizziness and headache. Benzene results in blood cell apoptosis and depression of hematopoiesis (142, 143). It also alters expression and secretion of cytokines and immunological growth factors. For example, benzene increases CYP4F3A expression, which results in increased leukotriene B4 (LTB_4) hydrolase concentrations, LTB_4 inaction, and neutropenia (144, 145). This immunosuppression results in increased susceptibility to respiratory infections, historically observed among workers exposed to high doses of benzene. Inhaled benzene also directly results in lung injury by inducing apoptosis in lung cells (146, 147). This injury is exacerbated by airway remodeling and accumulation of ECM without repair in the lung parenchyma. Benzene induces this remodeling by altering the physiological profile of MMP, endopeptidases that normally cleave one or more aspects of ECM. Imbalance in MMP and their regulators, inhibitors of metalloproteinases, is associated with fibrosis pathogenesis (146, 148). Benzene is one of the most common PAH, but the health effects from exposure to benzene can greatly differ from effects resulting from exposure to other PAHs For instance, another PAH, toluene-diisocyanate (TDI), induces occupational asthma. TDI triggers airway remodeling through E-cadherin, and activates the NLRP3 inflammasome leading to inflammation and an imbalance in pro-inflammatory cytokines and ROS (149, 150). The role of PAH compounds in general is discussed further in the next section.

3.3.3 Respirable Particulate Matter

Respirable particulate matter is commonly classified by particulate size, which can impact the depth of instillation in the lung as well as the disease pathogenesis when inhaled. Particles larger than 10 microns (μm) in diameter are readily inhalable. These include sand and soil particulates, and some pollen and tend to settle in the nasal passages if inhaled. Particles less than 10 μm in diameter (coarse, PM_{10}) include asbestos fibers, organic and inorganic dusts, pollen, bacteria, and fungal spores. These particles tend to deposit in areas of the upper respiratory tract, trachea, bronchi, and bronchioles. This also makes PM_{10} easier to clear through ciliary action. Particles less than 2.5 μm in diameter (fine, $PM_{2.5}$) include combustion particles, some organic compounds, and certain metals. These compounds are capable of depositing deep into the respiratory tract, including the alveoli, and are even capable of diffusing into the blood stream (151). The chemical, biological, and physical properties of particulate matter constituents determine the physiological effects upon inhalation. We will discuss some of these differences in the following sections.

3.3.3.1 Inorganic Occupational Dusts: Coal Dust, Asbestos, Silica

Exposure to inorganic dusts commonly occurs in mining/quarrying and manufacturing industries (152–154). Inhalation of inorganic dusts can cause pneumoconiosis, a class of interstitial lung disease defined by dust inhalation and characterized by chronic pulmonary inflammation and fibrosis (153). Coal pneumoconiosis from coal dust and silicosis from crystalline silicon dioxide inhalation share similar etiologies. In fact, they are common co-exposures in coal mining populations (155, 156). The disease severity associated with these exposures is due to the total exposure concentration and duration of exposure, as well as properties of the mineral dust and presence of other minerals. For instance, acute silicosis presents as alveolar silicolipoproteinosis, whereas classic silicosis presents either as "simple" nodular silicosis or "complicated" conglomatous silicosis (157). However, there is some debate to the fibrogenecity of different silica types, with some suggesting that quartz (the most common form of crystalline silica) may be less fibrogenic than other silica types (e.g., tridymite and cristobalite), whereas others see little difference (158, 159). Moreover, the presence of other minerals may alter silica toxicity (153).

In general, the pathogenesis for coal pneumoconiosis or silicosis is as follows. Alveolar macrophages are injured by inability to completely phagocytose inhaled dusts. In response to this damage, macrophages release ROS and lysosomal enzymes, which results in tissue damage. They also release pro-inflammatory cytokines (TNF-α, IL-1, arachidonic acid metabolites, cytokine-induced neutrophil chemoattractant, and macrophage inflammatory protein). Some of these cytokines are responsible for recruiting inflammatory cells to the lung, including additional macrophages (differentiated from circulating monocytes), lymphocytes, and neutrophils. These recruited cells are responsible for the initiation of alveolitis, subsequent damage to pulmonary tissues, and development of a fibrotic scar. Following this tissue damage, alveolar macrophages initiate a reparative phase by releasing TNF, IL-1, FGF, platelet derived growth factor, and insulin-like growth factor. These growth factors promote the recruitment and proliferation of type II pneumocytes and fibroblasts, which in turn induce fibrosis by overproducing fibronectin and collagen molecules (160, 161).

A very similar pathophysiology is directed by inhalation of asbestos fibers. During asbestosis, macrophages injured by incomplete fiber phagocytosis initiate an inflammatory response, followed by a repair process that leads to development of fibrotic scars. As with coal and silica dusts, properties of asbestos fiber and presence of other minerals alters toxicity (162). Shorter fibers (e.g., chrysolite) are more easily resolved, whereas longer fibers (e.g., amosite and crocidolite) bio-persist (163). Additionally, the presence of iron leads to greater redox reactivity and potential for lung injury (162). Separate from asbestosis, asbestos can also promote carcinogenesis, including malignant mesothelioma and bronchogenic carcinoma cancer types. Asbestos exposure is associated with approximately 10,000 annual lung cancer deaths in the United States (164, 165). The carcinogenic mechanisms induced by asbestos inhalation are similar to asbestosis. Chronic inflammation and oxidative stress lead to DNA damage, especially through generation of ROS, including superoxide and hydrogen peroxide (166–168). ROS induces oxidation of nucleic acid bases, single-strand breaks, and apoptosis. Additionally, phagocytosed asbestos fibers can interfere with chromosomal segregation during mitosis, resulting in aneuploidy and micronuclei (169, 170). This can lead to genomic instability and eventually lung cancer. There is also some evidence for involvement of asbestos-derived mutations in EGFR, KRAS, and TP53, leading to lung carcinogenesis (171, 172).

3.3.3.2 Organic Occupational Dusts: Wood, Textile, and Plant Dusts

Exposure to organic dusts commonly occurs in industries handling woodworking, textiles, and any manipulation of plant material. Inhalation of these dusts has myriad effects dependent upon and complicated by mixed dust constituents. These constituents may have irritating, inflammatory, allergenic, fibrotic, and even carcinogenic effects if inhaled. For instance, dust from several wood types is considered to be carcinogenic, including dust from both oak and beech wood. Other wood dusts, from birch, mahogany, teak, and walnut, are also suspected to be carcinogenic if inhaled (173). The carcinogenic mechanisms remain unresolved, but it is hypothesized that lung and nasal cancers may arise from carcinogenic chemicals inherent in (e.g., tannins) or added to (e.g., aldehydes) wood products. Alternatively, carcinogenesis may result from interference with mucociliary action, thereby increasing exposure susceptibility to other exogenous carcinogens. Another carcinogenic factor may be prolonged inflammation, marked by heightened neutrophil and eosinophil recruitment as well as cytokine production (174–176). Eosinophil recruitment is also important in regulation of allergenic responses to wood dust. Wood dust inhalation can exacerbate T_H2–high asthma through eosinophil recruitment (primarily birch dust) and transcription of genes encoding for T_H2-specific cytokines (primarily oak dust) (174).

Plant type may be important to disease etiology, but so too may be microbial contamination in the inhaled dust. This is especially true for organic fibers found in many textile industries, including cotton, flax, hemp, and jute (177–180). Chronic inhalation of these fibers results in a type of interstitial lung disease known as byssinosis, characterized by airflow obstruction and airway inflammation. Contamination by Gram-negative and Gram-positive bacteria, actinomycetes, and fungi is not uncommon in these organic fibers, and bacterial-derived endotoxin contributes to byssinosis development. Endotoxin-contaminated organic fibers induce goblet cell hyperplasia, resulting in increased mucin production and airflow obstruction. Neutrophil recruitment also induces inflammation, which can result in fibrotic scarring after sustained exposure (173). This fibrosis is characteristic of byssinosis and other pneumoconioses.

3.3.3.3 Particulate Matter Less Than 2.5 μm in Diameter ($PM_{2.5}$)

As $PM_{2.5}$ can enter circulation and is otherwise difficult to clear, it represents significant health risks upon inhalation (151). $PM_{2.5}$ is also highly heterogeneous, comprising carbon particles, metal particulates, and bioaerosols (including microorganisms and microbially derived compounds) (181). Not only the aerodynamic diameter of particulates, but also the constituents, determine the toxicological potential when inhaled. Different mixtures of $PM_{2.5}$ are associated with lung cancer, asthma, COPD, and fibrosis pathology, driven primarily through DNA damage, oxidative stress, and inflammation.

Carbon particles include both elemental carbon (EC) and organic carbon (OC). EC exists in graphite or amorphous forms and is synonymous with black carbon (BC) or carbon black (CB) (182). Although there are differences in hazard potential for molecules classified independently as BC or CB, we will refer to both simply as EC when discussing associated health effects. OC references

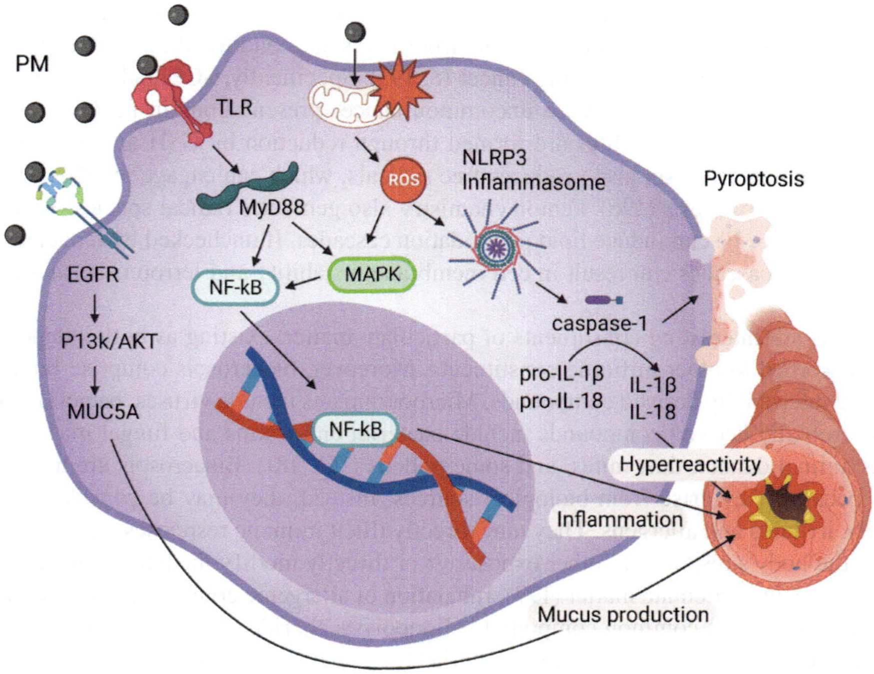

FIGURE 3.2 Cellular and molecular mechanisms of $PM_{2.5}$-induced airway injury. PM binds to TLRs on cellular membranes, which recruits MyD88 and activates the NF-κB and MAPK pathways. NF-κB promotes expression of pro-inflammatory cytokines, which lead to airway inflammation. PM also induces mitochondrial damage, leading to mitochondrial function disruption and generation of ROS. ROS further activate the MAPK pathway, which in turn may further upregulate the NF-κB pathway, heightening inflammation. ROS also activate the NLRP3 inflammasome, which releases caspase 1. Caspase 1 induces pyroptosis and cleaves pro–IL-1β and pro–IL-18 into IL-1β and IL-18, thus increasing airway inflammation and responsiveness. PM may also bind to EGFR on cell surface membranes, activating the P13K/AKT pathway and upregulating expression of MUC5A. This triggers mucus hyperproduction leading to airway obstruction. *EGFR*, epidermal growth factor receptor; *MAPK*, mitogen-activated protein kinase; *MUC5*, mucin-5AC; *MyD88*, myeloid differentiation primary response 88; *NF-κB*, nuclear factor kappa–light chain–enhancer of activated B cells; *NLRP3*, nod-like receptor family pyrin domain containing 3; *P13K/AKT*, phosphoinositide 3-kinase/protein kinase B; *PM*, particulate matter; *ROS*, reactive oxygen species; *TLR*, toll-like receptors.

Source: Adapted from Wang and Liu (2023) with modification, created with BioRender.com.

carbon in a myriad of organic molecules, where it is bound to other elements. OC includes PAHs, which use carbon as a cyclic backbone (181). EC activates the NLRP3 inflammasome and induces autophagy, both driving fibrosis in the lung (183). EC can also increase cell membrane permeability, thereby increasing pulmonary cell susceptibility to other harmful PM constituents (184). PAHs increase expression of CYP1A1 and OCG1, resulting in DNA damage and eventually lung carcinogenesis (185, 186). PAHs also induce inflammatory responses by increasing expression of IL-6 and IL-8, and activation of the NLRP3 inflammasome through ROS generated by mitochondrial damage (187, 188) (Figure 3.2). A specific class of PAHs, nitrated PAHs, also activate NF-κB, Akt phosphorylation, and MAPK pathways (189).

Metal signatures in particulate matter are also important in disease etiology. Some metals injure organ systems only after diffusing into blood circulation (lead [Pb^{2+}] affects the kidneys) but others (arsenic [As], chromium [Cr]) induce lung injury upon inhalation (190, 191). Arsenic, as well as zinc (Zn) and iron (Fe), can induce this injury through generation of ROS and by engaging Fenton chemistry (191). Arsenic exposure commonly occurs through ingestion of contaminated drinking water but can also occur through inhalation of airborne residues on PM (192). Both exposure routes are associated with development of lung cancer (8, 193). Specifically, SCC and SCLCs result from arsenic exposure. These are otherwise uncommon cancer presentations in never smokers (194). Highly carcinogenic As metabolites are formed through reduction by GSH and methylation processes (195). This metabolism also produces free radicals, which can engage in DNA damage and may result in carcinogenesis (196). Fenton chemistry also generates radical species. Which is why metals such as As or Fe can induce lipid peroxidation cascades. If unchecked by reducing enzymes (e.g., GSH), these cascades can result in cell membrane instability and ferroptosis, causing further lung injury (116).

Bioaerosols are interesting constituents of particulate matter, existing as both independent particles and absorbed to other airborne constituents. Moreover, bioaerosols comprise both microorganisms and microbially derived compounds. Microorganisms include viruses, bacteria, and fungal spores. Microbially derived compounds include bacterial endotoxins and fungal macromolecules, as well as animal dander, dust mites, and some pollens (151, 181). Bioaerosols are not considered toxicants because they arise from biological sources; instead, they may be considered toxins or, more likely, irritants and allergens. They may directly illicit immune responses; act as adjuvants by enhancing the body's response to other exposures; or directly modify the hazard potential of toxicant constituents in particulate matter (197). Inhalation of allergenic compounds can trigger asthma exacerbations. One such common compound is lipopolysaccharide (LPS), an endotoxin expressed on the outer membrane of Gram-negative bacteria (198). Exposure to high LPS concentrations can induce T_H17 asthma responses, defined by increased IL-17 and neutrophilic inflammation (199). Low doses of LPS can result in T_H2 responses associated with asthma development (200). In this way, LPS exposure has dose-dependent implications for asthma pathology. LPS can also induce chronic pulmonary inflammation, implicated in pathology of COPD and fibrosis (198). Respiratory infections by virulent viral or bacterial species found in PM also commonly exacerbate COPD pathogenesis by inducing macrophages and other phagocytic immune cells to generate ROS, which can damage pulmonary architecture (57, 58).

Sources of $PM_{2.5}$ are both natural and anthropogenic. Natural sources include incomplete combustion of carbon fuels, as in wildfires. Incomplete combustion also occurs anthropogenically in vehicle exhaust and in many industrial processes. However, an often-overlooked anthropogenic source of $PM_{2.5}$ is during cooking episodes in indoor environments. Different $PM_{2.5}$ constituents are released in different concentrations depending on fuel type (e.g., electric stoves, gas, wood, or other biomass) and cooking materials (different oils have different smoke points). These exposures are of special concern in developing countries, where alternative fuel sources are more commonly used in buildings. According to the World Health Organization, approximately 2.4 billion people cook using fuels such as coal, kerosene, or biomass (wood, animal dung, and crop waste). Worldwide, household air pollution generated by cooking episodes is responsible for nearly 3.2 million deaths each year, including 237,000 deaths among children under the age of 5 years old. Combined, ambient air pollution and household air pollution are associated with 6.7 million premature deaths each year (201). Thus, exposure to environmental toxicants is not just an ambient concern or occupational concern. It is a concern with global precedent mediated by both environmental and socioeconomic factors.

3.4 THE FUTURE OF ENVIRONMENTAL TOXICANTS IN LUNG DISEASE

3.4.1 Climate Changes

Both ambient and indoor exposure to respiratory toxicants are subject to environmental influences. Therefore, climate change may significantly impact diversity and concentration of common toxicants, as well as current patterns in exposure. For instance, ground-level ozone is generated from VOC and NO_X species in combination with heat and sunlight. Changes in meteorological patterns that increase heat and sunlight are estimated to increase ground-level ozone concentration. Moreover, increasing temperatures and decreased rainfall promotes drought conditions, which favor more frequent and intense wildfires. Wildfires are one source of ambient VOC and NO_X species, which contribute to ozone concentrations. Incomplete combustion of carbon in wildfires also releases coarse and fine particulate matter. Conversely, increased rainfall can promote vegetative growth, resulting in increased concentrations of ambient allergens. At the extreme, flooding from heavy rainfall can cause water damage in buildings. In turn, the increased moisture availability and indoor relative humidity can support growth of microbial species and the release of microbially derived VOC (202). Climate inextricably affects environmental exposure, which in turn may interact with lung physiology. Therefore, climate change has serious implications for the future of human lung health.

3.4.2 Regulation

The United States has passed several statutes aimed at reducing exposure to inhalable toxicants. Congress passed the Clean Air Act (CAA) in 1970 and has since amended its regulations, most recently in 1990. The CAA establishes national ambient air quality standards and requires states to adopt enforceable laws aimed at reducing airborne environmental and human health toxicants (3). The act also established six criteria of air pollutants to be monitored and regulated by the US Environmental Protection Agency (EPA): ground-level O_3, PM, CO, Pb, sulfur dioxide (SO_2), and nitrogen dioxide (NO_2) (203). These pollutants have significant implications for human health and/or environmental health. Additional regulation for inhalable toxicants came with the passage of the Toxic Substances Control Act in 1976, which empowers the EPA to require the testing, recording, and reporting on chemicals and mixtures by manufacturers, processors, and importers. It also gave the EPA authority to enforce chemical waste cleanup (4). In 1990, Congress furthered regulation of chemical waste with the Pollution Prevention (P2) Act. This statute aims to reduce and eliminate hazardous waste by modifying production processes. The EPA has since created a database on source reduction and is empowered to require toxic chemical release forms from commercial entities (5). These laws have succeeded in maintaining regulated release of ambient inhalable toxicants and environmental hazards. OSHA has also set standards to reduce hazardous exposure in the workplace (204). However, there are currently no statutes regulating air quality in indoor residential or public places. Strategies to improve indoor air quality are based on removing sources of exposure, increasing indoor air ventilation rates, and implementing air filtration (listed in descending order of general efficacy) (205).

3.5 CONCLUSIONS

We have discussed three main categories of inhalable, environmental toxicants: gaseous phase (e.g., O_3, CO), VOCs (e.g., formaldehyde, benzene), and respirable particulate matter (e.g., asbestos, $PM_{2.5}$). Even within these classes, resulting lung diseases are heterogenous and arise from varied etiology. Some commonalities are the induction and exacerbation of disease through chronic inflammation and oxidative stress. These pathways can result in DNA damage (lung cancer), airway hyperreactivity (asthma), tissue degradation (COPD), and tissue deposition (fibrosis). Neither the diseases nor the environmental toxicants listed here are meant to represent an exhaustive list. There are many other lung diseases and many more potentially hazardous substances. We have chosen to focus on a select and common few to elucidate important pathophysiology involved in toxicant-driven lung disease. Finally, as our environment changes, so too does our exposure. Current environmental regulations are aimed at reducing harmful exposures, but in a world increasingly impacted by climate change, revisions and new statutes may be necessary to regulate harmful environmental toxicants and ultimately, incidence and severity of lung disease.

ABBREVIATIONS

AAM	arachidonic acid metabolite
AHR	airway hyperresponsiveness
ASL	airway surface liquid
BC	black carbon
CAA	Clean Air Act
CB	carbon black
COPD	chronic obstructive pulmonary disease
CYP1A1	cytochrome P450, family 1, subfamily A, polypeptide 1
DAMP	damage-associated molecular pattern
DC	dendritic cell
EC	elemental carbon
ECM	extracellular matrix
EGFR	epidermal growth factor receptor
EPA	US Environmental Protection Agency
FGF	fibroblast growth factor
FVC	forced vital capacity
GM-CSF	granulocyte-macrophage colony–stimulating factor
GSH	glutathione
Hb	hemoglobin
HO-1	heme oxygenase-1
HTR	5-hydroxytryptamine receptor 4
IFN	interferon
Ig	immunoglobulin
IL	interleukin
MAPK	mitogen-activated protein kinase
M-CSF	macrophage colony-stimulating factor
MIP	macrophage inflammatory protein
MMP	matrix metalloproteinase
MPO	myeloperoxidase
NF-κB	nuclear factor kappa–light chain–enhancer of activated B cells
NLRP3	nod-like receptor family pyrin domain containing 3
NO_X	nitrogen oxides
OC	organic carbon
OCG1	oxoguanine glycosylase
P13/AKT	phosphoinositide 3-kinase/protein kinase B
PAH	polyacrylic hydrocarbon
PAMP	pattern-associated molecular pattern
$PM_{2.5}$	particulate matter with aerodynamic diameter <2.5 μm
POG	proto-oncogene
PRR	pattern recognition receptor
ROS	reactive oxygen species
SCC	squamous cell carcinoma
SCLC	small cell lung cancer
TDI	toluene-diisocyanate
TGF-β	transforming growth factor beta
T_H2	T helper 2
TLR	toll-like receptor
TNF-α	tumor necrosis factor alpha
TSP	tumor suppressor genes
VOC	volatile organic compound

REFERENCES

1. World Health Organization. (2022) Ambient (outdoor) air pollution. In *Air Pollution*. https://www.who.int/news-room/fact-sheets/detail/ambient-(outdoor)-air-quality-and-health
2. European Environment Agency. (2023) In *Europe's Air Quality Status 2023*. https://www.eea.europa.eu/publications/europes-air-quality-status-2023
3. United States Environmental Protection Agency. (2022) Clean air act requirements and history. In *Clean Air Act Overview*. https://www.epa.gov/clean-air-act-overview/clean-air-act-requirements-and-history
4. United States Environmental Protection Agency. (2022) Summary of the toxic substances control act. In *Laws & Regulations*. https://www.epa.gov/laws-regulations/summary-toxic-substances-control-act
5. United States Environmental Protection Agency. (2023) Pollution prevention law and policies. In *Pollution Prevention (P2)*. https://www.epa.gov/p2/pollution-prevention-law-and-policies
6. Parkin, D. M., Bray, F., Ferlay, J., and Pisani, P. (2005) Global cancer statistics, 2002. *CA Cancer J Clin.* **55**, 74–108
7. Brambilla, E., Travis, W. D., Colby, T. V., Corrin, B., and Shimosato, Y. (2001) The new World Health Organization classification of lung tumours. *Eur Respir J.* **18**, 1059–1068
8. Sun, S., Schiller, J. H., and Gazdar, A. F. (2007) Lung cancer in never smoker—A different disease. *Nat Rev Cancer.* **7**, 778–790
9. Compton, C. (2020) *Cancer the Enemy from Within*, Springer Nature Switzerland AG, Cham. ISBN 978-3-030-40650-9
10. Shepherd, F. A., Pereira, J. R., Ciuleanu, T., Tan, E. H., Hirsh, V., Thongprasert, S., Campos, D., Maoleekoonpiroj, S., Smylie, M., Martins, R., van Kooten, M., Dediu, M., Findlay, B., Tu, D., Johnston, D., Bezjak, A., Clark, G., Santabárbara, P., and Seymour, L. (2005) Erlotinib in previously treated non-small-cell lung cancer. *N Engl J Med.* **353**
11. Kosaka, T., Yatabe, Y., Endoh, H., Kuwano, H., Takahashi, T., and Mitsudomi, T. (2004) Mutations of the epidermal growth factor receptor gene in lung cancer biological and clinical implications. *Cancer Res.* **64**
12. Pham, D., Kris, M. G., Riely, G. J., Sarkaria, I. S., McDonough, T., Chuai, S., Venkatraman, E. S., Miller, V. A., Ladanyi, M., Pao, W., Wilson, R. K., Singh, B., and Rusch, V. W. (2006) Use of cigarette-smoking history to estimate the likelihood of mutations in epidermal growth factor receptor gene exons 19 and 21 in lung adenocarcinomas. *J Clin Oncol.* **24**, 1700–1704
13. Hill, W., Lim, E. L., Weeden, C. E., Lee, C., Augustine, M., Chen, K., Kuan, F. C., Marongiu, F., Evans, E. J., Jr., Moore, D. A., Rodrigues, *et al.* (2023) Lung adenocarcinoma promotion by air pollutants. *Nature.* **616**, 159–167
14. Johnson, L., Mercer, K., Greenbaum, D., Bronsonk, R. T., Crowley, D., Tuveson, D. A., and Jacks, T. (2001) Somatic activation of the K-ras oncogene causes early onset lung cancer in mice. *Nature.* **410**, 1111–1116
15. Politi, K., Zakowski, M. F., Fan, P. D., Schonfeld, E. A., Pao, W., and Varmus, H. E. (2006) Lung adenocarcinomas induced in mice by mutant EGF receptors found in human lung cancers respond to a tyrosine kinase inhibitor or to down-regulation of the receptors. *Genes Dev.* **20**, 1496–1510
16. Shigematsu, H., Lin, L., Takahashi, T., Nomura, M., Suzuki, M., Wistuba, II, Fong, K. M., Lee, H., Toyooka, S., Shimizu, *et al.* (2005) Clinical and biological features associated with epidermal growth factor receptor gene mutations in lung cancers. *J Natl Cancer Inst.* **97**, 339–346
17. Downward, J. (2003) Targeting RAS signalling pathways in cancer therapy. *Nat Rev Cancer.* **3**, 11–22
18. Sengupta, S., and Harris, C. C. (2005) p53: Traffic cop at the crossroads of DNA repair and recombination. *Nat Rev Mol Cell Biol.* **6**, 44–55
19. Le Calvez, F., Mukeria, A., Hunt, J. D., Kelm, O., Hung, R. J., Tanie're, P., Boffetta, P., Brennan, P., Zaridze, D., and Hainaut, P. (2005) TP53 and KRAS mutation load and types in lung cancers in relation to tobacco smoke distinct patterns in never, former, and current smokers. *Cancer Res.* **65**
20. Sharma, S. V., Bell, D. W., Settleman, J., and Haber, D. A. (2007) Epidermal growth factor receptor mutations in lung cancer. *Nat Rev Cancer.* **7**, 169–181
21. Hammad, H., and Lambrecht, B. N. (2021) The basic immunology of asthma. *Cell.* **184**, 1469–1485
22. Fahy, J. V. (2015) Type 2 inflammation in asthma—present in most, absent in many. *Nat Rev Immunol Perspect.* **15**
23. Green, R. H., Brightling, C. E., Woltmann, G., Parker, D., Wardlaw, A. J., and Pavord, I. D. (2002) Analysis of induced sputum in adults with asthma identification of subgroup isolated sputum neutrophilia and poor response to inhaled corticosteroids. *Thorax.* **57**, 875–879

24. Cayrol, C., Duval, A., Schmitt, P., Roga, S., Camus, M., Stella, A., Burlet-Schiltz, O., Gonzalez-de-Peredo, A., and Girard, J. P. (2018) Environmental allergens induce allergic inflammation through proteolytic maturation of IL-33. *Nat Immunol.* **19**, 375–385
25. Denney, L., Byrne, A. J., Shea, T. J., Buckley, J. S., Pease, J. E., Herledan, G. M., Walker, S. A., Gregory, L. G., and Lloyd, C. M. (2015) Pulmonary epithelial cell-derived cytokine TGF-beta1 is a critical cofactor for enhanced innate lymphoid cell function. *Immunity.* **43**, 945–958
26. Moon, H. G., Kim, S. J., Jeong, J. J., Han, S. S., Jarjour, N. N., Lee, H., Abboud-Werner, S. L., Chung, S., Choi, H. S., Natarajan, *et al.* (2018) Airway epithelial cell-derived colony stimulating factor-1 promotes allergen sensitization. *Immunity.* **49**, 275–287.e275
27. Willart, M. A., Deswarte, K., Pouliot, P., Braun, H., Beyaert, R., Lambrecht, B. N., and Hammad, H. (2012) Interleukin-1alpha controls allergic sensitization to inhaled house dust mite via the epithelial release of GM-CSF and IL-33. *J Exp Med.* **209**, 1505–1517
28. Halim, T. Y., Steer, C. A., Matha, L., Gold, M. J., Martinez-Gonzalez, I., McNagny, K. M., McKenzie, A. N., and Takei, F. (2014) Group 2 innate lymphoid cells are critical for the initiation of adaptive T helper 2 cell-mediated allergic lung inflammation. *Immunity.* **40**, 425–435
29. Locksley, R. M. (2010) Asthma and allergic inflammation. *Cell.* **140**, 777–783
30. Webb, L. M., Lundie, R. J., Borger, J. G., Brown, S. L., Connor, L. M., Cartwright, A. N., Dougall, A. M., Wilbers, R. H., Cook, P. C., Jackson-Jones, *et al.* (2017) Type I interferon is required for T helper (Th) 2 induction by dendritic cells. *EMBO J.* **36**, 2404–2418
31. Hoof, I., Schulten, V., Layhadi, J. A., Stranzl, T., Christensen, L. H., Herrera de la Mata, S., Seumois, G., Vijayanand, P., Lundegaard, C., Niss, K., Lund, A., Ahrenfeldt, *et al.* (2020) Allergen-specific IgG(+) memory B cells are temporally linked to IgE memory responses. *J Allergy Clin Immunol.* **146**, 180–191
32. Manise, M., Holtappels, G., Van Crombruggen, K., Schleich, F., Bachert, C., and Louis, R. (2013) Sputum IgE and cytokines in asthma: Relationship with sputum cellular profile. *PLoS ONE.* **8**, e58388
33. Redhu, N. S., and Gounni, A. S. (2013) The high affinity IgE receptor (FcepsilonRI) expression and function in airway smooth muscle. *Pulm Pharmacol Ther.* **26**, 86–94
34. Elieh Ali Komi, D., and Bjermer, L. (2019) Mast cell-mediated orchestration of the immune responses in human allergic asthma: Current insights. *Clin Rev Allergy Immunol.* **56**, 234–247
35. Ferreira, D. S., Carvalho-Pinto, R. M., Gregorio, M. G., Annoni, R., Teles, A. M., Buttignol, M., Araujo-Paulino, B. B., Katayama, E. H., Oliveira, B. L., Del Frari, *et al.* (2018) Airway pathology in severe asthma is related to airflow obstruction but not symptom control. *Allergy.* **73**, 635–643
36. Cinelli, M. A., Do, H. T., Miley, G. P., and Silverman, R. B. (2020) Inducible nitric oxide synthase: Regulation, structure, and inhibition. *Med Res Rev.* **40**, 158–189
37. Kanda, A., Yasutaka, Y., Van Bui, D., Suzuki, K., Sawada, S., Kobayashi, Y., Asako, M., and Iwai, H. (2020) Multiple biological aspects of eosinophils in host defense, eosinophil-associated diseases, immunoregulation, and homeostasis is their role beneficial, detrimental, regulator, or bystander? *Biol Pharm Bull.* **43**
38. Canas, J. A., Sastre, B., Rodrigo-Munoz, J. M., Fernandez-Nieto, M., Barranco, P., Quirce, S., Sastre, J., and Del Pozo, V. (2018) Eosinophil-derived exosomes contribute to asthma remodelling by activating structural lung cells. *Clin Exp Allergy.* **48**, 1173–1185
39. O'Byrne, P. M., Metev, H., Puu, M., Richter, K., Keen, C., Uddin, M., Larsson, B., Cullberg, M., and Nair, P. (2016) Efficacy and safety of a CXCR2 antagonist, AZD5069, in patients with uncontrolled persistent asthma: A randomised, double-blind, placebo-controlled trial. *Lancet Respir Med.* **4**, 797–806
40. Rossios, C., Pavlidis, S., Hoda, U., Kuo, C. H., Wiegman, C., Russell, K., Sun, K., Loza, M. J., Baribaud, F., Durham, A. L., Ojo, O., *et al.,* and Unbiased Biomarkers for the Prediction of Respiratory Diseases Outcomes Consortia Project, T. (2018) Sputum transcriptomics reveal upregulation of IL-1 receptor family members in patients with severe asthma. *J Allergy Clin Immunol.* **141**, 560–570
41. Li, X., Hastie, A. T., Peters, M. C., Hawkins, G. A., Phipatanakul, W., Li, H., Moore, W. C., Busse, W. W., Castro, M., Erzurum, *et al.,* and Blood Institute's Severe Asthma Research Program, N. (2020) Investigation of the relationship between IL-6 and type 2 biomarkers in patients with severe asthma. *J Allergy Clin Immunol.* **145**, 430–433
42. Tliba, O., and Panettieri, R. A., Jr. (2019) Paucigranulocytic asthma: Uncoupling of airway obstruction from inflammation. *J Allergy Clin Immunol.* **143**, 1287–1294
43. Guo, H., Callaway, J. B., and Ting, J. P. (2015) Inflammasomes: Mechanism of action, role in disease, and therapeutics. *Nat Med.* **21**, 677–687

44. Barnes, P. J. (2016) Inflammatory mechanisms in patients with chronic obstructive pulmonary disease. *J Allergy Clin Immunol*. **138**, 16–27
45. Wang, Y., Xu, J., Meng, Y., Adcock, I. M., and Yao, X. (2018) Role of inflammatory cells in airway remodeling in COPD. *Int J Chron Obstruct Pulmon Dis*. **13**, 3341–3348
46. Lareau, S. C., Fahy, B., Meek, P., and Wang, A. (2019) Chronic obstructive pulmonary disease (COPD). *Am J Respir Crit Care Med*. **199**
47. MacNee, W. (2006) ABC of chronic obstructive pulmonary disease: Pathology, pathogenesis, and pathophysiology. *Practice*. **332**
48. Gan, W. Q., Man, S. F., Senthilselvan, A., and Sin, D. D. (2004) Association between chronic obstructive pulmonary disease and systemic inflammation: A systematic review and a meta-analysis. *Thorax*. **59**, 574–580
49. Marchioni, A., Tonelli, R., Fantini, R., Tabbi, L., Castaniere, I., Livrieri, F., Bedogni, S., Ruggieri, V., Pisani, L., Nava, S., and Clini, E. (2019) Respiratory mechanics and diaphragmatic dysfunction in COPD patients who failed non-invasive mechanical ventilation. *Int J Chron Obstruct Pulmon Dis*. **14**, 2575–2585
50. Mathioudakis, A. G., Janssens, W., Sivapalan, P., Singanayagam, A., Dransfield, M. T., Jensen, J. S., and Vestbo, J. (2020) Acute exacerbations of chronic obstructive pulmonary disease: In search of diagnostic biomarkers and treatable traits. *Thorax*. **75**, 520–527
51. Meduri, G. U., Annane, D., Chrousos, G. P., Marik, P. E., and Sinclair, S. E. (2009) Activation and regulation of systemic inflammation in ARDS: Rationale for prolonged glucocorticoid therapy. *Chest*. **136**, 1631–1643
52. Sharma, B. B., and Singh, V. (2019) Diaphragmatic dysfunction in chronic obstructive pulmonary disease. *Lung India*. **36**
53. Rahman, I., and MacNee, W. (2012) Antioxidant pharmacological therapies for COPD. *Curr Opin Pharmacol*. **12**, 256–265
54. Schuliga, M. (2015) NF-kappaB signaling in chronic inflammatory airway disease. *Biomolecules*. **5**, 1266–1283
55. Fischer, B. M., Voynow, J. A., and Ghio, A. J. (2015) COPD: Balancing oxidants and antioxidants. *Int J Chron Obstruct Pulmon Dis*. **10**, 261–276
56. Rahman, I. (2012) Pharmacological antioxidant strategies as therapeutic interventions for COPD. *Biochim Biophys Acta*. **1822**, 714–728
57. Kim, V., and Aaron, S. D. (2018) What is a COPD exacerbation? Current definitions, pitfalls, challenges and opportunities for improvement. *Eur Respir J*. **52**
58. Papi, A., Bellettato, C. M., Braccioni, F., Romagnoli, M., Casolari, P., Caramori, G., Fabbri, L. M., and Johnston, S. L. (2006) Infections and airway inflammation in chronic obstructive pulmonary disease severe exacerbations. *Am J Respir Crit Care Med*. **173**, 1114–1121
59. Gao, W., Li, L., Wang, Y., Zhang, S., Adcock, I. M., Barnes, P. J., Huang, M., and Yao, X. (2015) Bronchial epithelial cells: The key effector cells in the pathogenesis of chronic obstructive pulmonary disease? *Respirology*. **20**, 722–729
60. Saetta, M., Mariani, M., Panina-Bordignon, P., Turato, G., Buonsanti, C., Baraldo, S., Bellettato, C. M., Papi, A., Corbetta, L., Zuin, R., Sinigaglia, F., and Fabbri, L. M. (2002) Increased expression of the chemokine receptor CXCR3 and its ligand CXCL10 in peripheral airways of smokers with chronic obstructive pulmonary disease. *Am J Respir Crit Care Med*. **165**, 1404–1409
61. Liu, A. N., Mohammed, A. Z., Rice, W. R., Fiedeldey, D. T., Liebermann, J. S., Whitsett, J. A., Braciale, T. J., and Enelow, R. I. (1999) Perforin-independent CD8(+) T-cell-mediated cytotoxicity of alveolar epithelial cells is preferentially mediated by tumor necrosis factor-alpha relative insensitivity to Fas ligand. *Am J Respir Cell Mol Biol*. **20**
62. Gao, D. K., Salomonis, N., Henderlight, M., Woods, C., Thakkar, K., Grom, A. A., Thornton, S., Jordan, M. B., Wikenheiser-Brokamp, K. A., and Schulert, G. S. (2021) IFN-gamma is essential for alveolar macrophage-driven pulmonary inflammation in macrophage activation syndrome. *JCI Insight*. **6**
63. Schroder, K., Hertzog, P. J., Ravasi, T., and Hume, D. A. (2004) Interferon-gamma: An overview of signals, mechanisms and functions. *J Leukoc Biol*. **75**, 163–189
64. Vlahos, R., Wark, P. A., Anderson, G. P., and Bozinovski, S. (2012) Glucocorticosteroids differentially regulate MMP-9 and neutrophil elastase in COPD. *PLoS ONE*. **7**, e33277

65. Rodrigues, S. O., Cunha, C., Soares, G. M. V., Silva, P. L., Silva, A. R., and Goncalves-de-Albuquerque, C. F. (2021) Mechanisms, pathophysiology and currently proposed treatments of chronic obstructive pulmonary disease. *Pharmaceuticals (Basel).* **14**
66. Saito, A., Horie, M., and Nagase, T. (2018) TGF-beta signaling in lung health and disease. *Int J Mol Sci.* **19**
67. Goldkorn, T., and Filosto, S. (2010) Lung injury and cancer: Mechanistic insights into ceramide and EGFR signaling under cigarette smoke. *Am J Respir Cell Mol Biol.* **43**, 259–268
68. Lin, V. Y., Kaza, N., Birket, S. E., Kim, H., Edwards, L. J., LaFontaine, J., Liu, L., Mazur, M., Byzek, S. A., Hanes, J., Tearney, G. J., Raju, S. V., and Rowe, S. M. (2020) Excess mucus viscosity and airway dehydration impact COPD airway clearance. *Eur Respir J.* **55**
69. Yang, T., Luo, F., Shen, Y., An, J., Li, X., Liu, X., Ying, B., Liao, Z., Dong, J., Guo, L., Wang, T., Xu, D., Chen, L., and Wen, F. (2012) Quercetin attenuates airway inflammation and mucus production induced by cigarette smoke in rats. *Int Immunopharmacol.* **13**, 73–81
70. Johns Hopkins Medicine. (2023) Interstitial lung disease: Pulmonary fibrosis. In *Health.* https://www.hopkinsmedicine.org/health/conditions-and-diseases/interstitial-lung-disease-pulmonary-fibrosis
71. CHEST Foundation. (2023) Conditions. In *Interstitial Lund Disease (ILD)*, American College of Chest Physicians. https://foundation.chestnet.org/lung-health-a-z/interstitial-lung-disease-ild/conditions/
72. CHEST Foundation. (2023) *Sarcoidosis*, American College of Chest Physicians. https://foundation.chestnet.org/lung-health-a-z/sarcoidosis/
73. Johns Hopkins Medicine. (2023) Pneumoconiosis. In *Health.* https://www.hopkinsmedicine.org/health/conditions-and-diseases/pneumoconiosis
74. Rose, J. J., Wang, L., Xu, Q., McTiernan, C. F., Shiva, S., Tejero, J., and Gladwin, M. T. (2017) Carbon monoxide poisoning: Pathogenesis, management, and future directions of therapy. *Am J Respir Crit Care Med.* **195**, 596–606
75. Motterlini, R., and Foresti, R. (2017) Biological signaling by carbon monoxide and carbon monoxide-releasing molecules. *Am J Physiol Cell Physiol.* **312**, C302–C313
76. Ryter, S. W. (2020) Therapeutic potential of heme oxygenase-1 and carbon monoxide in acute organ injury, critical illness, and inflammatory disorders. *Antioxidants (Basel).* **9**
77. Ryter, S. W., Alam, J., and Choi, A. M. (2006) Heme oxygenase-1/carbon monoxide: From basic science to therapeutic applications. *Physiol Rev.* **86**, 583–650
78. Wagener, F. A. D. T. G., Eggert, A., Boerman, O. C., Oyen, W. J. G., Verhofstad, A., Abraham, N. G., Adema, G., van Kooyk, Y., de Witte, A. T., and Figdor, C. G. (2001) Heme is a potent inducer of inflammation in mice and is counteracted by heme oxygenase. *Blood.* **96**
79. Hirayama, A., Noronha-Dutra, A. A., Gordge, M. P., Neild, G. H., and Hothersall, J. S. (199) S-nitrosothiols are stored by platelets and released during platelet—neutrophil interactions. *Nitric Oxide Biol Chem.* **3**, 95–104
80. Thom, S. R., Bhopale, V. M., Han, S. T., Clark, J. M., and Hardy, K. R. (2006) Intravascular neutrophil activation due to carbon monoxide poisoning. *Am J Respir Crit Care Med.* **174**, 1239–1248
81. Thom, S. R., Ohnishi, S. T., and Ischiropoulos, H. (1994) Nitric oxide released by platelets inhibits neutrophil B_2 integrin function following acute carbon monoxide poisoning. *Toxicol Appl Physiol.* **128**, 105–110
82. Thom, S. R., Bhopale, V. M., and Fisher, D. (2006) Hyperbaric oxygen reduces delayed immune-mediated neuropathology in experimental carbon monoxide toxicity. *Toxicol Appl Pharmacol.* **213**, 152–159
83. Otterbein, L. E., Mantell, L. L., and Choi, M. K. (1999) Carbon monoxide provides protection against hyperoxic lung injury. *Am J Physiol Lung Cell Mol Physiol.* **276**, L688–L694.
84. Mitchell, L. A., Channell, M. M., Royer, C. M., Ryter, S. W., Choi, A. M., and McDonald, J. D. (2010) Evaluation of inhaled carbon monoxide as an anti-inflammatory therapy in a nonhuman primate model of lung inflammation. *Am J Physiol Lung Cell Mol Physiol.* **299**, L891–L897
85. United States Environmental Protection Agency. (2023) Overview of greenhouse gases. In *Greenhouse Gas Emissions.* https://www.epa.gov/ghgemissions/overview-greenhouse-gases
86. Jacobson, T. A., Kler, J. S., Hernke, M. T., Braun, R. K., Meyer, K. C., and Funk, W. E. (2019) Direct human health risks of increased atmospheric carbon dioxide. *Nat Sustain.* **2**, 691–701
87. Jacobson, M. Z. (2010) Enhancement of local air pollution by urban CO_2 domes. *Environ Sci Technol.* **44**
88. Velasco, E., and Roth, M. (2010) Cities as net sources of CO2: Review of atmospheric CO2Exchange in urban environments measured by eddy covariance technique. *Geography Compass.* **4**, 1238–1259

89. Lu, R., and Turco, R. P. (1995) Air pollutant transport in a coastal environment—II. Three-dimensional simulations over Los Angeles basin. *Atmos Environ.* **29**
90. Rendón, A. M., Salazar, J. F., Palacio, C. A., Wirth, V., and Brötz, B. (2014) Effects of urbanization on the temperature inversion breakup in a mountain valley with implications for air quality. *J Appl Meteorol Climatol.* **53**, 840–858
91. Erdmann, C. A., Steiner, K. C., and Apte, M. G. (2002) *Indoor Carbon Dioxide Concentrations and Sick Building Syndrome Symptoms in the BASE Study Revisited: Analyses of the 100 Building Data*set, Lawrence Berkeley National Lab. (LBNL), Berkeley, CA. https://www.osti.gov/biblio/820782
92. Gall, E. T., Cheung, T., Luhung, I., Schiavon, S., and Nazaroff, W. W. (2016) Real-time monitoring of personal exposures to carbon dioxide. *Build Environ.* **104**, 59–67
93. Fisk, W. J. (2017) The ventilation problem in schools: Literature review. *Indoor Air.* **27**, 1039–1051
94. Thom, S. R., Bhopale, V. M., Hu, J., and Yang, M. (2017) Inflammatory responses to acute elevations of carbon dioxide in mice. *J Appl Physiol (1985).* **123**, 297–302
95. Zhang, X., Wargocki, P., and Lian, Z. (2017) Physiological responses during exposure to carbon dioxide and bioeffluents at levels typically occurring indoors. *Indoor Air.* **27**, 65–77
96. Miller, J., Edwards, L. D., Agusti, A., Bakke, P., Calverley, P. M., Celli, B., Coxson, H. O., Crim, C., Lomas, D. A., Miller, B. E., *et al.,* and Evaluation of, C. L. T. I. P. S. E. I. (2013) Comorbidity, systemic inflammation and outcomes in the ECLIPSE cohort. *Respir Med.* **107**, 1376–1384
97. Zappulla, D. (2008) Environmental stress, erythrocyte dysfunctions, inflammation, and the metabolic syndrome: Adaptations to CO2 increases? *J Cardiometabolic Syndr.* **3**, 30–34
98. Jenkin, M. E., and Clemitshaw, K. C. (2000) Ozone and other secondary photochemical pollutants chemical processes governing their formation in the planetary boundary layer. *Atmos Environ.* **34**
99. Su, F. C., Mukherjee, B., and Batterman, S. (2013) Determinants of personal, indoor and outdoor VOC concentrations: An analysis of the RIOPA data. *Environ Res.* **126**, 192–203
100. Nazaroff, W. W., and Weschler, C. J. (2022) Indoor ozone: Concentrations and influencing factors. *Indoor Air.* **32**, e12942
101. Schwarz, L., Hansen, K., Alari, A., Ilango, S. D., Bernal, N., Basu, R., Gershunov, A., and Benmarhnia, T. (2021) Spatial variation in the joint effect of extreme heat events and ozone on respiratory hospitalizations in California. *Proc Natl Acad Sci U S A.* **118**
102. Nicolas, M., Ramalho, O., and Maupetit, F. (2007) Reactions between ozone and building products: Impact on primary and secondary emissions. *Atmos Environ.* **41**, 3129–3138
103. Salonen, H., Salthammer, T., and Morawska, L. (2018) Human exposure to ozone in school and office indoor environments. *Environ Int.* **119**, 503–514
104. Al-Hegelan, M., Tighe, R. M., Castillo, C., and Hollingsworth, J. W. (2011) Ambient ozone and pulmonary innate immunity. *Immunol Res.* **49**, 173–191
105. Bromberg, P. A. (2016) Mechanisms of the acute effects of inhaled ozone in humans. *Biochim Biophys Acta.* **1860**, 2771–2781
106. Hatch, G. E., Duncan, K. E., Diaz-Sanchez, D., Schmitt, M. T., Ghio, A. J., Carraway, M. S., McKee, J., Dailey, L. A., Berntsen, J., and Devlin, R. B. (2014) Progress in assessing air pollutant risks from in vitro exposures: Matching ozone dose and effect in human airway cells. *Toxicol Sci.* **141**, 198–205
107. Sokolowska, M., Chen, L. Y., Eberlein, M., Martinez-Anton, A., Liu, Y., Alsaaty, S., Qi, H. Y., Logun, C., Horton, M., and Shelhamer, J. H. (2014) Low molecular weight hyaluronan activates cytosolic phospholipase A2alpha and eicosanoid production in monocytes and macrophages. *J Biol Chem.* **289**, 4470–4488
108. Morgan, M. J., and Liu, Z. G. (2011) Crosstalk of reactive oxygen species and NF-kappaB signaling. *Cell Res.* **21**, 103–115
109. Pantano, C., Reynaert, N. L., van der Vliet, A., and Janssen-Heininger, Y. M. W. (2006) Redox-sensitive kinases of the nuclear factor-κB signaling pathway. *Antioxid Redox Signal.* **8**, 1791–1806
110. Tighe, R. M., Wheeler, J., and Hollingsworth, J. W. (2015) *Air Pollution and Health Effects*, Humana Press, Springer-Verlag, London. ISBN 978-1-4471-6668-9
111. Yuan, S., Hollinger, M., Lachowicz-Scroggins, M. E., Kerr, S. C., Dunican, E. M., Daniel, B. M., Ghosh, S., Erzurum, S. C., Willard, B., Hazen, S. L., *et al.* (2015) Oxidation increases mucin polymer cross-links to stiffen airway mucus gels. *Lung Disease.* **7**

112. Dall'Acqua, M. C., Bonet, I. J., Zampronio, A. R., Tambeli, C. H., Parada, C. A., and Fischer, L. (2014) The contribution of transient receptor potential ankyrin 1 (TRPA1) to the in vivo nociceptive effects of prostaglandin E(2). *Life Sci.* **105**, 7–13
113. Passannante, A. N., Hazucha, M. J., Bromberg, P. A., Seal, E., Folinsbee, L., and Koch, G. (1998) Nociceptive mechanisms modulate ozone-induced human lung function decrements. *J Appl Physiol.* **85**, 1863–1870
114. Sharma, R., Yang, Y., Sharma, A., Awasthi, S., and Awasthi, Y. C. (2004) Antioxidant role of glutathione S-transferases protection against oxidant toxicity and regulation of stress-mediated apoptosis. *Antioxid Redox Signal.* **6**
115. Girotti, A. W. (1998) Lipid hydroperoxide generation, turnover, and effector action in biological systems. *J Lipid Res.* **39**, 1529–1542
116. Tang, D., and Kroemer, G. (2020) Ferroptosis. *Curr Biol.* **30**, R1292–R1297
117. Ghio, A. J., Turi, J. L., Madden, M. C., Dailey, L. A., Richards, J. D., Stonehuerner, J. G., Morgan, D. L., Singleton, S., Garrick, L. M., and Garrick, M. D. (2007) Lung injury after ozone exposure is iron dependent. *Am J Physiol Lung Cell Mol Physiol.* **292**, L134–L143
118. Lubin, J. H., Boice, J. D., Edling, C., Hornung, R. W., Howe, G. R., Kunz, E., Kusiak, R. A., Morrison, H. I., Radford, E. P., Samet, J. M., *et al.* (1995) Lung cancer in radon-exposed miners and estimation of risk from indoor exposure. *J Natl Cancer Inst.* **87**, 817–827
119. United States Environmental Protection Agency. (2023) Health effects of radon. In *Radon.* https://www.epa.gov/radon/health-risk-radon
120. Prise, K. M., Pinto, M., Newman, H. C., and Michael, B. D. (2001) A review of studies of ionizing radiation-induced double-strand break clustering. *Radiat Res.* **156**, 572–576
121. Riley, P. A. (1993) Free radicals in biology: Oxidative stress and the effects of ionizing radiation. *Int J Radiat Biol.* **65**, 27–33
122. Samet, J. M., Avila-Tang, E., Boffetta, P., Hannan, L. M., Olivo-Marston, S., Thun, M. J., and Rudin, C. M. (2009) Lung cancer in never smokers: Clinical epidemiology and environmental risk factors. *Clin Cancer Res.* **15**, 5626–5645
123. Chauhan, V., Howland, M., Kutzner, B., McNamee, J. P., Bellier, P. V., and Wilkins, R. C. (2012) Biological effects of alpha particle radiation exposure on human monocytic cells. *Int J Hyg Environ Health.* **215**, 339–344
124. United States Environmental Protection Agency. (2023) Technical overview of volatile organic compounds. In *Indoor Air Quality (IAQ).* https://www.epa.gov/indoor-air-quality-iaq/technical-overview-volatile-organic-compounds
125. Shi, Y., Lan, F., Matson, C., Mulligan, P., Whetstine, J. R., Cole, P. A., Casero, R. A., and Shi, Y. (2004) Histone demethylation mediated by the nuclear amine oxidase homolog LSD1. *Cell.* **119**, 941–953
126. Yu, P. H., Lai, C., and Zu, D. (1997) Formation of formaldehyde from adrenaline in vivo A potential risk factor for stress-related angiopathy. *Neurochem Res.* **22**
127. Yu, P. H., Lu, L. X., Fan, H., Kazachkov, M., Jiang, Z. J., Jalkanen, S., and Stolen, C. (2006) Involvement of semicarbazide-sensitive amine oxidase-mediated deamination in lipopolysaccharide-induced pulmonary inflammation. *Am J Pathol.* **168**, 718–726
128. Pontel, L. B., Rosado, I. V., Burgos-Barragan, G., Garaycoechea, J. I., Yu, R., Arends, M. J., Chandrasekaran, G., Broecker, V., Wei, W., Liu, L., *et al.* (2015) Endogenous formaldehyde is a hematopoietic stem cell genotoxin and metabolic carcinogen. *Mol Cell.* **60**, 177–188
129. Uotila, L., and Koivusalo, M. (1974) Purification and properties of S-formylglutathione hydrolase from human liver. *J Biolog Chem.* **249**, 7664–7672
130. Barbosa, E., Dos Santos, A. L. A., Peteffi, G. P., Schneider, A., Muller, D., Rovaris, D., Bau, C. H. D., Linden, R., Antunes, M. V., and Charao, M. F. (2019) Increase of global DNA methylation patterns in beauty salon workers exposed to low levels of formaldehyde. *Environ Sci Pollut Res Int.* **26**, 1304–1314
131. Katsnelson, B. A., Degtyareva, T. D., Privalova, L. I., Minigaliyeva, I. A., Slyshkina, T. V., Ryzhov, V. V., and Beresneva, O. Y. (2013) Attenuation of subchronic formaldehyde inhalation toxicity with oral administration of glutamate, glycine and methionine. *Toxicol Lett.* **220**, 181–186
132. O'Brien, P. J., Siraki, A. G., and Shangari, N. (2005) Aldehyde sources, metabolism, molecular toxicity mechanisms, and possible effects on human health. *Crit Rev Toxicol.* **35**, 609–662
133. Reingruber, H., and Pontel, L. B. (2018) Formaldehyde metabolism and its impact on human health. *Curr Opin Toxicol.* **9**, 28–34

134. Wu, Y., Duan, J., Li, B., Liu, H., and Chen, M. (2020) Exposure to formaldehyde at low temperatures aggravates allergic asthma involved in transient receptor potential ion channel. *Environ Toxicol Pharmacol.* **80**, 103469
135. Tan, S. L. W., Chadha, S., Liu, Y., Gabasova, E., Perera, D., Ahmed, K., Constantinou, S., Renaudin, X., Lee, M., Aebersold, R., and Venkitaraman, A. R. (2017) A class of environmental and endogenous toxins induces BRCA2 haploinsufficiency and genome instability. *Cell.* **169**, 1105–1118.e1115
136. Kang, J., Duan, J., Song, J., Luo, C., Liu, H., Li, B., Yang, X., Yu, W., and Chen, M. (2018) Exposure to a combination of formaldehyde and DINP aggravated asthma-like pathology through oxidative stress and NF-kappaB activation. *Toxicology.* **404–405**, 49–58
137. Wu, Y., Di, X., Zhao, M., Li, H., Bai, L., and Wang, K. (2022) The role of the NLRP3 inflammasome in chronic inflammation in asthma and chronic obstructive pulmonary disease. *Immun Inflamm Dis.* **10**, e750
138. Leso, V., Macrini, M. C., Russo, F., and Iavicoli, I. (2020) Formaldehyde exposure and epigenetic effects: A systematic review. *Appl Sci.* **10**
139. Leal, M. P., Brochetti, R. A., Ignacio, A., Camara, N. O. S., da Palma, R. K., de Oliveira, L. V. F., de Fatima Teixeira da Silva, D., and Lino-Dos-Santos-Franco, A. (2018) Effects of formaldehyde exposure on the development of pulmonary fibrosis induced by bleomycin in mice. *Toxicol Rep.* **5**, 512–520
140. Traboulsi, H., Cherian, M., Abou Rjeili, M., Preteroti, M., Bourbeau, J., Smith, B. M., Eidelman, D. H., and Baglole, C. J. (2020) Inhalation toxicology of vaping products and implications for pulmonary health. *Int J Mol Sci.* **21**
141. Wang, L., He, X., Bi, Y., and Ma, Q. (2012) Stem cell and benzene-induced malignancy and hematotoxicity. *Chem Res Toxicol.* **25**, 1303–1315
142. Aksoy, M. (1989) Hematotoxicity and carcinogenicity of benzene. *Environ Health Perspect.* **82**
143. White, W. C. A., and Gammon, A. M. (1914) The influence of benzol inhalation on experimental pulmonary tuberculosis in rabbits. *Trans Assoc Am Physicians.* **29**, 332–337
144. Snyder, R. (2002) Benzene and leukemia. *Crit Rev Toxicol.* **32**, 155–210
145. Zhao, Z., He, X., Bi, Y., Xia, Y., Tao, N., Li, L., and Ma, Q. (2009) Induction of CYP4F3 by benzene metabolites in human white blood cells in vivo in human promyelocytic leukemic cell lines and ex vivo in human blood neutrophils. *Drug Metab Dispos.* **37**, 282–291
146. Falzone, L., Marconi, A., Loreto, C., Franco, S., Spandidos, D. A., and Libra, M. (2016) Occupational exposure to carcinogens: Benzene, pesticides and fibers (Review). *Mol Med Rep.* **14**, 4467–4474
147. Weaver, C. V., Liu, S. P., Lu, J. F., and Lin, B. S. (2007) The effects of benzene exposure on apoptosis in epithelial lung cells: Localization by terminal deoxynucleotidyl transferase-mediated dUTP-biotin nick end labeling (TUNEL) and the immunocytochemical localization of apoptosis-related gene products. *Cell Biol Toxicol.* **23**, 201–220
148. Pardo, A., and Selman, M. (2002) Molecular mechanisms of pulmonary fibrosis. *Front Biosci.* **7**, 1743–1761
149. Liang, J., Zhao, H., Yao, L., Tang, H., Dong, H., Wu, Y., Liu, L., Zou, F., and Cai, S. (2015) Phosphatidylinositol 3-kinases pathway mediates lung caspase-1 activation and high mobility group box 1 production in a toluene-diisocyanate induced murine asthma model. *Toxicol Lett.* **236**, 25–33
150. Sulaiman, I., Tan, K., Mohtarrudin, N., Lim, J. C. W., and Stanslas, J. (2018) Andrographolide prevented toluene diisocyanate-induced occupational asthma and aberrant airway E-cadherin distribution via p38 MAPK-dependent Nrf2 induction. *Pulm Pharmacol Ther.* **53**, 39–51
151. Kim, K. H., Kabir, E., and Kabir, S. (2015) A review on the human health impact of airborne particulate matter. *Environ Int.* **74**, 136–143
152. Duarte, J., Castelo Branco, J., Rodrigues, F., Vaz, M., and Santos Baptista, J. (2022) Occupational exposure to mineral dust in mining and earthmoving works: A scoping review. *Safety.* **8**
153. Fujimura, F. (2000) Pathology and pathophysiology of pneumoconiosis. *Curr Opin Pulm Med.* **6**
154. Hubaux, R., Becker-Santos, D. D., Enfield, K. S., Lam, S., Lam, W. L., and Martinez, V. D. (2012) Arsenic, asbestos and radon: Emerging players in lung tumorigenesis. *Environ Health.* **11**, 89
155. Cohen, R. A., Rose, C. S., Go, L. H. T., Zell-Baran, L. M., Almberg, K. S., Sarver, E. A., Lowers, H. A., Iwaniuk, C., Clingerman, S. M., Richardson, *et al.* (2022) Pathology and mineralogy demonstrate respirable crystalline silica is a major cause of severe pneumoconiosis in U.S. coal miners. *Ann Am Thorac Soc.* **19**, 1469–1478

156. Weill, H., Jones, R. N., and Parkes, W. R. (1994) Silicosis and related diseases. In *Occupational Lung Disorders* (Parkes, W. R., ed.), 3rd ed., Butterworths, London
157. Rimala, B., Greenberg, A. K., and Rom, W. N. (2005) Basic pathogenetic mechanisms in silicosis current understanding. *Curr Opin Pulm Med.* **11**
158. Hughes, A. M., Weill, H., Checkoway, H., Jones, R. N., Henry, M. M., Heyer, N. J., Seixas, N. S., and Demers, P. A. (1997) Radiographic evidence of silicosis risk in the diatomaceous earth industry. *Am J Respir Crit Care Med.* **158**
159. Mossman, B. T., and Glenn, R. E. (2013) Bioreactivity of the crystalline silica polymorphs, quartz and cristobalite, and implications for occupational exposure limits (OELs). *Crit Rev Toxicol.* **43**, 632–660
160. Mossman, B. T., and Churg, A. (1997) Mechanisms in the pathogenesis of asbestosis and silicosis. *Am J Respir Crit Care Med.* **157**
161. Robledo, R., and Mossman, B. (1999) Cellular and molecular mechanisms of asbestos-induced fibrosis. *J Cell Physiol.* **180**, 158–166
162. Srivastava, R. K., Lohani, M., Pant, A. B., and Rahman, Q. (2010) Cyto-genotoxicity of amphibole asbestos fibers in cultured human lung epithelial cell line: Role of surface iron. *Toxicol Ind Health.* **26**, 575–582
163. Browne, K. (1994) Asbestos-related disorders. In *Occupational Lung Disorders* (Parkes, W. R., ed.), Butterworths, London
164. McCormack, V., Peto, J., Byrnes, G., Straif, K., and Boffetta, P. (2012) Estimating the asbestos-related lung cancer burden from mesothelioma mortality. *Br J Cancer.* **106**, 575–584
165. Zervos, M. D., Bizekis, C., and Pass, H. I. (2008) Malignant mesothelioma 2008. *Curr Opin Pulm Med.* **14**
166. Choe, N., Tanaka, S., and Kagan, E. (1998) Asbestos fibers and interleukin-1 upregulate the formation of reactive nitrogen species in rat pleural mesothelial cells. *Am J Respir Cell Mol Biol.* **19**
167. Hei, T. K., Xu, A., Huang, S. X., and Zhao, Y. (2006) Mechanism of fiber carcinogenesis: From reactive radical species to silencing of the beta igH3 gene. *Inhal Toxicol.* **18**, 985–990
168. Xu, A., Huang, X., Lien, Y., Bao, L., Yu, Z., and Hei, T. K. (2007) Genotoxic mechanisms of asbestos fibers role of extranuclear targets. *Chem Res Toxicol.* **20**
169. Aljandali, A., Pollack, H., Yeldandi, A., Li, Y., Weitzman, S. A., and Kamp, D. W. (2001) Asbestos causes apoptosis in alveolar epithelial cells: Role of iron-induced free radicals. *J Lab Clin Med.* **137**, 330–339
170. Panduri, V., Weitzman, S. A., Chandel, N. S., and Kamp, D. W. (2004) Mitochondrial-derived free radicals mediate asbestos-induced alveolar epithelial cell apoptosis. *Am J Physiol Lung Cell Mol Physiol.* **286**, L1220–L1227.
171. Husgafvel-Pursiainen, K., Karjalainen, A., Kannio, A., Anttila, S., Partanen, T., Ojajärvi, A., and Vainio, H. (199) Lung cancer and past occupational exposure to asbestos. *Am J Respir Cell Mol Biol.* **20**
172. Shukla, A., Flanders, T., Lounsbury, K. M., and B. T., M. (2004) The γ-glutamylcysteine synthetase and glutathione regulate asbestos-induced expression of activator protein-1 family members and activity. *Cancer Res.* **64**
173. Johnson, D. R. (2015) Organic dusts. In *Hamilton & Hardy's Industrial Toxicology*, pp. 969–996, Wiley, Hoboken.
174. Maatta, J., Lehto, M., Leino, M., Tillander, S., Haapakoski, R., Majuri, M. L., Wolff, H., Rautio, S., Welling, I., Husgafvel-Pursiainen, K., *et al.* (2006) Mechanisms of particle-induced pulmonary inflammation in a mouse model: Exposure to wood dust. *Toxicol Sci.* **93**, 96–104
175. Nylander, L. A., and Dement, J. M. (1993) Carcinogenic effects of wood dust: Review and discussion. *Am J Ind Med.* **24**, 619–647
176. Hancock, D. G., Langley, M. E., Chia, K. L., Woodman, R. J., and Shanahan, E. M. (2015) Wood dust exposure and lung cancer risk: A meta-analysis. *Occup Environ Med.* **72**, 889–898
177. Decuyper, II, Green, B. J., Sussman, G. L., Ebo, D. G., Silvers, W. S., Pacheco, K., King, B. S., Cohn, J. R., Zeiger, R. S., Zeiger, *et al.* (2020) Occupational allergies to Cannabis. *J Allergy Clin Immunol Pract.* **8**, 3331–3338
178. Bouhuys, A., Wolfson, R. L., Horner, D. W., Brain, J. D., and Zuskin, E. (1969) Byssinosis in cotton textile workers: Respiratory survey of a mill with rapid labor turnover. *Ann Intern Med.* **71**
179. Elwood, P. C., Pemberton, J., Merrett, J. D., Carey, G. C. R., and McAulay, I. R. (1965) Byssinosis and other respiratory symptoms in flax workers in Northern Ireland. *Br J Ind Med.* **22**

180. Niven, R. M., and Pickering, C. A. C. (1996) Byssinosis: A review. In *Occupational Lung Disease*, Thorax, London.
181. Wang, Q., and Liu, S. (2023) The effects and pathogenesis of PM2.5 and its components on chronic obstructive pulmonary disease. *Int J Chron Obstruct Pulmon Dis.* **18**, 493–506
182. Long, C. M., Nascarella, M. A., and Valberg, P. A. (2013) Carbon black vs. black carbon and other airborne materials containing elemental carbon: Physical and chemical distinctions. *Environ Pollut.* **181**, 271–286
183. Zhou, L., Li, P., Zhang, M., Han, B., Chu, C., Su, X., Li, B., Kang, H., Ning, J., Zhang, *et al.* (2020) Carbon black nanoparticles induce pulmonary fibrosis through NLRP3 inflammasome pathway modulated by miR-96 targeted FOXO3a. *Chemosphere.* **241**, 125075
184. Wang, L., Bao, S., Liu, X., Wang, F., Zhang, J., Dang, P., Wang, F., Li, B., and Lin, Y. (2021) Low-dose exposure to black carbon significantly increase lung injury of cadmium by promoting cellular apoptosis. *Ecotoxicol Environ Saf.* **224**, 112703
185. Fu, Y., Niu, Y., Pan, B., Liu, Y., Zhang, B., Li, X., Yang, A., Nie, J., Wang, R., and Yang, J. (2019) OGG1 methylation mediated the effects of cell cycle and oxidative DNA damage related to PAHs exposure in Chinese coke oven workers. *Chemosphere.* **224**, 48–57
186. Spitz, M. R., Wei, Q., Dong, Q., Amos, C. I., and Wu, X. (2003) Genetic susceptibility to lung cancer the role of DNA damage and repair. *Cancer Epidemiol Biomarkers Prev.* **12**
187. Lauer, F. T., Mitchell, L. A., Bedrick, E., McDonald, J. D., Lee, W. Y., Li, W. W., Olvera, H., Amaya, M. A., Berwick, M., Gonzales, M., *et al.* (2009) Temporal-spatial analysis of U.S.-Mexico border environmental fine and coarse PM air sample extract activity in human bronchial epithelial cells. *Toxicol Appl Pharmacol.* **238**, 1–10
188. Zheng, R., Song, P., Wu, Y., Wang, Y., Han, X., Yan, J., Wu, X., and Zhang, H. (2022) Property-activity relationship between physicochemical properties of PM(2.5) and their activation of NLRP3 inflammasome. *NanoImpact.* **25**, 100380
189. Hu, B., Tong, B., Xiang, Y., Li, S. R., Tan, Z. X., Xiang, H. X., Fu, L., Wang, H., Zhao, H., and Xu, D. X. (2020) Acute 1-NP exposure induces inflammatory responses through activating various inflammatory signaling pathways in mouse lungs and human A549 cells. *Ecotoxicol Environ Saf.* **189**, 109977
190. Bai, X., Liu, Y., Wang, S., Liu, C., Liu, F., Su, G., Peng, X., Yuan, C., Jiang, Y., and Yan, B. (2018) Ultrafine particle libraries for exploring mechanisms of PM(2.5)-induced toxicity in human cells. *Ecotoxicol Environ Saf.* **157**, 380–387
191. Xu, F., Qiu, X., Hu, X., Shang, Y., Pardo, M., Fang, Y., Wang, J., Rudich, Y., and Zhu, T. (2018) Effects on IL-1β signaling activation induced by water and organic extracts of fine particulate matter (PM2.5) in vitro. *Environ Pollut.* **237**, 592–600
192. Brunt, R., Vasak, L., and Griffioen, J. (2004) *Arsenic in Groundwater Probability of Occurrence of Excessive Concentration on Global Scale*, International Groundwater Resources Assessment Center, Delft
193. Subramanian, J., and Govindan, R. (2007) Lung cancer in never smokers: A review. *J Clin Oncol.* **25**, 561–570
194. Heck, J. E., Andrew, A. S., Onega, T., Rigas, J. R., Jackson, B. P., Karagas, M. R., and Duell, E. J. (2009) Lung cancer in a U.S. population with low to moderate arsenic exposure. *Environ Health Perspect.* **117**, 1718–1723
195. Ebert, F., Weiss, A., Bultemeyer, M., Hamann, I., Hartwig, A., and Schwerdtle, T. (2011) Arsenicals affect base excision repair by several mechanisms. *Mutat Res.* **715**, 32–41
196. Jomova, K., Jenisova, Z., Feszterova, M., Baros, S., Liska, J., Hudecova, D., Rhodes, C. J., and Valko, M. (2011) Arsenic: Toxicity, oxidative stress and human disease. *J Appl Toxicol.* **31**, 95–107
197. Samake, A., Uzu, G., Martins, J. M. F., Calas, A., Vince, E., Parat, S., and Jaffrezo, J. L. (2017) The unexpected role of bioaerosols in the Oxidative Potential of PM. *Sci Rep.* **7**, 10978
198. Yang, J., Kim, Y. K., Kang, T. S., Jee, Y. K., and Kim, Y. Y. (2017) Importance of indoor dust biological ultrafine particles in the pathogenesis of chronic inflammatory lung diseases. *Environ Health Toxicol.* **32**, e2017021
199. Yang, J., Kim, E. K., Park, H. J., McDowell, A., and Kim, Y. K. (2020) The impact of bacteria-derived ultrafine dust particles on pulmonary diseases. *Exp Mol Med.* **52**, 338–347

200. Soumelis, V., Reche, P. A., Kanzler, H., Yuan, W., Edward, G., Homey, B., Gilliet, M., Ho, S., Antonenko, S., Lauerma, A., *et al.* (2002) Human epithelial cells trigger dendritic cell mediated allergic inflammation by producing TSLP. *Nat Immunol.* **3**, 673–680
201. World Health Organization. (2022) Household air pollution. In *Air Pollution.* https://www.who.int/news-room/fact-sheets/detail/household-air-pollution-and-health
202. United States Environmental Protection Agency. (2022) Climate change impacts on air quality. In *Climate Change Impacts.* https://www.epa.gov/climateimpacts/climate-change-impacts-air-quality
203. United States Environmental Protection Agency. (2022) Criteria air pollutants. https://www.epa.gov/criteria-air-pollutants
204. United States Occupational Safety and Health Administration. Chemical hazards and toxic substances. In *Safety and Health Topics*, United States Department of Labor, Washington, DC. https://www.osha.gov/chemical-hazards
205. United States Environmental Protection Agency. (2023) Regulatory and guidance information by topic: Air. In *Regulatory Information by Topic.* https://www.epa.gov/regulatory-information-topic/regulatory-and-guidance-information-topic-air

4 Epigenetic Modifications of Histone in Lung Fibrosis

Yan Y. Sanders

4.1 INTRODUCTION

Lung fibrosis is an age-related, chronic fatal disease characterized by the scarring of lung tissues, which often results from dysregulated wound repair (1). Idiopathic pulmonary fibrosis (IPF) is the most common form of lung fibrosis, with unknown etiology (1). Although the cause of IPF is not clear, studies have shown that certain genetic alterations, such as telomere-related gene mutation (2) and MUC5B promoter polymorphism (3), are associated with lung fibrosis. However, some environmental factors, including cigarette smoke (4), inhalation exposure to silica and coal mine dusts (5), and changes of nutritional states (6) are contributing factors to the pathogenesis of lung fibrosis. Emerging studies have demonstrated that aging is a major risk factor for lung fibrosis (7).

All these factors outside of gene alterations are likely causing changes in gene expression favor for developing fibrotic lung tissue through epigenetic mechanisms (Figure 4.1) (8). Classically, epigenetic mechanisms that regulate gene expression include three main components: DNA methylation, noncoding RNA, and histone modifications (9). All these epigenetic mechanisms have been reported in the pathogenesis of lung fibrosis (8). These epigenetic alterations regulate gene expression without changing DNA sequences, through modifications of DNA bases and histone tails to change the accessibility of genes to regulate transcription. For example, Thy-1 is a cell surface marker that is lost in IPF, and Thy-1 negative fibroblasts are more fully differentiated myofibroblasts (10, 11). Thy-1 can be silenced by DNA methylation (12), and by histone modifications (13) in lung fibroblasts. These epigenetic modifications affect chromatin structures to regulate gene transcription. Chromatins are either euchromatin, with loose structure for active transcription, or heterochromatin, with dense structure to repress transcription. For example, histone modifications, especially acetylated histone, usually signal open chromatin and active transcription (14).

Previously, we have summarized changes of DNA methylation (15, 16) and anti-fibrotic function of histone deacetylases (HDACs) inhibitors (1) in lung fibrosis. In this review, we will focus on histone modification changes in lung fibrosis. We will discuss the roles of the writers, such as acetyl- or methyltransferases, and erasers, such as de-acetylases or -methylases, of histone modifications in lung fibrosis.

4.1.1 Overview of Histone Acetylation

Histone acetylation is usually associated with active transcription and is balanced by acetyltransferases and deacetylases (Figure 4.2A) (1). HATs are the acetyltransferases that transfer acetyl groups to various substrates, including histones. These enzymes are categorized into several families, including Gcn5-related *N*-acetyltransferases (GNATs), named for their similarity to the Gnc5 enzyme; the MYST family, named by the founding members Moz, Ybf2, Sas2 and Tip60; and the p300/CBP (CREB-binding protein) family (17). Some of the HATs, such as p300, are reported to be involved in pro-fibrotic gene upregulation in lung fibrosis (18, 19).

HDACs are the enzymes that remove the acetyl group from lysine residues, which are separated into three classes of zinc-dependent enzymes and one class of NAD^+-dependent sirtuins (17). There are various reports of the involvement of HADCs in the pathogenesis of lung fibrosis, and HDAC inhibitors have been proposed as therapeutic methods (1, 20, 21).

DOI: 10.1201/9781003355243-5

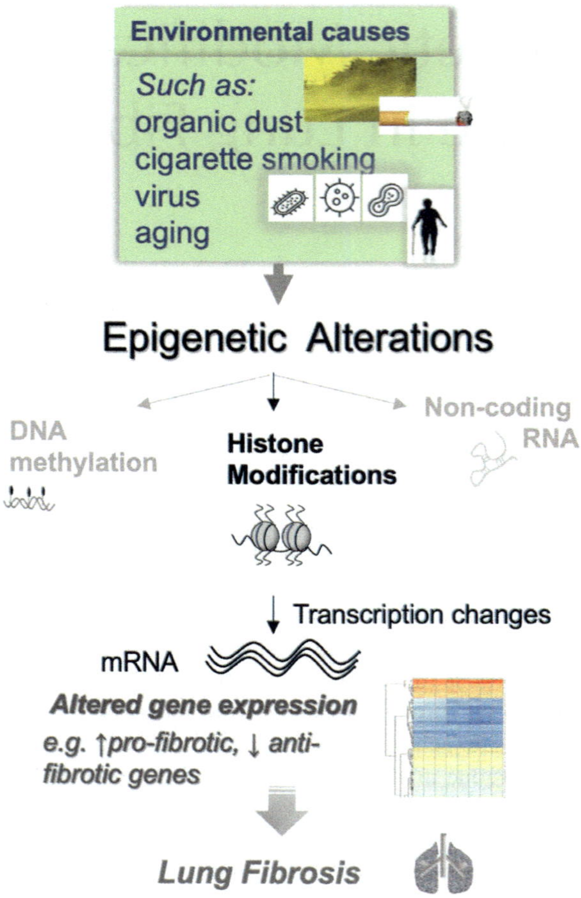

FIGURE 4.1 Epigenetic modifications are involved in the pathogenesis of lung fibrosis. Many environmental factors induce epigenetic alterations, including DNA methylation, noncoding RNA, and histone modifications, to change the gene expression profile and result in lung fibrosis.

4.1.2 Overview of Histone Methylation

Another extensively studied histone modification is histone methylation. Unlike histone acetylation, which is always associated with active transcription, histone methylation can be associated with either active or repressive transcription (Figure 4.2B). Histone methylation usually happens on lysine (K) or on arginine (R) and is mediated by lysine or arginine methyltransferases and demethylases. Although arginine methylation has been regarded as one of the common posttranslational modifications (PTM) in epigenetic regulation and is involved in many cellular processes, it has started to gain traction only in recent years as a result of the advancement of technology, providing techniques and animal models to support its related research (22). In lung fibrosis studies, the widely studied histone methylation modifications are lysine methylation; only few explored the role of arginine methylation in lung fibrosis (23, 24). Arginine can be mono- or dimethylated, whereas histone lysine can have a mono-, di-, or trimethyl-group (me1, me2, or me3) added by lysine methyltransferases (KMT). These different methylation statuses at different positions add another layer of regulation on gene expression. Usually, histone H3 lysine 4 mono-/di-/trimethylation (short for H3K4Me1/2/3, similar to the following), H3K36Me1/2/3, and H3K79Me1/2/3 are usually marked as *active* marks associated with actively transcribed genes; other methylated histone marks, such as H3K9Me3 and H3K27Me3, are regarded as repressive marks associated with closed chromatin (25).

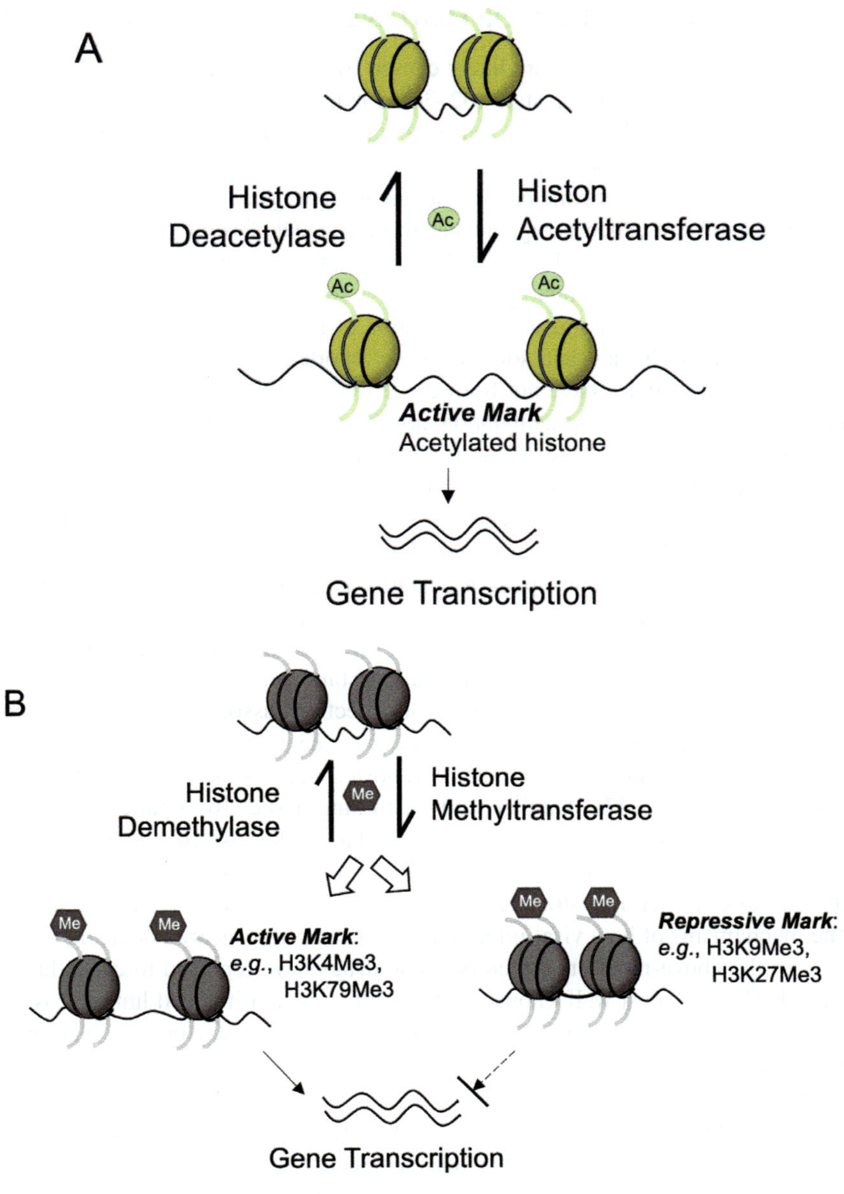

FIGURE 4.2 Histone modifications regulate gene transcription. (A) Histone acetylation is balanced by histone acetyltransferases and deacetylases; it is usually an active mark and is associated with active transcription. (B) Histone methylation is balanced by histone methyltransferases and demethylases; it can be either an active mark or a repressive mark. Some histone methylation, such as H3K4me3 or H3K79me3, is associated with active transcription; other histone methylation is a repressive mark, like H3K9me3 or H3K27me3, and is associated with transcriptional repression. Ac, acetylation; Me, methylation.

Besides the histone methylation changes, some methyltransferases such as G9a, EZH2, and DOT1L are reported to be involved in the epigenetic regulation of lung fibrosis (26–29). Lysine demethylases, including LSD1 (lysine-specific demethylase 1) (30) and JMJD3 (Jumonji domain-containing protein-3) (31) are also reported in the pathogenesis of lung fibrosis. As for arginine methylation–related enzymes, PRMT1 (protein arginine methyltransferase 1) was found to be elevated in IPF lungs and regulates lung fibroblast motility in IPF (24).

4.2 ENVIRONMENTAL CAUSES OF HISTONE MODIFICATIONS

The changes of histone modifications are likely caused by environmental alterations, such as the air breathed in, smoking, and metabolic state (32–34). Environmental factors could affect epigenetic makeup, including histone modifications to shape the transcriptome and metabolome to cause diseases epigenetically. Histone modification changes that responding to various environmental stimuli could form a specific pattern as epigenetic memory to pass on to future generations (32). Studies have shown that environmental exposures, such as organic dust from livestock, farming, wood, metal, and mineral, all contribute to the pathogenesis of IPF (35). A study found that dust from the rubble of the World Trade Center towers induced inflammatory and oxidative stress in mice lungs that are associated with epigenetic modifications, including histone methylation of H3K4, K27, and K36, in the lung tissues and altered pulmonary mechanics, and some of those reported changes were found in lung fibrosis (36). Although extensive studies have not been carried out, silicon dust has been reported to cause aberrant epigenetic changes in DNA methylation, noncoding RNA, and histone modifications (37).

Cigarette smoking has long been regarded as a risk factor associated with IPF (38). It is well documented that smoking can induce histone modification changes. Cigarette smoking causes histone hyperacetylation in human and rat lungs with increased inflammation; one of the effects of smoking is the reduced activity and expression of certain HDACs (39–41). One study found that smoking induced distinct histone modification pattern changes, including mono- and dimethylation of H3 and H4 in mouse lung and human bronchial epithelial cells (42). Increased risk of IPF was shown in both current and past tobacco smokers; interestingly, the study also found an increased risk for IPF in those who were exposed to maternal smoking around birth (43).

Besides certain types of dust and smoking, viral infections associated with IPF have been well documented (44–46). In one study, herpesviruses were detected in 97% of IPF subjects compared with 36% in controls (44). A recent study reported that about 45% of COVID-19 survivors have developed lung fibrosis (47). Another study with lung tissues from patients of non-resolvable COVID-19 showed increased fibrotic genes and cellular markers similar to IPF, indicating viral infection inducing lung fibrosis (48). Viral infection alters the epigenome, including histone modification, and methods of targeting histone changes to control viral infection have been proposed (49). However, the mechanisms of how viral infection causes lung fibrosis are not clear.

In addition to the above-mentioned factors, some drugs have reported to cause lung fibrosis as a side effect (50, 51). The most well-known of these is bleomycin-induced lung fibrosis, which has been widely used in animal models for lung fibrosis (52). Studies using the bleomycin-induced lung fibrosis models have shown histone modification changes, which are associated with altered fibrotic-related gene expression (33, 53, 54).

Another important factor related to environmental changes is aging (55). Bulk RNA sequencing data with human donor lung tissues revealed that cellular senescence and activated pro-fibrotic pathway both increase with age (56). The widely studied histone acetylation and methylation were known to impact the aging process (57). We have showed that the differential state of H4K16Ac in young and aged mice, which may be one of the important factors why aged mice have unresolved persistent lung fibrosis (53).

Overall, environmental exposure is a critical component in the pathogenesis of lung fibrosis. As common genetic variants are not always observed in IPF, the epigenetic changes in response to environmental stimuli may provide the critical link of the disease pathogenesis.

4.3 HISTONE ACETYLATION IN LUNG FIBROSIS

Though not many studies have demonstrated the direct impact of altered specific histone acetylation in the pathogenesis of lung fibrosis (13, 53), emerging studies showed that histone acetylation status is associated with specific fibrotic-related genes and controls the expression of these genes (1, 21). Most of these studies explored the alterations of related enzymes, HDACs and HATs, either directly regulate gene expression, or through change histone modification to regulate gene expression.

4.3.1 HDACs

In lung fibrosis, aberrant HDAC activities have been reported. HDAC inhibitors have shown promising results in cellular and animal lung fibrosis models, which were summarized in previous reviews (1, 21). Here we will summarize the main findings of dysregulated HDACs and their roles in the pathogenesis of IPF (Table 4.1). In one report, the authors found a strong induction of almost all class I and II HDACs in IPF lung tissues (58), which suggests HDAC overexpression is critical in the pathogenesis of IPF.

TABLE 4.1
HDACs in Lung Fibrosis

	Mechanisms	Material Used	Note	Ref
HDAC2, 4	HDAC2 activity increased in middle and late stages of lung fibrosis. HDAC4 activity similar to saline control group.	Bleo mice model C57BL/6J, 6–7 weeks old Lung tissue	Activities measured at 7, 14, and 21 days after bleo injury	(59)
	Reduced HDAC4 nuclear translocation by aberrant protein phosphatase 2A/HDAC4 axis in IPF fibroblasts, which reduces miR-29 and results in increased collagen I	Primary IPF and non-IPF fibroblasts	Lentiviral vector for HDAC4 knockdown	(60)
HDAC3	Upregulated, suppresses Nrf2	Whole lung tissue (IPF and 8- to 10-week-old bleo mice), alveolar epithelia cells	Inhibition with RGFP966, and flox mice improves fibrosis	(61)
	Regulate alveolar EMT via AKT pathway under hypoxia	Alveolar epithelial cell, IPF fibroblasts	siRNA HDAC3 relieves bleo-induced fibrosis in mice (7 weeks old) HDAC3 interacts with miR-224	(62)
	Upregulated Promotes EMT, inflammation, fibrosis development via Notch1 and STAT1 signaling	Bleo mice (8 weeks) MRC5 lung fibroblast	Inhibition with specific inhibitor RGFP966	(63)
	Decreased HDAC3 in IPF fibroblast nucleus, resulting in loss of heterochromatin	IPF-derived fibroblasts	Cells were grown on hydrogels to mimic stiffness of fibrotic lung	(64)
	Increased HDAC3 upregulates IL17RA by targeting miR-19a-3p	RA-ILD fibrosis patients lung tissue, RA-ILD mice model and isolated lung fibroblasts	RA-ILD fibrosis	(65)
HDAC6	Overexpressed in IPF lung tissue	IPF lung tissues	Immunohistochemistry stain	(66)
	HDAC6 increased in some IPF samples. However, HDAC6 KO mice were not protected from bleomycin injury.	IPF and non-IPF lung tissues and fibroblasts, 6- to 10-week-old mice bleo model, HDAC6 KO mice	C57/BL6 and HDAC6 KO mice (6–10 weeks), bleo model with lung tissues and primary fibroblasts	(67)
HDAC7	Attenuating TGF-β induced fibroblast activation	IMR-90 lung fibroblast with TGF-β	In vitro data	(69)

(Continued)

TABLE 4.1 (*Continued*)
HDACs in Lung Fibrosis

	Mechanisms	Material Used	Note	Ref
	ET-1 activates HDAC7 to initiate AP-1 transcriptional activity by recruiting p300 and promotes CTGF production	WI-38 lung fibroblast Ovalbumin-induced airway fibrosis mice model	In vitro data Mice model is in airway fibrosis, not lung fibrosis	(70)
HDAC8	HDAC8 increased in IPF lungs and is associated with α-smooth actin; inhibition of HDAC8 by NCC170 reduced loss of TGF-β H3K27Ac at PPARγ and reduced fibrosis in bleomycin mice model	IPF and non-IPF lung fibroblasts Bleo mice (6–10 weeks)		(71)
SIRT1	SIRT1 mediates Ku70-deacetylation, leads to Ku70/FLIP complex, stabilizing FLIP; reduced apoptosis and fibrosis occurs	IPF, non-IPF fibroblasts, mice 11–12 weeks, chimeric mice	Induced lung fibrosis in $Sirt1^{tm2.1Mcby}$ chimeric mice with SIRT1-deficient mesenchymal cells SIRT1 deficiency *reduces* apoptosis resistance	(83)
	Overexpression of SIRT1 attenuates lung fibrosis	MRC-5 lung fibroblasts, Bleo mice model (8 weeks old)	However, SIRT1 was found upregulated in fibrotic lung	(84)
	SIRT1 deacetylates H3K9/14ac to downregulate IL-11 expression	MRC-5, primary lung fibroblasts from 9-week-old mice	Used vitamin D–deficient mice, with or without Sirt1 overexpression	(80)
	SIRT1 reduced in alveolar epithelial cells; ZIP8/SIRT1 axis associated with alveolar epithelial renewal capacity and enhanced lung fibrosis	Human donor lung tissues, single cell, mice bleo model, 8–12 weeks and 18–20 months old		(82)
	SIRT1 blocks cigarette smoking–induced lung fibroblast activation	6-week-old mice primary lung fibroblast culture		(81)
SIRT2	Increased SIRT2 in response to TGF-β; SIRT2 inhibition reduced smad2/3 phosphorylation and fibrotic genes	MRC-5 with TGF-β, CCL-134, and Bleo mice model (6–8 weeks)	Improved fibrosis in bleo mice with SIRT2 inhibitor AGK2	(89)
SIRT3	Deficient AEC SIRT3 activity, with induced acetylation of MnSOD and OGG1; affects AEC mtDNA damage and apoptosis	A549 and MLE-12 human and murine epithelial, 129S WT and $SIRT3^{-/-}$ mice (8–10 weeks) asbestos or bleo-induced fibrosis		(85)
	SIRT3 reduced in AECs of IPF; increased SIRT3 is protective by reducing recruitment of profibrotic alveolar macrophages	A546 and MLE-12 epithelial cells, and 8- to 10-week-old WT mice, $Sirt3^{Tg}$ (whole-body SIRT3 overexpression) mice, bleo and asbestos models	$SIRT3^{-/-}$ mice asbestos, and bleo-induced fibrosis is augmented; overexpressed SIRT3 is protective	(78)
	Reduced SIRT3 in lung fibroblasts mediates senescence, apoptosis resistance, and increased pro-fibrotic genes	Human lung tissues/ fibroblasts, bleo mice (18–22 months)	Airway delivery SIRT3 cDNA facilitates fibrosis resolution through macrophage-fibroblast crosstalk via SIRT3-FOXO3 axis	(77)
	Reduced SIRT3 mediates increased IDH2 and SOD2 acetylation, results in deficient antioxidant response, promotes myofibroblast differentiation	IPF, SSc-PF patients' tissues/ fibroblasts Bleo mice model (2 or 22 months old), and Sirt3 knockout and control mice		(86)

	Mechanisms	Material Used	Note	Ref
	Probucol restored SIRT3 and relives lung fibrosis by reducing oxidative stress and HIF-1α activation	8-week-old mice bleo model, murine epithelial MLE-12, and macrophage RAW264.7	Probucol was given starting at the same time as bleo, for 2 weeks	(87)
	Baicalein reduces lung fibrosis, fibroblast senescence, and TGF-β1/Smad pathway by preserving SIRT3 expression	Bleo mice model (8-week-old mice), murine primary lung fibroblasts		(88)
SIRT6	Increased SIRT6 in IPF lung, and with TGF-β Overexpressed SIRT6 inhibited TGF-β–induced senescence via proteasomal degradation of p21	HBEC, fibroblasts from donors		(90)
	Reduced Sirt6 with aging; deficiency of SIRT6 activates TGF-β signaling and induces fibrosis	WT and SIRT6 heterozygous (SIRT6$^{+/-}$) mice	Multiple-organ fibrosis	(91)
	SIRT6 inhibits EMT in lung fibrosis via TGF-β/Smad3 signaling	A549 epithelial, bleo mice model (8-week-old mice)		(92)
SIRT7	SIRT7 is reduced in lung fibrosis. Overexpressed SIRT7 reduces pro-fibrotic markers.	Patients' lung tissues/fibroblasts Bleo mice model (10–12 weeks old)		(93)
	Increased GLS1 by TGF-β downregulates SIRT7 to target FOXO4 acetylation	Lung fibroblasts, bleo mice model (10 weeks old)		(95)
	SIRT7 deficiency suppresses inflammation, induces endomesenchymal transition	Endothelial cells, acute lung injury and bleo mice model (10–12 weeks old)		(96)

4.3.1.1 HDAC2 and HDAC4

One study measured HDAC2 and HDAC4 activities in a bleomycin-induced fibrosis model at 7, 14 and 21 days after bleomycin injury. The study found that HDAC2 peaked at day 14 and maintained till day 21 after the injury, but HDAC4 did not show much difference as compared to a saline control group (59). However, in another study exploring the regulation of increased collagen I in IPF fibroblasts, researchers found HDAC4 protein level is reduced in IPF fibroblasts, which regulates collagen I expression via miR-29 (60).

4.3.1.2 HDAC3

A study with fibrotic lung tissues, either in mice with bleomycin-induced lung fibrosis or in IPF patients, found that HDAC3 is upregulated, and no significant changes of HDAC1, 2, or 8 under the same condition were observed (61). The HDAC3 aberration resulted in inhibition of Nrf2, a master transcription factor of antioxidative stress, contributes significantly to pulmonary fibrosis. This group also showed that HDAC3-selective inhibitor could effectively alleviate Nrf2 suppression and reduced lung fibrosis in the bleomycin mice model (61). HDAC3 was also found regulating hypoxia-induced alveolar EMT via the AKT pathway; inhibition of HDAC3 with siRNA decreased the migration and invasiveness of IPF lung fibroblasts and inhibited lung fibrosis induced by bleomycin in 6- to 7-week-old mice (62). In addition to promoting EMT, HDAC3 was also reported to promote inflammation and lung fibrosis development by activating Notche1 and STAT1 signaling (63).

In another study, the role of HDAC3 in the IPF lung fibroblast nucleus was examined (64). The investigators grew the cells on hydrogels to mimic the stiffness of a fibrotic lung, and found that HDAC3 is reduced in the nucleus, which correlates with the loss of heterochromatin. When inhibiting HDAC3 with specific inhibitor RGFP966, they observed H3K9 hyperacetylation with increased Col1A1, ACTA2, and p21 (64). The results of increased expression of these pro-fibrotic markers with HDAC3-specific inhibitors are different than other studies using the same inhibitor (61, 63), which showed reduced expression. However, the other studies did not use IPF primary lung fibroblasts and did not culture the cells on a stiff matrix, which may contribute to the differences. These studies also indicate the complexity of epigenetic regulation under a different culture environment, as well as in different cells.

Besides IPF, in rheumatoid arthritis (RA)-associated interstitial lung disease (ILD), HDAC3 was found to be increased, which upregulates IL17RA expression by targeting miR-19a-3p to facilitate RA-ILD fibrosis development (65). Although more studies are needed to explicit the effects of HDAC3 in lung fibrosis, HDAC3 has shown its importance in the pathogenesis of lung fibrosis.

4.3.1.3 HDAC6

HDAC6 is reported to be overexpressed in IPF lung tissues (66). Studies have reported that HDAC6 is higher in some IPF samples and significantly higher in others (67). The bleomycin-induced lung fibrosis is not protected in HDAC6 knockout (KO) mice, and the protective effects from the HDAC6 "selective" inhibitor Tubastatin are likely via HDAC6-independent mechanisms, such as targeting the TGF-β–PI3K–Akt pathway (67). Some preliminary studies found that a novel and selective HDAC6 inhibitor, CDK-506, has strong anti-fibrotic effects by inhibiting SAMD2/3 phosphorylation (68). Many of the HDAC6-related effects are independent of histone acetylation.

4.3.1.4 HDAC7

HDAC7 is found to be required for TGF-β induced fibroblast activation; after knockdown by silencing HDAC7 it did not significantly change the total acetyl-H3 protein level, indicating HDAC7 may only specifically affect a subset of genomic loci (69). Another study in lung fibroblasts indicates that HDAC7 can be activated by ET-1 to initiate AP-1 transcriptional activity by recruiting p300 (70). Both in vitro studies with lung fibroblasts support the role of HDAC7 in lung fibrosis, and propose it as potential therapeutic target.

4.3.1.5 HDAC8

The key event in IPF pathogenesis is the fibroblast differentiation to myofibroblast, and HDAC8 is associated with α-smooth muscle actin in TGF-β–treated normal lung fibroblasts (71). HDAC8 selective inhibitor NCC170 or siRNA would repress TGF-β–induced pro-fibrotic proteins and would induce anti-fibrotic proteins; in addition, NCC170 significantly reduced fibrotic scores in bleomycin-induced lung fibrosis in mice (71).

Currently, not many studies have explored the roles of HDAC9 or HDAC10 in lung fibrosis. In one of the studies with HDAC6 (67), the investigators also knocked down HDAC10 in normal lung fibroblasts and did not observe any changes in TGF-β–induced collagen expression. However, the report that examined various HDACs expression in IPF by immunohistochemistry (IHC) found that HDAC10 is much higher in IPF when compared to the control (58). Therefore, more studies are needed to clarify the functions of these HDACs in lung fibrosis.

4.3.1.6 Sirtuins

Sirtuins (SIRTs) are NAD^+-dependent class III HDACs with seven family members (SIRT1–7) in mammalian cells; high NAD+/NADH ratio stimulates sirtuin activation, and a low ratio leads to sirtuin inactivation (72). Some sirtuins shuttle between nucleus and cytosol, like SIRT1 and SIRT 2;

some are mainly located in mitochondria, such as SIRT3, SIRT4, and SIRT 5; and some are mainly located in the nucleus such as SIRT6; SIRT7 is located in the nucleolus (73, 74). The targets of sirtuins are like other HDACs that are not only histones but also many non-histone proteins. Because sirtuin enzymatic activities are NAD^+ dependent, sirtuins act as energy sensors linking environmental stimuli and cellular metabolic status (72). Many studies have revealed the positive roles of sirtuins against aging and many diseases, including lung fibrosis (74, 75). Sirtuins are distinctly different from the other classes of HDACs that are Zn^{2+} dependent; thus those pan-HDAC inhibitors usually will not affect the sirtuins. Among the sirtuins, SIRT1, SIRT6, and SIRT7 are mostly in the nucleus and act as histone deacetylases; they also regulate key transcription factors and cofactors in response to cellular changes (72).

The sirtuins are involved in many cellular processes, such as cell cycle, DNA repair, and metabolism; they have emerged as a major regulator in aging (76). In lung fibrosis, sirtuins are involved in various inflammation and fibrotic related pathways (75, 76). Many of the studies on sirtuins are related with non-histone acetylation, whereas only few are associated with histone deacetylation in lung fibrosis. For example, several reports indicate that SIRT3 is reduced in lung fibrosis, and induced SIRT3 will ameliorate lung fibrosis in animal models through the SIRT3-FoxO3a signaling axis (77) or preserve alveolar epithelial cell mitochondrial DNA damage (78).

SIRT1 and SIRT3 showed anti-fibrotic effects in fibrosis-related diseases (76). A study found reduced serum levels of SIRT1 and SIRT3 are associated with a greater extent of cutaneous fibrosis, persistence of interstitial lung disease, and worse pulmonary function in systemic sclerosis patients (79). SIRT1 is reduced with aging, and its overexpression improves senescence-associated lung fibrosis induced by vitamin D deficiency through downregulating IL-11 transcription by reducing IL-11 promoter association with H3K9/14ac at the binding region of Smad2 (80). SIRT1 is found to prevent cigarette smoking–induced lung fibroblast activation by regulating lipid metabolism in an autophagy-dependent manner (81). SIRT1 was also found to be involved in type 2 alveolar epithelial cell regeneration with a zinc transporter, ZIP8 (82). Although several studies reported reduced SIRT1 in lung fibrosis (82, 83), a study indicated SIRT1 is overexpressed in lung fibrosis (84). Nonetheless, they all conclude that increased SIRT1 will increase resolution in lung fibrosis.

SIRT3 is mostly located in the mitochondria, where it mainly deacetylates mitochondrial proteins. In IPF patients, reduced SIRT3 was reported to increase acetylation of its targets MnSOD and OGG1, which are important for maintain mtDNA integrity and apoptosis (85). Reduced SIRT3 also promotes myofibroblast differentiation by inactivating the major oxidative stress response regulators SOD2 and isocitrate dehydrogenase 2 (IDH2) with acetylation (86). Overexpression of SIRT3 relieves asbestos-induced lung fibrosis and mt-damage, and it reduces lung fibrogenic monocyte recruitment (78). The cholesterol-lowering drug probucol increases SIRT3 by reducing oxidative stress and HIF-1 activation in the lung fibrosis model (87). Another widely used drug, baicalein, attenuates lung fibrosis, lung fibroblast senescence, and TGF-β1/Smad activation by preserving SIRT3 expression (88). SIRT3 is also reduced in IPF fibroblasts; its anti-fibrotic effects are associated with myofibroblast apoptosis susceptibility via macrophage-fibroblast crosstalk to activate SIRT3-FoxO3a signaling path in fibrosis (77).

Contrary to the reduced SIRT1 and SIRT3 in response to TGF-β1 stimulation, one study found SIRT2 is increased, and the siRNA SIRT2 reduced TGF-β1–induced pro-fibrotic marker expression (89). The roles of SIRT4 and SIRT5 in lung fibrosis are not clear.

SIRT6 is reported to be increased in IPF lung and in response to TGF-β in human bronchial epithelial cells, and was found regulating TGF-β–induced cellular senescence by suppressing p21/waf-1 expression via posttranslational modification for proteasome degradation (90). However, SIRT6 is reduced with aging, and its deficiency increased SMAD3 and H3K9, K56 acetylation that upregulates TGF-β, and induced multiple organ fibrosis, including lung, in mice (91). In lung epithelial cells, forced SIRT6 expression abrogated TGF-β–induced epithelial to mesenchymal transition (92).

SIRT7 mRNA is reduced in IPF patients' lung tissues, as well as in fibroblasts derived from IPF lungs (93). In addition, TGF-β upregulates glutaminase 1 (GLS1) in lung fibroblasts (31, 94), partially through downregulating SIRT7, whereas SIRT7 deacetylation of FOXO4 blocks GLS1 expression (95). The loss of SIRT7 during aging and acute lung injury is reported to be associated with inflammation and fibrosis (96).

4.3.2 HATs and Histone Acetylation

Histone acetylation is balanced by HDAC and HAT. HATs are enzymes that transfer an acetyl group from acetyl-CoA to histone lysine amino acid, which relaxes the chromatin structure and facilitates transcription (97). Compared to HDACs, there are fewer studies of HATs in lung fibrosis. Some acetyltransferases are also transcriptional factors, which play a crucial role in regulating gene expression in the pathogenesis of lung fibrosis (19, 53, 98).

Studies of HATs in lung fibrosis are largely on p300 (also called EP300), a transcriptional cofactor, and HAT. P300 is upregulated in IPF lung tissues; its overexpression would stimulate FN1 production and DDR1 (discoidin domain receptor 1) expression; synergistical inhibition of p300 and DDR1 would alleviate fibrosis in experimental models (99). A natural compound, plumbagin, was found to suppress lung fibrosis via inhibition of p300 activity, which reduced pro-fibrotic gene expression and cell proliferation (100). The recruitment of p300 is reduced on the promoter of pro-fibrotic gene in the presence of an inhibitor of reader protein bromodomain (19). The HAT inhibitor anacardic acid downregulates IL-6 via p300 in macrophages in a paraquat-induced pulmonary fibrosis (101).

In a detailed study of decreased Cox-2 expression in IPF, it was found that Cox-2 promoter is linked to reduced H3 and H4 acetylation due to the reduced recruitment of HAT, as well as the increased recruitment of transcriptional co-repressor complex (98). The researchers found that some HATs, like CBP, p300, GCN5 and PCAF, are required for Cox-2 gene transcription in normal lung fibroblasts. In IPF fibroblasts, the HAT recruitment to the Cox-2 promoter is significantly reduced, but the transcriptional corepressor complexes CoREST, NCoR, and mSin3a are increased, resulting in diminished Cox-2 transcription in IPF fibroblasts. Later, the same group further explored that G9a-mediated H3K9me3 and EZH2-mediated H3K27 are enriched at the Cox-2 promoter region, whereas the active mark H3K4me3 is reduced (28). The regulation of Cox-2 is complex. They also showed that DNA methylation, DNA methyltransferases, and binding proteins together contribute to the regulation of Cox-2 in IPF fibroblasts (28). This group also explored another anti-fibrotic gene, CXCL10 (also called IP-10) (102). They reported that the downregulation of CXCL10 is not only because of the histone deacetylation but also H3 hypermethylation in IPF. Reduced CXCL10 expression is because its promoter region decreases HAT recruitment, together with increased recruitment of the HDAC-containing transcriptional repressor complexes NCoR, CoRESt, and mSin3a; H3K9-specific methyltransferase G9a and SUV39H1; and HP1, the H3K9me3 binding protein (102).

There are many studies that explored individual gene expression that was regulated by specific histone acetylation, and many of these modifications are interactive. For example, one study treated the fibroblasts with SAHA, a pan-HDAC inhibitor, and not only observed increased histone acetylation (H3Ac, H4Ac, and H3K9ac) but also increased H3K9me3, a repressive mark (103). The changes of different histone marks at the general level may or may not reflect the situation at the specific gene promoter region. The SAHA treated cells showed increased association with H3K9ac with the pro-apoptosis gene Bak, but they are depleted in the anti-apoptosis gene Bcl-xL promoter, corresponding to increased apoptosis with upregulation of Bak and downregulation of Bcl-xL (103). Similarly, in the aging lung, H4K16ac is reported to be reduced (53) with increased Nox4 expression (104), which Nox4 promoter region is associated with increased H4K16Ac and p300 for its upregulation (19, 104).

Most of the above-mentioned studies explored the non-histone protein targets of the HDACs and HATs. Few studies explored general histone acetylation level changes in lung fibrosis. One study showed reduced H4K16Ac in normal aging mice lung tissues and fibroblasts, whereas increased H4K16ac was noticed in aged mice with persistent lung fibrosis and in some IPF lung tissues (53).

Knocking down H4K16Ac major acetyltransferase, Mof, decreased H4K16Ac levels, reduced profibrotic genes expression by depleting their promoter association with H4K16Ac, and mitigated lung fibrosis in the aged mice model (53).

4.4 HISTONE METHYLATION AND OTHER HISTONE MODIFICATION IN LUNG FIBROSIS

Unlike histone acetylation, which is usually associated with actively transcribed genes, histone methylation can be an active mark or a repressive mark (105). Studies in histone methylation–related roles in lung fibrosis are emerging. There are a few genes that are reported to be regulated by histone methylation. For example, an anti-fibrotic gene, Cox-2, was associated with repressive marks of H3K9me3 and H3K27me3 in lung fibrosis (28), which was discussed in detail in the earlier part of HAT and acetylated histone. PGC1α was reported to be inhibited in IPF, and repressor complex CBX5 (chromobox homolog 5) and G9a-mediated H3K9 methylation are critical for the repression of PGC1α (26). TGF-β enhances CBX5 recruitment and H3K9me2 binding to the PPARGC1A gene (encoding PGC1α) promoter region to downregulate its expression, which is essential for fibroblast activation during lung fibrosis (26).

Among the enzymes that are involved in histone methylation in lung fibrosis, H3K27me3 methyltransferase, EZH2, is one of the most studied. One report indicates that EZH2 is increased in IPF, which enhances fibroblast differentiation likely through p-SMAD2/3 nuclear translocation (106). EZH2 and H3K9me3 methyltransferase G9a together regulate the anti-fibrotic gene CXCL10 in IPF (107). The report indicated EZH2 and G9a physically interact with each other; reducing one will reduce the association with the other as well as the target histone H3K9me3 or H3K27me3, and will restore the CXCL10 expression in IPF. This report also demonstrated that H3K27me3 demethylase JMJD3 overexpression reduces H3K27me3 and impacts CXCL10 expression (107). Another report indicates that EZH2 is involved in pulmonary fibrosis associated with acute respiratory distress syndrome (108). Inhibition of EZH2 protects against lung fibrosis likely by regulating the crosstalk of alveolar macrophages and epithelial cells to prevent EMT through the TGF-β/Smad signaling pathway.

Besides being a methyltransferase of the repressive mark H3K27me3, a recent study revealed that EZH2 has an important non-canonical role, in which it acts as a transcriptional regulator driving aberrant repair in injured epithelial cells. Epithelial injury enriches nuclear TGF-β–activated kinase 1 (TAK1), which mediates EZH2 phosphorylation to facilitate its release from PRC2 and to establish a transcriptional complex to orchestrate aberrant epithelial repair programs (27). Loss of epithelial EZH2 or TAK1 attenuates the fibrotic cascade and improves lung fibrosis outcomes. The release of EZH2 from PRC2 is accompanied by an EZH2-EZH1 switch to preserve H3K27me3 deposition at non-target genes.

In addition to EZH2, the demethylases of H3K27me3, JMJD3, and UTX are also involved in lung fibrosis. A study reported that STAT6 regulates JMJD3, and the deficiency of JMJD3 abrogates profibrotic macrophage gene expression (109). JMJD3 and UTX are enzymes that use metabolite α-ketoglutarate (α-KG) as a cofactor for their activities; therefore, reduced glutaminolysis would reduce their activities and affect the levels of H3K27me3 (31). The altered levels of H3K27me3 would affect fibrosis-related gene expression in IPF lung fibroblasts (33). The expression of Col3A1 in lung fibroblasts is reported to be negatively correlated with H3K27me3 (110). In a study of lung fibrosis after nickel exposure, H3K27me3 is lost in genes that are persistently upregulated after the exposure (111). The report explored nickel-induced transcriptional changes and also found extensive genome-wide alterations relate to the active histone H3K4me3 after termination of nickel exposure. In addition to the studies of repressive mark H3K27me3 and active mark H3K4me3, H3K79me/me2 are linked to active gene transcription, and the methyltransferase DOT1L is reported to be a regulator of lung fibrosis (29). The study reported increased DOTL1 in mice bleomycin models of lung fibrosis, with increased H3K79me3, whereas inhibition of DOTL1 reduces related fibrosis. This group found that DOTL1 mediates fibrosis response the via Notch1 pathway through enriched H3K79me3 at the Jag1 promoter to upregulate its expression and binding to Notch1 receptor to modulate profibrotic gene expression and accelerate lung fibrosis. For the histone methylation in H4, the repressive H4K20me3 was reported to be inversely associated with the profibrotic gene Nox4 in senescent lung fibroblasts (104). Overall, histone methylation, active or repressive, coordinated with other regulatory mechanisms, participated in the pathogenesis of lung fibrosis.

Besides histone acetylation and methylation, there are other histone modifications reported in lung fibrosis. For example, increased glycolysis was reported in lung fibrosis, and the increased lactate induced histone lactylation in macrophages (112). Lactate increases histone lactylation at the promoters of profibrotic genes in macrophages, which is mediated by the acetyltransferase p300. However, because of technical limitations, such as valid antibodies, the exploration of histone modifications other than acetylation and methylation is very limited.

4.5 CONCLUSIONS

The studies of general levels of histone modification changes in lung fibrosis have just begun, and the exact mechanisms that control histone modifications have yet to be elucidated. The regulation of specific genes by specific histone modification would require in-depth study. With the development of new technology, it is now possible to define the regulatory role of specific histone modification. For example, in mouse embryonic stem cells, the role of H3K27 in regulating gene expression can be addressed by using CRISPR base editing (113). Such an approach demonstrated the feasibility of elucidating that the function of histone modification can be applied in lung cells, which would provide better understanding of the functions of histone modification in different lung cells. We should always examine cautiously the functions of histone modification in a gene- and context-specific way in response to the microenvironment, because this is a dynamic process coordinating with other factors to regulate gene expression.

As histone modification changes are relatively easy to detect and are abundant, using histone modification changes as biomarkers has been proposed in other diseases. For example, H3K27me3 and H4K20me have been proposed as diagnostic indicators of melanoma (114). More studies are needed to establish histone modification change as possible biomarkers in lung fibrosis. Current studies are mainly explored specific genes that are regulated by various histone modifications, their related enzymes, and other epigenetic regulators, such as noncoding RNA, all coordinated to regulate the gene expression. Using epigenetic modifiers to alter targeted gene expression is usually not specific, especially not organ or cell type specific. This provides challenges for clinical applications for targeted therapy.

Although there are many obstacles to applying histone modifiers in current therapeutic strategies, histone modification marks could be used alone or in combination, or with other markers to determine prognosis or to stratify different treatment options for different categories of patients, to identify the ones may response better than others. At this stage, epigenetic modifiers that have been proposed as therapeutic drugs are mainly examined in pre-clinical models. Extensive studies to explore such drugs for their mechanisms, benefits, potential side effects and interactions with other drugs are needed before using in clinical settings. Although many unknowns remain to be addressed, it is clear that histone modifications provide an exciting new avenue to understand and to target as new therapeutic options for this fatal disease, lung fibrosis.

ABBREVIATIONS

Ac	acetylation
acetyl-CoA	acetyl coenzyme A
ACTA2	actin alpha 2; also called alpha-smooth muscle actin

AEC	alveolar epithelial cell
α-KG	α-ketoglutarate
Bcl-xL	B cell lymphoma–extra large
Bleo	bleomycin
CBP	CREB-binding protein
CBX5	chromobox protein homolog 5
Col1A1	collagen type 1
Col3A1	collagen type 3
CoREST	co-repressor of repressor element–1 silencing transcription
Cox-2	cyclooxygenase-2
CRISPR	clustered regularly interspaced short palindromic repeats
CXCL10	C-X-C motif chemokine ligand 10
DDR1	discoidin domain receptor 1
DOT1L	disruptor of telomeric silencing 1–like
EMT	epithelial to mesenchymal transition
EZH2	enhancer of zeste homolog 2
FN1	fibronectin
FoxO3a	forkhead box O3
GCN5	general control non-depressible 5
GLS1	glutaminase-1
GNAT	Gcn5-related *N*-acetyltransferase
H3	histone 3
H4	histone 4
HAT	histone acetyltransferase
HBEC	human bronchial epithelial cells
HDAC	histone deacetylase
HIF-1	hypoxia-inducible factor 1
IDH2	isocitrate dehydrogenase 2
IHC	immunohistochemistry
IL	interleukin
ILD	interstitial lung disease
IPF	idiopathic pulmonary fibrosis
JMJD3	Jumonji domain-containing protein-3
K	lysine
KMT	lysine methyltransferase
KO	knockout
LSD1	lysine-specific demethylase 1
Me	methylation
Me1, me2, me3	mono-, di-, trimethyl group
miR	micro-RNA
MnSOD	manganese-superoxide dismutase
Mof	males absent on the first
mt	mitochondria
NAD	nicotinamide adenine dinucleotide
NCoR	nuclear receptor corepressor
Nox4	nicotinamide adenine dinucleotide phosphate oxidase 4
Nrf2	nuclear factor–erythroid 2–related factor 2
p-	phosphorate-
PCAF	P300/CBP-associated factor
PGC1a	peroxisome proliferator–activated receptor–gamma coactivator
PRC2	polycomb repressive complex 2

PRMT1	protein arginine methyltransferase 1
PTM	posttranslational modification
R	arginine
RA	rheumatoid arthritis
SAHA	suberoylanilide hydroxamic acid
siRNA	small interfering RNA
SIRT	sirtuin
Smad	suppressor of mothers against decapentaplegic
SOD	superoxide dismutase
STAT1	signal transducer and activator of transcription 1
TAK1	transforming growth factor-β–activated kinase 1
TGF-β	transforming growth factor beta
UTX	ubiquitously transcribed tetratricopeptide repeat on chromosome X

REFERENCES

1. Lyu, X., Hu, M., Peng, J., Zhang, X., and Sanders, Y. Y. (2019) HDAC inhibitors as antifibrotic drugs in cardiac and pulmonary fibrosis. *Ther Adv Chronic Dis.* **10**. https://doi.org/10.1177/2040622319862697
2. Tirelli, C., Pesenti, C., Miozzo, M., Mondoni, M., Fontana, L., and Centanni, S. (2022) The genetic and epigenetic footprint in idiopathic pulmonary fibrosis and familial pulmonary fibrosis: A state-of-the-art review. *Diagnostics.* **12**, 3107. www.mdpi.com/2075-4418/12/12/3107
3. Seibold, M. A., Wise, A. L., Speer, M. C., Steele, M. P., Brown, K. K., Loyd, J. E., *et al.* (2011) A common MUC5B promoter polymorphism and pulmonary fibrosis. *N Engl J Med.* **364**, 1503–1512. https://doi.org/10.1056/NEJMoa1013660
4. Bae, W., Lee, C.-H., Lee, J., Kim, Y. W., Han, K., and Choi, S. M. (2022) Impact of smoking on the development of idiopathic pulmonary fibrosis: Results from a nationwide population-based cohort study. *Thorax.* **77**, 470–476. https://doi.org/10.1136/thoraxjnl-2020-215386
5. Wood, C., and Yates, D. (2020) Respiratory surveillance in mineral dust-exposed workers. *Breathe.* **16**, 190632. https://doi.org/10.1183/20734735.0362-2019
6. Hamanaka, R. B., and Mutlu, G. M. (2021) Metabolic requirements of pulmonary fibrosis: Role of fibroblast metabolism. *FEBS J.* **288**, 6331–6352. https://doi.org/10.1111/febs.15693
7. Gulati, S., and Thannickal, V. J. (2019) The aging lung and idiopathic pulmonary fibrosis. *Am J Med Sci.* **357**, 384–389. https://doi.org/10.1016/j.amjms.2019.02.008
8. Liu, Y., Wen, D., Ho, C., Yu, L., Zheng, D., O'Reilly, S., *et al.* (2023) Epigenetics as a versatile regulator of fibrosis. *J Trans Med.* **21**, 164. https://doi.org/10.1186/s12967-023-04018-5
9. Al Aboud, N. M., Tupper, C., and Jialal, I. (2022) Genetics, epigenetic mechanism. In *StatPearls*, StatPearls Publishing Copyright © 2022, StatPearls Publishing LLC, Treasure Island, FL
10. Hagood, J. S., Prabhakaran, P., Kumbla, P., Salazar, L., MacEwen, M. W., Barker, T. H., *et al.* (2005) Loss of fibroblast Thy-1 expression correlates with lung fibrogenesis. *Am J Pathol.* **167**, 365–379. https://doi.org/10.1016/s0002-9440(10)62982-3
11. Sanders, Y. Y., Kumbla, P., and Hagood, J. S. (2007) Enhanced myofibroblastic differentiation and survival in Thy-1(-) lung fibroblasts. *Am J Respir Cell Mol Biol.* **36**, 226–235. https://doi.org/10.1165/rcmb.2006-0178OC
12. Sanders, Y. Y., Pardo, A., Selman, M., Nuovo, G. J., Tollefsbol, T. O., Siegal, G. P., *et al.* (2008) Thy-1 promoter hypermethylation: A novel epigenetic pathogenic mechanism in pulmonary fibrosis. *Am J Respir Cell Mol Biol.* **39**, 610–618. https://doi.org/10.1165/rcmb.2007-0322OC
13. Sanders, Y. Y., Tollefsbol, T. O., Varisco, B. M., and Hagood, J. S. (2011) Epigenetic regulation of thy-1 by histone deacetylase inhibitor in rat lung fibroblasts. *Am J Respir Cell Mol Biol.* **45**, 16–23. https://doi.org/10.1165/rcmb.2010-0154OC
14. Li, B., Carey, M., and Workman, J. L. (2007) The role of chromatin during transcription. *Cell.* **128**, 707–719. https://doi.org/10.1016/j.cell.2007.01.015
15. Zhang, X., Hu, M., Lyu, X., Li, C., Thannickal, V. J., and Sanders, Y. Y. (2017) DNA methylation regulated gene expression in organ fibrosis. *Biochim Biophys Acta Mol Basis Dis.* **1863**, 2389–2397. https://doi.org/10.1016/j.bbadis.2017.05.010

16. Duan, J., Zhong, B., Fan, Z., Zhang, H., Xu, M., Zhang, X., *et al.* (2022) DNA methylation in pulmonary fibrosis and lung cancer. *Expert Rev Respir Med.* **16**, 519–528. https://doi.org/10.1080/17476348.2022.2085091
17. Yang, X. J., and Seto, E. (2007) HATs and HDACs: From structure, function and regulation to novel strategies for therapy and prevention. *Oncogene.* **26**, 5310–5318. https://doi.org/10.1038/sj.onc.1210599
18. Deng, X., Zhou, X., Deng, Y., Liu, F., Feng, X., Yin, Q., *et al.* (2017) Thrombin induces CCL2 expression in human lung fibroblasts via p300 mediated histone acetylation and NF-KappaB activation. *J Cell Biochem.* **118**, 4012–4019. https://doi.org/10.1002/jcb.26057
19. Sanders, Y. Y., Lyv, X., Zhou, Q. J., Xiang, Z., Stanford, D., Bodduluri, S., *et al.* (2020) Brd4-p300 inhibition downregulates Nox4 and accelerates lung fibrosis resolution in aged mice. *JCI Insight.* **5**. https://doi.org/10.1172/jci.insight.137127
20. Yoon, S., Kang, G., and Eom, G. H. (2019) HDAC inhibitors: Therapeutic potential in fibrosis-associated human diseases. *Int J Mol Sci.* **20**. https://doi.org/10.3390/ijms20061329
21. Korfei, M., Mahavadi, P., and Guenther, A. (2022) Targeting histone deacetylases in idiopathic pulmonary fibrosis: A future therapeutic option. *Cells.* **11**, 1626. www.mdpi.com/2073-4409/11/10/1626
22. Blanc, R. S., and Richard, S. (2017) Arginine methylation: The coming of age. *Mol Cell.* **65**, 8–24. https://doi.org/10.1016/j.molcel.2016.11.003
23. Kitowska, K., Zakrzewicz, D., Königshoff, M., Chrobak, I., Grimminger, F., Seeger, W., *et al.* (2008) Functional role and species-specific contribution of arginases in pulmonary fibrosis. *Am J Physiol Lung Cell Mol Physiol.* **294**, L34–L45. https://doi.org/10.1152/ajplung.00007.2007
24. Zakrzewicz, D., Zakrzewicz, A., Didiasova, M., Korencak, M., Kosanovic, D., Schermuly, R. T., *et al.* (2015) Elevated protein arginine methyltransferase 1 expression regulates fibroblast motility in pulmonary fibrosis. *Biochim Biophys Acta.* **1852**, 2678–2688. https://doi.org/10.1016/j.bbadis.2015.09.008
25. Black, J. C., Van Rechem, C., and Whetstine, J. R. (2012) Histone lysine methylation dynamics: Establishment, regulation, and biological impact. *Mol Cell.* **48**, 491–507. https://doi.org/10.1016/j.molcel.2012.11.006
26. Ligresti, G., Caporarello, N., Meridew, J. A., Jones, D. L., Tan, Q., Choi, K. M., *et al.* (2019) CBX5/G9a/H3K9me-mediated gene repression is essential to fibroblast activation during lung fibrosis. *JCI Insight.* **5**. https://doi.org/10.1172/jci.insight.127111
27. Le, H. Q., Hill, M. A., Kollak, I., Keck, M., Schroeder, V., Wirth, J., *et al.* (2021) An EZH2-dependent transcriptional complex promotes aberrant epithelial remodelling after injury. *EMBO Rep.* **22**, e52785. https://doi.org/10.15252/embr.202152785
28. Coward, W. R., Feghali-Bostwick, C. A., Jenkins, G., Knox, A. J., and Pang, L. (2014) A central role for G9a and EZH2 in the epigenetic silencing of cyclooxygenase-2 in idiopathic pulmonary fibrosis. *FASEB J.* **28**, 3183–3196. https://doi.org/10.1096/fj.13-241760
29. Yang, D., Xu, P., Su, H., Zhong, W., Xu, J., Su, Z., *et al.* (2022) The histone methyltransferase DOT1L is a new epigenetic regulator of pulmonary fibrosis. *Cell Death Dis.* **13**, 60. https://doi.org/10.1038/s41419-021-04365-5
30. Pan, X., Li, J., Tu, X., Wu, C., Liu, H., Luo, Y., *et al.* (2020) Lysine-specific demethylase-1 regulates fibroblast activation in pulmonary fibrosis via TGF-β1/Smad3 pathway. *Pharmacol Res.* **152**, 104592. https://doi.org/10.1016/j.phrs.2019.104592
31. Bai, L., Bernard, K., Tang, X., Hu, M., Horowitz, J. C., Thannickal, V. J., *et al.* (2019) Glutaminolysis epigenetically regulates antiapoptotic gene expression in idiopathic pulmonary fibrosis fibroblasts. *Am J Respir Cell Mol Biol.* **60**, 49–57. https://doi.org/10.1165/rcmb.2018-0180OC
32. Dai, H., and Wang, Z. (2014) Histone modification patterns and their responses to environment. *Curr Environ Health Rep.* **1**, 11–21. https://doi.org/10.1007/s40572-013-0008-2
33. Xiang, Z., Bai, L., Zhou, J. Q., Cevallos, R. R., Sanders, J. R., Liu, G., *et al.* (2023) Epigenetic regulation of IPF fibroblast phenotype by glutaminolysis. *Mol Metab.* **67**, 101655. https://doi.org/10.1016/j.molmet.2022.101655
34. Yang, J., Liang, C., Liu, L., Wang, L., and Yu, G. (2023) High-fat diet related lung fibrosis-epigenetic regulation matters. *Biomolecules.* **13**, 558. www.mdpi.com/2218-273X/13/3/558
35. Trethewey, S. P., and Walters, G. I. (2018) The role of occupational and environmental exposures in the pathogenesis of idiopathic pulmonary fibrosis: A narrative literature review. *Medicina (Kaunas).* **54**. https://doi.org/10.3390/medicina54060108

36. Sunil, V. R., Vayas, K. N., Fang, M., Zarbl, H., Massa, C., Gow, A. J., *et al.* (2017) World trade center (WTC) dust exposure in mice is associated with inflammation, oxidative stress and epigenetic changes in the lung. *Exp Mol Pathol.* **102**, 50–58. https://doi.org/10.1016/j.yexmp.2016.12.005
37. Yin, H., Xie, Y., Gu, P., Li, W., Zhang, Y., Yao, Y., *et al.* (2022) The emerging role of epigenetic regulation in the progression of silicosis. *Clin Epigenetics.* **14**, 169. https://doi.org/10.1186/s13148-022-01391-8
38. Baumgartner, K. B., Samet, J. M., Stidley, C. A., Colby, T. V., and Waldron, J. A. (1997) Cigarette smoking: A risk factor for idiopathic pulmonary fibrosis. *Am J Respir Crit Care Med.* **155**, 242–248. https://doi.org/10.1164/ajrccm.155.1.9001319
39. Borgas, D., Chambers, E., Newton, J., Ko, J., Rivera, S., Rounds, S., *et al.* (2016) Cigarette smoke disrupted lung endothelial barrier integrity and increased susceptibility to acute lung injury via histone deacetylase 6. *Am J Respir Cell Mol Biol.* **54**, 683–696. https://doi.org/10.1165/rcmb.2015-0149OC
40. Winkler, A. R., Nocka, K. N., and Williams, C. M. (2012) Smoke exposure of human macrophages reduces HDAC3 activity, resulting in enhanced inflammatory cytokine production. *Pulm Pharmacol Ther.* **25**, 286–292. https://doi.org/10.1016/j.pupt.2012.05.003
41. Marwick, J. A., Kirkham, P. A., Stevenson, C. S., Danahay, H., Giddings, J., Butler, K., *et al.* (2004) Cigarette smoke alters chromatin remodeling and induces proinflammatory genes in rat lungs. *Am J Respir Cell Mol Biol.* **31**, 633–642. https://doi.org/10.1165/rcmb.2004-0006OC
42. Sundar, I. K., Nevid, M. Z., Friedman, A. E., and Rahman, I. (2014) Cigarette smoke induces distinct histone modifications in lung cells: Implications for the pathogenesis of COPD and lung cancer. *J Proteome Res.* **13**, 982–996. https://doi.org/10.1021/pr400998n
43. Bellou, V., Belbasis, L., and Evangelou, E. (2021) Tobacco smoking and risk for pulmonary fibrosis: A prospective cohort study from the UK biobank. *Chest.* **160**, 983–993. https://doi.org/10.1016/j.chest.2021.04.035
44. Tang, Y. W., Johnson, J. E., Browning, P. J., Cruz-Gervis, R. A., Davis, A., Graham, B. S., *et al.* (2003) Herpesvirus DNA is consistently detected in lungs of patients with idiopathic pulmonary fibrosis. *J Clin Microbiol.* **41**, 2633–2640. https://doi.org/10.1128/jcm.41.6.2633-2640.2003
45. Lawson, W. E., Crossno, P. F., Polosukhin, V. V., Roldan, J., Cheng, D. S., Lane, K. B., *et al.* (2008) Endoplasmic reticulum stress in alveolar epithelial cells is prominent in IPF: Association with altered surfactant protein processing and herpesvirus infection. *Am J Physiol Lung Cell Mol Physiol.* **294**, L1119–L1126. https://doi.org/10.1152/ajplung.00382.2007
46. Stewart, J. P., Egan, J. J., Ross, A. J., Kelly, B. G., Lok, S. S., Hasleton, P. S., *et al.* (1999) The detection of Epstein-Barr virus DNA in lung tissue from patients with idiopathic pulmonary fibrosis. *Am J Respir Crit Care Med.* **159**, 1336–1341. https://doi.org/10.1164/ajrccm.159.4.9807077
47. Hama Amin, B. J., Kakamad, F. H., Ahmed, G. S., Ahmed, S. F., Abdulla, B. A., Mohammed, S. H., *et al.* (2022) Post COVID-19 pulmonary fibrosis: A meta-analysis study. *Ann Med Surg (Lond).* **77**, 103590. https://doi.org/10.1016/j.amsu.2022.103590
48. Jyothula, S. S. K., Peters, A., Liang, Y., Bi, W., Shivshankar, P., Yau, S., *et al.* (2022) Fulminant lung fibrosis in non-resolvable COVID-19 requiring transplantation. *EBioMedicine.* **86**, 104351. https://doi.org/10.1016/j.ebiom.2022.104351
49. Nehme, Z., Pasquereau, S., and Herbein, G. (2020) Targeting histone epigenetics to control viral infections. *Histone Modificat Ther.* 255–292. https://doi.org/10.1016/b978-0-12-816422-8.00011-8
50. Hubbard, R., Venn, A., Smith, C., Cooper, M., Johnston, I., and Britton, J. (1998) Exposure to commonly prescribed drugs and the etiology of cryptogenic fibrosing alveolitis: A case-control study. *Am J Respir Crit Care Med.* **157**, 743–747. https://doi.org/10.1164/ajrccm.157.3.9701093
51. Macneal, K., and Schwartz, D. A. (2012) The genetic and environmental causes of pulmonary fibrosis. *Proc Am Thorac Soc.* **9**, 120–125. https://doi.org/10.1513/pats.201112-055AW
52. Liu, T., De Los Santos, F. G., and Phan, S. H. (2017) The bleomycin model of pulmonary fibrosis. *Methods Mol Biol.* **1627**, 27–42. https://doi.org/10.1007/978-1-4939-7113-8_2
53. Zhang, X., Liu, H., Zhou, J. Q., Krick, S., Barnes, J. W., Thannickal, V. J., *et al.* (2022) Modulation of H4K16Ac levels reduces pro-fibrotic gene expression and mitigates lung fibrosis in aged mice. *Theranostics.* **12**, 530–541. https://doi.org/10.7150/thno.62760
54. Sanders, Y. Y., Liu, H., Scruggs, A. M., Duncan, S. R., Huang, S. K., and Thannickal, V. J. (2017) Epigenetic regulation of caveolin-1 gene expression in lung fibroblasts. *Am J Respir Cell Mol Biol.* **56**, 50–61. https://doi.org/10.1165/rcmb.2016-0034OC

55. Meiners, S., Eickelberg, O., and Königshoff, M. (2015) Hallmarks of the ageing lung. *Eur Respir J.* **45**, 807–827. https://doi.org/10.1183/09031936.00186914
56. Lee, S., Islam, M. N., Boostanpour, K., Aran, D., Jin, G., Christenson, S., *et al.* (2021) Molecular programs of fibrotic change in aging human lung. *Nat Commun.* **12**, 6309. https://doi.org/10.1038/s41467-021-26603-2
57. Sidler, C., Kovalchuk, O., and Kovalchuk, I. (2017) Epigenetic regulation of cellular senescence and aging. *Front Genet.* **8**, 138. https://doi.org/10.3389/fgene.2017.00138
58. Korfei, M., Skwarna, S., Henneke, I., MacKenzie, B., Klymenko, O., Saito, S., *et al.* (2015) Aberrant expression and activity of histone deacetylases in sporadic idiopathic pulmonary fibrosis. *Thorax.* **70**, 1022–1032. https://doi.org/10.1136/thoraxjnl-2014-206411
59. Li, M., Zheng, Y., Yuan, H., Liu, Y., and Wen, X. (2017) Effects of dynamic changes in histone acetylation and deacetylase activity on pulmonary fibrosis. *Int Immunopharmacol.* **52**, 272–280. https://doi.org/10.1016/j.intimp.2017.09.020
60. Khalil, W., Xia, H., Bodempudi, V., Kahm, J., Hergert, P., Smith, K., *et al.* (2015) Pathologic regulation of collagen I by an aberrant protein phosphatase 2A/histone deacetylase C4/MicroRNA-29 signal axis in idiopathic pulmonary fibrosis fibroblasts. *Am J Respir Cell Mol Biol.* **53**, 391–399. https://doi.org/10.1165/rcmb.2014-0150OC
61. Chen, F., Gao, Q., Zhang, L., Ding, Y., Wang, H., and Cao, W. (2021) Inhibiting HDAC3 (Histone Deacetylase 3) aberration and the resultant Nrf2 (nuclear factor erythroid-derived 2-related factor-2) repression mitigates pulmonary fibrosis. *Hypertension.* **78**, e15–e25. https://doi.org/10.1161/hypertensionaha.121.17471
62. Jeong, S. H., Son, E. S., Lee, Y. E., Kyung, S. Y., Park, J.-W., and Kim, S.-H. (2022) Histone deacetylase 3 promotes alveolar epithelial—mesenchymal transition and fibroblast migration under hypoxic conditions. *Exp Mol Med.* **54**, 922–931. https://doi.org/10.1038/s12276-022-00796-y
63. Zheng, Q., Lei, Y., Hui, S., Tong, M., and Liang, L. (2022) HDAC3 promotes pulmonary fibrosis by activating NOTCH1 and STAT1 signaling and up-regulating inflammasome components AIM2 and ASC. *Cytokine.* **153**, 155842. https://doi.org/10.1016/j.cyto.2022.155842
64. Toscano-Marquez, F., Romero, Y., Espina-Ordoñez, M., and Cisneros, J. (2023) Absence of HDAC3 by matrix stiffness promotes chromatin remodeling and fibroblast activation in idiopathic pulmonary fibrosis. *Cells.* **12**, 1020. www.mdpi.com/2073-4409/12/7/1020
65. Yuan, H., Jiao, L., Yu, N., Duan, H., Yu, Y., and Bai, Y. (2020) Histone deacetylase 3-mediated inhibition of microRNA-19a-3p facilitates the development of rheumatoid arthritis-associated interstitial lung disease. *Front Physiol.* **11**. https://doi.org/10.3389/fphys.2020.549656
66. Campiani, G., Cavella, C., Osko, J. D., Brindisi, M., Relitti, N., Brogi, S., *et al.* (2021) Harnessing the role of HDAC6 in idiopathic pulmonary fibrosis: Design, synthesis, structural analysis, and biological evaluation of potent inhibitors. *J Med Chem.* **64**, 9960–9988. https://doi.org/10.1021/acs.jmedchem.1c00184
67. Saito, S., Zhuang, Y., Shan, B., Danchuk, S., Luo, F., Korfei, M., *et al.* (2017) Tubastatin ameliorates pulmonary fibrosis by targeting the TGFβ-PI3K-Akt pathway. *PLoS ONE.* **12**, e0186615. https://doi.org/10.1371/journal.pone.0186615
68. Lee, S. J., Kim, M. C., Baek, J., Suh, D., Song, J. W., and Ha, N. (2022) CKD-506, a selective HDAC6 inhibitor, is a first-in-class drug candidate for the treatment of idiopathic pulmonary fibrosis (IPF). In *D29. Mechanisms in Lung Injury, Repair, and Fibrosis*, pp. A5233–A5233, American Thoracic Society
69. Jones, D. L., Haak, A. J., Caporarello, N., Choi, K. M., Ye, Z., Yan, H., *et al.* (2019) TGFβ-induced fibroblast activation requires persistent and targeted HDAC-mediated gene repression. *J Cell Sci.* **132**. https://doi.org/10.1242/jcs.233486
70. Hua, H.-S., Wen, H.-C., Weng, C.-M., Lee, H.-S., Chen, B.-C., and Lin, C.-H. (2021) Histone deacetylase 7 mediates endothelin-1-induced connective tissue growth factor expression in human lung fibroblasts through p300 and activator protein-1 activation. *J Biomed Sci.* **28**, 38. https://doi.org/10.1186/s12929-021-00735-5
71. Saito, S., Zhuang, Y., Suzuki, T., Ota, Y., Bateman, M. E., Alkhatib, A. L., *et al.* (2019) HDAC8 inhibition ameliorates pulmonary fibrosis. *Am J Physiol Lung Cell Mol Physiol.* **316**, L175–L186. https://doi.org/10.1152/ajplung.00551.2017
72. Kupis, W., Pałyga, J., Tomal, E., and Niewiadomska, E. (2016) The role of sirtuins in cellular homeostasis. *J Physiol Biochem.* **72**, 371–380. https://doi.org/10.1007/s13105-016-0492-6

73. Hamaidi, I., and Kim, S. (2022) Sirtuins are crucial regulators of T cell metabolism and functions. *Exp Mol Med.* **54**, 207–215. https://doi.org/10.1038/s12276-022-00739-7
74. Zhao, L., Cao, J., Hu, K., He, X., Yun, D., Tong, T., *et al.* (2020) Sirtuins and their biological relevance in aging and age-related diseases. *Aging Dis.* **11**, 927–945. https://doi.org/10.14336/ad.2019.0820
75. Mazumder, S., Barman, M., Bandyopadhyay, U., and Bindu, S. (2020) Sirtuins as endogenous regulators of lung fibrosis: A current perspective. *Life Sci.* **258**, 118201. https://doi.org/10.1016/j.lfs.2020.118201
76. Zullo, A., Mancini, F. P., Schleip, R., Wearing, S., and Klingler, W. (2021) Fibrosis: Sirtuins at the checkpoints of myofibroblast differentiation and profibrotic activity. *Wound Repair Regen.* **29**, 650–666. https://doi.org/10.1111/wrr.12943
77. Rehan, M., Kurundkar, D., Kurundkar, A. R., Logsdon, N. J., Smith, S. R., Chanda, D., *et al.* (2021) Restoration of SIRT3 gene expression by airway delivery resolves age-associated persistent lung fibrosis in mice. *Nat Aging.* **1**, 205–217. https://doi.org/10.1038/s43587-021-00027-5
78. Cheresh, P., Kim, S. J., Jablonski, R., Watanabe, S., Lu, Z., Chi, M., *et al.* (2021) SIRT3 overexpression ameliorates asbestos-induced pulmonary fibrosis, mt-DNA damage, and lung fibrogenic monocyte recruitment. *Int J Mol Sci.* **22**. https://doi.org/10.3390/ijms22136856
79. Manetti, M., Rosa, I., Fioretto, B. S., Matucci-Cerinic, M., and Romano, E. (2022) Decreased serum levels of SIRT1 and SIRT3 correlate with severity of skin and lung fibrosis and peripheral microvasculopathy in systemic sclerosis. *J Clin Med.* **11**. https://doi.org/10.3390/jcm11051362
80. Zhou, J., Chen, H., Wang, Q., Chen, S., Wang, R., Wang, Z., *et al.* (2022) Sirt1 overexpression improves senescence-associated pulmonary fibrosis induced by vitamin D deficiency through downregulating IL-11 transcription. *Aging Cell.* **21**, e13680. https://doi.org/10.1111/acel.13680
81. Zhang, Y., Li, T., Pan, M., Wang, W., Huang, W., Yuan, Y., *et al.* (2022) SIRT1 prevents cigarette smoking-induced lung fibroblasts activation by regulating mitochondrial oxidative stress and lipid metabolism. *J Transl Med.* **20**, 222. https://doi.org/10.1186/s12967-022-03408-5
82. Liang, J., Huang, G., Liu, X., Taghavifar, F., Liu, N., Wang, Y., *et al.* (2022) The ZIP8/SIRT1 axis regulates alveolar progenitor cell renewal in aging and idiopathic pulmonary fibrosis. *J Clin Invest.* **132**. https://doi.org/10.1172/jci157338
83. Bulvik, R., Breuer, R., Dvir-Ginzberg, M., Reich, E., Berkman, N., and Wallach-Dayan, S. B. (2020) Sirt1 deficiency, specifically in fibroblasts, decreases apoptosis resistance and is associated with resolution of lung-fibrosis. *Biomolecules.* **10**, 996. www.mdpi.com/2218-273X/10/7/996
84. Zeng, Z., Cheng, S., Chen, H., Li, Q., Hu, Y., Wang, Q., *et al.* (2017) Activation and overexpression of Sirt1 attenuates lung fibrosis via P300. *Biochem Biophys Res Commun.* **486**, 1021–1026. https://doi.org/10.1016/j.bbrc.2017.03.155
85. Jablonski, R. P., Kim, S. J., Cheresh, P., Williams, D. B., Morales-Nebreda, L., Cheng, Y., *et al.* (2017) SIRT3 deficiency promotes lung fibrosis by augmenting alveolar epithelial cell mitochondrial DNA damage and apoptosis. *FASEB J.* **31**, 2520–2532. https://doi.org/10.1096/fj.201601077R
86. Sosulski, M. L., Gongora, R., Feghali-Bostwick, C., Lasky, J. A., and Sanchez, C. G. (2016) Sirtuin 3 deregulation promotes pulmonary fibrosis. *J Gerontol Series A.* **72**, 595–602. https://doi.org/10.1093/gerona/glw151
87. Zhang, H.-X., Li, Y.-N., Wang, X.-L., Ye, C.-L., Zhu, X.-Y., Li, H.-P., *et al.* (2019) Probucol ameliorates EMT and lung fibrosis through restoration of SIRT3 expression. *Pulm Pharmacol Ther.* **57**, 101803. https://doi.org/10.1016/j.pupt.2019.101803
88. Ji-hong, Y., Yu, M., Ling-hong, Y., Jing-jing, G., Ling-li, X., Lv, W., *et al.* (2023) Baicalein attenuates bleomycin-induced lung fibroblast senescence and lung fibrosis through restoration of Sirt3 expression. *Pharm Biol.* **61**, 288–297. https://doi.org/10.1080/13880209.2022.2160767
89. Gong, H., Zheng, C., Lyu, X., Dong, L., Tan, S., and Zhang, X. (2021) Inhibition of Sirt2 alleviates fibroblasts activation and pulmonary fibrosis via Smad2/3 pathway. *Front Pharmacol.* **12**, 756131. https://doi.org/10.3389/fphar.2021.756131
90. Minagawa, S., Araya, J., Numata, T., Nojiri, S., Hara, H., Yumino, Y., *et al.* (2011) Accelerated epithelial cell senescence in IPF and the inhibitory role of SIRT6 in TGF-β-induced senescence of human bronchial epithelial cells. *Am J Physiol Lung Cell Mol Physiol.* **300**, L391–L401. https://doi.org/10.1152/ajplung.00097.2010
91. Maity, S., Muhamed, J., Sarikhani, M., Kumar, S., Ahamed, F., Spurthi, K. M., *et al.* (2020) Sirtuin 6 deficiency transcriptionally up-regulates TGF-β signaling and induces fibrosis in mice. *J Biol Chem.* **295**, 415–434. https://doi.org/10.1074/jbc.RA118.007212

92. Tian, K., Chen, P., Liu, Z., Si, S., Zhang, Q., Mou, Y., *et al.* (2017) Sirtuin 6 inhibits epithelial to mesenchymal transition during idiopathic pulmonary fibrosis via inactivating TGF-β1/Smad3 signaling. *Oncotarget.* **8**, 61011–61024. https://doi.org/10.18632/oncotarget.17723
93. Wyman, A. E., Noor, Z., Fishelevich, R., Lockatell, V., Shah, N. G., Todd, N. W., *et al.* (2017) Sirtuin 7 is decreased in pulmonary fibrosis and regulates the fibrotic phenotype of lung fibroblasts. *Am J Physiol Lung Cell Mol Physiol.* **312**, L945–L958. https://doi.org/10.1152/ajplung.00473.2016
94. Bernard, K., Logsdon, N. J., Benavides, G. A., Sanders, Y., Zhang, J., Darley-Usmar, V. M., *et al.* (2018) Glutaminolysis is required for transforming growth factor-β1-induced myofibroblast differentiation and activation. *J Biol Chem.* **293**, 1218–1228. https://doi.org/10.1074/jbc.RA117.000444
95. Choudhury, M., Yin, X., Schaefbauer, K. J., Kang, J.-H., Roy, B., Kottom, T. J., *et al.* (2020) SIRT7-mediated modulation of glutaminase 1 regulates TGF-β-induced pulmonary fibrosis. *FASEB J.* **34**, 8920–8940. https://doi.org/10.1096/fj.202000564R
96. Wyman, A. E., Nguyen, T. T. T., Karki, P., Tulapurkar, M. E., Zhang, C.-O., Kim, J., *et al.* (2020) SIRT7 deficiency suppresses inflammation, induces EndoMT, and increases vascular permeability in primary pulmonary endothelial cells. *Sci Rep.* **10**, 12497. https://doi.org/10.1038/s41598-020-69236-z
97. Fang, Z., Wang, X., Sun, X., Hu, W., and Miao, Q. R. (2021) The role of histone protein acetylation in regulating endothelial function. *Front Cell Dev Biol.* **9**, 672447. https://doi.org/10.3389/fcell.2021.672447
98. Coward, W. R., Watts, K., Feghali-Bostwick, C. A., Knox, A., and Pang, L. (2009) Defective histone acetylation is responsible for the diminished expression of cyclooxygenase 2 in idiopathic pulmonary fibrosis. *Mol Cell Biol.* **29**, 4325–4339. https://doi.org/10.1128/mcb.01776-08
99. Tao, J., Zhang, M., Wen, Z., Wang, B., Zhang, L., Ou, Y., *et al.* (2018) Inhibition of EP300 and DDR1 synergistically alleviates pulmonary fibrosis in vitro and in vivo. *Biomed Pharmacother.* **106**, 1727–1733. https://doi.org/10.1016/j.biopha.2018.07.132
100. Lee, S. Y., Kim, M. J., Jang, S., Lee, G. E., Hwang, S. Y., Kwon, Y., *et al.* (2020) Plumbagin suppresses pulmonary fibrosis via inhibition of p300 histone acetyltransferase activity. *J Med Food.* **23**, 633–640. https://doi.org/10.1089/jmf.2019.4670
101. Hu, L., Yu, Y., Huang, H., Fan, H., Hu, L., Yin, C., *et al.* (2016) Epigenetic regulation of interleukin 6 by histone acetylation in macrophages and its role in paraquat-induced pulmonary fibrosis. *Front Immunol.* **7**, 696. https://doi.org/10.3389/fimmu.2016.00696
102. Coward, W. R., Watts, K., Feghali-Bostwick, C. A., Jenkins, G., and Pang, L. (2010) Repression of IP-10 by interactions between histone deacetylation and hypermethylation in idiopathic pulmonary fibrosis. *Mol Cell Biol.* **30**, 2874–2886. https://doi.org/10.1128/mcb.01527-09
103. Sanders, Y. Y., Hagood, J. S., Liu, H., Zhang, W., Ambalavanan, N., and Thannickal, V. J. (2014) Histone deacetylase inhibition promotes fibroblast apoptosis and ameliorates pulmonary fibrosis in mice. *Eur Respir J.* **43**, 1448–1458. https://doi.org/10.1183/09031936.00095113
104. Sanders, Y. Y., Liu, H., Liu, G., and Thannickal, V. J. (2015) Epigenetic mechanisms regulate NADPH oxidase-4 expression in cellular senescence. *Free Radic Biol Med.* **79**, 197–205. https://doi.org/10.1016/j.freeradbiomed.2014.12.008
105. Lawrence, M., Daujat, S., and Schneider, R. (2016) Lateral thinking: How histone modifications regulate gene expression. *Trends Genet.* **32**, 42–56. https://doi.org/10.1016/j.tig.2015.10.007
106. Xiao, X., Senavirathna, L. K., Gou, X., Huang, C., Liang, Y., and Liu, L. (2016) EZH2 enhances the differentiation of fibroblasts into myofibroblasts in idiopathic pulmonary fibrosis. *Physiol Rep.* **4**. https://doi.org/10.14814/phy2.12915
107. Coward, W. R., Brand, O. J., Pasini, A., Jenkins, G., Knox, A. J., and Pang, L. (2018) Interplay between EZH2 and G9a regulates CXCL10 gene repression in idiopathic pulmonary fibrosis. *Am J Respir Cell Mol Biol.* **58**, 449–460. https://doi.org/10.1165/rcmb.2017-0286OC
108. Bao, X., Liu, X., Liu, N., Zhuang, S., Yang, Q., Ren, H., *et al.* (2021) Inhibition of EZH2 prevents acute respiratory distress syndrome (ARDS)-associated pulmonary fibrosis by regulating the macrophage polarization phenotype. *Respir Res.* **22**, 194. https://doi.org/10.1186/s12931-021-01785-x
109. He, C., Larson-Casey, J. L., Gu, L., Ryan, A. J., Murthy, S., and Carter, A. B. (2016) Cu,Zn-superoxide dismutase-mediated redox regulation of jumonji domain containing 3 modulates macrophage polarization and pulmonary fibrosis. *Am J Respir Cell Mol Biol.* **55**, 58–71. https://doi.org/10.1165/rcmb.2015-0183OC
110. Zhang, X., Liu, H., Hock, T., Thannickal, V. J., and Sanders, Y. Y. (2013) Histone deacetylase inhibition downregulates collagen 3A1 in fibrotic lung fibroblasts. *Int J Mol Sci.* **14**, 19605–19617. https://doi.org/10.3390/ijms141019605

111. Jose, C. C., Wang, Z., Tanwar, V. S., Zhang, X., Zang, C., and Cuddapah, S. (2019) Nickel-induced transcriptional changes persist post exposure through epigenetic reprogramming. *Epigenetics Chromatin.* **12**, 75. https://doi.org/10.1186/s13072-019-0324-3
112. Cui, H., Xie, N., Banerjee, S., Ge, J., Jiang, D., Dey, T., *et al.* (2021) Lung myofibroblasts promote macrophage profibrotic activity through lactate-induced histone lactylation. *Am J Respir Cell Mol Biol.* **64**, 115–125. https://doi.org/10.1165/rcmb.2020-0360OC
113. Sankar, A., Mohammad, F., Sundaramurthy, A. K., Wang, H., Lerdrup, M., Tatar, T., *et al.* (2022) Histone editing elucidates the functional roles of H3K27 methylation and acetylation in mammals. *Nat Genet.* **54**, 754–760. https://doi.org/10.1038/s41588-022-01091-2
114. Davis, L. E., Shalin, S. C., and Tackett, A. J. (2020) Utility of histone H3K27me3 and H4K20me as diagnostic indicators of melanoma. *Melanoma Res.* **30**, 159–165. https://doi.org/10.1097/cmr.0000000000000648

5 Translational Potential of Hypoxia-Inducible Factor-2α Signaling in Pulmonary Hypertension

Zhiyu Dai

5.1 INTRODUCTION

Pulmonary hypertension (PH) is a progressive and often life-threatening condition characterized by elevated blood pressure in the pulmonary arteries (PAs), which leads to right heart failure and death (1). PH is currently defined by a mean pulmonary arterial pressure of more than 20 mm Hg at rest measured by right heart catheterization (2). In the context of PH, hypoxia-inducible factor (HIF) has been shown to play a role in the regulation of pulmonary vascular tone, and pulmonary vascular and right heart remodeling (3). Human and animal genetic studies demonstrated HIF-2α might play a critical role in contributing to the PH development compared to HIF-1α. Targeting the HIF-2α pathway has been proposed as a potential therapeutic strategy for PH. In preclinical studies, inhibition of HIF-2α has been shown to reduce pulmonary vascular remodeling and improve hemodynamic parameters in animal models of PH. However, more research is needed to determine the safety and efficacy of HIF-2α inhibitors in humans with PH. In this review, I will summarize the current knowledge of HIF-2α in PH and provide the opinion and perspective based on the understanding from our research findings.

5.2 HIF SIGNALING

HIFs are transcription factors that play a crucial role in the cellular response to hypoxia. Under normal oxygen conditions, HIF-α is rapidly degraded by oxygen-dependent prolyl hydroxylase enzymes (PHDs), including PHD1, PHD2, and PHD3. However, when in hypoxic conditions, the PHDs are inhibited, allowing HIF-α to accumulate and bind to HIF-β, also known as aryl hydrocarbon receptor nuclear translocator, to form an active HIF complex. This complex then translocates to the nucleus where it binds to specific DNA sequences known as hypoxia-responsive elements (HREs) to activate the transcription of genes involved in a variety of cellular processes such as angiogenesis, erythropoiesis, glucose metabolism, and cell survival (4–6) (Figure 5.1). There are three main isoforms of HIF-α: HIF-1α, HIF-2α (EPAS1), and HIF-3α. HIF-1α is the most extensively studied isoform and is widely expressed in various tissues. HIF-2α is expressed in a more restricted pattern, including endothelial cells (ECs) (7). HIF-3a is the least studied isoform, and it is shown as a negative regulator of HIF-1α and HIF-2α. HIF has been implicated in a number of physiological and pathological conditions, including cancer, ischemic diseases, and PH, among others. As a result, HIF has become an attractive therapeutic target for the development of drugs aimed at treating these diseases (8, 9).

DOI: 10.1201/9781003355243-6

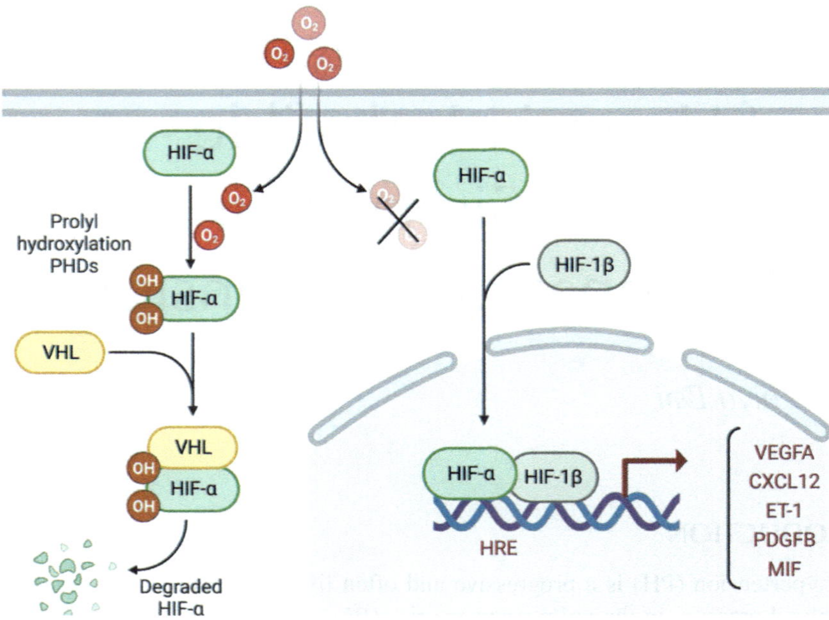

FIGURE 5.1 Schematic overview of the HIF pathway.

5.3 EXPRESSION PATTERN OF HIF-2α SIGNALING MOLECULES IN THE LUNG

HIF is expressed in many tissue and cell types throughout the body, including epithelial cells, ECs, smooth muscle cells (SMCs), and immune cells. In the lung, HIF-1α (HIF1A) expression is widespread and can be observed in a variety of cell types, including monocytes/macrophages, neutrophils, ECs, and SMCs based on recent human lung single-cell RNA sequencing (scRNA-seq) analysis. HIF-2α mRNA is highly expressed in the ECs and presented in pericytes, SMCs, and epithelial cells on the human scRNA-seq data (Figure 5.2A). In situ hybridization showed that *Epas1* strongly expresses in the mouse lung ECs compared to alveolar epithelial cells and fibroblasts (Figure 5.2B). In the human developmental lung, HIF-2α proteins localized to the epithelium including airway and alveolar epithelium as well as vascular ECs (10, 11). Our previous studies show that HIF-2α and PHD2 are highly expressed in the pulmonary arterial ECs (PAECs). Upregulation of HIF-2α protein were observed in the PAECs of idiopathic PAH patients and PH patients with congenital diaphragmatic hernia (11, 12). Loss of PHD2 was evident in the PAECs of occlusive lesions from patients with idiopathic pulmonary arterial hypertension (IPAH), and other PHDs like PHD1 and PHD3 were also downregulated in isolated PAECs from IPAH patients (13, 14). Many HIF-2α downstream target genes including NDRG1, CXCL12, CSF2, IL6, PDGFB, TGFA, and BCL2 were upregulated in the dissected PAs from PAH patients (15). Taken together, these data suggest that activation of HIF-2α signaling is evident in the PAECs of PAH patients.

5.4 HUMAN GENETIC EVIDENCE OF HIF-2α SIGNALING IN PH

The presence of human populations residing at high altitudes for extended periods of time offers a unique opportunity to investigate genetic adaptations enabling them to cope with the inherent low-oxygen conditions (16, 17). Among these populations, Tibetans and Andeans have been extensively studied and exhibit a certain level of defense against developing PAH resulting from

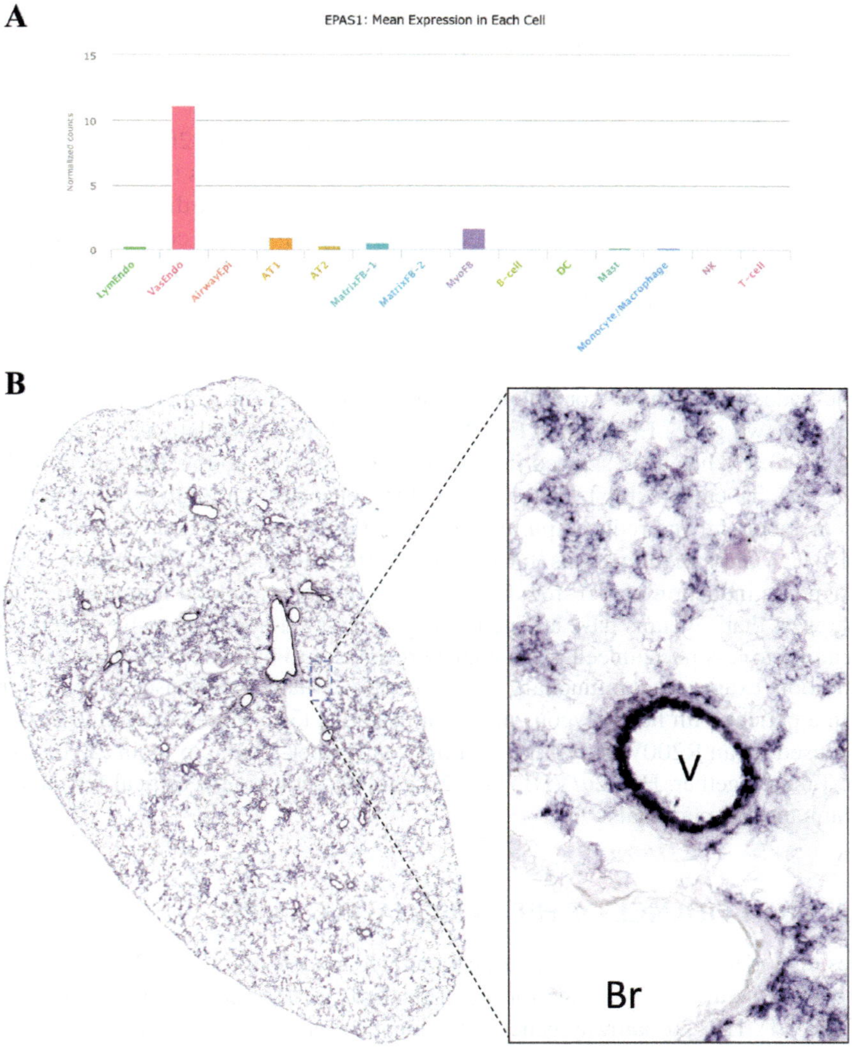

FIGURE 5.2 Expression pattern of HIF-2α in the lungs. (A) Sc-RNA-seq data showed that HIF-2α (EPAS1) is highly expressed in the lung vascular ECs. (B) In situ hybridization data showed that *Epas1* strongly expresses in the mouse lung ECs.

Source: Data was generated by LungMAP (A) from a 24-year-old human subject and (B) from a postnatal mouse lung.

prolonged exposure to hypoxia. Conversely, individuals from lowland populations who have recently relocated to high altitudes face a risk of PAH because of the overall hypoxic environment (18). The researchers sequenced the exomes of 50 Tibetans and identified the EPAS1 gene, which encodes HIF-2α. One particular single nucleotide polymorphism at EPAS1 showed a substantial frequency difference of 78% between Tibetan and Han populations, representing the most rapid allele frequency change observed in any human gene so far (19). Another group conducted genetic and functional analyses, finding that specific variants of EPAS1 in Tibetans led to a downregulation of gene expression in human ECs and placentas. Additionally, the Tibetan version of EPAS1 was associated not only with lower hemoglobin levels, serving as a protective factor against polycythemia, but also with reduced pulmonary vasoconstriction in Tibetans.

These data suggest that the downregulation of EPAS1 contributes to the molecular foundation of Tibetans' adaptation to high-altitude hypoxia (20). The genetic changes of EPAS1 underlying adaptation to hypoxia environments have also been identified in Sherpas (21) and multiple animals including horses (22) and dogs (23).

Genetic HIF-2α activation due to a G → A substitution at position 2097 (a glycine-to-arginine change at residue 537) has been found to cause severe PH in humans (24). Patients with HIF-2α gain-of-function mutations were associated with PH, increased cardiac output, and increased heart rate (25). By using whole-exome sequencing, Newman and colleagues discovered two non-synonymous substitutions (p.A606T and p.G610S) in the EPAS1 gene, specifically in the oxygen degradation domain, in Angus cattle with high-altitude PH. This finding was confirmed in two independent herds, with affected cattle exhibiting a mean PA pressure exceeding 50 mm Hg (26).

Genome-wide studies have consistently identified compelling genetic signatures of natural selection in the two genes EPAS1 and EGLN1 in high-altitude populations. A variant in EGLN1, c.[12C>G; 380G>C], has been showed to contribute functionally to the Tibetan high-altitude phenotype (27, 28). Tibetans exhibit lower hemoglobin concentration, higher pulmonary ventilation relative to metabolism, and reduced pulmonary vascular responses to acute and sustained hypoxia compared to Han Chinese individuals (lowlanders). At the cellular level, Tibetans show lower expression and hypoxic induction of HIF-regulated genes in peripheral blood lymphocytes. Overall, the findings suggest that Tibetans have adapted to hypoxia with a hyporesponsive HIF transcriptional system, leading to less pronounced physiological responses to hypoxic challenges even at sea level (29). In addition, exaggerated pulmonary vascular and ventilatory responses to acute hypoxia were reported in a patient with heterozygous mutation in PHD2 (30). Moreover, Chuvash polycythemia patients, caused by an R200W substitution in the von Hippel–Lindau protein (VHL), also develop PAH (31). Taken together, HIF-2α/VHL/PHD2 signaling activation is critical for the development of PH in humans and mammals.

5.5 RODENT EVIDENCE OF HIF-2α SIGNALING IN PH

Many lines of evidence from genetic modification mice models of the HIF-2α pathway (*Egln1*, *Vhl*, *Hif2a*) provide unequivocal evidence for the critical role of HIF-2α signaling in the pathogenesis of PH (Table 5.1). HIF-2α pathway genes are critical for embryonical development. For example, *Egln1*$^{-/-}$, *Vhl*$^{-/-}$, and *Hif2a*$^{-/-}$ mice are all embryonically lethal (7, 32, 33). Earlier studies employing both *Hif1a*$^{+/-}$ and *Hif2a*$^{+/-}$ mice suggest both HIF-1α and HIF-2α activation contribute to PH development (34, 35). Mice with genetic knock-in of an activation mutation (G536W) of Hif2a develop severe PH with right ventricular systolic pressure (RVSP) of 66 mm Hg (36). Recent studies also showed the superior role of endothelial HIF-2α versus HIF-1α in mediating chronic hypoxia-induced or *Vhl* or *Egln1* deficiency–induced pulmonary vascular remodeling and PH. Mice lacking HIF-2α but not HIF-1α in lung endothelium (L1Cre) are completely resistant to chronic hypoxia-induced PH (37). Mice with Vhl^{R200W} mutation (loss of function) develop PH, which is rescued by a heterozygous deletion of *Hif2a* but not *Hif1a* (38). Our studies generated *Egln1*^{Tie2Cre} mice, which develop progressive PH with RVSP levels ranging from 60 to 90 mm Hg at the age of 3.5 months. These mice die progressively (i.e., 80% mortality by the age of 6 months). Remarkably, these mice also develop unprecedented RV hypertrophy (right ventricle/left ventricle plus septum [RV/LV+S] ratio ranging from 0.75 to 1.1), RV fibrosis, and RV failure (14, 39). Kapitsinou and colleagues and Chen and coworkers showed that *Egln1*^{Cdh5Cre} mice exhibit much weaker PH compared to *Egln1*^{Tie2Cre} mice (40, 41). Endothelial deletion of *Hif2a* but not *Hif1a in Egln1*^{Tie2Cre} or *Egln1*^{Cdh5Cre} mice completely inhibited the development of PH (14, 40). Taken together, these studies suggest that HIF hydroxylase signaling, specifically the PHD2/HIF-2α axis, is critical in regulating pulmonary vascular homeostasis.

TABLE 5.1
Summary of Genetic Mouse Model of HIF-2 and Regulators in PH

Gene	Gene Modification	Disease Model, Time	Tissue/Cell Location	RVSP (mm Hg)	RV Remodeling/RV/LV+S Ratio	PA Remodeling	Reference
Hif2a	$Hif2a^{+/-}$	Hx, 4 weeks	Global	↓	↓	↓	Brusselmans 2003
Vhl	Vhl^{200W}	Nx, 7 months	Global	37	↑	↑	Hickey 2010
Vhl/Hif2a	$Vhl^{200W}/Hif2a^{+/-}$	Nx, 7 months	Global	↓	↓	↓	Hickey 2010
Hif2a	$Hif2a^{G536w/G536w}$	Nx, 4–6 months	Global	66	↑	↑	Tan 2013
Egln1	$Egln1^{Tie2Cre}$	Nx, 3 months	EC/constitutive	72	0.85/RV failure	↑ Musc, occlusive big and small PA, complex plexiform lesions	Dai 2016
Egln1/Hif2a	$Egln1/Hif2a^{Tie2Cre}$	Nx, 3 months	EC/constitutive	↓	↓	↓	Dai 2016
Egln1	$Egln1^{Cdh5Cre}$	Nx, 3 months	EC/constitutive	50	0.5	↑	Kapitsinou 2016
Egln1	$Egln1^{Cdh5Cre}$	Nx, 15 months	EC/constitutive	↑	↑	↑	Chen 2016
Egln1/Hif2a	$Egln1/Hif2a^{Cdh5Cre}$	Nx, 3 months	EC/constitutive	↓	↓	↓	Kapitsinou 2016
Hif2a	$Hif2a^{L1Cre}$	Hx, 3 weeks	EC/constitutive	↓	↓	↓	Cowburn 2016
Hif2a	$Hif2a^{Tie2CreER}$	Hx, 3 weeks	EC/inducible	↓	↓	↓	Tang 2018
Hif2a	$Hif2a^{fl/wt;UbcCreER}$	Hx, 5 weeks	Global/inducible	—	↓	—	Hu 2019
Hif1a/Hif2a	$Hif1a/Hif2a^{Myh6Cre}$	Hx, 4 weeks	Cardiomyocyte/constitutive	—	↓	—	Smith 2020

EC, endothelial cell; Hx, hypoxia; Nx, normoxia. ↑, higher than compared with controls; ↓ lower than compared with controls; —, no change compared with controls.

5.6 MOLECULAR AND CELLULAR MECHANISMS OF HIF-2 IN PH DEVELOPMENT

HIF plays a critical role in regulating vascular function, blood flow, and vascular remodeling through its effects on vasomotor tone, angiogenesis, metabolism, endothelial-mesenchymal transition (EndoMT), paracrine effect on other vascular cells such as pulmonary artery smooth muscle cells (PASMCs), fibroblasts, and leukocytes. Recently, studies highlighted the important role of HIF-2α in the pulmonary vascular ECs and the underlying molecular mechanisms in inducing the pathogenesis of PH (Figure 5.3).

HIF-2α has been shown to regulate the expression of a number of genes involved in pulmonary vasomodulation, including those encoding endothelin-1 (ET-1), endothelial nitric oxide synthase (eNOS or NOS2), and arginase-1 and arginase-2 (ARG1 and 2). Under hypoxic conditions, HIF induces the expression of ET-1, a potent vasoconstrictor that increases vascular resistance and contributes to the maintenance of blood pressure. Previous studies have characterized a HRE site in the human ET-1 promoter that contributed to the activation of ET-1 expression in EC (42). ET-1 binds to two types of receptors, ETA and ETB; both ETA and ETB receptors are found in SMCs. Activation of ETA and ETB receptors on SMCs mediates the vasoconstrictive and mitogenic effects of ET-1. An activation of the ET-1 system has been documented in PAH patients as well as in animal models of PH (43). Our previous studies also demonstrated that ET-1 was upregulated in *Egln1*^{Tie2Cre} mice in a HIF-2α but not HIF-1α dependent manner in vivo (14, 44). ET-1 system antagonists like bosentan and ambrisentan have been approved for PAH patients globally. HIF also regulates the expression of eNOS, which produces nitric oxide (NO), a potent vasodilator. eNOS is a HIF-2 downstream target, and two contiguous HRE sites located at -5375 to -5366 of the human eNOS promoter were previously identified (45). Our recent studies also demonstrated that eNOS activation was induced in the lungs of PH mice regulated by PHD2 and HIF-2α (46). eNOS activation resulted in formation

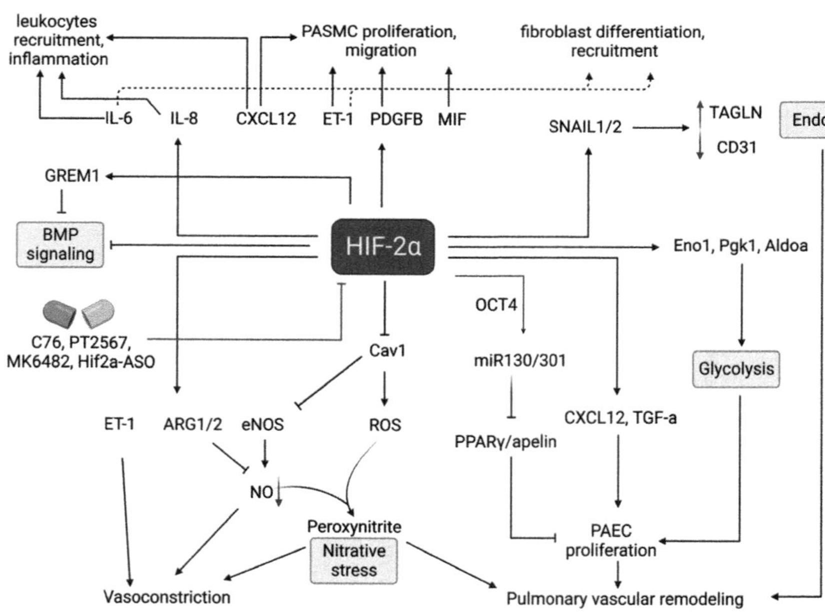

FIGURE 5.3 Downstream modulator of HIF-2α in pulmonary ECs. Lung endothelial HIF-2α control many downstream target genes and regulates signaling pathways related to vasoconstriction, proliferation, glycolysis, nitrative stress, the paracrine effect on perivascular cells including PASMCs, fibroblasts, and leukocytes. ↑, upregulation; ↓ downregulation.

of peroxynitrite, nitrative stress, and PH development. In addition, increased ARG2 was observed in ECs of patients with PAH (47). Preclinic studies also identified Argl and Arg2 were upregulated in hypoxic or *Egln1*^{Tie2Cre} mice lungs, which was abolished in Hif2a EC deletion mice (37). Hypoxic upregulation of ARG2 in human lung ECs was mediated by HIF-2α in vitro (48). Upregulation of arginase limits the availability of arginine for NO synthesis in cells, which further contributes to the development of PH (47).

PAECs hyperproliferation is a hallmark of PAH. HIF-2α has been shown to drive its specific target gene octamer-binding transcription factor 4 expression and subsequent upregulation of miR-130/301, leading to the downregulation of PPARγ/apelin signaling and PAECs proliferation in vitro (49). Our studies demonstrated that lung endothelial HIF-2α activation induced autocrine effect via upregulation of the CXCL12/CXCR4/FOXM1 axis and contributed to PAEC hyperproliferation and PH development in mice (50). CXCL12 is a classic HIF target. Elevated plasma CXCL12 is associated with an unfavorable prognosis in PAH patients (51). Our data also demonstrate a marked increase of CXCL12 expression in lungs of patients with IPAH (44). HIF-2α but not HIF-1α knockdown in vitro or knockout in vivo completely inhibited CXCL12 expression in lung ECs (14). Hu et al. showed that HIF-2α also mediated acute hypoxia induced the expression of transforming growth factor TGF-a, a ligand of EGFR, in normal human ECs (52).

The effect of HIF-1 on glycolytic metabolism is well defined (53). Previous studies suggest that HIF-2α does not regulate glycolytic flux. For example, hypoxia fails to induce glycolytic gene expression (LDHA, PGK-1, PGM-1, and PKM) in human renal carcinoma cells (786-O) and 293 cells (54). However, our recent unpublished data showed that HIF-2α activation secondary to *Egln1* deficiency induces the glycolytic gene programming, including upregulation of *Eno1, Tpi1, Pkm, Pfkl, Gapdh, Aldoa,* and *Pgk1* in lung ECs in mice. Further studies are warranty to investigate the role of HIF-2α on lung endothelial glycolysis in the context of PH development. In addition, recent studies highlighted the role of EndoMT in the PAH. In the lungs of *Egln1*^{Tie2Cre} mice, there is an increase in the expression of EndoMT markers and snail family transcriptional repressors (SNAI)1/2, master factors for epithelial mesenchymal transition, and this upregulation is dependent on HIF-2α but not HIF-1α (13). To further investigate the role of EndoMT in occlusive vascular remodeling and the development of severe PAH, future studies should use genetic lineage tracing methods in different animal models such as *Egln1*^{Tie2Cre} mice and sugen5416 plus hypoxia (SuHx) rats.

The lung vascular system maintains its homeostasis by producing various factors to act on peripheral cells through paracrine effect. Our previous studies using RNA-sequencing analysis demonstrated that ECs from severe PH mice (*Egln1*^{Tie2Cre}) expressed many genes (123 of 1680 upregulated genes) related to secretive proteins, including angiocrine factors *Cxcl12, Tgfb1, Edn1, Mif,* and *Pdgfb* because of the activation of endothelial HIF-2α (50). Indeed, these angiocrine factors (CXCL12, PDGFB, EDN1 and MIF) promote PASMC proliferation via induction of transcriptional factor FOXM1 and PH development (44). HIF-2α activation driven EC also induce pro-inflammatory cytokines including IL-6, TNF-α, IL-8, IL-1β, CXCL12, and adhesion molecule ICAM1 to promote leukocyte recruitment and accumulation (14, 37, 52). EC-derived factors include ET-1, PDGFB, and CXCL12, which play a role in stimulating the migration, recruitment, and proliferation of fibroblasts. Additionally, ECs can release factors like ET-1 and IL-6, which promote the differentiation of fibroblasts into myofibroblasts (55).

There is accumulating evidence of oxidative and/or nitrative stress in the lungs of patients with IPAH. Nitrative stress was markedly elevated, whereas endothelial caveolin-1 (Cav1) expression was suppressed in the lungs of *Egln1*^{Tie2Cre} mice. The nitrative stress and Cav1 expression were normalized when *Hif2a* was co-deleted in *Egln1*^{Tie2Cre} mice (46). Bone morphogenetic protein (BMP) signaling deficiency is evident in the lungs of patients with PAH. Expression of several components of BMP signaling, including *Bmpr1b, Bmp3, Bmpr2, Acvrl1, Bmp6, Eng, Acvr2b,* and *Smad9* was decreased, whereas *Grem1*, the negative regulator of BMP signaling was markedly increased in *Egln1*^{Tie2Cre} lungs. Alteration of BMP signaling molecules was also normalized in *Egln1/Hif2a*^{Tie2Cre}

mice. Further experiments showed that GREM1 is an HIF-2α direct target gene with evidence that human GREM1 promoter has HRE sites (56).

Overall, HIF plays a critical role in the pathogenesis of vasoconstriction and pulmonary vascular remodeling and PH. Targeting HIF signaling pathways may therefore represent a promising therapeutic approach for the treatment of these conditions.

5.7 HIF AND RV REMODELING AND FAILURE DUE TO PH

Progressive PAH results in increased RV afterload, which can lead to RV hypertrophy, fibrosis, and eventually RV failure and death. Mice with gain-of-function mutations in *Hif2a* display RV hypertrophy without evidence of RV dilatation (36). Endothelial HIF-2α activation in *Egln-1*^{Tie2Cre} mice leads to severe RV hypertrophy (RV/LV+S ratio ranging from 0.75 to 1.1), RV fibrosis, and RV failure (14, 39). Interesting, Myh6Cre-mediated loss of cardiomyocytes HIF-2α abolished cardiomyocyte HIF-1α deletion–induced RV hypertrophy under hypoxia conditions (57). These data suggest activation of HIF-2α in both ECs and cardiomyocytes contribute to RV remodeling. Previous studies show that the attenuation of PH in chronic hypoxia fails to prevent RV remodeling, suggesting that PH-independent factors regulate RV remodeling. Further studies are warranted to investigate the role of HIF-2α in RV remodeling employing inducible and cell-specific deletion or overexpression of HIF-2α and HIF pathway molecules such as PHD2 and VHL in fixed-afterload models of RV hypertrophy and failure such as the PA banding model.

5.8 TRANSLATIONAL POTENTIAL OF ANTI-HIF-2 THERAPY IN PH

Accumulative evidence including human genetic, preclinic animal studies, and cellular and molecular mechanism findings indicate PHD2/VHL/HIF-2α axis is critical in regulating pulmonary vascular homeostasis and targeting abnormal HIF-2α signaling, representing a novel and potential impactful approach for PAH therapy. Multiple preclinical trials have been evaluated in PH mice and rat PH models (Table 5.2). Our group was the first to show that pharmacological inhibition of HIF-2α using HIF-2α translational inhibitor C76 (58) can effectively inhibit obliterative pulmonary vascular remodeling and right heart failure in three distinct rodent models (*Egln1*^{Tie2Cre} mice, monocrotaline [MCT]-rats, and SuHx-rats) of severe PH. We also for the first time observed clear survival benefits in two of the three models through HIF-2α inhibition (12). Hu and colleagues demonstrated that antisense oligonucleotides against Hif2a reduced pulmonary vascular muscularization and RV hypertrophy in mice exposed to hypoxia (52). The same group also showed that a small-molecule HIF2 inhibitor (PT2567) significantly attenuated monocyte recruitment and vascular cell proliferation, vessel muscularization, and PH development in rats exposed to hypoxia. Macias and colleagues further support the finding that global inhibition of HIF-2α reduced pulmonary vascular hemodynamics and pulmonary vascular remodeling in both su5416/hypoxia prevention and intervention models and MCT-exposed rodents (59). Additionally, Ghosh and colleagues demonstrated that inactivating HIF-2α with the second-generation allosteric HIF-2α inhibitor, MK-6482 attenuated polycythemia, and PH in Vhl^{R200W} mice, *Irp1*-KO mice, and double mutant Vhl^{R200W}; *Irp1*-KO mice with PH (60). Taken together, inhibition of HIF-2 signaling is a promising therapeutic approach for PAH.

Although these preclinical studies are promising, the translation of anti–HIF-2 therapy to clinical practice requires careful consideration of potential safety and efficacy concerns. HIF-2 regulates a wide range of physiological processes, including erythropoiesis, angiogenesis, barrier integrity, and carotid body development and growth. Therefore, the systemic inhibition of HIF-2 could have unintended effects on these processes, leading to off-target effects and potential toxicity. In addition, it is important to note that PH is a heterogeneous condition with multiple subtypes, each with distinct pathogenic mechanisms. The efficacy of anti–HIF-2 therapy may vary depending on the

TABLE 5.2
Summary of the Pharmacological Studies in Preclinical PH Treatment

Agent	Disease Model	Treatment Protocol	Treatment Dose	RVSP (mm Hg)	RV Remodeling/ RV/LV+S ratio	PA Remodeling	Reference
C76	*Egln1*^{Tie2Cre} mice	i.p. injection, daily, 3 months, reverse	12.5 mg/kg BW	↓	↓	↓	Dai 2016
C76	MCT rats	2 weeks, MCT, i.p. injection, daily, 2 weeks, reverse	12.5 mg/kg BW	↓	↓	↓	Dai 2016
C76	SuHx rats	3 weeks SuHx, 3 weeks rest, i.p. injection, daily, 3 weeks, reverse	12.5 mg/kg BW	↓	↓	↓	Dai 2016
Hif2a-ASO	Hx mice, 5 weeks	i.p. injection, twice a week, prevention	50 mg/kg BW	↓	↓	↓	Hu 2019
PT2567	Hx rats, 4 weeks	Oral gavage daily, prevention	300 mg/kg BW	↓	↓	↓	Hu 2019
PT2567	SuHx rats	Oral gavage, twice daily, prevention	100 or 300 mg/kg BW	↓	↓	↓	Macias 2021
PT2567	SuHx rats	3 weeks SuHx, oral gavage, twice daily for 3 weeks, reverse	100 mg/kg BW	↓	↓	↓	Macias 2021
MK-6842	*Irp1*$^{-/-}$ mice, low-iron diet	Oral gavage, daily, 5 weeks	100 mg/kg BW	↓	↓	↓	Ghosh 2020
MK-6842	*Vhl*^{200W} mice	Oral gavage, daily, 5 weeks	100 mg/kg BW	↓	↓	↓	Ghosh 2020
MK-6842	*Irp1*$^{-/-}$;*Vhl*^{200W} mice	Oral gavage, daily, 5 weeks	100 mg/kg BW	↓	↓	↓	Ghosh 2020

BW, body weight; Hx, hypoxia; SuHx, sugen5416 plus hypoxia. ↑, higher than compared with controls; ↓ lower than compared with controls; —, no change compared with controls.

underlying cause of PH, and further research is needed to determine which patients may benefit most from this approach.

5.9 CONCLUSIONS

In conclusion, HIF-2α plays a major role in the initiation and progression of PH development. Anti–HIF-2 therapy holds promise as a potential therapeutic approach for the treatment of PH.

SOURCES OF FUNDING

This work was supported in part by NIH Grants R00HL138278, R01HL158596, R01HL62794, R01HL169509, R01HL170096; AHA Career Development Award 20CDA35310084, The Cardiovascular Research and Education Foundation, Arizona Biomedical Research Centre funding (RFGA2022–01–06), and the University of Arizona start-up funding to Z.D.

ABBREVIATIONS

ARG	arginase
BMPs	bone morphogenetic proteins
Cav1	caveolin-1
EC	endothelial cell
EndoMT	endothelial-mesenchymal transition
eNOS	endothelial nitric oxide synthase
HIF	hypoxia-inducible factor
IPAH	idiopathic pulmonary arterial hypertension
MCT	monocrotaline
PA	pulmonary artery
PAEC	pulmonary arterial endothelial cell
PH	pulmonary hypertension
PHD	prolyl hydroxylase enzyme
RV/LV+S	right ventricle/left ventricle plus septum
scRNA-seq	single-cell RNA sequencing
SMC	smooth muscle cell
SuHx	sugen5416 plus hypoxia
VHL	von Hippel–Lindau protein

REFERENCES

1. Hassoun, P. M. (2021) Pulmonary arterial hypertension. *N Engl J Med.* **385**, 2361–2376
2. Simonneau, G., Montani, D., Celermajer, D. S., Denton, C. P., Gatzoulis, M. A., Krowka, M., Williams, P. G., and Souza, R. (2019) Haemodynamic definitions and updated clinical classification of pulmonary hypertension. *Eur Respir J.* **53**, 1801913. https://doi.org/10.1183/13993003.01913-2018
3. Pullamsetti, S. S., Mamazhakypov, A., Weissmann, N., Seeger, W., and Savai, R. (2020) Hypoxia-inducible factor signaling in pulmonary hypertension. *J Clin Investig.* **130**, 5638–5651
4. Majmundar, A. J., Wong, W. J., and Simon, M. C. (2010) Hypoxia-inducible factors and the response to hypoxic stress. *Mol Cell.* **40**, 294–309
5. Semenza, G. L. (2012) Hypoxia-inducible factors in physiology and medicine. *Cell.* **148**, 399–408
6. Lee, K. E., and Simon, M. C. (2015) SnapShot: Hypoxia-inducible factors. *Cell.* **163**, 1288–1288.e1
7. Compernolle, V., Brusselmans, K., Acker, T., Hoet, P., Tjwa, M., Beck, H., Plaisance, S., Dor, Y., Keshet, E., Lupu, F., Nemery, B., Dewerchin, M., Van Veldhoven, P., Plate, K., Moons, L., Collen, D., and Carmeliet, P. (2002) Loss of HIF-2α and inhibition of VEGF impair fetal lung maturation, whereas treatment with VEGF prevents fatal respiratory distress in premature mice. *Nat Med.* **8**, 702–710

8. Bishop, T., and Ratcliffe, P. J. (2015) HIF hydroxylase pathways in cardiovascular physiology and medicine. *Circ Res.* **117**, 65–79
9. Ackerman, D., and Simon, M. C. (2014) Hypoxia, lipids, and cancer: Surviving the harsh tumor microenvironment. *Trends Cell Biol.* **24**, 472–478
10. Groenman, F., Rutter, M., Caniggia, I., Tibboel, D., and Post, M. (2007) Hypoxia-inducible factors in the first trimester human lung. *J Histochem Cytochem.* **55**, 355–363
11. Huang, Y., Boerema-de Munck, A., Buscop-van Kempen, M., Sluiter, I., de Krijger, R., Tibboel, D., and Rottier, R. J. (2018) Hypoxia inducible factor 2α (HIF2α/EPAS1) is associated with development of pulmonary hypertension in severe congenital diaphragmatic hernia patients. *Pulm Circ.* **8**, 204589401878373
12. Dai, Z., Zhu, M. M., Peng, Y., Machireddy, N., Evans, C. E., Machado, R., Zhang, X., and Zhao, Y.-Y. (2018) Therapeutic targeting of vascular remodeling and right heart failure in pulmonary arterial hypertension with a HIF-2α inhibitor. *Am J Respir Crit Care Med.* **198**, 1423–1434
13. Tang, H., Babicheva, A., McDermott, K. M., Gu, Y., Ayon, R. J., Song, S., Wang, Z., Gupta, A., Zhou, T., Sun, X., Dash, S., Wang, Z., Balistrieri, A., Zheng, Q., Cordery, A. G., Desai, A. A., Rischard, F., Khalpey, Z., Wang, J., Black, S. M., Garcia, J. G. N., Makino, A., and Yuan, J. X.-J. (2018) Endothelial HIF-2α contrib6482utes to severe pulmonary hypertension due to endothelial-to-mesenchymal transition. *Am J Physiol Lung Cell Mol Physiol.* **314**, L256–L275
14. Dai, Z., Li, M., Wharton, J., Zhu, M. M., and Zhao, Y. Y. (2016) Prolyl-4 hydroxylase 2 (PHD2) deficiency in endothelial cells and hematopoietic cells induces obliterative vascular remodeling and severe pulmonary arterial hypertension in mice and humans through hypoxia-inducible factor-2α. *Circulation.* **133**, 2447–2458
15. Hu, C.-J., Laux, A., Gandjeva, A., Wang, L., Li, M., Brown, D., Riddle, S., Kheyfets, V. O., Tuder, R. M., Zhang, H., and Stenmark, K. R. (2023) The effect of HIF inhibition on the phenotype of fibroblasts in human and bovine pulmonary hypertension. *Am J Respir Cell Mol Biol.* **69**, 73–86. https://doi.org/10.1165/rcmb.2022-0114OC
16. Beall, C. M. (2014) Adaptation to high altitude: Phenotypes and genotypes. *Annu Rev Anthropol.* **43**, 251–272
17. Bigham, A. W., and Lee, F. S. (2014) Human high-altitude adaptation: Forward genetics meets the HIF pathway. *Genes Dev.* **28**, 2189–2204
18. Simonson, T. S., Yang, Y., Huff, C. D., Yun, H., Qin, G., Witherspoon, D. J., Bai, Z., Lorenzo, F. R., Xing, J., Jorde, L. B., Prchal, J. T., and Ge, R. (2010) Genetic evidence for high-altitude adaptation in Tibet. *Science.* **329**, 72–75
19. Yi, X., Liang, Y., Huerta-Sanchez, E., Jin, X., Cuo, Z. X. P., Pool, J. E., Xu, X., Jiang, H., Vinckenbosch, N., Korneliussen, T. S., Zheng, H., Liu, T., He, W., Li, K., Luo, R., Nie, X., Wu, H., Zhao, M., Cao, H., *et al.* (2010) Sequencing of 50 human exomes reveals adaptation to high altitude. *Science.* **329**, 75–78
20. Peng, Y., Cui, C., He, Y., Ouzhuluobu, Zhang, H., Yang, D., Zhang, Q., Bianbazhuoma, Yang, L., He, Y., Xiang, K., Zhang, X., Bhandari, S., Shi, P., Yangla, Dejiquzong, Baimakangzhuo, Duojizhuoma, Pan, Y., *et al.* (2017) Down-regulation of EPAS1 transcription and genetic adaptation of Tibetans to high-altitude hypoxia. *Mol Biol Evol.* **34**, 818–830
21. Hanaoka, M., Droma, Y., Basnyat, B., Ito, M., Kobayashi, N., Katsuyama, Y., Kubo, K., and Ota, M. (2012) Genetic variants in EPAS1 contribute to adaptation to high-altitude hypoxia in Sherpas. *PLoS ONE.* **7**, e50566
22. Liu, X., Zhang, Y., Li, Y., Pan, J., Wang, D., Chen, W., Zheng, Z., He, X., Zhao, Q., Pu, Y., Guan, W., Han, J., Orlando, L., Ma, Y., and Jiang, L. (2019) EPAS1 gain-of-function mutation contributes to high-altitude adaptation in Tibetan horses. *Mol Biol Evol.* **36**, 2591–2603
23. Gou, X., Wang, Z., Li, N., Qiu, F., Xu, Z., Yan, D., Yang, S., Jia, J., Kong, X., Wei, Z., Lu, S., Lian, L., Wu, C., Wang, X., Li, G., Ma, T., Jiang, Q., Zhao, X., Yang, J., Liu, B., Wei, D., Li, H., Yang, J., Yan, Y., Zhao, G., Dong, X., Li, M., Deng, W., Leng, J., Wei, C., Wang, C., Mao, H., Zhang, H., Ding, G., and Li, Y. (2014) Whole-genome sequencing of six dog breeds from continuous altitudes reveals adaptation to high-altitude hypoxia. *Genome Res.* **24**, 1308–1315
24. Gale, D. P., Harten, S. K., Reid, C. D. L., Tuddenham, E. G. D., and Maxwell, P. H. (2008) Autosomal dominant erythrocytosis and pulmonary arterial hypertension associated with an activating HIF2α mutation. *Blood.* **112**, 919–921

25. Formenti, F., Beer, P. A., Croft, Q. P. P., Dorrington, K. L., Gale, D. P., Lappin, T. R. J., Lucas, G. S., Maher, E. R., Maxwell, P. H., McMullin, M. F., O'Connor, D. F., Percy, M. J., Pugh, C. W., Ratcliffe, P. J., Smith, T. G., Talbot, N. P., and Robbins, P. A. (2011) Cardiopulmonary function in two human disorders of the hypoxia-inducible factor (HIF) pathway: Von Hippel-Lindau disease and HIF-2α gain-of-function mutation. *FASEB J.* **25**, 2001–2011
26. Newman, J. H., Holt, T. N., Cogan, J. D., Womack, B., Phillips, J. A., Li, C., Kendall, Z., Stenmark, K. R., Thomas, M. G., Brown, R. D., Riddle, S. R., West, J. D., and Hamid, R. (2015) Increased prevalence of EPAS1 variant in cattle with high-altitude pulmonary hypertension. *Nat Commun.* **6**, 6863
27. Lorenzo, F. R., Huff, C., Myllymäki, M., Olenchock, B., Swierczek, S., Tashi, T., Gordeuk, V., Wuren, T., Ri-Li, G., McClain, D. A., Khan, T. M., Koul, P. A., Guchhait, P., Salama, M. E., Xing, J., Semenza, G. L., Liberzon, E., Wilson, A., Simonson, T. S., Jorde, L. B., Kaelin, W. G., Koivunen, P., and Prchal, J. T. (2014) A genetic mechanism for Tibetan high-altitude adaptation. *Nat Genet.* **46**, 951–956
28. Xiang, K., Ouzhuluobu, Peng, Y., Yang, Z., Zhang, X., Cui, C., Zhang, H., Li, M., Zhang, Y., Bianba, Gonggalanzi, Basang, Ciwangsangbu, Wu, T., Chen, H., Shi, H., Qi, X., and Su, B. (2013) Identification of a Tibetan-specific mutation in the hypoxic gene EGLN1 and its contribution to high-altitude adaptation. *Mol Biol Evol.* **30**, 1889–1898
29. Petousi, N., Croft, Q. P. P., Cavalleri, G. L., Cheng, H.-Y., Formenti, F., Ishida, K., Lunn, D., McCormack, M., Shianna, K. V., Talbot, N. P., Ratcliffe, P. J., and Robbins, P. A. (2014) Tibetans living at sea level have a hyporesponsive hypoxia-inducible factor system and blunted physiological responses to hypoxia. *J Appl Physiol.* **116**, 893–904
30. *Cardiopulmonary Phenotype Associated with Human PHD2 Mutation—Talbot—2017 — Physiological Reports—Wiley Online Library* [online]. https://physoc.onlinelibrary.wiley.com/doi/10.14814/phy2.13224 (Accessed May 17, 2023)
31. Bond, J., Gale, D. P., Connor, T., Adams, S., de Boer, J., Gascoyne, D. M., Williams, O., Maxwell, P. H., and Ancliff, P. J. (2011) Dysregulation of the HIF pathway due to VHL mutation causing severe erythrocytosis and pulmonary arterial hypertension. *Blood.* **117**, 3699–3701
32. Takeda, K., Ho, V. C., Takeda, H., Duan, L.-J., Nagy, A., and Fong, G.-H. (2006) Placental but not heart defects are associated with elevated hypoxia-inducible factor α levels in mice lacking prolyl hydroxylase domain protein 2. *Mol Cell Biol.* **26**, 8336–8346
33. Gnarra, J. R., Ward, J. M., Porter, F. D., Wagner, J. R., Devor, D. E., Grinberg, A., Emmert-Buck, M. R., Westphal, H., Klausner, R. D., and Linehan, W. M. (1997) Defective placental vasculogenesis causes embryonic lethality in VHL-deficient mice. *Proc Natl Acad Sci U S A.* **94**, 9102–9107
34. Brusselmans, K., Compernolle, V., Tjwa, M., Wiesener, M. S., Maxwell, P. H., Collen, D., and Carmeliet, P. (2003) Heterozygous deficiency of hypoxia-inducible factor-2alpha protects mice against pulmonary hypertension and right ventricular dysfunction during prolonged hypoxia. *J Clin Investig.* **111**, 1519–1527
35. Yu, A. Y., Shimoda, L. A., Iyer, N. V., Huso, D. L., Sun, X., McWilliams, R., Beaty, T., Sham, J. S., Wiener, C. M., Sylvester, J. T., and Semenza, G. L. (1999) Impaired physiological responses to chronic hypoxia in mice partially deficient for hypoxia-inducible factor 1alpha. *J Clin Investig.* **103**, 691–696
36. Tan, Q., Kerestes, H., Percy, M. J., Pietrofesa, R., Chen, L., Khurana, T. S., Christofidou-Solomidou, M., Lappin, T. R. J., and Lee, F. S. (2013) Erythrocytosis and pulmonary hypertension in a mouse model of human HIF2A gain of function mutation. *J Biol Chem.* **288**, 17134–17144
37. Cowburn, A. S., Crosby, A., Macias, D., Branco, C., Colaço, R. D. D. R., Southwood, M., Toshner, M., Crotty Alexander, L. E., Morrell, N. W., Chilvers, E. R., and Johnson, R. S. (2016) HIF2α—arginase axis is essential for the development of pulmonary hypertension. *Proc Natl Acad Sci.* **113**, 8801–8806
38. Hickey, M. M., Richardson, T., Wang, T., Mosqueira, M., Arguiri, E., Yu, H., Yu, Q.-C., Solomides, C. C., Morrisey, E. E., Khurana, T. S., Christofidou-Solomidou, M., and Simon, M. C. (2010) The von Hippel-Lindau Chuvash mutation promotes pulmonary hypertension and fibrosis in mice. *J Clin Investig.* **120**, 827–839
39. Dai, Z., and Zhao, Y.-Y. (2017) Discovery of a murine model of clinical PAH: Mission impossible? *Trends Cardiovasc Med.* **27**, 229–236
40. Kapitsinou, P. P., Rajendran, G., Astleford, L., Michael, M., Schonfeld, M. P., Fields, T., Shay, S., French, J. L., West, J., and Haase, V. H. (2016) The endothelial PHD2/HIF-2 axis regulates pulmonary artery pressure in mice. *Mol Cell Biol.* **36**, 1584–1594

41. Wang, S., Zeng, H., Xie, X.-J., Tao, Y.-K., He, X., Roman, R. J., Aschner, J. L., and Chen, J.-X. (2016) Loss of prolyl hydroxylase domain protein 2 in vascular endothelium increases pericyte coverage and promotes pulmonary arterial remodeling. *Oncotarget*. **7**, 58848–58861
42. Yamashita, K., Discher, D. J., Hu, J., Bishopric, N. H., and Webster, K. A. (2001) Molecular regulation of the endothelin-1 gene by hypoxia: Contributions of hypoxia-inducible factor-1, activator protein-1, GATA-2, and p300/CBP *. *J Biol Chem*. **276**, 12645–12653
43. Galié, N., Manes, A., and Branzi, A. (2004) The endothelin system in pulmonary arterial hypertension. *Cardiovasc Res*. **61**, 227–237
44. Dai, Z., Zhu, M. M., Peng, Y., Jin, H., Machireddy, N., Qian, Z., Zhang, X., and Zhao, Y. Y. (2018) Endothelial and smooth muscle cell interaction via FoxM1 signaling mediates vascular remodeling and pulmonary hypertension. *Am J Respir Crit Care Med*. **198**, 788–802
45. Coulet, F., Nadaud, S., Agrapart, M., and Soubrier, F. (2003) Identification of hypoxia-response element in the human endothelial nitric-oxide synthase gene promoter. *J Biol Chem*. **278**, 46230–46240
46. Liu, B., Peng, Y., Yi, D., Machireddy, N., Dong, D., Ramirez, K., Dai, J., Vanderpool, R., Zhu, M. M., Dai, Z., and Zhao, Y.-Y. (2022) Endothelial PHD2 deficiency induces nitrative stress via suppression of caveolin-1 in pulmonary hypertension. *Eur Respir J*. **22**, 2102643. https://doi.org/10.1183/13993003.02643-2021
47. Xu, W., Kaneko, F. T., Zheng, S., Comhair, S. A. A., Janocha, A. J., Goggans, T., Thunnissen, F. B. J. M., Farver, C., Hazen, S. L., Jennings, C., Dweik, R. A., Arroliga, A. C., and Erzurum, S. C. (2004) Increased arginase II and decreased NO synthesis in endothelial cells of patients with pulmonary arterial hypertension. *FASEB J*. **18**, 1746–1748
48. Krotova, K., Patel, J. M., Block, E. R., and Zharikov, S. (2010) Hypoxic upregulation of arginase II in human lung endothelial cells. *Am J Physiol Cell Physiol*. **299**, C1541–C1548
49. Bertero, T., Lu, Y., Annis, S., Hale, A., Bhat, B., Saggar, R., Saggar, R., Wallace, W. D., Ross, D. J., Vargas, S. O., Graham, B. B., Kumar, R., Black, S. M., Fratz, S., Fineman, J. R., West, J. D., Haley, K. J., Waxman, A. B., Chau, B. N., Cottrill, K. A., and Chan, S. Y. (2014) Systems-level regulation of microRNA networks by miR-130/301 promotes pulmonary hypertension. *J Clin Investig*. **124**, 3514–3528
50. Yi, D., Liu, B., Wang, T., Liao, Q., Zhu, M. M., Zhao, Y. Y., and Dai, Z. (2021) Endothelial autocrine signaling through cxcl12/cxcr4/foxm1 axis contributes to severe pulmonary arterial hypertension. *Int J Mol Sci*. **22**, 1–11
51. McCullagh, B. N., Costello, C. M., Li, L., O'Connell, C., Codd, M., Lawrie, A., Morton, A., Kiely, D. G., Condliffe, R., Elliot, C., McLoughlin, P., and Gaine, S. (2015) Elevated plasma CXCL12α is associated with α poorer prognosis in pulmonary arterial hypertension. *PLoS ONE*. **10**, e0123709
52. Hu, C.-J., Poth, J. M., Zhang, H., Flockton, A., Laux, A., Kumar, S., McKeon, B., Frid, M. G., Mouradian, G., Li, M., Riddle, S., Pugliese, S. C., Brown, R. D., Wallace, E. M., Graham, B. B., and Stenmark, K. R. (2019) Suppression of HIF2 signalling attenuates the initiation of hypoxia-induced pulmonary hypertension. *Eur Respir J*. **54**, 1900378
53. Kierans, S. J., and Taylor, C. T. (2021) Regulation of glycolysis by the hypoxia-inducible factor (HIF): Implications for cellular physiology. *J Physiol*. **599**, 23–37
54. Hu, C.-J., Wang, L.-Y., Chodosh, L. A., Keith, B., and Simon, M. C. (2003) Differential roles of hypoxia-inducible factor 1α (HIF-1α) and HIF-2α in hypoxic gene regulation. *Mol Cell Biol*. **23**, 9361–9374
55. Evans, C. E., Cober, N. D., Dai, Z., Stewart, D. J., and Zhao, Y.-Y. (2021) Endothelial cells in the pathogenesis of pulmonary arterial hypertension. *Eur Respir J*. **58**, 2003957
56. Liu, B., Yi, D., Pan, J., Dai, J., Zhu, M. M., Zhao, Y., Oh, S. P., Fallon, M. B., and Dai, Z. (2022) Suppression of BMP signaling by PHD2 deficiency in pulmonary arterial hypertension. *Pulm Circ*. **12**, 0–2
57. Smith, K. A., Waypa, G. B., Dudley, V. J., Budinger, G. R. S., Abdala-Valencia, H., Bartom, E., and Schumacker, P. T. (2020) Role of hypoxia-inducible factors in regulating right ventricular function and remodeling during chronic hypoxia—induced pulmonary hypertension. *Am J Respir Cell Mol Biol*. **63**, 652–664
58. Zimmer, M., Ebert, B. L., Neil, C., Brenner, K., Papaioannou, I., Melas, A., Tolliday, N., Lamb, J., Pantopoulos, K., Golub, T., and Iliopoulos, O. (2008) Small-molecule inhibitors of HIF-2α translation link its 5′UTR iron-responsive element to oxygen sensing. *Mol Cell*. **32**, 838–848

59. Macias, D., Moore, S., Crosby, A., Southwood, M., Du, X., Tan, H., Xie, S., Vassallo, A., Wood, A. J. T., Wallace, E. M., and Cowburn, A. S. (2021) Targeting HIF2α-ARNT hetero-dimerisation as a novel therapeutic strategy for pulmonary arterial hypertension. *Eur Respir J*. **57**, 1902061
60. Ghosh, M. C., Zhang, D., Ollivierre, H., Noguchi, A., Springer, D., Linehan, W. M., and Rouault, T. A. (2020) Therapeutic inhibition of HIF2α reverses polycythemia and pulmonary hypertension in murine models of two human diseases. *Blood*. **137**, 2424. https://doi.org/10.1182/blood.2020009138

Part II

Research Strategies and Methods in Lung Biology and Pathophysiology

6 Utility of Lung Organoids to Accelerate Our Understanding of Human Pulmonary Development and Diseases

Nicole Pek and Mingxia Gu

6.1 INTRODUCTION

The lung is one of the most complex organs; it is home to at least 40 different cell types and is segregated into proximal and distal regions, each comprising different cell types and serving unique functions (1–4). It is therefore a challenge to develop a suitable in vitro system for pulmonary research. Therefore, rodent models are typically used to study lung organogenesis, as well as disease initiation and progression, despite apparent differences between mice and humans. Equipped with our increasing knowledge of both developmental and organoid biology, the field has undoubtedly witnessed groundbreaking developments over the past few years—efforts to generate 3D models called lung organoids in vitro have yielded success. Although numerous methods of generating lung organoids in vitro have been reported, the final products consist of lung epithelial cells to varying extents and in various proportions. This essentially means that "mini-lungs" can be readily made in the lab from various sources of cells and can be used instantaneously for multiple biomedical applications including but not limited to disease modeling, drug discovery and testing, personalized medicine, and transplantation therapy.

Animals have been instrumental in allowing us to gain insights into the intricacies of lung development. By performing genetic manipulations and gain- and loss-of-function studies in mice, we were able to identify key signaling pathways that underscore different stages of lung development (5). Mice studies have also enabled us to perform in-depth analyses of the complex process of branching morphogenesis (6). However, fundamental differences between rodent and human lung development warrant the need for human-specific models/tissues. Obtaining fetal tissues to study human lung development, particularly for certain developmental stages, is difficult given ethical considerations. Additionally, primary lung cells are terminally differentiated and are not well suited for developmental studies. Therefore, human lung organoids (HLOs) are valuable resources in that they have the potential to serve as "organ surrogates," allowing us to gain cellular and molecular insights into various stages of human lung development. Coupled with powerful single-cell RNA sequencing (scRNA-Seq) and bioinformatic technologies, we can predict the developmental paths of all cell populations regardless of their abundance. HLOs are also poised to change the way we approach disease modeling. Conventional methods include the use of rodent and 2D cellular models—rodent models are unable to fully replicate many facets of human lung diseases (7), and 2D cell cultures are homogenous (only constitute one cell type), which means that they are ineffective in capturing whole-organ phenotypes (structural, cellular, architectural) caused by the disease. In striking contrast, HLOs offer a significant advantage by faithfully recapitulating features of the native organs, rendering them more physiologically relevant for modeling complex human diseases.

DOI: 10.1201/9781003355243-8

In this chapter, we introduce the current lung organoid models and explore the utility of these 3D in vitro systems to better understand the complexities associated with human lung development and disease.

6.2 LUNG ORGANOID MODELS—CURRENT STATE OF TECHNOLOGY

6.2.1 Organoids Are Three-Dimensional "Miniature Organs"

Organoids are 3D structures that consist of multiple cell types that can self-organize, resulting in profound cellular structures, tissue-like architecture, and even physiological functions that resemble that of the native organ (8, 9). Therefore, organoids are considered by some to essentially be "miniature organs" that are made and cultured in the dish. The self-organization process is a result of cellular processes such as cell sorting and spatially restricted lineage commitment that are largely driven by cell-cell and cell-matrix interactions (10–12). Today, organoids representing at least ten organ systems have been described, including the intestines, lungs, pancreas, brain, heart, liver, and stomach (13, 14). The list will continue to expand. One common feature among all types of organoids is that they are generated by following key principles of mammalian organogenesis and tissue regeneration; information gathered from in vivo studies is vital in informing the field of organoid biology (15, 16). In turn, the process of organoid formation is also able to provide cellular and molecular insights into human development and tissue regeneration (13). The utility of organoids transcends that of just basic research; they hold tremendous potential for translational research including but not limited to disease modeling, toxicology studies, drug discovery and testing, personalized medicine, regenerative medicine, and cell therapies (8, 17–20).

6.2.2 The Inadequacies of Conventional Lung Cell Cultures

Because of the highly complex cellular and structural features of organoids, they are considered by many to be a "superior" in vitro model system compared to their 2D counterparts. Conventional 2D monolayer cell cultures were typically obtained from donor tissues or tumors and they have been used to study cellular physiology, model genetic diseases, and cancers, and perform pharmaceutical and therapeutic discovery/testing. These cells grow homogenously on flat surfaces of a cell culture dish, and because of the simplicity of these cultured cells, they serve as attractive pre-clinical model systems for various biological studies.

In the context of lung research, primary lung cells such as airway epithelial cells and nasal epithelial cells have been successfully used to study diseases such as cystic fibrosis (CF) (21). Alveolar epithelial cells were either obtained from tumors or immortalized from cells isolated from healthy/donor tissues (22). Immortalizing alveolar epithelial type 2 cells (AT2) has been particularly useful because they have the tendency to differentiate into alveolar epithelial type 1 cells (AT1) and have a limited life span ex vivo. Immortalized AT2 cell lines were used to study alveolar injury and repair and mechanisms of viral infections (23, 24). Also, airway epithelial cells obtained from a lung adenocarcinoma were shown to demonstrate many features of the airway epithelium, such as barrier integrity, and were used to study bronchial epithelium functions (25). Often, the lung epithelial cells were cultured in an air-liquid interface (ALI) that recapitulates the actual lung environment by exposing the apical side of cells to the air and the basolateral side to liquid (culture media). Studies have shown that ALI induces mucociliary differentiation of airway epithelial cells, resulting in a formation of a pseudostratified layer composed of not just ciliated cells, but also goblet/secretory cells and basal cells (26). ALI also promotes the maturation of stem cell–derived alveolar epithelium (27). Additionally, non-epithelial cell types such as pulmonary endothelial cells have been successfully isolated and cultured in vitro, providing a viable source of cells to perform studies related to pulmonary vascular diseases (28). Clearly, there are merits of conventional lung cell cultures—many important discoveries pertaining to lung diseases, cancers, epithelial functions, respiration, and epithelial repair stemmed from these 2D models (21, 29–33). Another added advantage is that primary lung cells or cell lines can be adapted for personalized medicine because they are isolated from an individual (21, 34). However, there are major limitations associated with the use of 2D cell cultures (35).

A significant drawback of 2D cell cultures is that because the cells are cultured on a flat surface, the natural cell shape is not preserved (36). Most notably, 2D cell cultures often lack extracellular matrix (ECM) that is naturally present in the tissue environment; therefore these cultures are unable to replicate the structural and even functional features of the native organ. ECM is essential in driving cellular organization, enabling cell-cell/cell-ECM communication, which collectively allows cells to adopt a 3D conformation and to self-organize, differentiate, and thrive (37). Next, most of the 2D lung cell cultures are homogenous; thus they do not represent the cellular diversity and heterogeneity in the organ. Also, although immortalized human lung epithelial cells provide unlimited quantities of starting material for various biomedical applications, each cell line only represents one donor, raising questions on the generalizability of findings. Additionally, because these immortalized cell lines were generated by manipulating the expression of oncogenes, they differ significantly from cells found in vivo and are not suitable for studies on cell cycle regulation/proliferation.

6.2.3 Different Strategies of Deriving Lung Organoids

Compared to other organs, the mammalian lung is particularly challenging to "reconstruct" in an organoid system. This is largely because of the highly sophisticated nature of the organ that has been adapted for breathing. First, because of its function, the lungs interact with both air and liquid. Next, the mammalian lungs house more than 40 different cell types, including epithelial, interstitial, vascular, and immune cells (1–4). These cells partake in forming a variety of complex interconnected structures (Figure 6.1a) that work in tandem to efficiently perform gaseous exchange. To add to the complexity, the lungs can be broadly segregated into two regions—the proximal and distal lungs, each comprising vastly different epithelial cells and housing a precise combination of different cell types (Figure 6.1a, b). The proximal lung comprises the large airways that conduct air from the external environment into the organ; this includes goblet cells (secretory), basal cells, ciliated cells, club cells, neuroendocrine cells, tuft cells, and even ionocytes (Figure 6.1b). The distal lung comprises the distal airways (small/medium airways) and the alveoli, which are air sacs that facilitate gaseous exchange. The alveoli consist of AT1 and AT2 cells (also known as AEC1 and AEC2 cells, respectively), alveolar macrophages, and mesenchymal cell types such as fibroblasts, alveolar capillary endothelial cells, and pericytes (not shown in Figure 6.1b). Today, researchers are still uncovering new cell types, particularly stem cells/progenitors in the human lungs (38, 39). Hence, the field of lung organoids is plagued by the seemingly impossible task of recapitulating all these complexities in an in vitro system.

However, over the past decade, great strides have been made since the first lung organoid culture was described in 1987 (40). Currently, a variety of lung organoids representing different lung regions can be generated in vitro from various sources of cells (Figure 6.2) (41). Regardless of the source of the starting material, the success of generating structurally and functionally sound lung organoids relies on two important factors: (1) the incorporation of ECM and (2) the adaptation of the ALI to the organoid culture system. First, the ECM mimics the stem cell niche present in vivo by providing an environment rich in biochemical and biomechanical signals that support cellular growth, differentiation, self-organization, and maturation (29). Matrigel, a natural basement membrane extract obtained from Engelbreth–Holm–Swarm mouse sarcoma cells, is commonly used for lung organoid formation (42), and it primarily contains laminin and collagen IV. Although other forms of ECM have been used for lung organoid generation, Matrigel is still viewed as the gold standard ECM. Second, the use of ALI, as mentioned earlier, recapitulates the respiratory environment (27, 43). When used in combination, both factors support the growth and differentiation of the lung epithelium in a 3D organoid system.

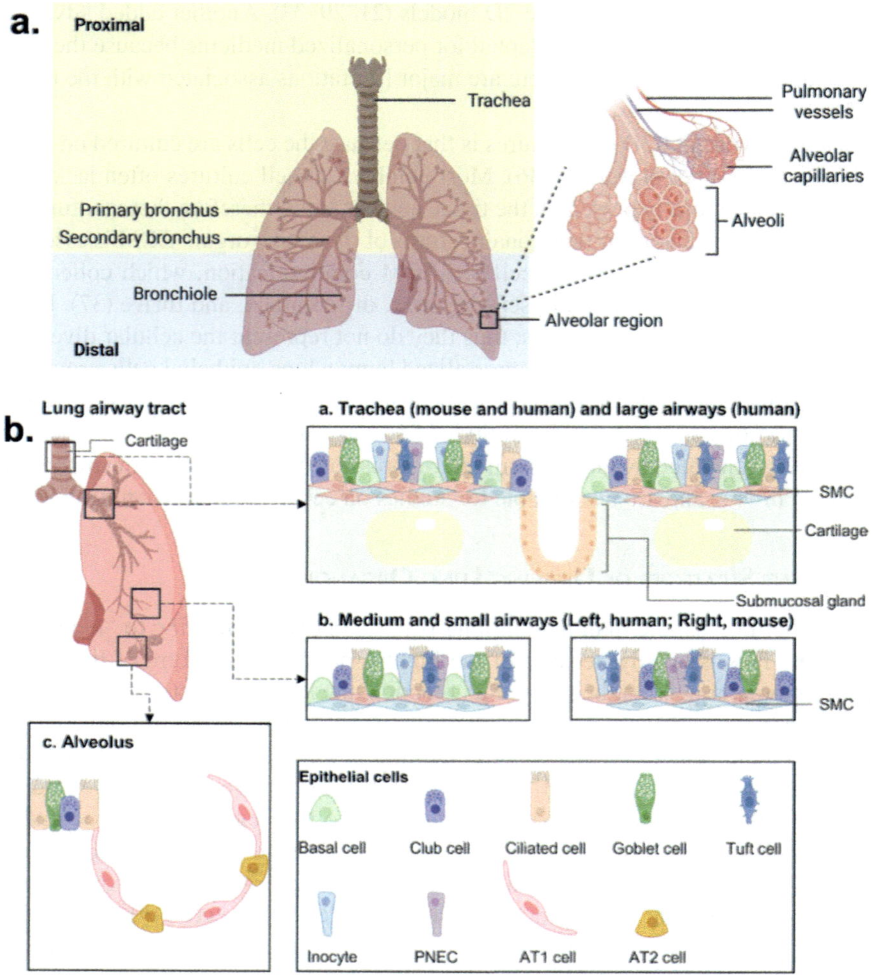

FIGURE 6.1 Structural and cellular organization of the mammalian lungs. (a) Anatomy of the human lungs. (b) Cellular architecture of the mammalian lungs.

Source: Image obtained from Meng et al. (2023) (38).

At present, HLOs can be generated from four different sources of cells:fetal progenitor cells, adult stem cells (ASCs), tumor cells, and human pluripotent stem cells (hPSCs).

6.2.3.1 Fetal Progenitor–Derived Lung Organoids

There are five main stages of human lung development: embryonic (4–7 weeks), pseudoglandular (5–17 weeks), canicular (16–26 weeks), saccular (24 weeks to term), and alveolar stage (36 weeks–21 years) (44). During the pseudoglandular stage, the lung buds undergo regulated branching into the surrounding mesenchyme via a process called branching morphogenesis to generate an extensive network of airways with several thousand terminal branches (analogous to a tree). During the canicular stage, the terminal branches narrow and form epithelial sacs that will later give rise to the alveoli (45). At a cellular level, the distal epithelial tip cells (branching tips), which are the multipotent progenitor cell population, are responsible for branching morphogenesis in both the pseudoglandular and canicular stages of pulmonary development. As they differentiate, they remain

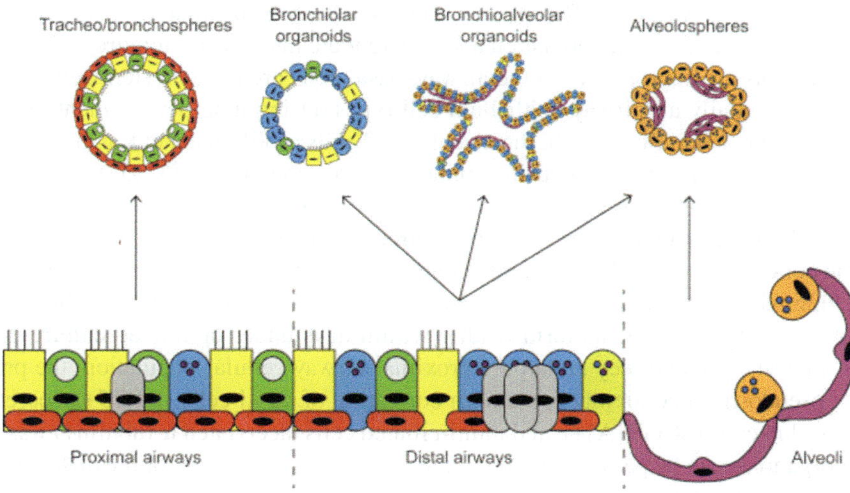

FIGURE 6.2 Various types of lung organoids representing different lung regions. A schematic diagram illustrating the various types of lung epithelial organoids that represents different lung regions.

Source: Image obtained and modified from Nadkarni et al. (2016) (41).

behind, forming stalk cells while the self-renewing tip cells remain at the branching tips (46, 47). Several researchers have tapped into the multipotent potential of these tip epithelial cells to generate lung organoids in vitro.

In 2017, Nikolić and colleagues described a method to generate organoids from SOX9, SOX2, and NKX2.1-expressing human embryonic lung epithelial tip cells isolated from 5–9 pcw lungs (48). The epithelial tip cells form organoids readily and acquire branching morphologies over time. Also, these organoids can self-renew and maintain SOX2 and SOX9 expression for over nine passages (4 months). Moreover, the organoids were able to differentiate into bronchiolar structures that contain goblet, basal, and ciliated cells when cultured in human airway differentiation media. The organoids were also capable of differentiating to the alveolar fate when exposed to the alveolar differentiation media; pro-SFTPC+ cells (but no progenitor markers) were detected in the organoids. Similarly, Lamers and colleagues used human fetal lung cells (16–17 pcw) to generate SOX2 and SOX9-expressing lung bud-tip organoids (49). They subsequently established a differentiated bronchioalveolar-like model from the progenitor cells obtained from lung bud-tip organoids compared to AT1, AT2, and basal cells.

Fetal tip progenitors (EPCAM+ CD44+ CD36+) were isolated by Lim and colleagues from 17–21 pcw human lung tissues and used to form epithelial tip organoids (50). After 3 weeks in culture, two distinct populations of organoids were observed: cystic (Lin^{NEG}) organoids that express tip progenitor markers and folded (LIN^{POS}) organoids that express progenitor and AT2 markers. The authors found that LIN^{POS} organoids differentiate readily to form alveolar organoids that robustly express AT2 markers SFTPC, LAMP3, and HOPX. Furthermore, electron microscopy (EM) confirmed the presence of few immature lamellar bodies (LBs) that consist of a surfactant projection core in the LIN^{POS} organoids. However, significant production and secretion of the LBs were observed in the differentiated alveolar organoids, suggesting that the LIN^{POS} organoids were able to differentiate into a mature AT2 cell fate.

6.2.3.2 ASC-Derived Lung Organoids

Lung organoids recapitulating the proximal and distal regions of the lungs have been successfully generated from adult lung cells from healthy donors. This strategy yielded success because of the presence of multipotent ASCs in different lung regions, which means that they are capable of self-renewal to maintain the stem cell population and differentiate into various mature cell types in vitro when the conditions are appropriate. In the airway, basal cells are resident stem cells that are able to give rise to essentially all airway epithelium cell types during lung development, homeostasis, and even epithelial regeneration (51). In the alveoli, AT2 cells proliferate and function as local stem cells that give rise to AT1, especially during epithelial repair upon injury (52–54). By exploiting the multipotent potential of these ASCs, coupled with the optimization of culture and differentiation conditions, adult lung tissues serve as a suitable source of cells for lung organoid generation.

Sachs and colleagues were able to generate human airway organoids from epithelial cells isolated from adult lung tissue that can be expanded and cultured in vitro for an extended duration of time (55). The human airway organoids form readily within days under optimized culture conditions, and further analysis revealed they are a pure proximal-airway cellular population; the presence of a polarized, pseudostratified airway epithelium consisting of basal cells (KRT5+), secretory cells (MUC5AC+), club cells (SCGB1A1+), and multi-ciliated cells (acetylated α-tubulin+) was detected. The airway organoids can be passaged while retaining similar proportions of the different proximal airway epithelial cell types. Next, recent work by Salahudeen and colleagues, reported the generation of AT2 organoids (SFTPC+ HTII-280+) and basal organoids (KRT5+) from adult distal lung cells (56). Upon further differentiation, basal organoids developed lumens lined with club and ciliated cells, whereas AT2 organoids consist of mature AT2 cells, as well as AT1 cells.

A method to generate pure alveolospheres from HTII-280+ AT2 cells purified from healthy human lungs was reported by Katsura and colleagues (57). The alveolospheres consisted only of AT2 cells (no airway or AT1 markers were detected). The culture conditions of alveolospheres were able to support AT2 maintenance, but upon induction, alveolospheres were able to differentiate readily into AT1 cells (express AT1 marker AGER). Additionally, the generation of bronchiolar and alveolar organoids from human adult lung cells was described by Ekanger and colleagues (58). The bronchiolar organoids consist of basal cells (KRT5) and secretory cells (MUC5AC), and EM also confirmed the presence of ultra-structures such as pseudo-stratified epithelium and microtubule arrangements that resembles the airway respiratory epithelium. The alveolar organoids consist of both alveolar epithelial cells—AT1 (AQP5) and AT2 cells (pro-SFTPC).

The protocols described so far described the independent derivation of either proximal airway or alveolar organoids. Tindle and colleagues reported on a method to generate adult lung organoids (ALOs) that comprise both proximal and distal airway epithelium from adult human lung cells (59). The ALOs consist of all six major epithelial cell types: AT2, AT1, basal, goblet, ciliated, and club cells. Via RNA sequencing, the authors confirmed that the cellular composition of their ALOs reflected that of the human lung tissues. Moreover, this cellular diversity was maintained even after several passages.

6.2.3.3 Tumor-Derived Organoids/Lung Cancer Organoids

Cancer organoids are organoids that consist of neoplastic cells derived from tissue samples of cancer patients (60). The derivation and utility of cancer organoids have gained traction in the cancer community in recent years as a pre-clinical model because they are able to mimic phenotypic features of the parent tumor and successfully model cancer heterogeneity (60). This success was also observed for lung cancers. An extensive list of lung cancer organoids (LCOs) models was compiled by Ma and colleagues recently (61). LCOs will not only serve as tools to study carcinogenesis, but because LCOs from individual patients can be readily generated, LCOs have immense potential in personalized medicine and individualized drug testing (62).

One such model was described by Kim and colleagues, in which surgically resected lung cancer tissues were dissociated into individual cells or cell clusters that were subsequently embedded in Matrigel and cultured to form LCOs (63). The LCOs were able to maintain the histology of the original cancer tissues. For instance, LCOs generated from adenocarcinoma exhibited acinar or glandular features and express lung adenocarcinoma marker expression, whereas LCOs made from squamous cell carcinoma displayed cytoplasmic keratinization. The adenocarcinoma LCOs consisted of goblet and ciliated cells, whereas the squamous LCOs consisted of basal and some ciliated cells. Another group of researchers described establishing 12 LCOs lines from 15 patients with different subtypes of adenocarcinoma using a similar method (64). These LCOs were also expandable in culture and preserved characteristics of the parental tumors. All the LCOs express CK7, confirming an epithelial origin; the LCOs also express SFTPC, suggesting that AT2 cells could be the cell of origin.

In a similar vein, immortalized lung epithelial cells that were transformed from cells isolated from healthy donors (see Section 6.2.3.2) were also used to form lung organoids. This was described by Tran and colleagues (65), where alveolar-like organoids were generated from a combination of immortalized AT2 cells and neonatal mouse lung fibroblasts. The immortalized AT2 cells express progenitor markers SOX2 and SOX9 and are void for mature alveolar markers. However, when the alveolar-like organoids were formed, they expressed mature alveolar epithelial markers such as aquaporin 5 (AQP5), G-protein-coupled receptor class C group 5 member A, and even to a certain extent HOPX, which are AT1 markers.

6.2.3.4 hPSC-Derived Lung Organoids

As its name suggests, hPSCs are "pluripotent," which means that they are able to give rise to most human cell types in vitro. hPSCs must be carefully and methodically differentiated to generate the desired cell type(s) in a dish by mimicking natural organ development. Generating lung organoids via directed differentiation of hPSCs closely follows the development trajectory of the lungs in vivo. In brief, the mammalian lungs are derived from the anterior foregut endoderm (AFE), where the lung buds begin forming and future airways develop via branching morphogenesis. Then as the primordial lung continues to develop further, alveolar tissues will form. By understanding the molecular and cellular drivers of embryonic lung development, particularly through the study of animal models, this knowledge has been applied to the creation of lung organoid differentiation protocols.

The first strategy of generating hPSC-derived lung organoids relies on the stepwise generation of multipotent progenitor cell populations, which can then be expanded and further differentiated into more mature lung epithelial cell types. A clear advantage of this method is the continual propagation of the progenitors, which can be either cryopreserved or differentiated when needed. This method was employed by McCauley and colleagues, where hPSCs were first differentiated into the definitive endoderm (DE) lineage (CXCR4+ SOX17+ FOXA2+) and subsequently patterned into the AFE fate (NKX2.1+ PAX9+ SOX2+) before being differentiated into EPCAM-expressing NKX2.1+ lung epithelial progenitors (66). The NKX2.1+ progenitors were then purified and further differentiated into proximal airway organoids or alveolar organoids by manipulating the dose of Wnt signals in culture (Figure 6.3a). By exposing aggregates of NKX2.1+ progenitors to a "low-Wnt" protocol, proximal airway organoids expressing high levels of proximal airway markers such as *TP63*, *SCGB3A2*, *SCGB1A1*, *MUC5B*, *CFTR*, *FOXJ1*, and *SFTPB* (66). These proximal airway organoids were found to contain secretory, basal, and goblet lineages (Figure 6.4). When the same progenitor population was exposed to a "high-Wnt" protocol (presence of Wnt activator CHIR99021), the organoids were pattern toward the distal fate; levels of distal lung markers such as ETV5 and SFTPC were observed. Instead of isolating NKX2.1+ progenitors, Jacob and colleagues further differentiated the NKX2.1+ progenitors into NKX2.1+ SFTC+ AT2 cells (termed iAEC2s) that were known facultative progenitors in the alveolus. These iAECs are multipotent and displayed the ability to proliferate and self-renew as spheroids (termed alveolospheres). These alveolospheres express epithelial marker EPCAM, as well as AT2 markers such as SFTPB and SFTC (Figure 6.4).

Hein and colleagues (2022) recently reported the derivation of bud-tip progenitor (BTP) cells from hiPSCs that express not only NKX2.1 but human bud-tip progenitor markers such as SOX9, SOX2, and CPM (67). The NKX2.1+ CPM+ BTP cells were purified and further cultured into BTP organoids (iBTOs) that were shown to bear resemblance to native bud-tip progenitors found in human fetal lungs. These iBTOs were expandable and able to maintain their multipotency in prolonged culture. Hein and

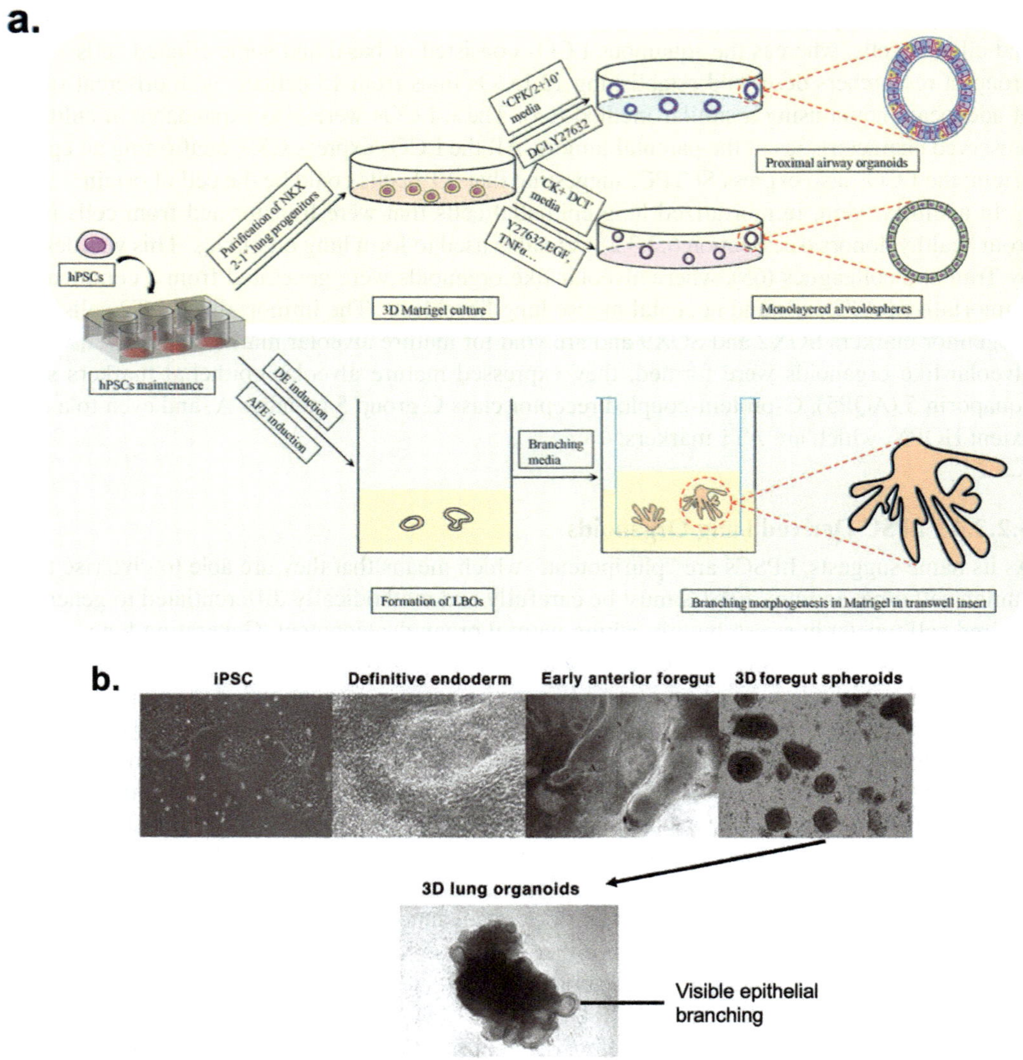

FIGURE 6.3 Brief overview of generating lung organoids from human pluripotent stem cells (hPSCs). (a) A schematic diagram illustrating the different methods of generating hPSC-derived lung organoids. (b) Brightfield images showing changes in cellular morphology at different stages of lung organoid differentiation from human-induced pluripotent stem cells.

Source: (a) Image obtained from Heo et al. (2022) (37). (b) Image obtained and modified from Bosáková et al. (2022) (29).

colleagues were then able to robustly differentiate these iBTOs into either airway or alveolar organoids. Similarly, Yamamoto and colleagues differentiated hPSCs into CPM+ NKX2.1+ progenitors, which were subsequently differentiated into alveolar organoids (AOs) (68). However, they co-cultured the CPM+ NKX2.1+ progenitors with human fetal lung fibroblast to form the AOs.

The second strategy involves the continual differentiation of hPSCs to lung organoids; no purification of progenitor cell populations was performed (Figure 6.3a, b). This method was adopted by Dye and colleagues where human embryonic stem cells were differentiated into DE, and subsequently, to SOX2+ ventral anterior foregut spheroids (69). The ventral anterior foregut spheroids were then embedded in Matrigel and subsequently differentiated into HLOs by stimulating the Hedgehog signaling pathway, which significantly induces NKX2.1 expression. A high dose of FGF10 was also added to the differentiation media, which induced branching morphogenesis and maintenance of lung progenitors. At day 15, the HLOs contained high percentages of ECAD+ epithelial cells that express both proximal and distal lung markers. After culturing these HLOs for 2 months, however, they display more prominent proximal airway-like features such as the presence of ciliated cells marked by ACTTUB and FOXJ1; notably, the proximal-like airway epithelium was surrounded by smooth muscle actin-expressing mesenchyme, similar to that of the bronchi/bronchioles (69). Building on this protocol, Miller and colleagues reported on the generation of lung bud-tip progenitor–like organoids using this approach; the NKX2.1+ ventral anterior foregut spheroids were encapsulated in Matrigel droplets and further differentiated into "patterned lung organoids" (PLOs) that possess multiple bud structures and exhibit proximal airway-like (express club cell and goblet cell markers) and bud tip–like characteristics (70). Furthermore, they showed that by passaging the PLOs, they were able to obtain and further propagate proliferative bud-tip organoids (express progenitor markers SOX2 and SOX9). These bud-tip organoids displayed multi-lineage potency and were able to differentiate into proximal airway-like organoids and alveolar-like organoids. Another similar protocol was also described by Leibel and colleagues where "whole lung organoids" (WLOs) were generated from hPSC-derived NKX2.1+ lung progenitor cells (71, 72). These WLOs were reported to consist of airway and alveolar epithelial cells, as well as mesenchymal cells.

A similar strategy was used by Chen and colleagues where hPSCs were differentiated to DE. Then, the DE cells were dissociated into clumps, which were further patterned toward the AFE fate before specification toward the lung lineage. By day 25, organoids with budding epithelial structures were formed ("lung bud organoids" [LBOs]; Figure 6.3a, b) (73, 74). These LBOs consist of EPCAM+ epithelial cells that express lung lineage markers including FOXA2 and NKX2.1 (Figure 6.4). The LBOs can be further matured in vitro where more extensive branching structures will

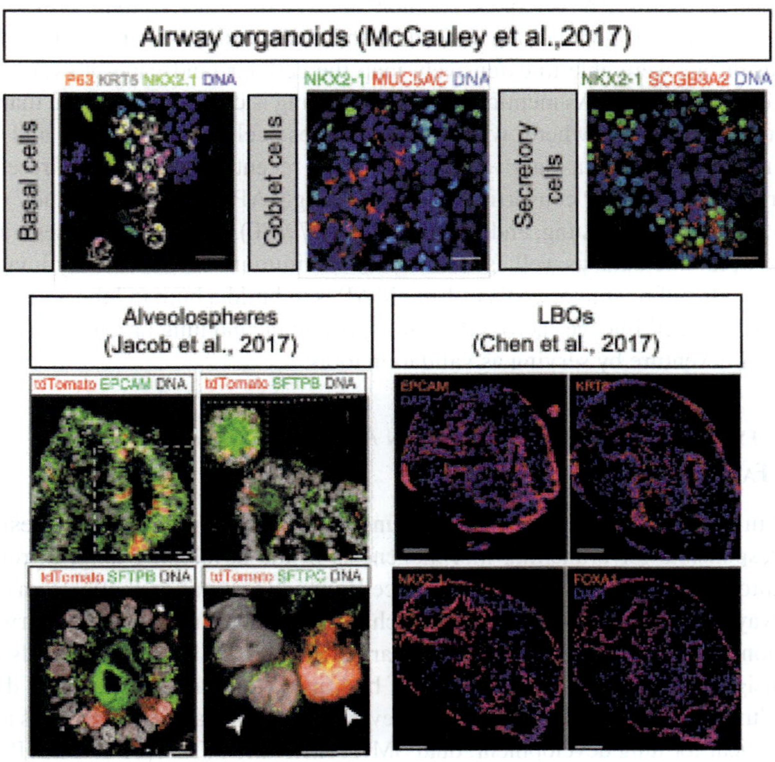

FIGURE 6.4 Cellular and molecular features of various hPSC-derived lung organoids.

Source: Immunofluorescence images obtained and modified from McCauley et al., 2017 (66), Jacob et al., 2017 (76), and Chen et al., 2017 (73).

form ("branching organoids"). The branching organoids expressed higher levels of AT2 markers (e.g., MUC1, SFTPB, SFTPC); HOPX, an AT1 marker; and some airway markers. Another similar protocol was described by Han and colleagues recently (75).

All in all, the field of lung organoids has seen great progress over the past decade; organoids encapsulating features of different lung regions that are functional and share similarities to the native organ can now be readily generated in the lab. A collection of published HLO systems was also recently reviewed by Yang and colleagues (77). These lung organoids have since been used to better understand the intricacies of lung organogenesis, as well as the pathogenesis of complex human diseases. In the next two sections, we will explore the applications of lung organoids in accelerating our understanding of pulmonary development and disease.

6.3 DISSECTING THE COMPLEXITIES OF LUNG DEVELOPMENT IN VITRO

Lung organoids, when combined with in vivo studies, can greatly advance our understanding of human lung development. Because organoids are generated in vitro, cellular samples from various stages of differentiation can be obtained, allowing us to gain access to numerous forms of information at different stages of lung development holistically—mRNA transcript, protein, epigenetics, chromatin accessibility, structural, intercellular dynamics, cell-ECM interactions, and so forth. In this section, we delve deeper into how lung organoids have been used hitherto to fill some gaps in our knowledge of human lung development.

6.3.1 Charting Lung Organogenesis at a Single-Cell Resolution

The advent of scRNA-Seq has brought about massive waves of discoveries and knowledge across multiple disciplines. With scRNA-Seq, we can now analyze the transcriptome profile of individual cells, identify new/rare cell populations in tissues, and characterize cells regardless of their abundance. As a result, we were able to extensively map the cellular landscape of every human organ and construct cell atlases (78). As mentioned earlier, the lung is composed of more than 40 different cell types, and understanding where, when, and how those cells arise, mature, and interact with the tissue microenvironment is a massive undertaking. In recent years, groups of researchers came together to form the LungMAP consortium to create LungMAP (www.lungmap.net/), an initiative funded by the National Heart, Lung, and Blood Institute (79, 80). LungMAP houses transcriptomic, proteomic, and lipidomic data, as well as histological and immunohistochemistry images of human and mouse lungs. One of the major goals of LungMAP is to build a human lung atlas that maps and characterizes all lung cell lineages across the human life span, and lung organoids have contributed significantly to this venture by serving as validation tools.

6.3.2 Identifying Mechanisms That Govern Airway and Alveolar Cell Fate Specification

As mentioned in Section 6.2.3.1, bud-tip cells (found at the tips of epithelial branches) are progenitors that are responsible for branching morphogenesis in the developing lung. In rodents, Sox9+ bud-tip progenitors give rise to TP63+ basal stem cells, which serve as epithelial stem cells that give rise to the airway epithelia (81). However, the mechanisms underlying the human airway and basal cell specification from bud-tip cells remain unclear. hPSC-derived bud-tip organoids were used to gather more insights into the molecular drivers of bud tip–to–airway and basal cell differentiation during human lung development (82). By testing several signaling pathway activators and inhibitors known to be critical for lung development, dual SMAD activation (TGFβ-1 and BMP4) (DSA) was discovered to significantly induce TP63 in the bud-tip organoids. Culturing DSA-induced organoids in media containing TGFβ-1 and BMP4 inhibitors (DSI) resulted in the derivation of airway cells such as secretory, goblet, ciliated, and club cells, suggesting that DSI stimulated the bud-tip organoids to differentiate into the proximal airway fate. scRNA-Seq confirmed that the DSI-treated organoids share high levels of similarity to the in vivo airway. Hence, this study exemplifies the utility of lung organoids (bud-tip organoids) to identify regulators and elucidate mechanisms responsible for airway epithelial fate specification in a human-specific in vitro system.

Furthermore, fetal lung organoids formed from late-stage (15–22 pcw) tip epithelial cells were found to exhibit alveolar lineage signatures, signifying that late-stage tip progenitors were committed to the alveolar fate. By adding signals to the culture media, tip organoids grown in the presence of Wnt signals had thin epithelium and expressed high levels of AT2 markers. On the other hand, tip organoids grown in the presence of FGF had thicker epithelium and expressed high levels of basal cell marker TP63 (50). Also, by performing bulk ATAC-seq (assay for transposase-accessible chromatin using sequencing), transcription factor NKX2.1 was found to be a major driving force for alveolar differentiation in these tip epithelial organoids (50). NKX2.1 directly binds to and activates the expression of genes related to surfactant production and secretion in the alveolar-fated cells. Altogether, these tip epithelial organoids were able to uncover insights into alveolar cell fate determination.

6.3.3 Uncovering Regulators of Human Lung Progenitors

As previously mentioned, lung epithelial progenitors give rise to all epithelial lineages from the proximal to the distal regions of the lungs. How these multipotent progenitors self-renew in the developing human lungs, however, remains unknown. CRISPRi screen in human fetal lung organoids generated from fetal tip progenitors was performed and transcription factor SOX9 was found to promote proliferation and suppress differentiation in the epithelial progenitors by inhibiting expression of secretory lineage-related metabolic genes as well as genes that mark non-lung lineages (83). Moreover, SOX9 directly regulates the expression of ETV4 and ETV5, thereby enhancing FGF signaling. SOX9 also directly regulates WNT pathway activators. Hence, fetal lung organoids facilitated the identification of a key transcription factor that governs the self-renewing properties of lung epithelial progenitors.

6.3.4 Elucidating the Role of the Mesenchyme in Epithelial Differentiation

During lung development, the pulmonary mesenchyme regulates epithelial differentiation, branching morphogenesis, and alveolar maturation by producing essential signals (84). The role of myofibroblasts in spatially patterning the developing alveolus is explored using human fetal lung organoids are critical in spatially patterning human AT2 cells (50). When freshly isolated WNT2-expressing fibroblasts were co-cultured with LIN^{POS} organoids, AT2 markers were maintained. However, AT2 marker expression was not supported in LIN^{POS} organoids that were cultured with $PDGFRA^{+}CD141^{+}$ myofibroblasts that express high levels of NOTUM, a secreted Wnt inhibitor. Then, when LIN^{POS} organoids were co-cultured with fibroblast that overexpress NOTUM, SFTPC expression was drastically lost. Therefore, Wnt-responsive SFTPC+ cells are restricted to the epithelial tips because of the presence of NOTUM-expressing myofibroblasts that surrounds the differentiated stalk cells. At the tips, WNT2-expressing fibroblasts continue to promote and sustain AT2 marker expression in the absence of the myofibroblast. Therefore, LIN^{POS} organoids were effectively used to better understand the role of the mesenchyme, particularly the myofibroblasts in regulating epithelial differentiation and branching.

6.4 MODELING AND INVESTIGATING HUMAN PULMONARY DISEASES

Lung organoids hold great promise in translational research. Aside from the use of lung cancer organoids to model and study a variety of lung cancers, HLOs have been employed to model various other pulmonary diseases (summarized in Table 6.1).

TABLE 6.1
Summary of Pulmonary Diseases Modeled by Lung Organoids

	Name of Disease	Lung Organoid Model System	Reference
Viral infections	Respiratory syncytial virus	hPSC-derived LBOs hPSC-derived HLOs	Chen et al., 2017 Harford et al., 2022
	Influenza	ASC-derived AT2 organoids ASC-derived airway organoids ASC-derived airway organoids	Salahudeen et al., 2020 Zhou et al., 2018 Hysenaj et al., 2023
	COVID-19	ASC-derived AT2 organoids ASC-derived bronchiolar and alveolar organoids ASC-derived airway organoids hPSC-derived proximal and alveolar organoids	Salahudeen et al., 2020 Ekanger et al., 2022 Hysenaj et al., 2023 Pei et al., 2021
Genetic diseases	Cystic fibrosis	hPSC-derived proximal airway organoids ASC-derived airway organoids	McCauley et al., 2017 Sachs et al., 2019
	Idiopathic pulmonary fibrosis	hPSC-derived LBOs Pulmospheres hPSC-derived alveolar organoids	Chen et al., 2017 Surolia et al., 2017 Kim et al., 2021
Congenital lung disorders	Neonatal respiratory distress syndrome	hPSC-derived alveolospheres	Jacob et al., 2017
	Congenital lung malformations	ASC-derived lung organoids	Swarr et al., 2018
Chronic lung disorders	Bronchopulmonary dysplasia	Fetal progenitor–derived lung organoids	Sucre et al., 2016 (103)

6.4.1 Viral Infections

6.4.1.1 Respiratory Syncytial Virus

Respiratory syncytial virus (RSV) is a common respiratory virus that causes upper respiratory infection resulting in mild, cold-like symptoms in people. Whereas most people recover from RSV infections readily, some vulnerable patients may develop bronchiolitis with obstruction of small airways, which may progress to pneumonia and even respiratory failure (85). When hPSC-derived LBOs were infected with RSV, infected cells were observed to shed into the lumen of the branching epithelium, closely resembling the clinical symptom of bronchiolitis (73). Another group performed a similar study where hPSC-derived HLOs were micro-injected with RSV (86). They showed that RSV-infected HLOs showed an increased expression of CC10 (club cell marker), a significant decreased expression of FOXJ1 (ciliated cell marker), and disruption of F-active organization, suggesting that RSV infection causes drastic alterations to the cellular architecture of fetal lungs.

6.4.1.2 H1N1 Influenza

H1N1 influenza is a subtype of influenza A virus that causes the H1N1 swine flu; it infects the upper and respiratory tracts resulting in cold-like symptoms (87). The H1N1 virus was able to infect AT2 organoids, which expressed influenza receptors (56). H1N1 infection was also successfully modeled in human airway organoids (comprising ciliated, goblet, club, and basal cells) generated from normal adult lung tissues; the H1N1 viruses were able to infect and replicate in the airway monolayers generated from the airway organoids (42). Similarly, ASC-derived airway organoids were also able to be infected by H1N1 viruses. Through this disease model, the authors observed that H1N1 predominantly infected ciliated and goblet cells (88).

6.4.1.3 SARS-CoV-2

The recent coronavirus disease 2019 (COVID-19) global pandemic has highlighted the imminent need for viable and physiologically relevant in vitro human lung models to understand how SARS-CoV-2, the causative agent of COVID-19, infects cells and causes disease. COVID-19 is a highly contagious infectious disease that causes symptoms such as cough, fever, and shortness of breath. COVID-19 can result in severe illness and even death (89).

The use of airway and alveolar organoids in modeling COVID-19 has been reviewed by Oliveria and colleagues (2021) (90). HLOs serve as a robust pre-clinical model to not only understand the mode of action of SARS-CoV-2 and how the virus impacts lung epithelial cells, but also for drug discovery and testing. SARS-CoV-2 was found to be able to infect distal lungs, resulting in alveolar damage and respiratory failure (91). To model this, apical-out epithelial organization was induced in ASC-derived AT2 organoids to expose the luminal cells that express ACE2, the SARS-CoV-2 receptor. Then, the organoids were infected with SARS-CoV-2. The SARS-CoV-2 virus was able to replicate in the organoids, and primarily infected SFTPC+ AT2 cells and SCGB1A1+ club cells (56). SARS-CoV-2 was also able to infect and replicate in ASC-derived bronchiolar and alveolar organoids; multi-nucleated cells were observed in the organoids, similar to that of human COVID-19 patients (58). Interestingly, a mediator called Tetraspanin-8 was discovered to facilitate SARS-CoV-2 infection using ASC-derived lung organoids (88). Additionally, the efficacy of the antiviral drug remdesivir was evaluated using the hPSC-derived human proximal and alveolar organoids (92). Remdesivir was shown to be effective in inhibiting SARS-CoV-2 replication in both types of lung organoids.

6.4.2 Genetic Diseases

6.4.2.1 Cystic Fibrosis

Cystic fibrosis (CF) is an inherited disease caused by mutations in the *CF transmembrane conductance regulator* (*CFTR*) gene. Mutated CFTR results in the production of thickened mucus secretions on epithelial linings, thereby affecting organoids such as the lungs, pancreas, intestines, and hepatic and biliary systems (93). Lung disease remains the main cause of morbidity and mortality in patients with CF. In the lungs, the thickened mucus forms plugs and plaques that obstruct airways, ultimately resulting in respiratory failure if left untreated (94). HLOs have been used to model the pulmonary features of CF. McCauley and colleagues generated proximal airway organoids from CF patient hiPSCs and were able to show that CF organoids exhibited CFTR-dependent swelling in the presence of forskolin (66). Also, Sachs and colleagues reported on the derivation of ASC-derived airway organoids from CF patients, which were able to recapitulate certain disease phenotypes such as forskolin-induced swelling (55).

6.4.2.2 Idiopathic Pulmonary Fibrosis

Idiopathic pulmonary fibrosis (IPF) is a severe and lethal genetic disease characterized by progressive scarring of the lungs. Lungs of IPF patients have excessive deposition of ECM, resulting in structural alterations and overall decline in organ function (95). At a cellular level, lung epithelial cells of IPF patients undergo repeated injury resulting in the recruitment of inflammatory cells, fibrosis, and pronounced ECM deposition. hPSC-derived LBOs were used to model pulmonary fibrosis associated with Hermansky-Pudlak syndrome by mutating *HPS1* gene via CRISPR/Cas9. The mutant LBOs displayed reduced epithelial branching, as well as abnormal accumulation of ECM and mesenchymal cells (73). 3D pulmospheres generated from IPF lung tissue were found to be surrounded by ECM proteins, resembling disease phenotype. Also, significant upregulation of smooth muscle marker α-SMA was observed in IPF pulmospheres (96). hPSC-derived AOs were also used to model TGF-β1-induced fibrosis—TGF-β1 treated hPSC-AOs were able to exhibit fibrogenic responses and collagen accumulation (97). Additionally, the fibrotic hPSC-AOs were then used to for drug testing; NP-011, a truncated form of milk fat globule-EGF factor 8 protein was effective in ameliorating fibrosis of TGF-β1 treated hPSC-AOs.

6.4.3 Congenital Lung Disorders

6.4.3.1 Neonatal Respiratory Distress Syndrome

Neonatal respiratory distress syndrome (RDS) is a serious condition that primarily occurs in premature newborns because of insufficient development of the lungs. The immature lungs are unable to function appropriately because of the severe deficiency of surfactant, resulting in respiratory distress (98). Mutations in the SFTPB gene that encodes for surfactant B proteins are found to cause RDS (99). Alveolospheres were derived from hiPSCs of an RDS patient (SP212) to model RDS caused by a particular *SFTPB* gene mutation termed 121ins2, where two base pairs were inserted at codon 121 in exon 4. SP212 alveolospheres did not express any SFTPB protein. By correcting the mutation using CRISPR/Cas9, SFTPB production was successfully restored (76).

6.4.3.2 Congenital Lung Malformations

Congenital lung malformations (CLM) consist of a broad range of conditions that affects the proper development of the lungs. The most common form of CLM is congenital pulmonary airway malformation (CPAM). CPAM is caused by the cessation of lung development during various stages of embryogenesis (100). Lung organoids were generated by combining lung epithelium isolated from CPAM lesions with human lung fibroblast cells to model CPAM in vitro (101). CPAM organoids were found to be consistently larger and consist of a nearly occlusive lumen surrounded by unusually thick layers of basal cells. CPAM organoids displayed profound epithelial defects akin to phenotypes observed in patients (101).

6.4.4 Chronic Lung Disorders

6.4.4.1 Bronchopulmonary Dysplasia

Bronchopulmonary dysplasia (BPD) is a chronic lung disease that primarily affects premature infants. BPD is the most common cause of respiratory morbidity in preterm infants and is caused by interrupted saccular and alveolar stages of lung development. Some of the key features of BPD include alveolar simplification and dysmorphic pulmonary vascularization (102). Human fetal lung organoids derived by combining fetal lung cells and human fetal lung fibroblasts were exposed to alternating cycles of hypoxia and hyperoxia to model BPD (103). The human fetal lung organoids exposed to hypoxia-hyperoxia exhibited fibroblast activation as evidenced by the drastic upregulation of α-SMA, which was also observed in BPD patients. Additionally, the diseased organoids also showed increased expression of several genes such as *ELASTIN, COL1A1, PDGFRa*, *VIM*, which are known to be upregulated in BPD patients (103).

6.5 ADDRESSING LIMITATIONS AND FUTURE STRATEGIES

6.5.1 Limitations

Without a doubt, the emergence of lung organoids propagated by the advancements in stem cell and organoid technologies has accelerated pulmonary research. However, there are still existing barriers to be overcome. In this section, we will explore some of these challenges and discuss plausible strategies to overcome these barriers to achieve lung organoids that not only exhibit profound levels of epithelial maturity but are also of higher quality and complexity.

6.5.1.1 Reliance on Tumor-Derived ECM

One major limitation of the current lung organoid model system is the heavy reliance on commercial sources of ECM such as Matrigel. As mentioned in Section 6.2.3, Matrigel is produced from mouse tumor cells. Therefore, it is poorly defined, has lot-to-lot variability, and most importantly, do not consist of any organ-specific ECM composition. The mechanical properties of Matrigel were also found to be heterogenous throughout a given area of substrate (104). Due to these limitations, researchers have begun exploring alternatives to generate Matrigel-free organoids. Some of these alternatives include decellularized tissues, natural matrices (e.g., collagen, alginate, laminin, fibrin), and even synthetic hydrogel polymers such as polyethylene glycol (PEG) and poly(lactic-co-glycolic acid) (PGLA). Synthetic hydrogels are especially promising because they are tuneable, which means that the stiffness and mechanical properties of the matrix can be altered to best mimic in vivo conditions (37, 104, 105). A synthetic hydrogel called PEG-4MAL was used to embed HLOs. The HLOs were able to maintain cell viability and consisted of well-organized lung epithelium comprising NKX2.1 cells, as well as airway basal cells (106).

6.5.1.2 Culture Conditions Are Not Chemically Defined

Many lung organoid protocols use components that are not chemically defined; this includes Matrigel (elaborated in Section 6.5.1.1). Some rely on the use of bovine pituitary extract (BPE) or fetal bovine serum (FBS). Whether BPE or FBS can artificially alter the self-renewal, proliferation, differentiation, and growth of lung progenitors and/or epithelial cells in vitro remains unknown, thus posing a question of whether the lung organoids generated in vitro using these components (1) are reproducible and (2) are physiologically relevant and can be used to accurately study human development and disease. Nonetheless, more researchers are increasingly pivoting toward using chemically defined media for lung organoid derivation (55, 56). This change, coupled with the elimination of Matrigel, will be crucial in ensuring that lung organoids are biologically relevant and meaningful.

6.5.1.3 Limited Cell Types

As mentioned earlier, the lung is composed of more than 40 types of cells, including a variety of stromal and immune cells. Although several lung organoid model systems described thus far have reported the presence of numerous different lung epithelial cell types (59, 70, 72, 73), they are still unable to fully mirror the cellular makeup of a human lung. The human lung also consists of non-epithelial cells such as fibroblasts and vascular cells, as well as tissue-resident immune cells such as macrophages (alveolar and interstitial macrophages), and infiltrating immune cells such as neutrophils, monocytes, and T cells. However, lung organoids lack the presence of these disease-relevant cells, rendering them unsuitable for studies pertaining to pulmonary inflammation and diseases whereby immune cells play a significant role. A feasible solution to the lack of immune cells is to incorporate immune cells into lung organoids via co-culture. This will be further discussed in Section 6.5.2.1.

As mentioned in Section 6.3.4, the lung mesenchyme is necessary to support the growth and differentiation of the lung epithelium. At the same time, the lung mesenchyme receives signals from multiple cell types to give rise to cells including but not limited to airway and vascular smooth muscle cells (SMCs), endothelial cells (ECs), and pericytes, giving rise to the overall lung vasculature. In adult lungs, mesenchymal stromal/stem cells (MSCs) continue to play a role in maintaining lung epithelium homeostasis and repair. Most lung organoid models are devoid of these essential populations of cells. However, simultaneous differentiation of both lung epithelial and mesenchymal cells in lung organoids has been described (69, 70, 76). In an attempt to address this concern, researchers have begun co-culturing lung organoids with mesenchymal cells; increased lung organoid formation was observed when lung organoids were co-cultured with mesenchymal stem cells (MSCs) (107). Also, when lung progenitors were co-cultured with hiPSC-derived mesenchymal cells, alveolar epithelial cell fates were induced during organoid formation (108).

6.5.1.4 Lack of Vascularization

As mentioned in Section 6.5.1.3 because of the lack of vascular cell types such as ECs, pericytes, and SMCs in most of the current lung organoid model systems, the lung organoids lack vascularization. This remains to be a major hurdle in the field. Vascularization of lung organoids has only been accomplished by grafting the organoids into animals, which promoted significant epithelial differentiation and maturation (69, 73). This is unsurprising because in vivo, the host vasculature penetrates and integrates into the organoid grafts. Thereafter, oxygen, nutrients, and even paracrine factors from the host will be delivered to the grafts, allowing the cells in the organoids to thrive and mature (109). Without robust organoid vascularization in vitro, the duration of organoid culture, as well as growth and maturation of the lung epithelium in lung organoids, is greatly limited.

Efforts are currently being made to enhance vascular cell populations in organoids. Incorporating mesodermal progenitor cells into organoids has been shown to allow vessel formation in tumor and neural organoids (110). Lung organoid vascularization could also potentially be achieved through endothelial co-cultures and organoid-on-chip systems (which will be elaborated in Section 6.5.2.1).

6.5.2 Future Strategies—"Next-Gen" Lung Organoids?

6.5.2.1 Recapitulating the Organ Niche

One of the most critical considerations toward better recapitulating the complexity of the organ niche includes vascularizing the lung organoids and incorporating immune cells in the culture system as discussed in Section 6.5.1.3. Endothelial co-cultures have been used to vascularize liver, kidney, and brain organoids (111). For instance, endothelial networks were able to form in hPSC-derived cerebral organoids when adding hPSC-derived ECs were added to differentiating organoids (112). For lung organoids, a similar observation was described when bronchio-alveolar stem cells were co-cultured with mouse lung ECs, generating multi-lineage lung organoids (bronchiolar and alveolar lineages) (113). Currently, the vascularization of HLOs via endothelial co-cultures has not been extensively reported.

Organoid-on-chip has also recently emerged to be a potential solution for vascularizing lung organoids. In brief, organoid-on-chip involves a microfabricated chip that is designed to contain different functional units (114). One proposed technique to vascularize organoids using the organoid-on-chip technology is to first engineer a perfusable vascular bed on a microfluidic chip. Then, a pre-vascularized organoid will be added on a vascular bed to allow functional connections to form between the vascular bed and organoid (115). The feasibility of this method in generating vascularized HLOs remains untested.

Next, sophisticated multi-lineage organoids could be a plausible solution, particularly into incorporate the mesodermal derivatives such as the vascular and stromal cells into the lung organoids and have them coexist, interact, and even communicate with the lung epithelium akin to natural organ development. This will be elaborated in Section 6.5.2.2.

Also, organoid-immune co-cultures have been proposed for onco-immunological studies (116). Tumor organoid–T cell co-cultures have been reported to test the tumor reactivity and tumoricidal capacity of patient T cells (117). Recently, human alveolar organoids were co-cultured with hiPSC-derived macrophages (iMACs) to generate iMACs-alveolar organoids (iMACs-AOs) that were suitable for modeling inflammatory diseases (118). iMACs coexisted in the iMACs-AOs for up to 2 weeks and were able to respond readily to inflammatory stimuli. More studies will have to be done to develop and test more organoid-immune co-cultures for tumor and non-tumor lung organoids.

Finally, cells in the lungs experience mechanical stretching due to the physical forces exerted during breathing movements. The importance of biomechanical signals in regulating lung epithelial differentiation has also been well documented in animal studies (119, 120). In vitro, inducing mechanical stretch in primary cultures of AT2 cells by incorporating cyclic stretch has been shown to promote AT2 differentiation and maturation (121), illustrating the importance of mechano-signals in the development of lung cell types. Therefore, the incorporation of mechanical stretch into organoid cultures will likely aid in better recapitulating organ niche.

6.5.2.2 Complex Multi-Lineage Organoids

During the process of natural organogenesis, organs do not develop in isolation; they co-develop intimately with other adjacent organs/tissues. Cell-cell communication between cells of different lineages is a critical driver of proper organ development. However, this intricate interplay is not reflected in current lung organoid systems; lung epithelial cells differentiate and mature in the absence of cells of other lineages/organs. Recently, however, organoid scientists have been exploring the concept of multi-lineage organoids to better encapsulate the dynamic cell-cell interactions that occur during development in vivo. This can be accomplished by two techniques—assembloid formation or multi-lineage co-development.

First, assembloids are defined as self-organizing 3D systems that are formed from the integration of multiple organoids that captures inter-tissue or inter-lineage cellular interactions (122). The assembloid technique was founded in the field of neurobiology where hiPSC-derived cortical and spinal spheroids were "fused" together with skeletal muscle spheroids to form corticomotor assembloids. The cortical neurons were able to project and connect with the spinal spheroids, and the spinal-derived motor neurons were able to connect with muscle (123). Multi-lineage assembloids have also been reported—neuro-mesodermal assembloids that mimic peripheral nervous system development were constructed by integrating neuroepithelial and mesodermal aggregates (124). Also, complex hepato-biliary-pancreatic organoids were generated using the assembloid technique (125, 126). Perhaps vasculo-lung assembloids made from integrating vessel organoids with lung organoids and/or lung progenitor spheroids with mesodermal spheroids can be explored as a potential method to either vascularize lung organoids or incorporate diverse mesenchymal cell types into lung organoids.

Second, in contrast to the assembloid technique, the concept of multi-lineage co-development involves carefully inducing the differentiation of multiple cell lineage from hPSCs within a single structure. The feasibility of this method was exemplified by the co-development of cardio-pulmonary micro-tissues consisting of both cardiac and lung epithelial components (127). The presence of the cardiac lineages was found to promote alveolar maturation. Hence, the application of co-development to achieve better-vascularized lung organoids or lung organoids with viable immune cells can be further explored.

6.6 CONCLUSIONS

Here, we explored the current lung organoid models derived from a variety of cell sources—fetal progenitors, ASCs, hPSCs, as well as tumor/cancer cells. To date, great progress has been made in the field; several types of HLOs with varying cellular and molecular features can now be readily generated in the lab. Many of the lung organoid model systems reported success in obtaining either proximal or distal epithelial cell types, others have described identifying both or even mesenchymal cells within their organoids. Regardless, these lung organoids have served as indispensable tools for the study of human organogenesis and disease pathogenesis. However, there are several notable limitations that must be addressed to improve the reproducibility, relevancy, and quality of current lung organoid models to permit downstream biomedical applications. Therefore, it is paramount to explore new strategies to overcome these limitations (Figure 6.5). By leveraging the advancement of cell culture technologies and incorporating cutting-edge bioengineering strategies alongside our increasing knowledge of developmental biology, we are optimistic that the next generation of HLOs will propel us toward more discoveries and breakthroughs in pulmonary research.

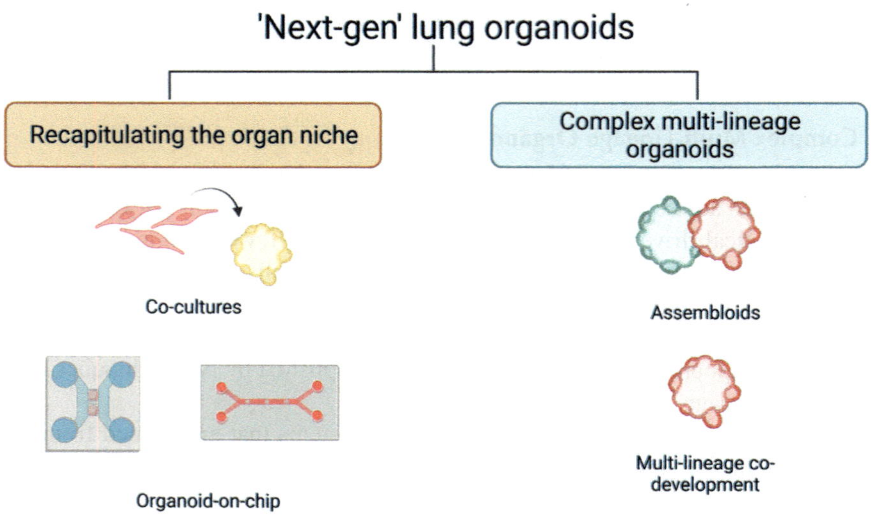

FIGURE 6.5 Considerations toward "next-gen" lung organoids. Schematic diagram summarizing the considerations toward improving current lung organoid models.

6.7 ACKNOWLEDGMENTS

M.G. is funded by NIH R01HL166283 and Cincinnati Children's Research Foundation Endowed Scholar Award. N.P. is funded by the American Heart Association Pre-doctoral Fellowship (1013861), LungMAP2 Pilot Award, the University of Cincinnati Graduate Student Government Research Fellowship, and the National University of Singapore Development Grant 2022–2023. We acknowledge and attribute the figures included in this book chapter to their respective authors. Other figures were generated using www.biorender.com.

ABBREVIATIONS

AEC1	alveolar type 1 epithelial cell
AEC2	alveolar type 2 epithelial cell
AFE	anterior foregut endoderm
ALI	air-liquid interface
ALOs	adult lung organoids
AOs	alveolar organoids
AQP5	aquaporin 5
ASCs	adult stem cells
AT1	alveolar type 1 epithelial cell
AT2	alveolar type 2 epithelial cell
ATAC-Seq	assay for transposase-accessible chromatin using sequencing
BPD	bronchopulmonary dysplasia
BPE	bovine pituitary extract
BTP	bud-tip progenitors
CF	cystic fibrosis
CLM	congenital lung malformations
COVID-19	coronavirus disease 2019
CPAM	congenital pulmonary airway malformation
DE	definitive endoderm
DSA	dual SMAD activation

DSI	dual SMAD activation inhibitor
ECM	extracellular matrix
ECs	endothelial cells
EM	electron microscopy
FBS	fetal bovine serum
hiPSCs	human-induced pluripotent stem cells
HLOs	human lung organoids
hPSCs	human pluripotent stem cells
iBTOs	bud-tip organoids
iMACs	hiPSC-derived macrophages
IPF	idiopathic pulmonary fibrosis
LB	lamellar bodies
LBOs	lung bud organoids
MSC	mesenchymal stem cells
PEG	polyethylene glycol
PGLA	poly(lactic-co-glycolic acid)
PLO	patterned lung organoid
RDS	respiratory distress syndrome
RSV	respiratory syncytial virus
scRNA-Seq	single-cell RNA sequencing
SMC	smooth muscle cell
WLO	whole lung organoid

REFERENCES

1. Li, F., He, J., Wei, J., Cho, W. C., and Liu, X. (2015) Diversity of epithelial stem cell types in adult lung. *Stem Cells Int.* **2015**, 728307
2. Basil, M. C., Katzen, J., Engler, A. E., Guo, M., Herriges, M. J., Kathiriya, J. J., Windmueller, R., Ysasi, A. B., Zacharias, W. J., Chapman, H. A., Kotton, D. N., Rock, J. R., Snoeck, H.-W., Vunjak-Novakovic, G., Whitsett, J. A., and Morrisey, E. E. (2020) The cellular and physiological basis for lung repair and regeneration: Past, present, and future. *Cell Stem Cell.* **26**, 482–502
3. Cardoso, W. V., and Whitsett, J. A. (2008) Resident cellular components of the lung: Developmental aspects. *Proc Am Thorac Soc.* **5**, 767
4. Whitsett, J. A., Kalin, T. V., Xu, Y., and Kalinichenko, V. V. (2019) Building and regenerating the lung cell by cell. *Physiol Rev.* **99**, 513–554
5. Cardoso, W. V. (2001) Molecular regulation of lung development. *Annu Rev Physiol.* **63**, 471–494
6. Metzger, R. J., Klein, O. D., Martin, G. R., and Krasnow, M. A. (2008) The branching programme of mouse lung development. *Nature.* **453**, 745–750
7. Baron, R. M., Choi, A. J. S., Owen, C. A., and Choi, A. M. K. (2012) Genetically manipulated mouse models of lung disease: Potential and pitfalls. *Am J Physiol Lung Cell Mol Physiol.* **302**, L485–L497
8. Clevers, H. (2016) Modeling development and disease with organoids. *Cell.* **165**, 1586–1597
9. Lancaster, M. A., and Knoblich, J. A. (2014) Generation of cerebral organoids from human pluripotent stem cells. *Nat Protoc.* **9**, 2329–2340
10. Lancaster, M. A., and Knoblich, J. A. (2014) Organogenesis in a dish: Modeling development and disease using organoid technologies. *Science.* **345**, 1247125
11. Marsee, A., Roos, F. J. M., Verstegen, M. M. A., Marsee, A., Roos, F., Verstegen, M., Clevers, H., Vallier, L., Takebe, T., Huch, M., Peng, W. C., Forbes, S., Lemaigre, F., de Koning, E., Gehart, H., van der Laan, L., *et al.* (2021) Building consensus on definition and nomenclature of hepatic, pancreatic, and biliary organoids. *Cell Stem Cell.* **28**, 816–832
12. Unagolla, J. M., and Jayasuriya, A. C. (2022) Recent advances in organoid engineering: A comprehensive review. *Appl Mater Today.* **29**, 101582

13. Kim, J., Koo, B.-K., and Knoblich, J. A. (2020) Human organoids: Model systems for human biology and medicine. *Nat Rev Mol Cell Biol.* **21**, 571–584
14. Kaluthantrige Don, F., and Huch, M. (2021) Organoids, where we stand and where we go. *Trends Mol Med.* **27**, 416–418
15. McCauley, H. A., and Wells, J. M. (2017) Pluripotent stem cell-derived organoids: Using principles of developmental biology to grow human tissues in a dish. *Development.* **144**, 958–962
16. Huch, M., and Koo, B.-K. (2015) Modeling mouse and human development using organoid cultures. *Development.* **142**, 3113–3125
17. Lancaster, M. A., and Huch, M. (2019) Disease modelling in human organoids. *Dis Model Mech.* **12**, dmm039347
18. Ho, B. X., Pek, N. M. Q., and Soh, B.-S. (2018) Disease modeling using 3D organoids derived from human induced pluripotent stem cells. *Int J Mol Sci.* https://doi.org/10.3390/ijms19040936
19. Zhou, Z., Cong, L., and Cong, X. (2021) Patient-derived organoids in precision medicine: Drug screening, organoid-on-a-chip and living organoid biobank. *Frontiers in Oncology* [online]. www.frontiersin.org/articles/10.3389/fonc.2021.762184 (Accessed May 30, 2023)
20. Driehuis, E., Kretzschmar, K., and Clevers, H. (2020) Establishment of patient-derived cancer organoids for drug-screening applications. *Nat Protoc.* **15**, 3380–3409
21. Awatade, N. T., Wong, S. L., Hewson, C. K., Fawcett, L. K., Kicic, A., Jaffe, A., and Waters, S. A. (2018) Human primary epithelial cell models: Promising tools in the era of cystic fibrosis personalized medicine. *Front Pharmacol* [online]. www.frontiersin.org/articles/10.3389/fphar.2018.01429 (Accessed May 28, 2023)
22. Katsumiti, A., Ruenraroengsak, P., Cajaraville, M. P., Thorley, A. J., and Tetley, T. D. (2020) Immortalisation of primary human alveolar epithelial lung cells using a non-viral vector to study respiratory bioreactivity in vitro. *Sci Rep.* **10**, 20486
23. Sengupta, A., Roldan, N., Kiener, M., Froment, L., Raggi, G., Imler, T., de Maddalena, L., Rapet, A., May, T., Carius, P., Schneider-Daum, N., Lehr, C.-M., Kruithof-de Julio, M., Geiser, T., Marti, T. M., Stucki, J. D., Hobi, N., and Guenat, O. T. (2022) A new immortalized human alveolar epithelial cell model to study lung injury and toxicity on a breathing lung-on-chip system. *Front Toxicol.* https://doi.org/10.3389/ftox.2022.840606
24. Grau-Expósito, J., Perea, D., Suppi, M., Massana, N., Vergara, A., Soler, M. J., Trinite, B., Blanco, J., García-Pérez, J., Alcamí, J., Serrano-Mollar, A., Rosado, J., Falcó, V., Genescà, M., and Buzon, M. J. (2022) Evaluation of SARS-CoV-2 entry, inflammation and new therapeutics in human lung tissue cells. *PLoS Pathog.* **18**, e1010171
25. Kreft, M. E., Jerman, U. D., Lasič, E., Hevir-Kene, N., Rižner, T. L., Peternel, L., and Kristan, K. (2015) The characterization of the human cell line Calu-3 under different culture conditions and its use as an optimized in vitro model to investigate bronchial epithelial function. *Eur J Pharm Sci.* **69**, 1–9
26. Jiang, D., Schaefer, N., and Chu, H. W. (2018) Air-liquid interface culture of human and mouse airway epithelial cells. *Methods Mol Biol.* **1809**, 91–109
27. Abo, K. M., Sainz de Aja, J., Lindstrom-Vautrin, J., Alysandratos, K.-D., Richards, A., Garcia-de-Alba, C., Huang, J., Hix, O. T., Werder, R. B., Bullitt, E., Hinds, A., Falconer, I., Villacorta-Martin, C., Jaenisch, R., Kim, C. F., Kotton, D. N., and Wilson, A. A. (2022) Air-liquid interface culture promotes maturation and allows environmental exposure of pluripotent stem cell-derived alveolar epithelium. *JCI Insight.* **7**, e155589
28. Comhair, S. A. A., Xu, W., Mavrakis, L., Aldred, M. A., Asosingh, K., and Erzurum, S. C. (2012) Human primary lung endothelial cells in culture. *Am J Respir Cell Mol Biol.* **46**, 723–730
29. Bosáková, V., De Zuani, M., Sládková, L., Garlíková, Z., Jose, S. S., Zelante, T., Hortová Kohoutková, M., and Frič, J. (2022) Lung organoids—The ultimate tool to dissect pulmonary diseases? *Front Cell Dev Biol* [online]. www.frontiersin.org/articles/10.3389/fcell.2022.899368 (Accessed May 29, 2023)
30. Rothen-Rutishauser, B., Gibb, M., He, R., Petri-Fink, A., and Sayes, C. M. (2023) Human lung cell models to study aerosol delivery—Considerations for model design and development. *Eur J Pharm Sci.* **180**, 106337
31. Sato, M., Shay, J. W., and Minna, J. D. (2020) Immortalized normal human lung epithelial cell models for studying lung cancer biology. *Respir Investig.* **58**, 344–354
32. Crosby, L. M., and Waters, C. M. (2010) Epithelial repair mechanisms in the lung. *Am J Physiol Lung Cell Mol Physiol.* **298**, L715–L731
33. Gras, D., Bourdin, A., Vachier, I., de Senneville, L., Bonnans, C., and Chanez, P. (2012) An ex vivo model of severe asthma using reconstituted human bronchial epithelium. *J Allergy Clin Immunol.* **129**, 1259–1266.e1

34. Kelly, M., Dreano, E., Hatton, A., Lepissier, A., Golec, A., Sermet-Gaudelus, I., and Pranke, I. (2022) Primary human nasal epithelial cells: Biobanking in the context of precision medicine. *JoVE (J Vis Exp)*. https://doi.org/10.3791/63409
35. Jensen, C., and Teng, Y. (2020) Is it time to start transitioning from 2D to 3D cell culture? *Front Mol Biosci*. **7**, 33
36. Costa, E. C., Moreira, A. F., de Melo-Diogo, D., Gaspar, V. M., Carvalho, M. P., and Correia, I. J. (2016) 3D tumor spheroids: An overview on the tools and techniques used for their analysis. *Biotechnol Adv*. **34**, 1427–1441
37. Heo, J. H., Kang, D., Seo, S. J., and Jin, Y. (2022) Engineering the extracellular matrix for organoid culture. *Int J Stem Cells*. **15**, 60–69
38. Meng, X., Cui, G., and Peng, G. (2023) Lung development and regeneration: Newly defined cell types and progenitor status. *Cell Regen*. **12**, 5
39. Basil, M. C., Cardenas-Diaz, F. L., Kathiriya, J. J., Morley, M. P., Carl, J., Brumwell, A. N., Katzen, J., Slovik, K. J., Babu, A., Zhou, S., Kremp, M. M., McCauley, K. B., Li, S., Planer, J. D., Hussain, S. S., Liu, X., *et al.* (2022) Human distal airways contain a multipotent secretory cell that can regenerate alveoli. *Nature*. **604**, 120–126
40. Zimmermann, B. (1987) Lung organoid culture. *Differentiation*. **36**, 86–109
41. Nadkarni, R. R., Abed, S., and Draper, J. S. (2016) Organoids as a model system for studying human lung development and disease. *Biochem Biophys Res Commun*. **473**, 675–682
42. Kong, J., Wen, S., Cao, W., Yue, P., Xu, X., Zhang, Y., Luo, L., Chen, T., Li, L., Wang, F., Tao, J., Zhou, G., Luo, S., Liu, A., and Bao, F. (2021) Lung organoids, useful tools for investigating epithelial repair after lung injury. *Stem Cell Res Ther*. **12**, 1–13
43. Silva, S., Bicker, J., Falcão, A., and Fortuna, A. (2023) Air-liquid interface (ALI) impact on different respiratory cell cultures. *Eur J Pharm Biopharm*. **184**, 62–82
44. Schittny, J. C. (2017) Development of the lung. *Cell Tissue Res*. **367**, 427–444
45. Herriges, M., and Morrisey, E. E. (2014) Lung development: Orchestrating the generation and regeneration of a complex organ. *Development*. **141**, 502–513
46. Rawlins, E. L. (2008) Lung epithelial progenitor cells. *Proc Am Thorac Soc*. **5**, 675–681
47. Varghese, B., Ling, Z., and Ren, X. (2022) Reconstructing the pulmonary niche with stem cells: A lung story. *Stem Cell Res Ther*. **13**, 161
48. Nikolić, M. Z., Caritg, O., Jeng, Q., Johnson, J.-A., Sun, D., Howell, K. J., Brady, J. L., Laresgoiti, U., Allen, G., Butler, R., Zilbauer, M., Giangreco, A., and Rawlins, E. L. (2017) Human embryonic lung epithelial tips are multipotent progenitors that can be expanded in vitro as long-term self-renewing organoids. *eLife*. **6**, e26575
49. Lamers, M. M., van der Vaart, J., Knoops, K., Riesebosch, S., Breugem, T. I., Mykytyn, A. Z., Beumer, J., Schipper, D., Bezstarosti, K., Koopman, C. D., Groen, N., Ravelli, R. B. G., Duimel, H. Q., Demmers, J. A. A., Verjans, G. M. G. M., Koopmans, M. P. G., Muraro, M. J., Peters, P. J., Clevers, H., and Haagmans, B. L. (2021) An organoid-derived bronchioalveolar model for SARS-CoV-2 infection of human alveolar type II-like cells. *EMBO J*. **40**, e105912
50. Lim, K., Donovan, A. P. A., Tang, W., Sun, D., He, P., Pett, J. P., Teichmann, S. A., Marioni, J. C., Meyer, K. B., Brand, A. H., and Rawlins, E. L. (2023) Organoid modeling of human fetal lung alveolar development reveals mechanisms of cell fate patterning and neonatal respiratory disease. *Cell Stem Cell*. **30**, 20–37.e9
51. Wu, M., Zhang, X., Lin, Y., and Zeng, Y. (2022) Roles of airway basal stem cells in lung homeostasis and regenerative medicine. *Respir Res*. **23**, 122
52. Barkauskas, C. E., Cronce, M. J., Rackley, C. R., Bowie, E. J., Keene, D. R., Stripp, B. R., Randell, S. H., Noble, P. W., and Hogan, B. L. M. (2013) Type 2 alveolar cells are stem cells in adult lung. *J Clin Invest*. **123**, 3025–3036
53. Desai, T. J., Brownfield, D. G., and Krasnow, M. A. (2014) Alveolar progenitor and stem cells in lung development, renewal and cancer. *Nature*. **507**, 190–194
54. Chan, M., and Liu, Y. (2022) Function of epithelial stem cell in the repair of alveolar injury. *Stem Cell Res Ther*. **13**, 170
55. Sachs, N., Papaspyropoulos, A., Zomer-van Ommen, D. D., Heo, I., Böttinger, L., Klay, D., Weeber, F., Huelsz-Prince, G., Iakobachvili, N., Amatngalim, G. D., de Ligt, J., van Hoeck, A., Proost, N., Viveen, M. C., *et al.* (2019) Long-term expanding human airway organoids for disease modeling. *EMBO J*. **38**, e100300

56. Salahudeen, A. A., Choi, S. S., Rustagi, A., Zhu, J., van Unen, V., de la O, S. M., Flynn, R. A., Margalef-Català, M., Santos, A. J. M., Ju, J., Batish, A., Usui, T., Zheng, G. X. Y., Edwards, C. E., Wagar, L. E., *et al.* (2020) Progenitor identification and SARS-CoV-2 infection in human distal lung organoids. *Nature*. **588**, 670–675
57. Katsura, H., Sontake, V., Tata, A., Kobayashi, Y., Edwards, C. E., Heaton, B. E., Konkimalla, A., Asakura, T., Mikami, Y., Fritch, E. J., Lee, P. J., Heaton, N. S., Boucher, R. C., Randell, S. H., Baric, R. S., and Tata, P. R. (2020) Human lung stem cell-based alveolospheres provide insights into SARS-CoV-2-mediated interferon responses and pneumocyte dysfunction. *Cell Stem Cell*. **27**, 890–904.e8
58. Ekanger, C. T., Zhou, F., Bohan, D., Lotsberg, M. L., Ramnefjell, M., Hoareau, L., Røsland, G. V., Lu, N., Aanerud, M., Gärtner, F., Salminen, P. R., Bentsen, M., Halvorsen, T., Ræder, H., Akslen, L. A., Langeland, N., Cox, R., Maury, W., Stuhr, L. E. B., Lorens, J. B., and Engelsen, A. S. T. (2022) Human organotypic airway and lung organoid cells of bronchiolar and alveolar differentiation are permissive to infection by influenza and SARS-CoV-2 respiratory virus. *Front Cell Infect Microbiol* [online]. www.frontiersin.org/articles/10.3389/fcimb.2022.841447 (Accessed June 2, 2023)
59. Tindle, C., Fuller, M., Fonseca, A., Taheri, S., Ibeawuchi, S.-R., Beutler, N., Katkar, G. D., Claire, A., Castillo, V., Hernandez, M., Russo, H., Duran, J., Crotty Alexander, L. E., Tipps, A., Lin, G., Thistlethwaite, P. A., Chattopadhyay, R., Rogers, T. F., Sahoo, D., Ghosh, P., and Das, S. (2021) Adult stem cell-derived complete lung organoid models emulate lung disease in COVID-19. *eLife*. **10**, e66417
60. LeSavage, B. L., Suhar, R. A., Broguiere, N., Lutolf, M. P., and Heilshorn, S. C. (2022) Next-generation cancer organoids. *Nat Mater*. **21**, 143–159
61. Ma, H., Zhu, Y., Zhou, R., Yu, Y., Xiao, Z., and Zhang, H. (2022) Lung cancer organoids, a promising model still with long way to go. *Crit Rev Oncol/Hematol*. **171**, 103610
62. Li, Y., Gao, X., Ni, C., Zhao, B., and Cheng, X. (2023) The application of patient-derived organoid in the research of lung cancer. *Cell Oncol*. **46**, 503–519
63. Kim, M., Mun, H., Sung, C. O., Cho, E. J., Jeon, H.-J., Chun, S.-M., Jung, D. J., Shin, T. H., Jeong, G. S., Kim, D. K., Choi, E. K., Jeong, S.-Y., Taylor, A. M., Jain, S., Meyerson, M., and Jang, S. J. (2019) Patient-derived lung cancer organoids as in vitro cancer models for therapeutic screening. *Nat Commun*. **10**, 3991
64. Li, Z., Qian, Y., Li, W., Liu, L., Yu, L., Liu, X., Wu, G., Wang, Y., Luo, W., Fang, F., Liu, Y., Song, F., Cai, Z., Chen, W., and Huang, W. (2020) Human lung adenocarcinoma-derived organoid models for drug screening. *iScience*. **23**, 101411
65. Tran, E., Shi, T., Li, X., Chowdhury, A. Y., Jiang, D., Liu, Y., Wang, H., Yan, C., Wallace, W. D., Lu, R., Ryan, A. L., Marconett, C. N., Zhou, B., Borok, Z., and Offringa, I. A. (2022) Development of human alveolar epithelial cell models to study distal lung biology and disease. *iScience*. https://doi.org/10.1016/j.isci.2022.103780
66. McCauley, K. B., Hawkins, F., Serra, M., Thomas, D. C., Jacob, A., and Kotton, D. N. (2017) Efficient derivation of functional human airway epithelium from pluripotent stem cells via temporal regulation of Wnt signaling. *Cell Stem Cell*. **20**, 844–857.e6
67. Hein, R. F. C., Conchola, A. S., Fine, A. S., Xiao, Z., Frum, T., Brastrom, L. K., Akinwale, M. A., Childs, C. J., Tsai, Y.-H., Holloway, E. M., Huang, S., Mahoney, J., Heemskerk, I., and Spence, J. R. (2022) Stable iPSC-derived NKX2-1+ lung bud tip progenitor organoids give rise to airway and alveolar cell types. *Development*. **149**, dev200693
68. Yamamoto, Y., Korogi, Y., Hirai, T., and Gotoh, S. (2020) A method of generating alveolar organoids using human pluripotent stem cells. *Methods Cell Biol*. **159**, 115–141
69. Dye, B. R., Hill, D. R., Ferguson, M. A., Tsai, Y.-H., Nagy, M. S., Dyal, R., Wells, J. M., Mayhew, C. N., Nattiv, R., Klein, O. D., White, E. S., Deutsch, G. H., and Spence, J. R. (2015) In vitro generation of human pluripotent stem cell derived lung organoids. *eLife*. **4**, e05098
70. Miller, A. J., Dye, B. R., Ferrer-Torres, D., Hill, D. R., Overeem, A. W., Shea, L. D., and Spence, J. R. (2019) Generation of lung organoids from human pluripotent stem cells in vitro. *Nat Protoc*. **14**, 518–540
71. Leibel, S. L., McVicar, R. N., Winquist, A. M., Niles, W. D., and Snyder, E. Y. (2020) Generation of complete multi–cell type lung organoids from human embryonic and patient-specific induced pluripotent stem cells for infectious disease modeling and therapeutics validation. *Curr Protoc Stem Cell Biol*. **54**, e118
72. Leibel, S. L., McVicar, R. N., Winquist, A. M., and Snyder, E. Y. (2021) Generation of 3D whole lung organoids from induced pluripotent stem cells for modeling lung developmental biology and disease. *J Vis Exp*. https://doi.org/10.3791/62456

73. Chen, Y.-W., Huang, S. X., de Carvalho, A. L. R. T., Ho, S.-H., Islam, M. N., Volpi, S., Notarangelo, L. D., Ciancanelli, M., Casanova, J.-L., Bhattacharya, J., Liang, A. F., Palermo, L. M., Porotto, M., Moscona, A., and Snoeck, H.-W. (2017) A three-dimensional model of human lung development and disease from pluripotent stem cells. *Nat Cell Biol.* **19**, 542–549
74. Matkovic Leko, I., Schneider, R. T., Thimraj, T. A., Schrode, N., Beitler, D., Liu, H.-Y., Beaumont, K., Chen, Y.-W., and Snoeck, H.-W. (2023) A distal lung organoid model to study interstitial lung disease, viral infection and human lung development. *Nat Protoc.* https://doi.org/10.1038/s41596-023-00827-6
75. Han, L., Zhao, S., Yu, F., Rong, Z., Lin, Y., and Chen, Y. (2022) Generation of human embryonic stem cell-derived lung organoids. *STAR Protoc.* **3**, 101270
76. Jacob, A., Morley, M., Hawkins, F., McCauley, K. B., Jean, J. C., Heins, H., Na, C.-L., Weaver, T. E., Vedaie, M., Hurley, K., Hinds, A., Russo, S. J., Kook, S., Zacharias, W., Ochs, M., Traber, K., Quinton, L. J., *et al.* (2017) Differentiation of human pluripotent stem cells into functional lung alveolar epithelial cells. *Cell Stem Cell.* **21**, 472–488.e10
77. Yang, W., Li, Y., Shi, F., and Liu, H. (2023) Human lung organoid: Models for respiratory biology and diseases. *Dev Biol.* **494**, 26–34
78. Quake, S. R. (2022) A decade of molecular cell atlases. *Trends Genet.* **38**, 805–810
79. Ardini-Poleske, M. E., Clark, R. F., Ansong, C., Carson, J. P., Corley, R. A., Deutsch, G. H., Hagood, J. S., Kaminski, N., Mariani, T. J., Potter, S. S., Pryhuber, G. S., Warburton, D., Whitsett, J. A., Palmer, S. M., and Ambalavanan, N. (2017) LungMAP: The molecular atlas of lung development program. *Am J Physiol Lung Cell Mol Physiol.* **313**, L733–L740
80. Sun, X., Perl, A.-K., Li, R., Bell, S. M., Sajti, E., Kalinichenko, V. V., Kalin, T. V., Misra, R. S., Deshmukh, H., Clair, G., Kyle, J., Crotty Alexander, L. E., Masso Silva, J., Kitzmiller, J. A., Wikenheiser-Brokamp, K. A., *et al.* (2022) A census of the lung: CellCards from LungMAP. *Dev Cell.* **57**, 112–145.e2
81. Yang, Y., Riccio, P., Schotsaert, M., Mori, M., Lu, J., Lee, D.-K., García-Sastre, A., Xu, J., and Cardoso, W. V. (2018) Spatial-temporal lineage restrictions of embryonic p63+ progenitors establish distinct stem cell pools in adult airways. *Dev Cell.* **44**, 752–761.e4
82. Miller, A. J., Yu, Q., Czerwinski, M., Tsai, Y.-H., Conway, R. F., Wu, A., Holloway, E. M., Walker, T., Glass, I. A., Treutlein, B., Camp, J. G., and Spence, J. R. (2020) In vitro and in vivo development of the human airway at single-cell resolution. *Dev Cell.* **53**, 117–128.e6
83. Sun, D., Llora Batlle, O., van den Ameele, J., Thomas, J. C., He, P., Lim, K., Tang, W., Xu, C., Meyer, K. B., Teichmann, S. A., Marioni, J. C., Jackson, S. P., Brand, A. H., and Rawlins, E. L. (2022) SOX9 maintains human foetal lung tip progenitor state by enhancing WNT and RTK signalling. *EMBO J.* **41**, e111338
84. McCulley, D., Wienhold, M., and Sun, X. (2015) The pulmonary mesenchyme directs lung development. *Curr Opin Genet Dev.* **32**, 98–105
85. Jain, H., Schweitzer, J. W., and Justice, N. A. (2023) Respiratory syncytial virus infection. In *StatPearls*, StatPearls Publishing, Treasure Island, FL [online]. www.ncbi.nlm.nih.gov/books/NBK459215/ (Accessed June 5, 2023)
86. Harford, T. J., Rezaee, F., Dye, B. R., Fan, J., Spence, J. R., and Piedimonte, G. (2022) RSV-induced changes in a 3-dimensional organoid model of human fetal lungs. *PLoS ONE.* **17**, e0265094
87. Jilani, T. N., Jamil, R. T., and Siddiqui, A. H. (2023) H1N1 influenza. In *StatPearls*, StatPearls Publishing, Treasure Island, FL [online]. www.ncbi.nlm.nih.gov/books/NBK513241/ (Accessed June 5, 2023)
88. Hysenaj, L., Little, S., Kulhanek, K., Magnen, M., Bahl, K., Gbenedio, O. M., Prinz, M., Rodriguez, L., Andersen, C., Rao, A. A., Shen, A., Lone, J.-C., Lupin-Jimenez, L. C., Bonser, L. R., Serwas, N. K., Mick, E., *et al.* (2023) SARS-CoV-2 infection of airway organoids reveals conserved use of Tetraspanin-8 by Ancestral, Delta, and Omicron variants. *Stem Cell Rep.* **18**, 636–653
89. Esakandari, H., Nabi-Afjadi, M., Fakkari-Afjadi, J., Farahmandian, N., Miresmaeili, S.-M., and Bahreini, E. (2020) A comprehensive review of COVID-19 characteristics. *Biol Proced Online.* **22**, 19
90. de Oliveira, M., De Sibio, M. T., Costa, F. A. S., and Sakalem, M. E. (2021) Airway and alveoli organoids as valuable research tools in COVID-19. *ACS Biomater Sci Eng.* **7**, 3487–3502
91. Zhu, N., Zhang, D., Wang, W., Li, X., Yang, B., Song, J., Zhao, X., Huang, B., Shi, W., Lu, R., Niu, P., Zhan, F., Ma, X., Wang, D., Xu, W., Wu, G., Gao, G. F., and Tan, W. (2020) A novel coronavirus from patients with pneumonia in China, 2019. *N Engl J Med.* **382**, 727–733

92. Pei, R., Feng, J., Zhang, Y., Sun, H., Li, L., Yang, X., He, J., Xiao, S., Xiong, J., Lin, Y., Wen, K., Zhou, H., Chen, J., Rong, Z., and Chen, X. (2021) Host metabolism dysregulation and cell tropism identification in human airway and alveolar organoids upon SARS-CoV-2 infection. *Protein Cell.* **12**, 717–733
93. Yu, E., and Sharma, S. (2023) Cystic fibrosis. In *StatPearls*, StatPearls Publishing, Treasure Island, FL [online]. www.ncbi.nlm.nih.gov/books/NBK493206/ (Accessed June 5, 2023)
94. Turcios, N. L. (2020) Cystic fibrosis lung disease: An overview. *Respir Care.* **65**, 233–251
95. Mei, Q., Liu, Z., Zuo, H., Yang, Z., and Qu, J. (2022) Idiopathic pulmonary fibrosis: An update on pathogenesis. *Front Pharmacol.* **12**, 797292
96. Surolia, R., Li, F. J., Wang, Z., Li, H., Liu, G., Zhou, Y., Luckhardt, T., Bae, S., Liu, R.-M., Rangarajan, S., de Andrade, J., Thannickal, V. J., and Antony, V. B. (2017) 3D pulmospheres serve as a personalized and predictive multicellular model for assessment of antifibrotic drugs. *JCI Insight.* **2**, e91377
97. Kim, J.-H., An, G. H., Kim, J.-Y., Rasaei, R., Kim, W. J., Jin, X., Woo, D.-H., Han, C., Yang, S.-R., Kim, J.-H., and Hong, S.-H. (2021) Human pluripotent stem cell-derived alveolar organoids for modeling pulmonary fibrosis and drug testing. *Cell Death Discov.* **7**, 1–12
98. Yadav, S., Lee, B., and Kamity, R. (2023) Neonatal respiratory distress syndrome. In *StatPearls*, StatPearls Publishing, Treasure Island, FL [online]. www.ncbi.nlm.nih.gov/books/NBK560779/ (Accessed June 5, 2023)
99. Jo, H. S. (2014) Genetic risk factors associated with respiratory distress syndrome. *Korean J Pediatr.* **57**, 157–163
100. Mehta, P. A., and Sharma, G. (2023) Congenital pulmonary airway malformation. In *StatPearls*, StatPearls Publishing, Treasure Island, FL [online]. www.ncbi.nlm.nih.gov/books/NBK551664/ (Accessed June 5, 2023)
101. Swarr, D. T., Peranteau, W. H., Pogoriler, J., Frank, D. B., Adzick, N. S., Hedrick, H. L., Morley, M., Zhou, S., and Morrisey, E. E. (2018) Novel molecular and phenotypic insights into congenital lung malformations. *Am J Respir Crit Care Med.* **197**, 1328–1339
102. Thekkeveedu, R. K., Guaman, M. C., and Shivanna, B. (2017) Bronchopulmonary dysplasia: A review of pathogenesis and pathophysiology. *Respir Med.* **132**, 170–177
103. Sucre, J. M. S., Wilkinson, D., Vijayaraj, P., Paul, M., Dunn, B., Alva-Ornelas, J. A., and Gomperts, B. N. (2016) A three-dimensional human model of the fibroblast activation that accompanies bronchopulmonary dysplasia identifies Notch-mediated pathophysiology. *Am J Physiol Lung Cell Mol Physiol.* **310**, L889–L898
104. Kozlowski, M. T., Crook, C. J., and Ku, H. T. (2021) Towards organoid culture without Matrigel. *Commun Biol.* **4**, 1–15
105. Zhao, Z., Chen, X., Dowbaj, A. M., Sljukic, A., Bratlie, K., Lin, L., Fong, E. L. S., Balachander, G. M., Chen, Z., Soragni, A., Huch, M., Zeng, Y. A., Wang, Q., and Yu, H. (2022) Organoids. *Nat Rev Methods Primers.* **2**, 1–21
106. Cruz-Acuña, R., Quirós, M., Farkas, A. E., Dedhia, P. H., Huang, S., Siuda, D., García-Hernández, V., Miller, A. J., Spence, J. R., Nusrat, A., and García, A. J. (2017) Synthetic hydrogels for human intestinal organoid generation and colonic wound repair. *Nat Cell Biol.* **19**, 1326–1335
107. Leeman, K. T., Pessina, P., Lee, J.-H., and Kim, C. F. (2019) Mesenchymal stem cells increase alveolar differentiation in lung progenitor organoid cultures. *Sci Rep.* **9**, 6479
108. Tamai, K., Sakai, K., Yamaki, H., Moriguchi, K., Igura, K., Maehana, S., Suezawa, T., Takehara, K., Hagiwara, M., Hirai, T., and Gotoh, S. (2022) iPSC-derived mesenchymal cells that support alveolar organoid development. *Cell Rep Methods.* **2**, 100314
109. Grebenyuk, S., and Ranga, A. (2019) Engineering organoid vascularization. *Frontiers in Bioengineering and Biotechnology* [online]. www.frontiersin.org/articles/10.3389/fbioe.2019.00039 (Accessed June 4, 2023)
110. Wörsdörfer, P., Dalda, N., Kern, A., Krüger, S., Wagner, N., Kwok, C. K., Henke, E., and Ergün, S. (2019) Generation of complex human organoid models including vascular networks by incorporation of mesodermal progenitor cells. *Sci Rep.* **9**, 15663
111. Vargas-Valderrama, A., Messina, A., Mitjavila-Garcia, M. T., and Guenou, H. (2020) The endothelium, a key actor in organ development and hPSC-derived organoid vascularization. *J Biomed Sci.* **27**, 67
112. Pham, M. T., Pollock, K. M., Rose, M. D., Cary, W. A., Stewart, H. R., Zhou, P., Nolta, J. A., and Waldau, B. (2018) Generation of human vascularized brain organoids. *Neuroreport.* **29**, 588–593
113. Lee, J.-H., Bhang, D. H., Beede, A., Huang, T. L., Stripp, B. R., Bloch, K. D., Wagers, A. J., Tseng, Y.-H., Ryeom, S., and Kim, C. F. (2014) Lung stem cell differentiation in mice directed by endothelial cells via a BMP4-NFATc1-thrombospondin-1 axis. *Cell.* **156**, 440–455
114. Park, S. E., Georgescu, A., and Huh, D. (2019) Organoids-on-a-chip. *Science.* **364**, 960–965

115. Zhang, S., Wan, Z., and Kamm, R. D. (2021) Vascularized organoids on a chip: Strategies for engineering organoids with functional vasculature. *Lab Chip*. **21**, 473–488
116. Magré, L., Verstegen, M. M. A., Buschow, S., Laan, L. J. W. van der, Peppelenbosch, M., and Desai, J. (2023) Emerging organoid-immune co-culture models for cancer research: From oncoimmunology to personalized immunotherapies. *J Immunother Cancer*. **11**, e006290
117. Cattaneo, C. M., Dijkstra, K. K., Fanchi, L. F., Kelderman, S., Kaing, S., van Rooij, N., van den Brink, S., Schumacher, T. N., and Voest, E. E. (2020) Tumor organoid—T cell co-culture systems. *Nat Protoc*. **15**, 15–39
118. Seo, H.-R., Han, H.-J., Lee, Y., Noh, Y.-W., Cho, S.-J., and Kim, J.-H. (2022) Human pluripotent stem cell-derived alveolar organoid with macrophages. *Int J Mol Sci*. **23**, 9211
119. Flecknoe, S., Harding, R., Maritz, G., and Hooper, S. B. (2000) Increased lung expansion alters the proportions of type I and type II alveolar epithelial cells in fetal sheep. *Am J Physiol Lung Cell Mol Physiol*. **278**, L1180–L1185
120. Higuchi, M., Kato, T., Yoshino, H., Matsuda, K., Gotoh, K., Hirano, H., Koyama, K., and Maki, M. (1991) The influence of experimentally produced oligohydramnios on lung growth and pulmonary surfactant content in fetal rabbits. *J Dev Physiol*. **16**, 223–227
121. Sanchez-Esteban, J., Cicchiello, L. A., Wang, Y., Tsai, S. W., Williams, L. K., Torday, J. S., and Rubin, L. P. (2001) Mechanical stretch promotes alveolar epithelial type II cell differentiation. *J Appl Physiol (1985)*. **91**, 589–595
122. Kanton, S., and Paşca, S. P. (2022) Human assembloids. *Development*. **149**, dev201120
123. Andersen, J., Revah, O., Miura, Y., Thom, N., Amin, N. D., Kelley, K. W., Singh, M., Chen, X., Thete, M. V., Walczak, E. M., Vogel, H., Fan, H. C., and Paşca, S. P. (2020) Generation of functional human 3D cortico-motor assembloids. *Cell*. **183**, 1913–1929.e26
124. Rockel, A. F., Wagner, N., Spenger, P., Ergün, S., and Wörsdörfer, P. (2023) Neuro-mesodermal assembloids recapitulate aspects of peripheral nervous system development in vitro. *Stem Cell Rep*. **18**, 1155–1165
125. Koike, H., Iwasawa, K., Ouchi, R., Maezawa, M., Giesbrecht, K., Saiki, N., Ferguson, A., Kimura, M., Thompson, W. L., Wells, J. M., Zorn, A. M., and Takebe, T. (2019) Modelling human hepato-biliary-pancreatic organogenesis from the foregut—midgut boundary. *Nature*. **574**, 112–116
126. Koike, H., Iwasawa, K., Ouchi, R., Maezawa, M., Kimura, M., Kodaka, A., Nishii, S., Thompson, W. L., and Takebe, T. (2021) Engineering human hepato-biliary-pancreatic organoids from pluripotent stem cells. *Nat Protoc*. **16**, 919–936
127. Ng, W. H., Johnston, E. K., Tan, J. J., Bliley, J. M., Feinberg, A. W., Stolz, D. B., Sun, M., Wijesekara, P., Hawkins, F., Kotton, D. N., and Ren, X. (2022) Recapitulating human cardio-pulmonary co-development using simultaneous multilineage differentiation of pluripotent stem cells. *eLife*. **11**, e67872

7 Establishment of Animal Models of Pulmonary Fibrosis

Maolin Yao, Mo Chen, and Lian Li

7.1 INTRODUCTION

Pulmonary fibrosis (PF) is a component of interstitial lung diseases (ILDs), which is characterized clinically by persistent dry cough, progressive dyspnea, restricted ventilation dysfunction, hypoxemia, and respiratory failure (1). The genesis of pulmonary fibrosis may be attributed to an array of known pulmonary insults such as autoimmune diseases, drugs, radiation exposure, infectious or traumatic injuries, genetic variants, and occupational or environmental exposure to dusts and fibers (e.g., silica, asbestos, coal) (2). In a multitude of instances, the pulmonary fibrotic remodeling presents itself as idiopathic, and its etiology remaining unknown. This type of pulmonary fibrosis is dubbed idiopathic pulmonary fibrosis (IPF), which is categorized as idiopathic interstitial pneumonias (3). IPF is a chronic, progressive, and usually fatal fibrotic interstitial lung disease that is characterized by a distortion of the pulmonary architecture and an accumulation of extracellular matrix (ECM) (4). IPF patients have a poor prognosis with a median survival of 3–5 years after diagnosis, even as pirfenidone and nintedanib have exhibited the potential to retard the progression of IPF in clinical trials (5). To better understand the underlying mechanisms and identify more efficient clinical therapeutic targets for pulmonary fibrosis, establishment of appropriate experimental models is the first and most important consideration.

To date, multiple animal models of pulmonary fibrosis have been reported, mainly including cytokine (TGF-β, TGF-α, IL-13, TNF-α, IL-1β) overexpression models (6–10), targeted type II alveolar epithelial cell (AEC) injury models (11), familial IPF (SFTPC, SFTPA2, TERT, and TERC) models (12–14), models induced by environmental exposure (silica, asbestos, radiation) (15–17), drug/toxic (bleomycin, fluorescein isothiocyanate [FITC], acid, paraquat) injury models (18, 19), age-related models (20), and humanized models (21). Although no single animal model fully recapitulates all features of progressive pulmonary fibrosis until now, each of these models has successfully reproduced portions of patient manifestations, thereby providing invaluable insights into disease pathobiology. Among currently used animal models of pulmonary fibrosis, the bleomycin-induced model is the best characterized and the most prevalently used murine model (22). The silica-induced model, mimicking silicosis, is frequently employed to investigate the innate immune regulation of pulmonary fibrosis (23).

7.1.1 Bleomycin-Induced Model

Bleomycin, a multicomponent complex of basic glycopeptide antibiotics produced by *Streptomyces verticillus*, has traditionally been used as an antineoplastic drug for the treatment of various carcinomas and lymphomas (24). It has been employed for establishment of experimental models of pulmonary fibrosis for over four decades. Mechanism of bleomycin of action involves the induction of single- and double-stranded DNA breaks, thereby triggering the production of free radicals and sparking off oxidative stress, which contributes to pulmonary injury, inflammation, and the subsequent onset of interstitial fibrosis (25).

Species, genetics, and gender: Bleomycin has been used to induce pulmonary fibrosis in different animal species, including mice, rats, hamsters, guinea pigs, rabbits, dogs, and primates, but

 DOI: 10.1201/9781003355243-9

mice are the most commonly used (26). They are relatively less expensive associated with purchase and husbandry, and there are many available transgenic strains. The fibrotic response to bleomycin in mice is strain dependent, with C57BL/6, DBA/2, and 129 strains being more susceptible than BALB/c, white A/J, and C3Hf/Kam strains (27). Additionally, the severity of bleomycin-induced pulmonary fibrosis varies by gender difference. Male mice are susceptible than females to bleomycin-induced injury (28).

Route of bleomycin delivery: Bleomycin have been delivered by several different methods, including intravenous, intraperitoneal, subcutaneous, intratracheal, and inhalational. Among them, intratracheal instillation is the most prevalent method at present, further subdivided into traumatic administration of the trachea, trans-oral administration, and intranasal instillation (Figure 7.1). When bleomycin is delivered systemically (intravenous, intraperitoneal, or subcutaneous), the pulmonary vascular endothelium is the initial site of injury, succeeded by alveolar epithelial cell (AEC) damage, thereby simulating the process of pulmonary toxicity experienced by humans undergoing bleomycin chemotherapy. Intratracheal administration of bleomycin to the lung (generally 1.25–4 U/kg) is the most prevalent route, which produces many variable fibroblastic foci and better reflects human pathology with areas of subpleural fibrosis. Intratracheal administration results in direct injury initially to AECs. This event is followed by release of proinflammatory cytokines and chemokines and the influx of inflammatory cells. Peak inflammation occurs in the first week post-bleomycin administration (Figure 7.2). Subsequently, inflammatory response gradually subsides accompanying with proliferation of fibroblast, emergence of myofibroblast, and synthesis of ECM (29). Following a single intratracheal or trans-oral administration, the development of pulmonary fibrosis can be histologically and biochemically observed by the second week, with maximal fibrotic response typically noted around weeks 3–4 (Figure 7.3). Certain studies demonstrate fibrotic response can resolve between 6 and 8 weeks post-challenge (30).

Dose selection: Notable variances across bleomycin manufacturers and batches significantly influence the potency of bleomycin-induced pulmonary fibrosis, thereby accounting, in part, for the diverse doses observed in literature for the establishment of fibrosis models. Consequently, it becomes incumbent to ascertain the optimal dose for each bleomycin batch before embarking on research to ensure the desired outcome. From the existing literature, we have garnered the reported bleomycin administration doses for distinctive mouse strains. For CBA/J mice, the intratracheal administration dose is 1 U/kg; for C57BL/6 mice, it is 2 U/kg; and for BALB/c mice, it escalates

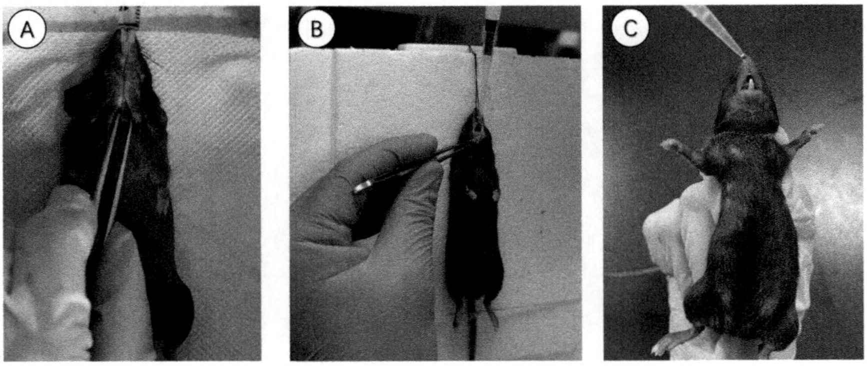

FIGURE 7.1 Intratracheal administration of bleomycin. (A) The trachea traumatic administration of bleomycin. The mice are anesthetized with avertin, and a vertical incision is made on the front of the neck. A needle is inserted into the trachea and bleomycin is injected into the lung. (B) Trans-oral administration of bleomycin. The tongue was controlled with sterile forceps and bleomycin was dripped orally with a pipette. (C) Intranasal instillation of bleomycin. Bleomycin is slowly dripped into the nose using a pipette.

Source: Photograph by Li laboratory.

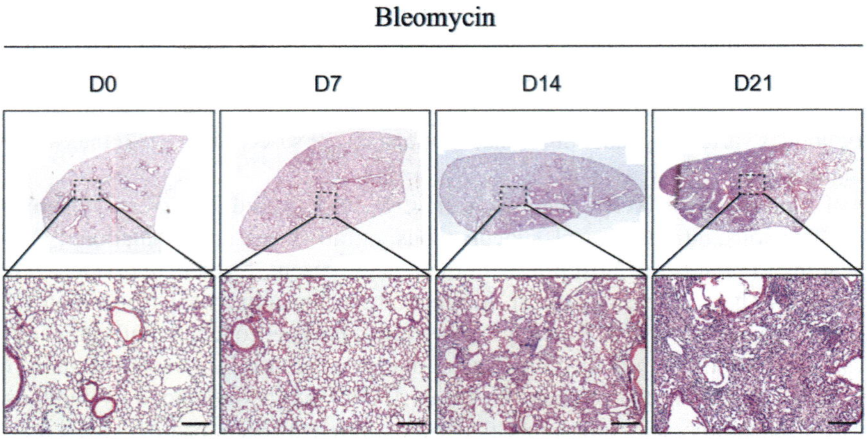

FIGURE 7.2 Lung histopathology after a single dose of bleomycin treatment. The mice (C57BL/6, 8 weeks old) receive an intratracheal instillation of bleomycin (2 U/kg) on day 0. Subsequently, the mice are sacrificed at intervals of 0, 7, 14, and 21 days post-bleomycin instillation. Representative H&E-stained lung tissue sections are showed. Bars, 100 μm.

Source: Photograph by Li laboratory.

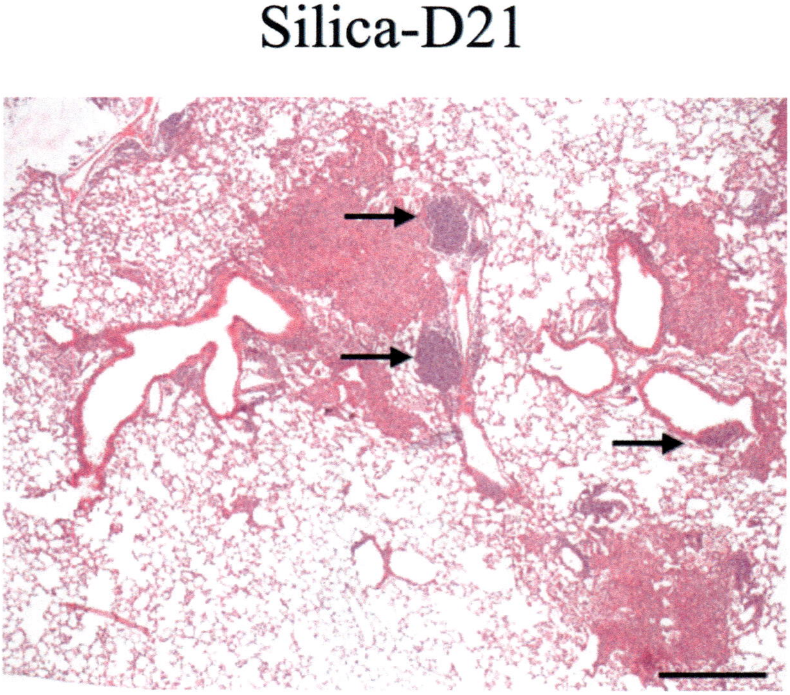

FIGURE 7.3 H&E staining of lung sections of mice 21 days after silica (200 mg/Kg) injury. Representative images of the staining are shown. Arrows show inflammatory cells. Bars, 100 μm.

Source: Photograph by Li laboratory.

to 10 U/kg (31). At these specified dosages, mortality rates in mice remain minimal, whereas a noticeable elevation in doses can precipitate significant fatalities, typically occurring during the acute injury phase and before the full-blown manifestation of chronic fibrosis. Regarding systemic administration, the dose escalates by 10- to 20-fold, and for repeated administrations, the dose generally equates to half of the single-dose regimen.

7.1.2 Silica-Induced Model

Silica particles measuring less than 5 μm are defined as "respirable" because of their capacity to penetrate distal airways and alveoli (32). Crystalline (quartz) silica particles, recognized as the etiological agents in occupational respiratory disease (silicosis), exhibit greater toxicity than their amorphous (diatom) silica counterparts (33). Instillation of crystalline silica into the lungs of rodents induces fibrous nodules (Figure 7.3) that closely simulate the development of silicosis seen in certain occupationally exposed populations. Silica is known to induce the activation of macrophage NLRP3, thus serving as an important model for studying innate immune modulation of pulmonary fibrosis.

Crystalline silica can be delivered to rodents predominantly via intratracheal administration and aerosol inhalation (34). Baking of silica is indispensable step before in vivo instillation to remove endotoxin and water. Intratracheal administration models typically take 14–28 days to develop, whereas aerosol inhalation models require 40–120 days and specialized equipment (35). The intratracheal administration models are easier, more cost-efficient, and often used. Low doses of silica can be effectively removed from the lung, thus minimizing toxicity to cells. In contrast, high doses of silica are retained in the lung, leading to a persistent inflammatory response and the progression of pulmonary fibrosis (36). The fibrotic response to silica in mice is strain dependent, with C3H/HeN, NZB, and MRL/MpJ strains being more susceptible than BALB/c strain in aerosol inhalation models (37). In intratracheal administration models, C57BL/6 mice are more susceptible than CBA/J mice (38).

The details on design, methodologies, and protocols for establishing bleomycin- or silica-induced mouse models of pulmonary fibrosis will be presented in the following sections.

7.2 MATERIALS

7.2.1 Animals and Reagents

1. Animals: C57BL/6J mouse (6–8 weeks old).
2. Anesthetic drug: 100% stock concentration of avertin is made by dissolving 10 g 2,2,2-tribromoethanol (Sigma-Aldrich, T48402) into 10 mL tertiary amyl ethanol. The stock solution is further diluted to 2.5% working solution with 0.9% sterile saline. Note: The working solution should be stored at 4°C without light.
3. Bleomycin: Bleomycin is purchased from Nippon Kayaku company (Nippon Kayaku Co., Ltd., Tokyo, Japan) in our lab and diluted in sterile saline to 5 U stock solution. Notes: Bleomycin should be kept on ice until administration. The bleomycin stock should be stored in single-use aliquots at 4°C.
4. Silica: The ultrafine silicon powder is purchased from US SILICA (MIN-U-SIL 5, Mill Creek, OK, USA). The content of silica is over 99.2%, and the median diameter of particle is 1.4 μm. The silica particles are baked in an oven at 180°C for 2 hours, then weighed and suspended in phosphate-buffered saline (PBS) with a final concentration of 200 mg/mL. The suspensions should be sonicated for 30 minutes and vortexed for 5 seconds before use.

7.3 METHODS

7.3.1 Intratracheal Instillation of Bleomycin or Silica

1. Age-, gender-, and weight-matched mice are randomly divided into bleomycin- or silica-treated and control groups.
2. The mice are weighed and recorded. Bleomycin or silica suspended solutions should be prepared according to body weight.
3. The mice are anesthetized by intraperitoneal injection of 100–200 μL/10 g 2.5% avertin working solution.
4. Once anesthesia is adequate, the mouse is placed on the operating board, and the surgical area is swabbed by cotton soaked in 75% ethanol. The skin on the front of mouse neck is cut a vertical incision about 1 cm length, and the mouse trachea is exposed.
5. Bleomycin (2–2.5 U/kg) or silica (200 mg/kg) is administered intratracheally by using a 0.3-mL syringe with a 31G needle (bleomycin) or 22G needle (silica) inserted between the cartilaginous rings of the trachea. The control group receives an equal volume of saline or PBS via the same approach. The entire procedure should take less than 5 minutes. Note: Air (50 μL) is sucked into the syringe before taking bleomycin, which contributes to pushing the bleomycin liquid out of the syringe.
6. The wound is closed with a 9-mm wound clip. The mice are allowed to recover from anesthesia by placing under a warming lamp and monitoring them until fully awake.
7. The mice are observed daily, and body weight and survival may be recorded.

7.4 ASSESSMENT OF PULMONARY FIBROSIS

7.4.1 H&E Staining

1. The trachea is cannulated, and the left lung is inflated with about 0.4 mL of 10% neutral buffered formalin.
2. The lungs are then fixed overnight with 10% neutral buffered formalin.
3. The lungs are dehydrated in 30%, 50%, 70%, 95% (twice), and 100% (twice) ethanol for 1 hour, treated with xylene for 10 minutes twice, and embedded in paraffin.
4. The lungs are cut into 5-μm thick slides.
5. The slides are deparaffinized with xylene for 10 minutes twice and washed in 100% (twice), 95% (twice), 85%, 70% ethanol, and distilled water for 10 minutes each.
6. The slides are stained in Mayer's hematoxylin (G1080, Solarbio) for 3 minutes, followed by a 10-minute rinse in water.
7. The slides are pretreated with 95% ethanol twice and then stained in eosin solution (G1120, Solarbio) for 30 seconds.
8. The slides are dehydrated with 95% and 100% ethanol twice, each time for 2 minutes, and treated with xylene twice, each for 5 minutes. Finally, the slides are mounted with neutral balsam (G8590, Solarbio).

7.4.2 Hydroxyproline Assay

The collagen content of the right lung is measured with a conventional hydroxyproline assay.

1. The lungs are washed with PBS, and then dried in an oven at 120°C overnight.
2. The dry weight of each lung is recorded.
3. The lung tissues in 3 mL 6 N HCl are heated at 120°C for 16 hours for acid hydrolysis.
4. The samples are cooled down. The pH of samples is adjusted by using 10 N NaOH to 6.5–8.0.

5. Filter out particles through a syringe filter (Millipore, 5-μm pore size) and make a total of 10 mL for each lung sample with PBS. Note: At this step, the sample can be stored at 4°C for later hydroxyproline assay.
6. Make a series of hydroxyproline standard dilutions (stock 100 μg/mL) of 10 μg/mL, 5 μg/mL, 2.5 μg/mL, 1.25 μg/mL, 0.625 μg/mL, and 0 μg/mL.
7. Add a 50-μL sample with 350 μL H_2O into each tube; add this 400 μL standard into each standard tube.
8. Add 200 μL chloramine T into each tube; mix and incubate at room temperature for 20 minutes.
9. Add 200 μL perchloric acid into each tube; mix and incubate at room temperature for 5 minutes.
10. Add 200 μL p-dimethylaminobenzaldehyde into each tube; mix and incubate at 60°C in a water bath for 20 minutes.
11. Cool down the samples; transfer 200 μL of each sample into 96-well plates in triplicate.
12. Determine the hydroxyproline content at 557 nm with microplate reader.

ABBREVIATIONS

AEC alveolar epithelial cell
ECM extracellular matrix
ILD interstitial lung disease
IPF idiopathic pulmonary fibrosis
PBS phosphate-buffered saline
PF pulmonary fibrosis

REFERENCES

1. Schafer, M. J., White, T. A., Iijima, K., Haak, A. J., Ligresti, G., Atkinson, E. J., Oberg, A. L., Birch, J., Salmonowicz, H., Zhu, Y., Mazula, D. L., Brooks, R. W., Fuhrmann-stroissnigg, H., Pirtskhalava, T., Prakash, Y. S., Tchkonia, T., Robbins, P. D., Passos, F., Kirkland, J. L., Tschumperlin, D. J., Kita, H., Aubry, M. C., and Lebrasseur, N. K. (2017) Cellular senescence mediates fibrotic pulmonary disease. *Nat Commun.* **23**, 14532
2. Ogawa, T., Shichino, S., Ueha, S., and Matsushima, K. (2021) Macrophages in lung fibrosis. *Int Immunol.* **33**, 665–671
3. Olson, A., Hartmann, N., Patnaik, P., and Wallace, L. (2021) Estimation of the prevalence of progressive fibrosing interstitial lung diseases: Systematic literature review and data from a physician survey. *Adv Ther.* **38**, 854–867
4. Raghu, G., Collard, H. R., Egan, J. J., Martinez, F. J., Behr, J., Brown, K. K., Colby, T. V., Flaherty, K. R., Lasky, J. A., Lynch, D. A., Ryu, J. H., Swigris, J. J., Wells, A. U., Ancochea, J., Bouros, D., Carvalho, C., Costabel, U., Ebina, M., Hansell, D. M., Johkoh, T., Kim, D. S., King, T. E., Kondoh, Y., Dudden, R. F., Nicholson, A. G., Richeldi, L., Myers, J., and Mu, N. L. (2011) An official ATS/ERS/JRS/ALAT statement: Idiopathic pulmonary fibrosis: Evidence-based guidelines for diagnosis and management. *Am J Respir Crit Care Med.* **183**, 788–824
5. Glaspole, I., Ph, D., Glassberg, M. K., Gorina, E., Hopkins, P. M., Kardatzke, D., Ph, D., Lancaster, L., Lederer, D. J., Nathan, S. D., Pereira, C. A., Sahn, S. A., Sussman, R., Swigris, J. J., Noble, P. W., and Study, A. (2014) A phase 3 trial of pirfenidone in patients with idiopathic pulmonary fibrosis. *N Engl J Med.* **370**, 2083–2092
6. Fernandez, I. E., and Eickelberg, O. (2012) The impact of TGF-β on lung fibrosis: From targeting to biomarkers. *Proc Am Thorac Soc.* **9**, 111–116
7. Baughman, R. P., Lower, E. E., Miller, M. A., Bejarano, P. A., and Heffelfinger, S. C. (1999) Overexpression of transforming growth factor-alpha and epidermal growth factor-receptor in idiopathic pulmonary fibrosis. *Sarcoidosis Vasc Diffuse Lung Dis.* **16**, 57–61
8. Sime, P. J., Marr, R. A., Gauldie, D., Xing, Z., Hewlett, B. R., Graham, F. L., and Gauldie, J. (1998) Transfer of tumor necrosis factor-alpha to rat lung induces severe pulmonary inflammation and patchy interstitial fibrogenesis with induction of transforming growth factor-beta1 and myofibroblasts. *Am J Pathol.* **153**, 825–832

9. Kolb, M., Margetts, P. J., Anthony, D. C., Pitossi, F., and Gauldie, J. (2001) Transient expression of IL-1beta induces acute lung injury and chronic repair leading to pulmonary fibrosis. *J Clin Invest.* **107**, 1529–1536
10. Lee, C. G., Homer, R. J., Zhu, Z., Lanone, S., Wang, X., Koteliansky, V., Shipley, J. M., Gotwals, P., Noble, P., Chen, Q., Senior, R. M., and Elias, J. A. (2001) Interleukin-13 induces tissue fibrosis by selectively stimulating and activating transforming growth factor beta(1). *J Exp Med.* **194**, 809–821
11. Sisson, T. H., Mendez, M., Choi, K., Subbotina, N., Courey, A., Cunningham, A., Dave, A., Engelhardt, J. F., Liu, X., White, E. S., Thannickal, V. J., Moore, B. B., Christensen, P. J., and Simon, R. H. (2010) Targeted injury of type II alveolar epithelial cells induces pulmonary fibrosis. *Am J Respir Crit Care Med.* **181**, 254–263
12. Wang, Y., Kuan, P. J., Xing, C., Cronkhite, J. T., Torres, F., Rosenblatt, R. L., DiMaio, J. M., Kinch, L. N., Grishin, N. V., and Garcia, C. K. (2009) Genetic defects in surfactant protein A2 are associated with pulmonary fibrosis and lung cancer. *Am J Hum Genet.* **84**, 52–59
13. Thomas, A. Q., Lane, K., Phillips, J., Prince, M., Markin, C., Speer, M., Schwartz, D. A., Gaddipati, R., Marney, A., Johnson, J., Roberts, R., Haines, J., Stahlman, M., and Loyd, J. E. (2002) Heterozygosity for a surfactant protein C gene mutation associated with usual interstitial pneumonitis and cellular nonspecific interstitial pneumonitis in one kindred. *Am J Respir Crit Care Med.* **165**, 1322–1328
14. Armanios, M. Y., Chen, J. J.-L., Cogan, J. D., Alder, J. K., Ingersoll, R. G., Markin, C., Lawson, W. E., Xie, M., Vulto, I., Phillips, J. A., Lansdorp, P. M., Greider, C. W., and Loyd, J. E. (2007) Telomerase mutations in families with idiopathic pulmonary fibrosis. *N Engl J Med.* **356**, 1317–1326
15. Karvonen, R. L., Fernandez-Madrid, F., Maughan, R. L., Palmer, K. C., and Fernandez-Madrid, I. (1987) An animal model of pulmonary radiation fibrosis with biochemical, physiologic, immunologic, and morphologic observations. *Radiat Res.* **111**, 68–80
16. Roggli, V. L., Gibbs, A. R., Attanoos, R., Churg, A., Popper, H., Cagle, P., Corrin, B., Franks, T. J., Galateau-Salle, F., Galvin, J., Hasleton, P. S., Henderson, D. W., and Honma, K. (2010) Pathology of asbestosis- An update of the diagnostic criteria: Report of the asbestosis committee of the college of american pathologists and pulmonary pathology society. *Arch Pathol Lab Med.* **134**, 462–480
17. Oberdörster, G. (1996) Significance of particle parameters in the evaluation of exposure-dose-response relationships of inhaled particles. *Inhal Toxicol.* **8**, 73–89
18. Moore, B. B., and Hogaboam, C. M. (2008) Murine models of pulmonary fibrosis. *Am J Physiol Lung Cell Mol Physiol.* **294**, L152–L160
19. Roberts, S. N., Howie, S. E. M., Wallace, W. A. H., Brown, D. M., Lamb, D., Ramage, E. A., and Donaldson, K. (1995) A novel model for human interstitial lung disease: Hapten-driven lung fibrosis in rodents. *J Pathol.* **176**, 309–318
20. Sueblinvong, V., Neujahr, D. C., Todd Mills, S., Roser-Page, S., Ritzenthaler, J. D., Guidot, D., Rojas, M., and Roman, J. (2012) Predisposition for disrepair in the aged lung. *Am J Med Sci.* **344**, 41–51
21. Pierce, E. M., Carpenter, K., Jakubzick, C., Kunkel, S. L., Flaherty, K. R., Martinez, F. J., and Hogaboam, C. M. (2007) Therapeutic targeting of CC ligand 21 or CC chemokine receptor 7 abrogates pulmonary fibrosis induced by the adoptive transfer of human pulmonary fibroblasts to immunodeficient mice. *Am J Pathol.* **170**, 1152–1164
22. Sleijfer, S. (2001) Bleomycin-induced pneumonitis. *Chest.* **120**, 617–624
23. Barbarin, V., Nihoul, A., Misson, P., Arras, M., Delos, M., Leclercq, I., Lison, D., and Huaux, F. (2005) The role of pro- and anti-inflammatory responses in silica-induced lung fibrosis. *Respir Res.* **6**, 112–125
24. Chen, J., and Stubbe, J. A. (2005) Bleomycins: Towards better therapeutics. *Nat Rev Cancer.* **5**, 102–112
25. Tounekti, O., Kenani, A., Foray, N., and Orlowski, S. (2001) The ratio of single- to double-strand DNA breaks and their absolute values determine cell death pathway. *Br J Cancer.* **84**, 1272–1279
26. Zhang, X., Jiang, L., Zhang, W., Wu, J., and Lu, X. (2008) Effect of Feixian recipe on laminin, collagen I and III in rats with pulmonary fibrosis induced by bleomycin. *Front Med.* **2**, 314–316
27. Walkin, L., Herrick, S. E., Summers, A., Brenchley, P. E., Hoff, C. M., Korstanje, R., and Margetts, P. J. (2013) The role of mouse strain differences in the susceptibility to fibrosis: A systematic review. *Fibrogenesis Tissue Repair.* **6**, 18–30
28. Gharaee-Kermani, M., Hatano, K., Nozaki, Y., and Phan, S. H. (2005) Gender-based differences in bleomycin-induced pulmonary fibrosis. *Am J Pathol.* **166**, 1593–1606
29. Chaudhary, N. I., Schnapp, A., and Park, J. E. (2012) Pharmacologic differentiation of inflammation and fibrosis in the rat bleomycin model. *Am J Respir Crit Care Med.* **173**, 769–776

30. Lawson, W. E., Polosukhin, V. V., Stathopoulos, G. T., Zoia, O., Han, W., Lane, K. B., Li, B., Donnelly, E. F., Holburn, G. E., Lewis, K. G., Collins, R. D., Hull, W. M., Glasser, S. W., Whitsett, J. A., and Blackwell, T. S. (2005) Increased and prolonged pulmonary fibrosis in surfactant protein C-deficient mice following intratracheal bleomycin. *Am J Pathol.* **167**, 1267–1277
31. Liu, T., Jin, H., Ullenbruch, M., Hu, B., Hashimoto, N., Moore, B., McKenzie, A., Lukacs, N. W., and Phan, S. H. (2004) Regulation of found in inflammatory zone 1 expression in bleomycin-induced lung fibrosis: Role of IL-4/IL-13 and mediation via STAT-6. *J Immunol.* **173**, 3425–3431
32. Căluțu, I. M., Smărăndescu, R. A., and Rașcu, A. (2022) Biomonitoring exposure and early diagnosis in silicosis: A comprehensive review of the current literature. *Biomedicines.* **11**, 100–117
33. Li, T., Yang, X., Xu, H., and Liu, H. (2022) Early identification, accurate diagnosis, and treatment of silicosis. *Can Respir J.* **25**, 3769134
34. Cassel, S. L., Eisenbarth, S. C., Iyer, S. S., Sadler, J. J., and Colegio, O. R. (2008) The Nalp3 inflammasome is essential for the development of silicosis. *Proc Natl Acad Sci U S A.* **105**, 9035–9040
35. Mercer, P. F., Abbott-banner, K., Adcock, I. M., and Knowles, R. G. (2015) Translational models of lung disease. *Clin Sci.* **128**, 235–256
36. Vorster, T., Mthombeni, J., and Phillips, J. I. (2022) The association between the histological subtypes of mesothelioma and asbestos exposure characteristics. *Int J Environ Res Public Health.* **19**, 14520
37. Davis, G. S., Leslie, K. O., and Hemenway, D. R. (1998) Silicosis in mice: Effects of dose, time, and genetic strain. *J Environ Pathol Toxicol Oncol.* **17**, 81–97
38. Ohtsuka, Y., Wang, X. T., Saito, J., Ishida, T., and Munakata, M. (2006) Genetic linkage analysis of pulmonary fibrotic response to silica in mice. *Eur Respir J.* **28**, 1013–1019

8 Research Strategies for Investigating Protein Ubiquitination in Lung Diseases

Yutong Zhao

8.1 INTRODUCTION

Ubiquitination is ubiquitin protein(s) or polyubiquitin chain(s) covalently conjugated to lysine (K) residues of target protein to regulate protein stability, cellular localization, and protein-protein interaction (1–7). Ubiquitination-dependent protein degradation has been well illustrated. Ubiquitination triggers protein degradation in proteasome or lysosome systems. Ubiquitin contains seven K residues (K6, K11, K27, K289, K33, K48, and K63). Polyubiquitin chains can be assembled using these K residues and the first methionine (Met1) residue (2–4, 7). In general, among these, K11-, K29- and K48-linked polyubiquitination is associated with proteasomal degradation, whereas mono- and K63-linked polyubiquitination regulates protein cellular translocation, protein-protein interaction, enzyme activity, and degradation in lysosomes (2–5, 8, 9). K63-linked polyubiquitination has been shown to increase protein stability possibly through counteracting its K48-linked polyubiquitination (10, 11).

A series of enzymatic reactions is necessary for ubiquitination. E1 ubiquitin-activating enzymes, E2 ubiquitin-conjugating enzymes, and E3 ubiquitin-protein ligases work concertedly to transfer ubiquitin to the target proteins to yield mono- or polyubiquitination. E3 ligases selectively target substrates and transfer ubiquitin or ubiquitin chains to the substrates (12–16). Each E3 ligase targets multiple substrates to regulate cellular functions. Each protein may be targeted by different E3 ligases to form different types of ubiquitination, resulting in distinct functions. The association of E3 ligases and substrates is dependent on other Post-translational modifications (PTMs) of substrates (17). For instance, phosphorylation of substrates may increase E3 ligase association with substrates (18–23). Proteasomal degradation of the inhibitor of nuclear factor kappa B (I-κB) is mediated by the Skp1-cullin 1-F-box (SCF) β-Trcp E3 ligase in a phosphorylation-dependent manner (23). In response to inflammatory stimuli, phosphorylated I-κB binds to SCF β-Trcp E3 ligase and thereby is polyubiquitinated (23). Ubiquitination is a reversible process. Deubiquitinating enzymes (DUBs) remove mono-ubiquitin or polyubiquitin chains from substrates (24–27). The balance between the substrate association with its E3 ligase and DUB determines the outcome of the substrate. The shift of substrate binding to E3 ligase or DUB may be regulated by other induced PTMs such as phosphorylation.

Lung diseases impose a huge global health burden. Since 2020, nearly 7 million deaths have been caused by COVID-19 worldwide. Disfunction of lung cells, including epithelial, endothelial, and fibroblasts, contributes to the pathogenesis of lung diseases. Several E3 ligases have been identified to regulate lung inflammation, cell death, and differentiation through targeting transcriptional factors, plasma membrane receptors, signaling proteins, and mitochondrial proteins (18, 21, 28–43). Recent studies have demonstrated that DUBs counteract the effects of E3 ligases and modulate lung cell function in lung injury and fibrosis (21, 26, 44–50). Thus, a better understanding of how

DOI: 10.1201/9781003355243-10

protein ubiquitination regulates physiological and pathological processes in lung cells is important to identify novel therapeutic strategies and targets for treating lung diseases. This chapter will focus on methodology and strategy of research in protein ubiquitination and deubiquitination in lung cells (Figure. 8.1).

8.2 DETERMINATION OF PROTEIN STABILITY IN PHYSIOLOGICAL AND PATHOLOGICAL CONDITIONS

Protein homeostasis is controlled by synthesis and degradation. Half-lives of proteins are highly variable, ranging from a few minutes to days in physiological condition. Cyclin D1 has a short half-life of around 30 minutes (51), whereas average half-lives of protein in HeLa cells were about 20 hours (52). To measure protein half-life, cycloheximide (CHX) chase assay is often performed. CHX is an inhibitor of protein biosynthesis because it inhibits translational elongation (53). Lung cells are treated with CHX (100–400 μg/ml) for different times (minutes to hours), and then cell lysates are analyzed by immunoblotting with a specific antibody against target protein. The concentration of CHX dependents on the cell types. To avoid the effect of cytotoxicity of CHX on protein degradation, cell viability needs to be monitored in the presence of CHX. Another traditional method to measure protein stability is the radioactive pulse-chase assay (54). Lung cells are incubated with ^{35}S-methionine for a short time to create the pulse of newly synthesized proteins. After designed period of time with "cold" media, cells are harvested and subjected to immunoprecipitation with an antibody against the target protein, followed by SDS-PAGE gel separation. Radioactive signals are detected by phosphoimager. The pulse-chase assay measures protein half-life without altering cellular activity; however, because of the use of radioactive materials, this method is not well used in the laboratory practice.

To determine if interest protein's stability is altered in the pathological conditions, such as infection, oxidative stress, and injury, first changes of protein levels are confirmed by immunoblotting analysis of interest protein after designed treatments, such as viral infection, endotoxin or cytokine treatments, or hypoxia exposure. In addition to protein stability, abnormal protein synthesis through transcription and translation machinery may alter protein levels in the pathological conditions. Thus, once changes of protein levels are confirmed by immunoblotting, mRNA levels of interest protein should be measured in the same condition by real-time polymerase chain reaction. In the case that changes in the mRNA levels are positively correlated with changes in protein levels, that suggests that protein level changes in a certain pathological condition are due to alteration of protein synthesis, not protein stability. If the protein level changes cannot be explained by changes in mRNA levels, it indicates that protein stability may be altered in the certain pathological condition. It is possible that under treatment, both synthesis and degradation of the interest protein are altered.

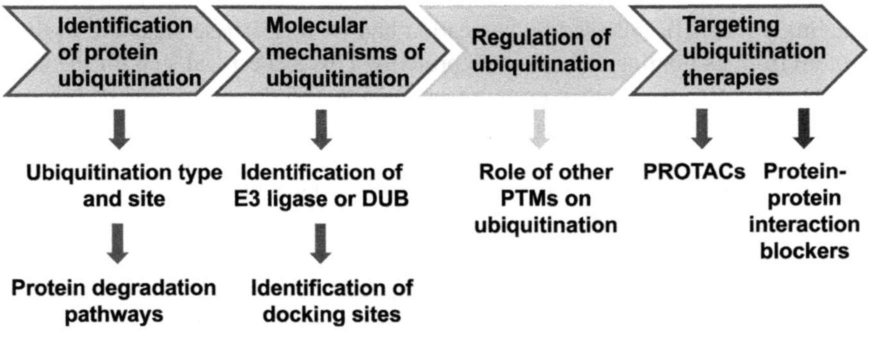

FIGURE 8.1 Research strategy to investigate protein ubiquitination.

8.3 DETERMINATION OF PROTEIN DEGRADATION PATHWAY

Once an interest protein's half-life or stability is confirmed in the physiological or pathological conditions, the next step is to investigate if the protein degradation is in the proteasome or lysosome system. The proteasome degrades damaged or unfolded proteins by proteolysis. The 20S proteasome contains β-subunits that have proteolytic active sites (55, 56). Specific inhibitors of the 20S proteasome, such as bortezomib and MG132, have been used to investigate if protein degrades in the proteasome (57). Lysosomes are membrane-bound organelles and degrade proteins in larger complexes, such as endosomes and mitochondria (58–60). Leupeptin, chloroquine, and bafilomycin A are often used to investigate if protein degrades in the lysosome (61, 62). To investigate which pathway regulates interest protein degradation in the physiological or pathological condition, lung cells are pretreated with inhibitors of the proteasome (MG132 or bortezomib) or lysosome (e.g., leupeptin) for 1 hour and then incubated with CHX (mimic physiological condition) or treatments (mimic pathological condition) for an additional designed period of time. If inhibitors of proteasome attenuate interest protein degradation, it suggests that interest protein degradation is by the proteasome system. The same adjustment is for using inhibitors of lysosome.

It has been shown that protein degradation in the proteasome is not totally dependent on ubiquitination. Misfolded and oxidized proteins may enter into the 20S proteasome and degrade (63). It is important to determine if the ubiquitination system regulates protein degradation. MLN7243 (TAK-243) is a selective E1-activating enzyme inhibitor (64) that inhibits the protein ubiquitination from the first step. Lung cells are treated with MLN7243 followed by CHX or pathological treatment. If protein degradation is prevented by MLN7243, it indicates that the interest protein degradation is regulated by the ubiquitination system.

8.4 DETERMINATION OF MONO- OR POLYUBIQUITINATION AND IDENTIFICATION OF UBIQUITIN SITES

After confirming the interest protein is degraded in the proteasome or lysosome, ubiquitination status of the interest protein needs to be determined. An in vivo ubiquitination assay can determine mono- or polyubiquitination of interest protein (21, 43, 65). Furthermore, different types of ubiquitination can be determined by using antibodies specific to K11-, K48-, or K63-linked polyubiquitin chains. Protein ubiquitination triggers protein degradation. In the pathological condition, loss of interest protein may reduce the overall ubiquitination degree of the substrate by in vivo ubiquitination assay, thus detecting the decreased ubiquitination cannot truly reflect the ubiquitination of the interest protein. To avoid this situation, cells are treated with the proteasome or lysosome inhibitor before pathological challenges. To increase the sensitivity of the assay, increases in ubiquitin levels in the cells can be achieved by overexpression of ubiquitin by plasmid or viral vector infection. After treatments, cells are collected in phosphate-buffered saline, and cell pellets are dissolved in 2% SDS-TBS buffer with deubiquitinating enzyme inhibitors (ubiquitin aldehyde and *N*-ethylmaleimide [NEM]), followed by sonication and boiling. These steps are to dissociate the interest protein from its binding or associated proteins. This is a critical step for the ubiquitination assay. Without separating the interest protein from its associated proteins, the detected ubiquitination bands may be due to ubiquitination of the associated proteins. After dissociation of the interest protein with its associated proteins, cell lysates are subjected to regulatory immunoprecipitation with an antibody against ubiquitin or the interest protein, followed by immunoblotting analysis with antibody against the interest protein or ubiquitin.

To further determine the types of ubiquitination, antibodies specific to K11-, K48-, or K63-linked polyubiquitin chains can be used for immunoprecipitation or immunoblotting analysis. Further, tagged ubiquitin K48 only, K63 only, K11 only, K48 deficient, K63 deficient, or K11 deficient plasmids can be used to transfect cells instead of pan-ubiquitin plasmid. Specific lysine-linked ubiquitination can be detected using these plasmids by in vivo ubiquitination assay.

K residues are common sites for accepting ubiquitin or polyubiquitin chains to the substrates. To initiate identifying the ubiquitin sites on interest proteins, computational methods may be applied to predict which K residues are potential ubiquitin-accepting sites. Computational machine learning methods have been well developed and employed for the ubiquitination site prediction based on protein sequences. However, researchers need to pay attention to the localization of the predicted K residues on the interest protein. For example, for the transmembrane protein, if the predicted K residues are on the extracellular membrane domain, the K sites do not need to be tested further. Experimental examination is needed to confirm whether the predicted K sites are ubiquitination sites. K residues are substituted with arginine (R) residues using site-directed mutagenesis to generate plasmid encoding K → R (K/R) mutants of the interest protein. The half-lives and degradation rates of wild type (Wt) and K/A mutants are compared by immunoblotting. The delay of degradation of K/R mutants suggests the K residue(s) are potential ubiquitination sites. To further validate if these K residues can be ubiquitinated or not, in vivo ubiquitination is used to measure the degree of ubiquitination of K/R mutants compared to Wt proteins. Multiple ubiquitination sites are possible within the interest proteins; thus several K residues need to be mutated. Proteomics have been employed to identify ubiquitination sites; however, it is challenging due to several factors, including small size of ubiquitination and low abundance of ubiquitinated proteins. Recently, improvement of tandem mass spectrometry (MS/MS) allows the detection of ubiquitination sites in enriched ubiquitinated proteins (66). Eventually, generation of K/R mutants and in vivo ubiquitination assay are needed to confirm the results from MS/MS.

8.5 IDENTIFICATION OF E3 LIGASES OR DUBs FOR THE INTEREST PROTEIN

Ubiquitination is regulated by a series of enzymatic reactions. E3 ligase is responsible for the ligation of ubiquitin or polyubiquitin chains to the substrate. There are major E3 ligase groups that have been well demonstrated, including RING-domain E3 ligases, HECT family E3 ligases, and RING-between-RING (RBR) E3 ligases (67–69). Like other PTMs, ubiquitination is negatively regulated by DUBs (70–72). There are an estimated 1000 E3 ligases and nearly 100 DUBs in human cells. The effects of E3 ligases and DUBs occurs in interaction with substrates. Identification of E3 ligase or DUB for the interest protein is important for further understanding the molecular regulation of substrate protein ubiquitination and deubiquitination. Enzyme screening library, protein interaction screening, and protein-protein interaction assays are used to identify E3 ligase or DUB for the interest protein.

Utilization of E3 ligase or DUB library assay is a common approach to identify E3 ligase or DUB in ubiquitination studies. A library that overexpresses or downregulates major E3 ligases or DUBs is conducted by plasmids, adenoviral or lentiviral vectors, small interfering (si)RNAs, or the CRISPR-cas9 system. The library is induced in the lung cells, and then the interest protein levels are examined by immunoblotting. If the interest protein levels are altered by the certain E3 ligase and DUB but the mRNA level of the interest protein is not affected, an E3 ligase or DUB is identified for the substrate.

Another common strategy to identify E3 ligase or DUB is based on unbiased protein-protein interaction assays. Endogenous or overexpressed interest protein as a bait can be co-immunoprecipitated (Co-IP) with its specific antibody. The precipitated complex is analyzed by mass spectrometry analysis (73, 74). The Co-IP–coupled MS provides a rapid, powerful, and unbiased screening to identifying the interest protein-associated partners including E3 ligases, DUBs, kinases, and other adaptors. Other methods including yeast two-hybridization (75, 76) and most recently developed proximity-based labeling techniques (BioID and TurboID) are often used for identifying ubiquitination-modifying enzymes (77–79).

Semi-unbiased strategies are be used to identify E3 ligase or DUB based on available knowledge about the interest protein. For instance, protein cellular localization is key information. If the interest protein is a membrane protein, the search for the ubiquitination modifying enzymes may be focused only on enzymes that have been known to be localized in the inner membrane. The sub-library for overexpressing or knocking down E3 ligases or DUBs can be developed based on cellular localization and functions. Protein-protein interaction involves protein motifs or domains. For instance, if the interest protein contains a PY motif, it is possible that a E3 ligase with WW domain has a ability to target and induce the interest protein ubiquitination. Thus, E3 ligases containing WW domains (Nedd4, Nedd4L, WWP1, etc.) (80, 81) are priority candidates to be considered.

After identifying the potential E3 ligase or DUB for the interest protein, in vivo or in vitro ubiquitination assay is used to confirm the effect of identified E3 ligase or DUB on substrate ubiquitination. For in vivo cellular ubiquitination assay, E3 ligase or DUB will be overexpressed or downregulated in lung cells. Cell lysates are subjected to denatured IP and ubiquitination assay as described above. For in vitro ubiquitination assay, recombinant or purified E3 ligase or DUB protein are incubated with substrate protein along with ATP, ubiquitin, and E1 and E2 enzymes, followed by immunoblotting with a specific antibody against the substrate (18, 82). To confirm E3 ligase or DUB directly targets the substrate, the association of the enzymes and substrate are confirmed by co-IP and co-immunofluorescence staining in lung cells. Evaluation of co-localization is an important process. Co-IP/MS may detect the association of two proteins that are not localized in the same fractions of cells because cell lysates are mixtures of proteins from all the fractions.

8.6 IDENTIFICATION OF DOCKING SITE

To further investigate the molecular regulation of E3 ligase or DUB mediation of interest protein ubiquitination, it is necessary to identify the docking site on the interest protein for E3 ligase or DUB. This information may benefit design blocking small molecules or peptide to prevent or promote the interest protein degradation. If the protein-docking motif has been predicted by bioinformatic tools, the docking motif can be mutated or deleted with site-directed mutagenesis. For example, if the interest protein is the substrate of a WW domain-containing E3 ligase, it is highly possible that PY motif in the interest protein is the docking site for the E3 ligase. To confirm the hypothesis, a PY motif-deleted mutant can be generated. Protein-protein interaction, substrate ubiquitination, and stability are be performed to confirm if the PY motif (PPXY or LPXY) is critical for interacting with the identified WW domain-containing E3 ligase (83).

If there is no certain motif or domain predicted as the docking site, generation of a series of N-terminal or C-terminal deletion mutants of the substrate will be performed to identify docking site by protein-protein interaction assay, such as co-IP. It should be cautioned that the deletion mutants may affect substrate's cellular localization and conformational changes. All these changes may disrupt the interaction with E3 ligase or DUB. Thus, generation and use of mutant with precise mutation or deletion of key amino acids for protein-protein interaction assay can identify the docking site. Furthermore, the ubiquitination and stability of the mutant are examined to confirm that the key amino acids are critical for interaction with E3 ligase or DUB.

Similar to the strategy for identifying docking site in the substrate, it will be of interest to identify the domain on E3 ligase or DUB for interaction with the substrate. The F-box protein in the SCF E3 ligase complex plays a critical role in targeting the substrate. The C-terminal domain of the F-box protein contains a substrate binding site. Based on the structure of the C-terminal domain, F-box proteins are characterized as leucine-rich domain-containing proteins (FBXL family), WD domain-containing proteins (FBXW family), and "other" domain-containing proteins (FBXO family) (84, 85). A series deletion mutant of the C-terminal of the F-box protein can be generated and subjected to a protein-protein interaction assay to determine the binding site for the substrate.

8.7 DETERMINATION OF REGULATORY MECHANISMS

Protein ubiquitination is not constant in the cells, whereas its level changes in different cellular states of flux and responds to external stimuli. Cell cycle–related proteins, such as cyclins and cyclin-dependent kinases, may be ubiquitinated and degraded as the cell grows and divides. In response to inflammatory stimuli, plasma receptors, such as interleukin-33R and lysophosphatidic acid R1, undergo ubiquitination and degradation in lung epithelial cells (18, 21). Ubiquitination is a regulatory process. Protein phosphorylation, acetylation, oxidization, or protein-protein interactions may influence ubiquitination (86), either promotion or reduction of ubiquitination through modulation of ubiquitination-related enzyme/substrate interaction, cellular localization, or ubiquitin conjugation. The chapter will introduce the research experiment design to investigate role of other PTMs and protein-protein interaction in ubiquitination and deubiquitination. Here, phosphorylation and acetylation are used as examples. First, phosphorylation or acetylation of substrate under physiological and pathological conditions should be examined by immunoblotting with specific phospho- or acetyl-antibodies, IP, or in vitro PTM assay. To investigate if the identified PTM regulates ubiquitination of the substrate, mutants of phosphorylation or acetylation sites are generated for investigating whether the site(s) are important for the substrate ubiquitination, degradation, or enzyme activity. Phosphorylation may enhance E3 ligase association with the substrate or reduce DUB/substrate interaction (21). To further understand the molecular mechanisms by which the PTMs influence ubiquitination, the effect of mutation on phosphorylation or acetylation sites on ubiquitination enzyme/substrate interaction and substrate's cellular localization should be considered. However, because both acetylation and ubiquitination use K residues, it is possible that the same K residue is responsible for both acetylation and ubiquitination. In this case, the K mutant may reduce both acetylation and ubiquitination. Thus, the conclusion from using K mutant needs be carefully considered. As discussed above, K residue is a common ubiquitin acceptor site. The ubiquitination site is not always exposed on the protein surface; this may prevent ubiquitination and degradation of the protein. During certain physiological and pathological changes, the K residue may be exposed to the protein surface because of changes in the substrate conformation. The protein structure conformation can be changed by PTMs including phosphorylation and protein-protein interaction. Substrate ubiquitination has been shown to be modulated by its association and disassociation with adapter proteins, such as 14–3–3 and calmodulin (87, 88). To investigate if protein ubiquitination is regulated by adapter protein interaction with a substrate, first modulate adapter protein levels by gene silencing or overexpression, and then determine substrate ubiquitination levels. If the binding site for adapter protein has been revealed, a mutant of the binding site can be generated and tested for ubiquitination.

In addition to the fact that PTMs of substrate regulate its ubiquitination, PTMs of E3 ligase or DUB may affect enzymatic activity or interaction with substrate (16, 19), therefore modulating substrate ubiquitination. Similar strategies as described above, such as mutation of phosphorylation or acetylation site, can be performed to investigate regulation of substrate ubiquitination.

8.8 DETERMINATION OF ROLE OF UBIQUITINATION IN LUNG CELLULAR FUNCTION AND DISEASES

Abnormal ubiquitination-dependent protein degradation has been shown to contribute to lung cellular disfunction and development of lung diseases (26, 89–91). Ubiquitination is a reversal and regulatory process (Figure 8.2). After confirming the interest protein is ubiquitinated, the next step is to investigate if the ubiquitination regulates protein-related cellular function. The mutant of ubiquitin acceptor site (K/R) or ubiquitination-related enzyme binding site is used to determine the substrate-related cellular functions, such as receptor-mediated signaling, cell proliferation, cytokine release, and cell death. The CRISPR-Cas9 system can be used to generate cell lines containing these mutants. As each E3 ligase or DUB targets and regulates multi-substrates, simple overexpression or knockdown of E3 ligase or DUB may result in unexpected and interest substrate-independent cellular response. The regulators, such as kinases or acetyltransferase, or adaptors (14–3–3 or calmodulin) may not only affect substrate ubiquitination but also ubiquitination-independent cellular functions. Thus, drawing a conclusion by modulation of E3 ligase, DUB, and regulators must be made carefully.

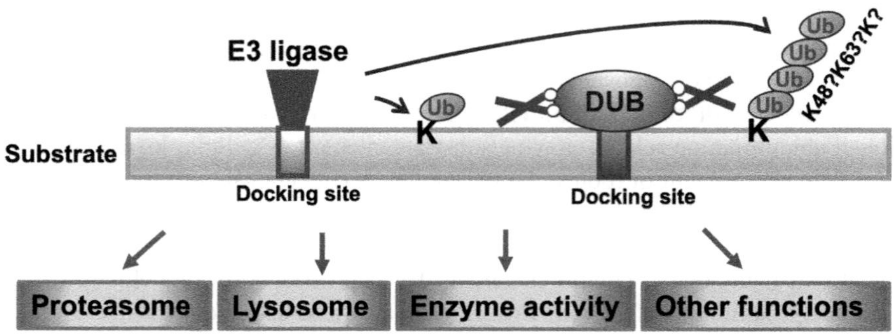

FIGURE 8.2 E3 ligase and DUB regulates protein ubiquitination.

To investigate the role of the ubiquitination of a substrate in the pathogenesis of lung diseases, the CRISPR-Cas9 system can be used to generate transgenic mice containing a mutant of ubiquitin acceptor site (K/R) or a ubiquitination-related enzyme binding site, followed by performing models of lung diseases. To precisely investigate the role of ubiquitination of interest protein in certain lung cell types during the development of lung diseases, cell type–specific transgenic mice may be generated.

8.9 UTILIZATION OF PROTEIN DEGRADATION PRINCIPLE TO DOWNREGULATE PROTEIN LEVELS BY PROTEOLYSIS TARGETING CHIMERAS

As discussed above, targeting a certain E3 ligase or DUB may affect multiple substrate-regulated cellular functions. Recently, a new biotechnology, proteolysis targeting chimeras (PROTACs), has been developed to target interest protein for its degradation based on modulation of protein ubiquitination (92–94). PROTAC technology does not need to identify specific E3 ligase for interest substrate, and it is suitable to degrade "undruggable proteins." Several PROTAC drugs have been tested in phase II clinical trials. The principle of PROTAC is to use an engineered small molecule comprising a E3 ligase binding site, a substrate binding site, and a linker between the two sites to recruit E3 ligase to substrate, therefore triggering substrate ubiquitination and proteasomal degradation. The substrate binding site can be designed based on known inhibitors of a substrate or a substrate's structure (Figure 8.3). The advantage of the design is not limited to a substrate's activation sites. Several endogenous E3 ligases, including von Hippel-Lindau (VHL) and cereblon (CRBN), have been widely used as baits to trap substrates (94). Small molecules binding to VHL or CRBN have been developed. Once the small molecule for substrate binding is revealed, the small molecule can be linked with a small molecule binding to E3 ligase with a linker. The new PROTACs can be used to test the substrate protein stability. However, the substrate may undergo various PTMs in the physiological and pathological conditions. The affinity of the small molecules binding to the substrate may vary; thus, it is causable that the PROTACs may have distinct efficiencies in different conditions.

Based on DUB-substrate or E3 ligase–substrate interaction, a docking site blocker can be designed and used to disrupt substrate interaction with DUB or E3 ligase to modulate its ubiquitination (95). For example, if the docking site for DUB is discovered, based on the docking site sequence and structure, a small molecule can be designed to prevent DUB interaction with the substrate. The small molecule will increase the protein ubiquitination. If the docking site for its specific E3 ligase is revealed, a small molecule can be used to reduce the interaction of E3 ligase, resulting in decreased ubiquitination.

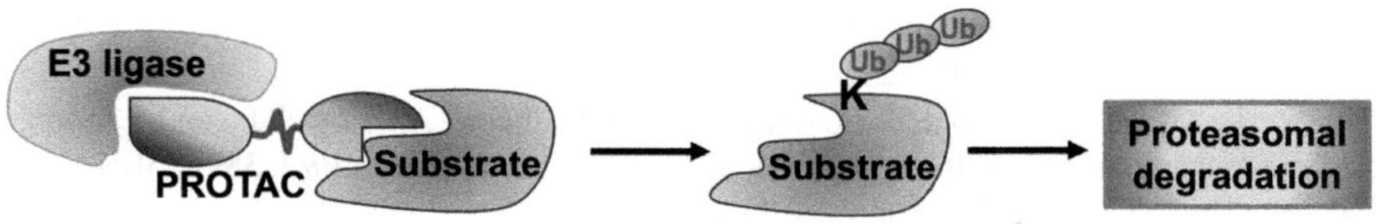

FIGURE 8.3 PROTAC target substrate for its ubiquitination and degradation.

8.10 CONCLUSIONS

In this chapter, we have discussed the principle of protein ubiquitination, research strategies, experimental design, and PROTACs in investigating the role of protein ubiquitination in lung diseases. This is not a guideline, but it provides a thoughtful idea to identify protein ubiquitination, its underlying mechanisms, and its potential applications in understanding lung diseases. This chapter focuses on interest substrate ubiquitination and biological functions. In addition to investigating substrate ubiquitination, a growing number of studies have been focused on determining the regulation of E3 ligases or DUBs. Changes of E3 ligases or DUBs in lung cells have been shown to contribute to disease development. Understanding the regulation of these enzymes' transcription, protein stability, and activities in physiological and pathological conditions is important to develop new therapies to treat lung diseases and other disorders. In this chapter, we do not provide detailed experimental information because different cell types exhibit distinct responses to the same stimulation or inhibitor treatment. Researchers need to optimize experimental conditions based on their own cell culture systems and disease models. Rapid technological discoveries and improvement, including artificial intelligence in protein structure, drug design, and development, will lead to develop more efficient approaches to target protein ubiquitination-mediated cellular responses and therapies.

8.11 ACKNOWLEDGMENT

The author would like to thank Sarah J. Taleb for editing.

ABBREVIATIONS

CHX	cycloheximide
Co-IP	co-immunoprecipitate
CRBN	cereblon
DUB	deubiquitinating enzyme
MS/MS	tandem mass spectrometry
NEM	*N*-ethylmaleimide
PROTAC	proteolysis-targeting chimera
PTM	post-translational modification
RBR	RING-between-RING
VHL	von Hippel-Lindau
Wt	wild type

REFERENCES

1. Ciechanover, A. (1994) The ubiquitin-proteasome proteolytic pathway. *Cell.* **79**, 13–21
2. Freiman, R. N., and Tjian, R. (2003) Regulating the regulators: Lysine modifications make their mark. *Cell.* **112**, 11–17
3. Pickart, C. M. (2001) Mechanisms underlying ubiquitination. *Annu Rev Biochem.* **70**, 503–533

4. Hodges, M., Tissot, C., and Freemont, P. S. (1998) Protein regulation: Tag wrestling with relatives of ubiquitin. *Curr Biol.* **8**, R749–R752
5. Mayer, R. J., Tipler, C., Arnold, J., Laszlo, L., Al-Khedhairy, A., Lowe, J., and Landon, M. (1996) Endosome-lysosomes, ubiquitin and neurodegeneration. *Adv Exp Med Biol.* **389**, 261–269
6. Jentsch, S. (1992) The ubiquitin-conjugation system. *Annu Rev Genet.* **26**, 179–207
7. Akutsu, M., Dikic, I., and Bremm, A. (2016) Ubiquitin chain diversity at a glance. *J Cell Sci.* **129**, 875–880
8. Cao, L., Liu, X., Zheng, B., Xing, C., and Liu, J. (2022) Role of K63-linked ubiquitination in cancer. *Cell Death Discov.* **8**, 410
9. Ohtake, F., Tsuchiya, H., Saeki, Y., and Tanaka, K. (2018) K63 ubiquitylation triggers proteasomal degradation by seeding branched ubiquitin chains. *Proc Natl Acad Sci U S A.* **115**, E1401–E1408
10. Choi, Y. B., and Harhaj, E. W. (2014) HTLV-1 tax stabilizes MCL-1 via TRAF6-dependent K63-linked polyubiquitination to promote cell survival and transformation. *PLoS Pathog.* **10**, e1004458
11. Zhang, J., Zhang, C., Cui, J., Ou, J., Han, J., Qin, Y., Zhi, F., and Wang, R. F. (2017) TRIM45 functions as a tumor suppressor in the brain via its E3 ligase activity by stabilizing p53 through K63-linked ubiquitination. *Cell Death Dis.* **8**, e2831
12. Hatakeyama, S., and Nakayama, K. I. (2003) Ubiquitylation as a quality control system for intracellular proteins. *J Biochem.* **134**, 1–8
13. Tanaka, K., Suzuki, T., and Chiba, T. (1998) The ligation systems for ubiquitin and ubiquitin-like proteins. *Mol Cells.* **8**, 503–512
14. Barghout, S. H., and Schimmer, A. D. (2021) E1 enzymes as therapeutic targets in cancer. *Pharmacol Rev.* **73**, 1–58
15. Kliza, K., and Husnjak, K. (2020) Resolving the complexity of ubiquitin networks. *Front Mol Biosci.* **7**, 21
16. Zheng, N., and Shabek, N. (2017) Ubiquitin ligases: Structure, function, and regulation. *Annu Rev Biochem.* **86**, 129–157
17. Lacoursiere, R. E., Hadi, D., and Shaw, G. S. (2022) Acetylation, phosphorylation, ubiquitination (Oh My!): Following post-translational modifications on the ubiquitin road. *Biomolecules.* **12**
18. Zhao, J., Wei, J., Mialki, R. K., Mallampalli, D. F., Chen, B. B., Coon, T., Zou, C., Mallampalli, R. K., and Zhao, Y. (2012) F-box protein FBXL19-mediated ubiquitination and degradation of the receptor for IL-33 limits pulmonary inflammation. *Nat Immunol.* **13**, 651–658
19. Chen, Y., Shao, X., Cao, J., Zhu, H., Yang, B., He, Q., and Ying, M. (2021) Phosphorylation regulates cullin-based ubiquitination in tumorigenesis. *Acta Pharm Sin B.* **11**, 309–321
20. Eiyama, A., and Okamoto, K. (2015) PINK1/Parkin-mediated mitophagy in mammalian cells. *Curr Opin Cell Biol.* **33**, 95–101
21. Zhao, J., Wei, J., Dong, S., Bowser, R. K., Zhang, L., Jacko, A. M., and Zhao, Y. (2016) Destabilization of lysophosphatidic acid receptor 1 reduces cytokine release and protects against lung injury. *EBioMedicine.* **10**, 195–203
22. Li, L., Wei, J., Suber, T. L., Ye, Q., Miao, J., Li, S., Taleb, S. J., Tran, K. C., Tamaskar, A. S., Zhao, J., and Zhao, Y. (2021) IL-37-induced activation of glycogen synthase kinase 3beta promotes IL-1R8/Sigirr phosphorylation, internalization, and degradation in lung epithelial cells. *J Cell Physiol.* **236**, 5676–5685
23. Kanarek, N., London, N., Schueler-Furman, O., and Ben-Neriah, Y. (2010) Ubiquitination and degradation of the inhibitors of NF-kappaB. *Cold Spring Harb Perspect Biol.* **2**, a000166
24. D'Andrea, A., and Pellman, D. (1998) Deubiquitinating enzymes: A new class of biological regulators. *Crit Rev Biochem Mol Biol.* **33**, 337–352
25. Kalderon, D. (1996) Protein degradation: De-ubiquitinate to decide your fate. *Curr Biol.* **6**, 662–665
26. Li, S., Zhao, J., Shang, D., Kass, D. J., and Zhao, Y. (2018) Ubiquitination and deubiquitination emerge as players in idiopathic pulmonary fibrosis pathogenesis and treatment. *JCI Insight.* **3**
27. Cai, J., Culley, M. K., Zhao, Y., and Zhao, J. (2018) The role of ubiquitination and deubiquitination in the regulation of cell junctions. *Protein Cell.* **9**, 754–769
28. Onose, A., Hashimoto, S., Hayashi, S., Maruoka, S., Kumasawa, F., Mizumura, K., Jibiki, I., Matsumoto, K., Gon, Y., Kobayashi, T., Takahashi, N., Shibata, Y., Abiko, Y., Shibata, T., Shimizu, K., and Horie, T. (2006) An inhibitory effect of A20 on NF-kappaB activation in airway epithelium upon influenza virus infection. *Eur J Pharmacol.* **541**, 198–204
29. Zou, C., Butler, P. L., Coon, T. A., Smith, R. M., Hammen, G., Zhao, Y., Chen, B. B., and Mallampalli, R. K. (2011) LPS impairs phospholipid synthesis by triggering beta-transducin repeat-containing protein (beta-TrCP)-mediated polyubiquitination and degradation of the surfactant enzyme acyl-CoA:lysophosphatidylcholine acyltransferase I (LPCAT1). *J Biol Chem.* **286**, 2719–2727

30. Chen, B. B., Coon, T. A., Glasser, J. R., McVerry, B. J., Zhao, J., Zhao, Y., Zou, C., Ellis, B., Sciurba, F. C., Zhang, Y., and Mallampalli, R. K. (2013) A combinatorial F box protein directed pathway controls TRAF adaptor stability to regulate inflammation. *Nat Immunol.* **14**, 470–479
31. Chen, B. B., Coon, T. A., Glasser, J. R., Zou, C., Ellis, B., Das, T., McKelvey, A. C., Rajbhandari, S., Lear, T., Kamga, C., Shiva, S., Li, C., Pilewski, J. M., Callio, J., Chu, C. T., Ray, A., Ray, P., Tyurina, Y. Y., Kagan, V. E., and Mallampalli, R. K. (2014) E3 ligase subunit Fbxo15 and PINK1 kinase regulate cardiolipin synthase 1 stability and mitochondrial function in pneumonia. *Cell Rep.* **7**, 476–487
32. Coon, T. A., McKelvey, A. C., Lear, T., Rajbhandari, S., Dunn, S. R., Connelly, W., Zhao, J. Y., Han, S., Liu, Y., Weathington, N. M., McVerry, B. J., Zhang, Y., and Chen, B. B. (2015) The proinflammatory role of HECTD2 in innate immunity and experimental lung injury. *Sci Transl Med.* **7**, 295ra109
33. Han, S., Lear, T. B., Jerome, J. A., Rajbhandari, S., Snavely, C. A., Gulick, D. L., Gibson, K. F., Zou, C., Chen, B. B., and Mallampalli, R. K. (2015) Lipopolysaccharide primes the NALP3 inflammasome by inhibiting its ubiquitination and degradation mediated by the SCFFBXL2 E3 ligase. *J Biol Chem.* **290**, 18124–18133
34. Liu, Y., Lear, T., Iannone, O., Shiva, S., Corey, C., Rajbhandari, S., Jerome, J., Chen, B. B., and Mallampalli, R. K. (2015) The proapoptotic F-box protein Fbxl7 regulates mitochondrial function by mediating the ubiquitylation and proteasomal degradation of survivin. *J Biol Chem.* **290**, 11843–11852
35. Zou, C., Li, J., Xiong, S., Chen, Y., Wu, Q., Li, X., Weathington, N. M., Han, S., Snavely, C., Chen, B. B., and Mallampalli, R. K. (2015) Mortality factor 4 like 1 protein mediates epithelial cell death in a mouse model of pneumonia. *Sci Transl Med.* **7**, 311ra171
36. Lear, T., McKelvey, A. C., Rajbhandari, S., Dunn, S. R., Coon, T. A., Connelly, W., Zhao, J. Y., Kass, D. J., Zhang, Y., Liu, Y., and Chen, B. B. (2016) Ubiquitin E3 ligase FIEL1 regulates fibrotic lung injury through SUMO-E3 ligase PIAS4. *J Exp Med.* **213**, 1029–1046
37. Letsiou, E., Sammani, S., Wang, H., Belvitch, P., and Dudek, S. M. (2017) Parkin regulates lipopolysaccharide-induced proinflammatory responses in acute lung injury. *Transl Res.* **181**, 71–82
38. Wei, J., Dong, S., Bowser, R. K., Khoo, A., Zhang, L., Jacko, A. M., Zhao, Y., and Zhao, J. (2017) Regulation of the ubiquitylation and deubiquitylation of CREB-binding protein modulates histone acetylation and lung inflammation. *Sci Signal.* **10**
39. Liu, S. S., Lv, X. X., Liu, C., Qi, J., Li, Y. X., Wei, X. P., Li, K., Hua, F., Cui, B., Zhang, X. W., Yu, J. J., Yu, J. M., Wang, F., Shang, S., Zhao, C. X., Hou, X. Y., Yao, Z. G., Li, P. P., Li, X., Huang, B., and Hu, Z. W. (2019) Targeting degradation of the transcription factor C/EBPbeta reduces lung fibrosis by restoring activity of the ubiquitin-editing enzyme A20 in macrophages. *Immunity.* **51**, 522–534.e527
40. Li, L., Wei, J., Mallampalli, R. K., Zhao, Y., and Zhao, J. (2019) TRIM21 Mitigates human lung microvascular endothelial cells' inflammatory responses to LPS. *Am J Respir Cell Mol Biol.* **61**, 776–785
41. Brazee, P. L., Morales-Nebreda, L., Magnani, N. D., Garcia, J. G., Misharin, A. V., Ridge, K. M., Budinger, G. R. S., Iwai, K., Dada, L. A., and Sznajder, J. I. (2020) Linear ubiquitin assembly complex regulates lung epithelial-driven responses during influenza infection. *J Clin Invest.* **130**, 1301–1314
42. Qian, Y., Wang, Z., Lin, H., Lei, T., Zhou, Z., Huang, W., Wu, X., Zuo, L., Wu, J., Liu, Y., Wang, L. F., Guan, X. H., Deng, K. Y., Fu, M., and Xin, H. B. (2022) TRIM47 is a novel endothelial activation factor that aggravates lipopolysaccharide-induced acute lung injury in mice via K63-linked ubiquitination of TRAF2. *Signal Transduct Target Ther.* **7**, 148
43. Li, S., Ye, Q., Wei, J., Taleb, S. J., Wang, H., Zhang, Y., Kass, D. J., Horowitz, J. C., Zhao, J., and Zhao, Y. (2023) Nedd4L suppression in lung fibroblasts facilitates pathogenesis of lung fibrosis. *Transl Res.* **253**, 1–7
44. Wang, D., Zhao, J., Li, S., Wei, J., Nan, L., Mallampalli, R. K., Weathington, N. M., Ma, H., and Zhao, Y. (2018) Phosphorylated E2F1 is stabilized by nuclear USP11 to drive Peg10 gene expression and activate lung epithelial cells. *J Mol Cell Biol.* **10**, 60–73
45. Nan, L., Jacko, A. M., Tan, J., Wang, D., Zhao, J., Kass, D. J., Ma, H., and Zhao, Y. (2016) Ubiquitin carboxyl-terminal hydrolase-L5 promotes TGFbeta-1 signaling by de-ubiquitinating and stabilizing Smad2/Smad3 in pulmonary fibrosis. *Sci Rep.* **6**, 33116
46. Jacko, A. M., Nan, L., Li, S., Tan, J., Zhao, J., Kass, D. J., and Zhao, Y. (2016) De-ubiquitinating enzyme, USP11, promotes transforming growth factor beta-1 signaling through stabilization of transforming growth factor beta receptor II. *Cell Death Dis.* **7**, e2474
47. Mialki, R. K., Zhao, J., Wei, J., Mallampalli, D. F., and Zhao, Y. (2013) Overexpression of USP14 protease reduces I-kappaB protein levels and increases cytokine release in lung epithelial cells. *J Biol Chem.* **288**, 15437–15441
48. Mitra, S., Epshtein, Y., Sammani, S., Quijada, H., Chen, W., Bandela, M., Desai, A. A., Garcia, J. G. N., and Jacobson, J. R. (2021) UCHL1, a deubiquitinating enzyme, regulates lung endothelial cell permeability in vitro and in vivo. *Am J Physiol Lung Cell Mol Physiol.* **320**, L497–L507

49. Li, S., Wang, D., Zhao, J., Weathington, N. M., Shang, D., and Zhao, Y. (2018) The deubiquitinating enzyme USP48 stabilizes TRAF2 and reduces E-cadherin-mediated adherens junctions. *FASEB J.* **32**, 230–242
50. Bednash, J. S., Weathington, N., Londino, J., Rojas, M., Gulick, D. L., Fort, R., Han, S., McKelvey, A. C., Chen, B. B., and Mallampalli, R. K. (2017) Targeting the deubiquitinase STAMBP inhibits NALP7 inflammasome activity. *Nat Commun.* **8**, 15203
51. Diehl, J. A., Zindy, F., and Sherr, C. J. (1997) Inhibition of cyclin D1 phosphorylation on threonine-286 prevents its rapid degradation via the ubiquitin-proteasome pathway. *Genes Dev.* **11**, 957–972
52. Boisvert, F. M., Ahmad, Y., Gierlinski, M., Charriere, F., Lamont, D., Scott, M., Barton, G., and Lamond, A. I. (2012) A quantitative spatial proteomics analysis of proteome turnover in human cells. *Mol Cell Proteomics.* **11**, M111 011429
53. Kao, S. H., Wang, W. L., Chen, C. Y., Chang, Y. L., Wu, Y. Y., Wang, Y. T., Wang, S. P., Nesvizhskii, A. I., Chen, Y. J., Hong, T. M., and Yang, P. C. (2015) Analysis of protein stability by the cycloheximide chase assay. *Bio Protoc.* **5**
54. Elgendy, M. (2017) Assessment of modulation of protein stability using pulse-chase method. *Bio Protoc.* **7**, e2443
55. Finley, D. (2009) Recognition and processing of ubiquitin-protein conjugates by the proteasome. *Annu Rev Biochem.* **78**, 477–513
56. Saeki, Y. (2017) Ubiquitin recognition by the proteasome. *J Biochem.* **161**, 113–124
57. Goldberg, A. L. (2012) Development of proteasome inhibitors as research tools and cancer drugs. *J Cell Biol.* **199**, 583–588
58. Luzio, J. P., Pryor, P. R., and Bright, N. A. (2007) Lysosomes: Fusion and function. *Nat Rev Mol Cell Biol.* **8**, 622–632
59. Watts, C. (2012) The endosome-lysosome pathway and information generation in the immune system. *Biochim Biophys Acta.* **1824**, 14–21
60. Lamming, D. W., and Bar-Peled, L. (2019) Lysosome: The metabolic signaling hub. *Traffic.* **20**, 27–38
61. Fedele, A. O., and Proud, C. G. (2020) Chloroquine and bafilomycin a mimic lysosomal storage disorders and impair mTORC1 signalling. *Biosci Rep.* **40**
62. Lee, S., Sato, Y., and Nixon, R. A. (2011) Lysosomal proteolysis inhibition selectively disrupts axonal transport of degradative organelles and causes an Alzheimer's-like axonal dystrophy. *J Neurosci.* **31**, 7817–7830
63. Jung, T., Hohn, A., and Grune, T. (2014) The proteasome and the degradation of oxidized proteins: Part II—protein oxidation and proteasomal degradation. *Redox Biol.* **2**, 99–104
64. Misra, M., Kuhn, M., Lobel, M., An, H., Statsyuk, A. V., Sotriffer, C., and Schindelin, H. (2017) Dissecting the specificity of adenosyl sulfamate inhibitors targeting the ubiquitin-activating enzyme. *Structure.* **25**, 1120–1129.e1123
65. Choo, Y. S., and Zhang, Z. (2009) Detection of protein ubiquitination. *J Vis Exp.* **30**, 1293.
66. Na, C. H., and Peng, J. (2012) Analysis of ubiquitinated proteome by quantitative mass spectrometry. *Methods Mol Biol.* **893**, 417–429
67. Toma-Fukai, S., and Shimizu, T. (2021) Structural diversity of ubiquitin E3 ligase. *Molecules.* **26**
68. Ardley, H. C., and Robinson, P. A. (2005) E3 ubiquitin ligases. *Essays Biochem.* **41**, 15–30
69. Berndsen, C. E., and Wolberger, C. (2014) New insights into ubiquitin E3 ligase mechanism. *Nat Struct Mol Biol.* **21**, 301–307
70. Harrigan, J. A., Jacq, X., Martin, N. M., and Jackson, S. P. (2018) Deubiquitylating enzymes and drug discovery: Emerging opportunities. *Nat Rev Drug Discov.* **17**, 57–78
71. Snyder, N. A., and Silva, G. M. (2021) Deubiquitinating enzymes (DUBs): Regulation, homeostasis, and oxidative stress response. *J Biol Chem.* **297**, 101077
72. McClurg, U. L., and Robson, C. N. (2015) Deubiquitinating enzymes as oncotargets. *Oncotarget.* **6**, 9657–9668
73. Richards, A. L., Eckhardt, M., and Krogan, N. J. (2021) Mass spectrometry-based protein-protein interaction networks for the study of human diseases. *Mol Syst Biol.* **17**, e8792
74. Vasilescu, J., and Figeys, D. (2006) Mapping protein-protein interactions by mass spectrometry. *Curr Opin Biotechnol.* **17**, 394–399
75. Miller, J., and Stagljar, I. (2004) Using the yeast two-hybrid system to identify interacting proteins. *Methods Mol Biol.* **261**, 247–262
76. Dove, S. L., and Hochschild, A. (2004) A bacterial two-hybrid system based on transcription activation. *Methods Mol Biol.* **261**, 231–246

77. Kanzler, C. R., Donohue, M., Dowdle, M. E., and Sheets, M. D. (2022) TurboID functions as an efficient biotin ligase for BioID applications in Xenopus embryos. *Dev Biol.* **492**, 133–138
78. Roux, K. J., Kim, D. I., Raida, M., and Burke, B. (2012) A promiscuous biotin ligase fusion protein identifies proximal and interacting proteins in mammalian cells. *J Cell Biol.* **196**, 801–810
79. Kim, D. I., Jensen, S. C., Noble, K. A., Kc, B., Roux, K. H., Motamedchaboki, K., and Roux, K. J. (2016) An improved smaller biotin ligase for BioID proximity labeling. *Mol Biol Cell.* **27**, 1188–1196
80. Huang, S. S., Hsu, L. J., and Chang, N. S. (2020) Functional role of WW domain-containing proteins in tumor biology and diseases: Insight into the role in ubiquitin-proteasome system. *FASEB Bioadv.* **2**, 234–253
81. Zou, X., Levy-Cohen, G., and Blank, M. (2015) Molecular functions of NEDD4 E3 ubiquitin ligases in cancer. *Biochim Biophys Acta.* **1856**, 91–106
82. Hussain, M., Saifi, S., Mohammed, A., and Sengupta, S. (2022) Protocol to detect in vitro and in cell ubiquitylation of mitochondrial DNA polymerase gamma by mitochondrial E3 ligase MITOL. *STAR Protoc.* **3**, 101710
83. Hatstat, A. K., Pupi, M. D., and McCafferty, D. G. (2021) Predicting PY motif-mediated protein-protein interactions in the Nedd4 family of ubiquitin ligases. *PLoS ONE.* **16**, e0258315
84. Nguyen, K. M., and Busino, L. (2020) The biology of F-box proteins: The SCF family of E3 ubiquitin ligases. *Adv Exp Med Biol.* **1217**, 111–122
85. Kipreos, E. T., and Pagano, M. (2000) The F-box protein family. *Genome Biol.* **1**, REVIEWS3002
86. Song, L., and Luo, Z. Q. (2019) Post-translational regulation of ubiquitin signaling. *J Cell Biol.* **218**, 1776–1786
87. Dar, A., Wu, D., Lee, N., Shibata, E., and Dutta, A. (2014) 14–3–3 proteins play a role in the cell cycle by shielding cdt2 from ubiquitin-mediated degradation. *Mol Cell Biol.* **34**, 4049–4061
88. Chen, B. B., Coon, T. A., Glasser, J. R., and Mallampalli, R. K. (2011) Calmodulin antagonizes a calcium-activated SCF ubiquitin E3 ligase subunit, FBXL2, to regulate surfactant homeostasis. *Mol Cell Biol.* **31**, 1905–1920
89. Vadasz, I., Weiss, C. H., and Sznajder, J. I. (2012) Ubiquitination and proteolysis in acute lung injury. *Chest.* **141**, 763–771
90. Li, T., and Zou, C. (2020) The role of deubiquitinating enzymes in acute lung injury and acute respiratory distress syndrome. *Int J Mol Sci.* **21**
91. Weiss, C. H., Budinger, G. R., Mutlu, G. M., and Jain, M. (2010) Proteasomal regulation of pulmonary fibrosis. *Proc Am Thorac Soc.* **7**, 77–83
92. Crew, A. P., Raina, K., Dong, H., Qian, Y., Wang, J., Vigil, D., Serebrenik, Y. V., Hamman, B. D., Morgan, A., Ferraro, C., Siu, K., Neklesa, T. K., Winkler, J. D., Coleman, K. G., and Crews, C. M. (2018) Identification and characterization of Von Hippel-Lindau-recruiting proteolysis targeting chimeras (PROTACs) of TANK-binding kinase 1. *J Med Chem.* **61**, 583–598
93. An, S., and Fu, L. (2018) Small-molecule PROTACs: An emerging and promising approach for the development of targeted therapy drugs. *EBioMed.* **36**, 553–562
94. Li, K., and Crews, C. M. (2022) PROTACs: Past, present and future. *Chem Soc Rev.* **51**, 5214–5236
95. Taleb, S. J., Wei, J., Mialki, R. K., Dong, S., Li, Y., Zhao, J., and Zhao, Y. (2021) A blocking peptide stabilizes lysophosphatidic acid receptor 1 and promotes lysophosphatidic acid-induced cellular responses. *J Cell Biochem.* **122**, 827–834

9 Peroxidative Metabolism of Lung Lipids in Pulmonary Health and Disease

Emphasis on Methods of Analysis of Oxidative Modification of Polyunsaturated Membrane Phospholipids

Austin C. Elliott, Lasya Kopuri, Surya Narla, Patrick J. Oliver, Anita Yenigalla, Elliott D. Crouser, Krishna Rao Maddipati, Thomas J. Hund, and Narasimham L. Parinandi

9.1 INTRODUCTION

Lipids are fundamental building blocks of living cells that play crucial roles in cellular structure and function. Among the cellular lipids, phospholipids (PLs) with a glycerol backbone and esterified fatty acids in the *Sn*-1 (saturated fatty acid) and *Sn*-2 (unsaturated fatty acid) positions are the principal integral components of cellular membranes, including the plasma membrane and subcellular (organelle) membranes. PLs contain both saturated and unsaturated fatty acids esterified to their glycerol backbones that contribute to the fluidity of biological membranes, as illustrated in Figure 9.1. The polyunsaturated fatty acids (PUFAs) of cell membrane PLs are susceptible to oxidative alteration/damage/deterioration that is mediated by oxidants including oxygen radicals, organic radicals, reactive oxygen species (ROS), and reactive nitrogen species (RNS) through the peroxidation of PUFA (broadly called *lipid peroxidation*). Examples of membrane PUFAs are shown in Figure 9.2. Currently, cell membrane lipid peroxidation in major organs of the mammalian body has been emphasized as a central metabolic process in nutrition, toxicology, pharmacology, aging, health, and disease. For the analysis or determination of PUFA peroxidation in cellular membranes of lung that are constantly exposed to oxygen and pathogens and toxicants capable of causing pulmonary oxidative stress leading to lipid peroxidation, a suitable and specific method of analysis is important. Therefore, the present chapter discusses (1) the lipidic nature of cellular membranes; (2) the nature, generation, and reactivity of oxygen radicals and ROS; (3) mechanisms of cellular lipid peroxidation and formation of oxidized PUFA molecules in living cells; (4) currently available methods of analysis/determination of cell membrane PUFA peroxidation and the oxidized PUFA species as applied to studies focused on lung disorders and disease; and (5) the suitable methods to determine PUFA peroxidation and peroxidatively modified/altered lipids in the lung during stress and disease conditions.

 DOI: 10.1201/9781003355243-11

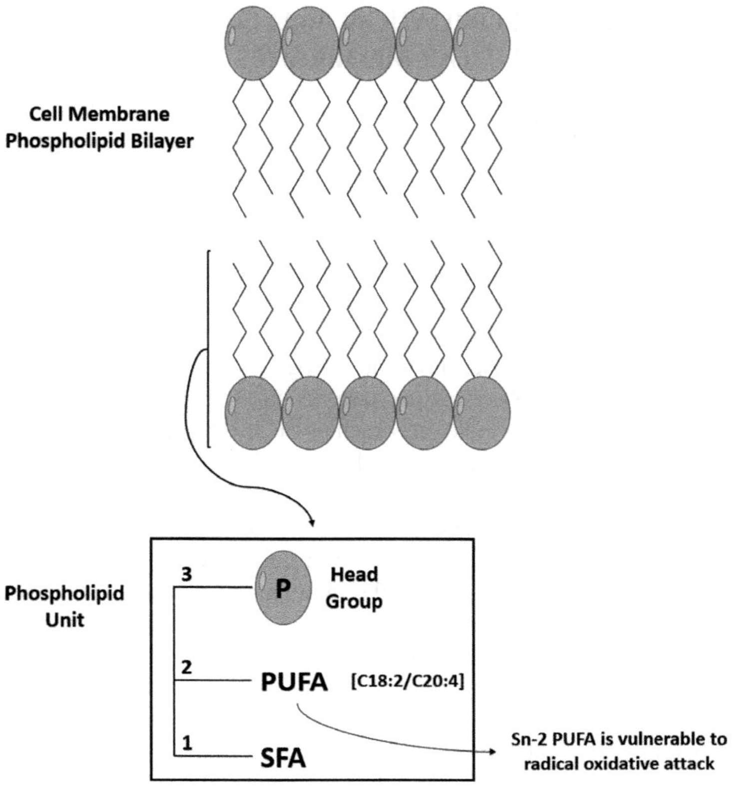

FIGURE 9.1 General structure of cell membrane phospholipid with esterified fatty acids, especially PUFA in the *Sn*-2 position of the glycerol backbone.

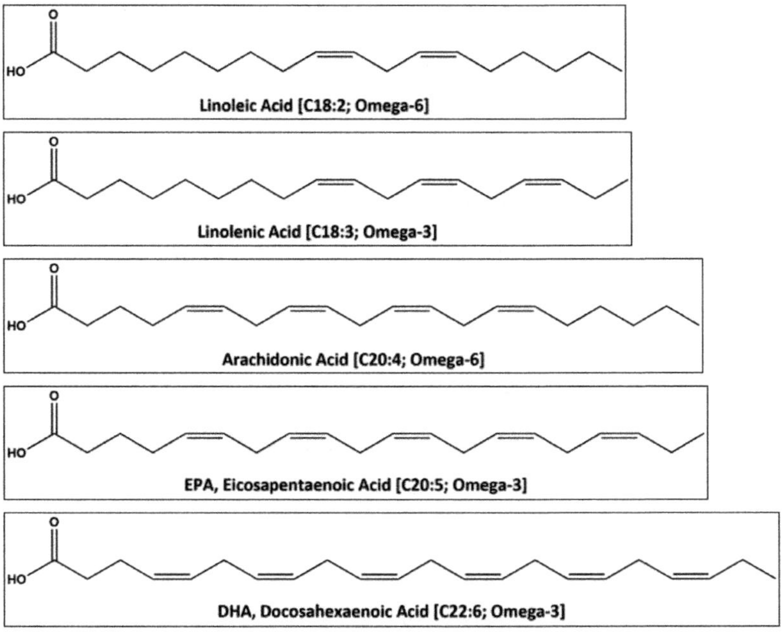

FIGURE 9.2 Membrane phospholipid esterified polyunsaturated fatty acids (PUFAs).

9.2 PATHWAY OF PUFA PEROXIDATION AND OXIDATIVE MODIFICATION IN CELLULAR MEMBRANES

Three major discoveries—the function and role of superoxide dismutase (SOD) (1), biosynthesis of prostaglandins from arachidonic acid (2), and oxidant chemical-induced toxicity through free radical reactions in the biological systems—have created a special and novel field of biochemistry known as *free radical biology and oxidative stress*. A free radical is any molecule that has an odd number of electrons (3). Free radicals participate in normal cellular metabolism, pathological/disease conditions, and biological aging (4). Free radicals are generated through the unimolecular hemolysis of molecules, radiolysis, photolysis, one-electron transfer from transition metals to organic species, and enzymatic catalysis in biological systems (xanthine oxidase, aldehyde oxidase, dihydroorotic acid dehydrogenase, galactose oxidase, flavin dehydrogenase, indoleamine dioxygenase, diamine oxidase, and ribulose-1,5-bisphosphate carboxylase) (5–8). Some example pathways are shown in Figure 9.3. Also, the autoxidation of molecules such as the hydroquinones, flavins, catecholamines, thiols, tetrahydropterins, ferrodoxins, and hemoglobin has been shown to generate free radicals (7, 9, 10). PUFA oxygenases, such as the cyclooxygenases (COXs) and lipoxygenases (LOXs), catalyze the formation of PUFA peroxides from PUFAs (mainly arachidonic acid) through the involvement of free radicals (11). Oxygen free radicals and ROS from the activation of molecular oxygen (O_2) are generated during the microbial infection, inflammation, and phagocytosis in neutrophils, monocytes, and macrophages (12). Toxic compounds including the agrochemicals (e.g., paraquat [PQ]), carbon tetrachloride, bromotrichloromethane, drugs (e.g., adriamycin and daunomycin), oxides of nitrogen, cigarette smoke, ozone, sulfur dioxide, sulfite, and transition metals induce or mediate the formation of the reactive free radical intermediates in biological systems including the lung tissue (13–18). Currently, free radical reactions in the biological systems have emerged in many biochemical and signaling pathways including the mitochondrial respiration and electron transport,

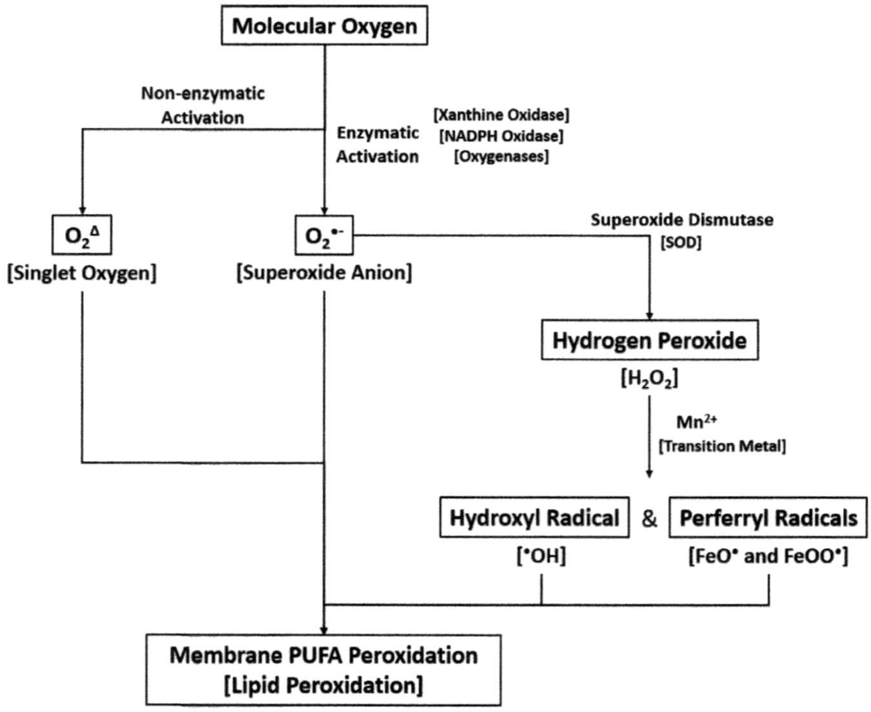

FIGURE 9.3 Non-enzymatic and enzymatic activation of molecular oxygen.

phagocytosis, arachidonic acid cascade of eicosanoid biosynthesis, exposure to and metabolism of toxic chemicals (e.g., ozone, drugs, herbicides, environmental toxicants, and transition metals), disease states (e.g., cancer, cardiovascular and cerebrovascular diseases, and ischemia-reperfusion damage), and biological aging (4, 18, 19).

Cellular membranes made up of PLs rich in PUFAs are constantly exposed to molecular oxygen in aerobes in the presence of transition metals (e.g., iron). Hence membrane PUFAs are susceptible to oxygen free radical–mediated oxidative attack, and those adverse biochemical events are seen in many pathophysiological conditions leading to membrane damage (20). The bilayer of the cell plasma membrane and intracellular organelles (e.g., the Golgi body, lysosomes, mitochondria, endoplasmic reticulum, and nucleus) are made up of PLs rich in PUFAs. PLs of the cellular membranes (either hydrophobic or amphipathic) facilitate creation of the radical centers in a non-aqueous environment, whereas the amphipathic lipids are important components of the cellular membranes (21). Therefore, PUFAs of the cellular membrane PLs tend to undergo free radical–mediated oxidative attack and deterioration in oxygen-rich environment (20), depicted as follows:

PUFA-H + R$^\bullet$ **PUFA$^\bullet$ + RH** [1]
PUFA$^\bullet$ + O_2 **PUFA-O-O$^\bullet$** [2]
PUFA-O-O$^\bullet$ + PUFA **PUFA-O-O-H + PUFA$^\bullet$** [3]

As shown in the above sequence of reactions, a free radical (R$^\bullet$) initiates attack on the membrane PUFA causing the formation of lipid (PUFA) hydroperoxides in the presence of molecular oxygen (O_2). The free radical R$^\bullet$ causes hydrogen (H) abstraction from the membrane PUFA at the methylene-interrupted double bond(s) leading to the formation of a PUFA radical (PUFA$^\bullet$), as shown in the initiation reaction [1]. The PUFA$^\bullet$ then reacts with molecular oxygen (O_2) forming a PUFA peroxy radical (PUFA-O-O$^\bullet$) [2]. The PUFA peroxy radical formed reacts with an unoxidized PUFA molecule leading to the formation of another PUFA free radical (PUFA$^\bullet$) and the PUFA hydroperoxide in a chain reaction [3]. This process is further illustrated in Figure 9.4.

9.3 CHEMICAL BASIS OF ANALYSIS AND METHODS OF DETERMINATION OF CELLULAR MEMBRANE LIPID PEROXIDATION AND OXIDATIVE MODIFICATION

Free radical (oxidant)-mediated cell membrane PUFA oxidation leads to the formation of carbonyls including malonaldehyde or malondialdehyde (MDA) (22) through the above generalized sequence of reactions via free radical-mediated hydrogen abstraction from membrane PUFA at the methylene-interrupted double bond(s), rearrangement of double bonds, oxygen consumption, and formation of a molecule with 5-membered ring that undergoes bond scission generating MDA. The structures of MDA and other reactive carbonyls are shown in Figure 9.5. More importantly, because molecular oxygen (O_2) is critical in the PUFA peroxidation process, the consumption of oxygen (cyanide-insensitive and non-mitochondrial oxygen consumption) has been measured as a biochemical and metabolic index of membrane PL PUFA peroxidation. Thus, the measurement of MDA (by spectrophotometric and chromatography methods) is widely exploited as an index of membrane PUFA peroxidation resulting from the oxidant attack. The conjugated diene (CD) rearrangement of methylene-interrupted double bonds in the PUFA is the sole basis for the increased absorption of ultraviolet (UV) light at 232–233 nm by the peroxidized PUFA. This unique feature of CD rearrangement of PUFA double bonds, a distinguishing property of the peroxidized PUFA-containing membrane lipids, is widely used in the analysis/determination of peroxidation of PUFA. The degradation molecules of cellular membrane peroxidized PUFAs, such as the reactive carbonyls, form adducts with other cellular macromolecules including peptides, proteins, PLs, and nucleic acids. MDA (CHO-CH_2-CHO), the widely examined lipid peroxidation end product, is shown to form Schiff base adducts with amines of proteins, PLs, and nucleic acids in biological systems (23).

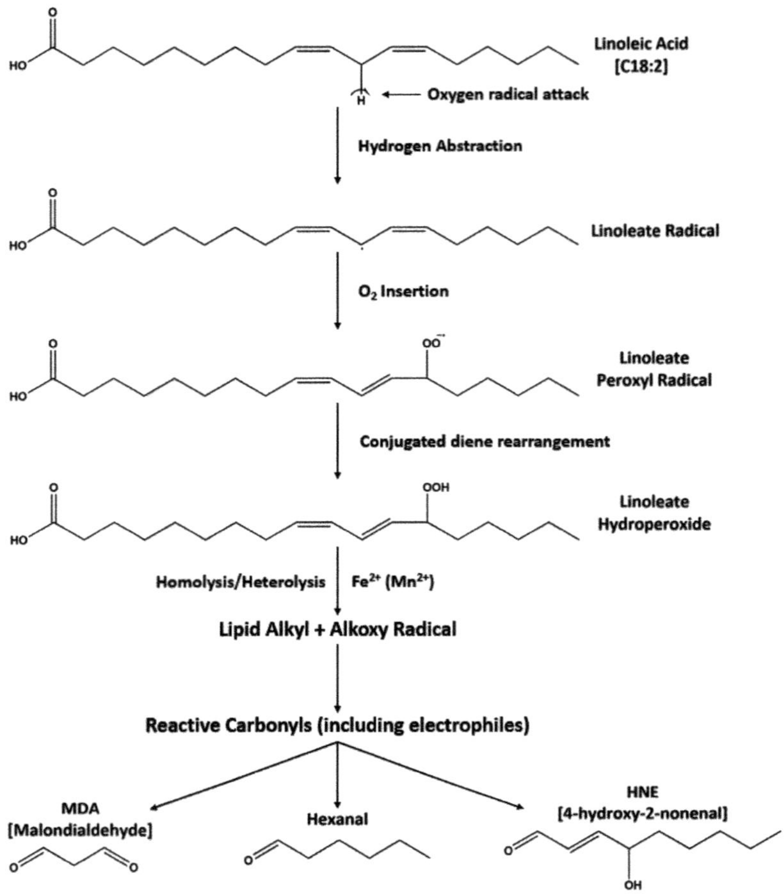

FIGURE 9.4 Oxygen radical-catalyzed PUFA peroxidation of cell membrane lipids.

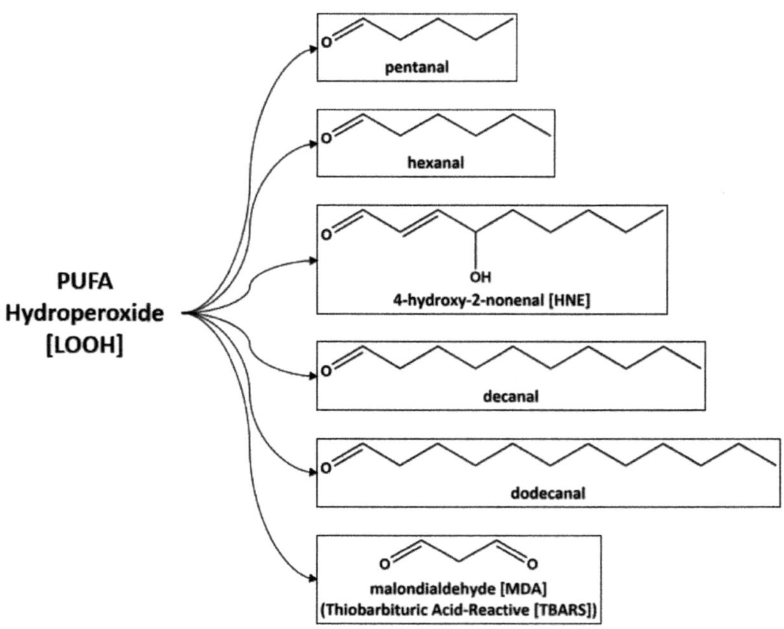

FIGURE 9.5 Membrane PUFA peroxide-formed reactive carbonyls and electrophiles.

These Schiff base adducts originating from cellular membrane lipid peroxidation display typical fluorescence at 360–380 nm excitation maxima and 440–470 nm emission maxima and are widely known as the *aging pigments, lipofuscin*, or *ceroid*; the buildup of these lipoperoxidative fluorescent pigments in tissues and organs serves as an index of cellular membrane PUFA peroxidation damage and biological aging in animals (23, 24). The cellular membrane PUFA peroxidation of PUFA also leads to the formation of volatile hydrocarbons such as ethane and pentane from the organs, including the lungs, and the release of those volatile hydrocarbons is used as a dependable index of noninvasive determination of in vivo lipid peroxidation (25). As the cellular membrane PL PUFAs undergo peroxidative deterioration, the loss of PUFAs in the membrane PLs also serves as a reliable analytical index of membrane lipid peroxidation. It is possible that certain peroxidized membrane PLs contain esterified PUFA hydroperoxide(s)/PUFA radical/reactive carbonyl(s) in the *Sn*-2 position and these peroxidatively modified cellular membrane PLs can be analyzed both qualitatively and quantitatively in cells and tissues undergoing lipid peroxidation. The biosynthesis of metabolic and bioactive lipids such as eicosanoids including leukotrienes (LTs) and prostaglandins (PGs) generated by the enzymatic catalysis of COXs and LOXs is crucial in inflammation (26). The lipid oxygenases such as COXs, LOXs, and cytochrome P450 (CYP) epoxygenases catalyze the oxidation of free and esterified PUFAs forming metabolites including the lipid peroxy radicals, lipid hydroperoxides, and lipid endoperoxides (27). The PUFA hydroperoxides also undergo reduced iron (Fe^{2+})-catalyzed decomposition leading to the formation of PUFA alkoxy radicals and oxidized iron (Fe^{3+})-catalyzed formation of PUFA peroxy radicals. Hence it is important to note that both the qualitative and quantitative analysis of eicosanoids (COX and LOX metabolites of arachidonic acid), PUFA radicals, PUFA peroxides, and PUFA epoxides formed from the peroxidized cellular membrane PLs will also provide suitable indices and specific biomarkers of peroxidative alteration of cellular membrane lipids under stress, pathophysiological, and disease conditions. Figure 9.6 provides the pathways of some of these enzymatic formations of oxidized metabolites.

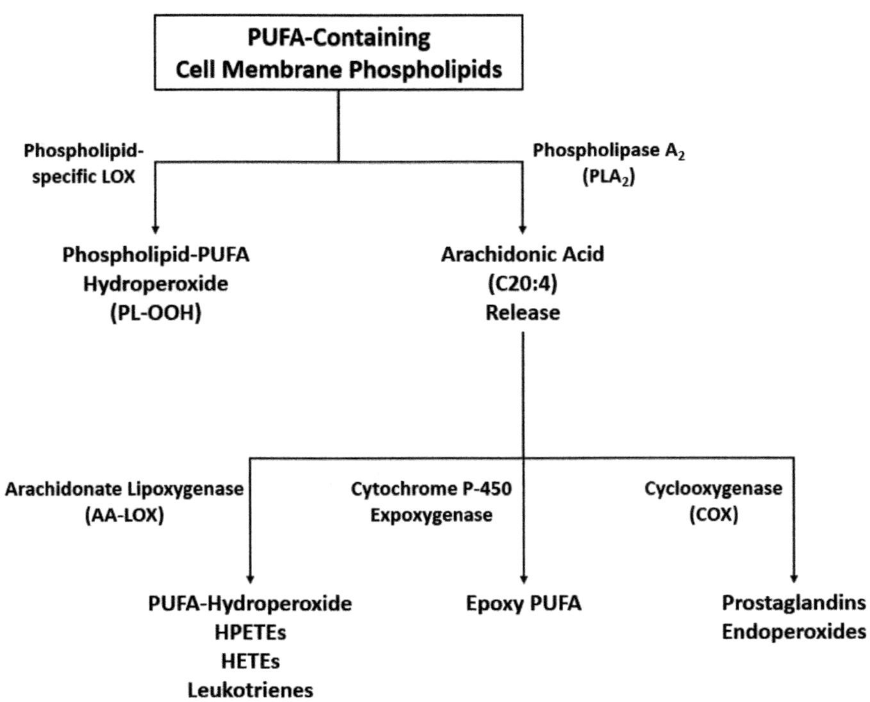

FIGURE 9.6 Lipid oxygenase-catalyzed formation of oxidized metabolites of PUFA.

9.4 SUITABLE METHODS OF ANALYSIS OF INDICES OF CELLULAR MEMBRANE LIPID PEROXIDATION AND OXIDATIVE MODIFICATION

The methods of qualitative and quantitative analysis of products/metabolites of cellular membrane PUFA peroxidation are extremely critical in the determination of extent of lipoperoxidative alteration of cells, tissues, organs, and organisms under the conditions of stress and pathophysiological states. The initial (primary) peroxidation products of cellular membrane PUFAs including CDs, PUFA radicals, and PUFA hydroperoxides are fundamental to the analysis of cellular membranes and determination of the physical and chemical alterations of the PUFAs, PLs, and lipoproteins (28). Furthermore, other important secondary peroxidation products arising from the peroxidative alterations of PUFAs, including F_2-isoprostanes, hydroxyalkenals (29), and 4-hydroxy-2-nonenal (4-NHE), have been analyzed as reliable indices of lipid peroxidation in several biological systems (30). Figure 9.7 summarizes the various analytes related to lipid peroxidation and their methods of detection. To analyze/determine the peroxidized lipids and lipid peroxidation in cellular membranes,

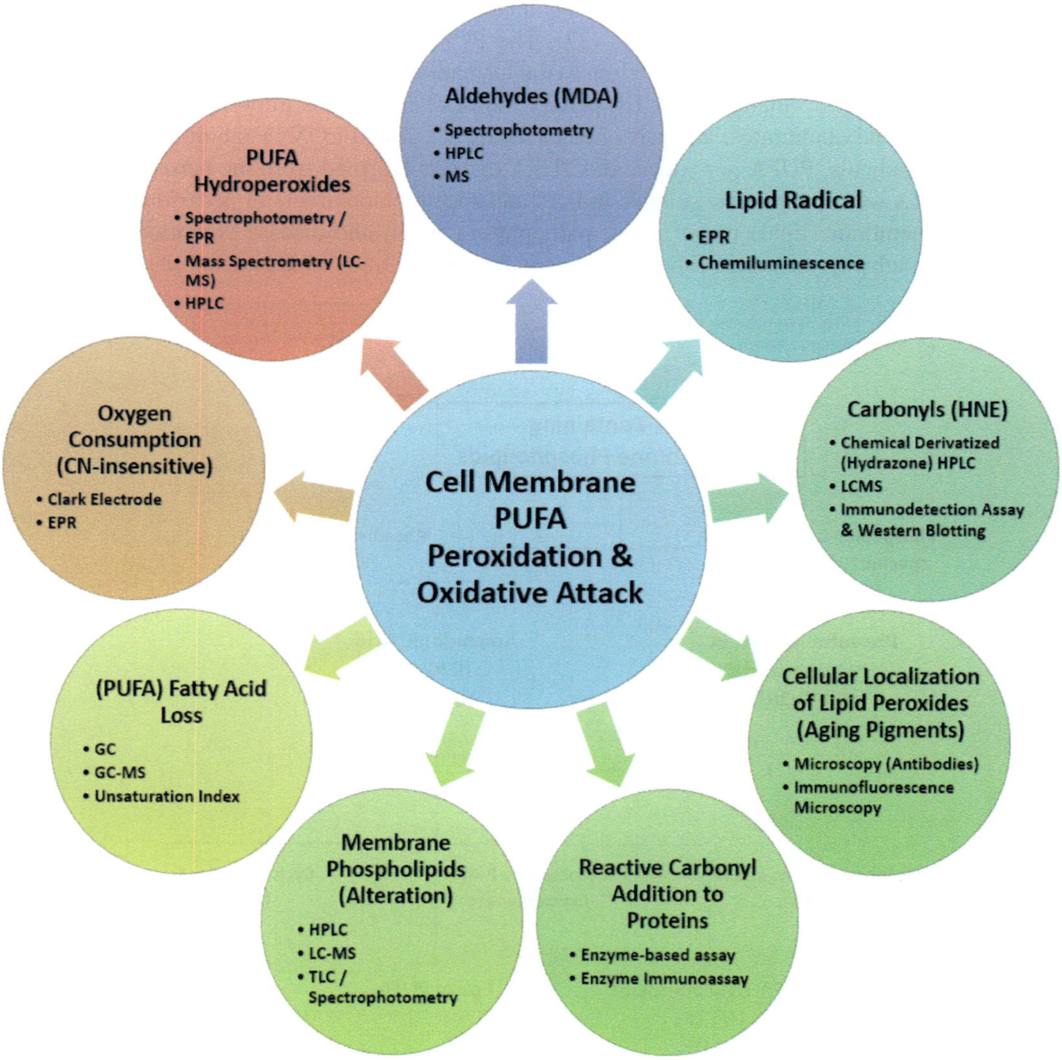

FIGURE 9.7 Methods of analysis of lipid (PUFA) peroxides and oxidatively modified lipoperoxide metabolites/products.

a suitable, reliable, and reproducible method of analysis focusing on a specific analyte (e.g., oxygen consumption, release of volatile hydrocarbons, PUFA or peroxidized PUFA, PL modifications, PUFA CDs, and PUFA hydroperoxides, reactive carbonyls and aldehydes, isoprostanes [IsoPs], Schiff base fluorescent adducts, and volatile hydrocarbons) has been chosen by the investigator. These analytical methods demand meticulous extraction and derivatization procedures with unique standards coupled with the use of techniques including polarography, chemiluminescence, UV-visible spectrophotometry, fluorescence spectrometry, column, chromatography, thin-layer chromatography (TLC), high-performance liquid chromatography (HPLC), gas chromatography–mass spectrometry (GC-MS), liquid chromatography–mass spectrometry (LC-MS), nuclear magnetic resonance (NMR), electron spin resonance (EPR), differential thermal calorimetry, immunoassay, sodium dodecyl sulfate-polyacrylamide gel electrophoresis (SDS-PAGE), and Western blotting, fluorescence microscopy, and radiochemical methods, to name a few.

The most simple and straightforward analytical methods of peroxidative degradation of PUFAs include determination of the acid value, determination of the peroxide value, the thiobarbituric acid (TBA) test, the Kreis test, analysis of total volatile carbonyl compounds, and determination of oxiranes (31, 32). Many of these analytical techniques are now available as commercial kits that are widely used to determine/assay lipid peroxidation in nutrition, pharmacology, toxicology, health, and disease. Assay of lipid peroxidation in biological systems uses the 2-TBA–based simple and sensitive method. TBA reacts with MDA formed from the PUFA peroxidation to form light red to pink TBA-reactive substances (TBARS). Stoichiometrically, one molecule of MDA reacts with two molecules of TBA resulting in a condensation product (33) with a distinct red color. Thus, the resultant TBARS (MDA-TBA_2 complex) is determined spectrophotometrically at 532 nm and spectrofluorimetrically by measuring the fluorescence excitation and emission at 515 nm and 553 nm, respectively (30). In addition, other reliable and accurate methods of determination of the lipid peroxidation–derived reactive aldehydes (carbonyls) such as MDA and 4-HNE by using the 2,4-dinitrophenylhydrazine derivatization and subsequent HPLC separation are available (34, 35).

The CD PUFA hydroperoxides formed from the oxidation of PUFAs (free and esterified) of cellular membranes are analyzed both qualitatively and quantitatively byspectrophotometric methods because the CDs exhibit maximum absorbance of the UV radiation (36) at 232–234 nm (37). In this analytical method, membrane lipids are extracted by Folch extraction (2:1 chloroform-methanol, vol/vol), dried under a stream of nitrogen gas, then reconstituted in cyclohexane (a solvent that has no UV absorbance at 232–234 nm), and the UV absorbance of CDs in the peroxidized PUFAs is determined (38, 39).

PUFA hydroperoxides (both free and esterified PUFAs) oxidize iodide (KI) and release stoichiometric iodine (I_2), which further reacts with iodide (I^-) to form triiodide (I_3^-) that can be determined spectrophotometrically at 353 nm as an index of lipid peroxidation (38). Ferrous (Fe^{2+}) oxidation-xylenol orange (FOX) assay is used to assay lipid peroxidation in several biological systems and in combination with the HPLC separation, and this method can determine different classes of lipid (PUFA) hydroperoxides. The FOX assay is dependent on the oxidation of Fe^{2+} to Fe^{3+} being catalyzed by the lipid (PUFA) hydroperoxide and then the subsequent reaction of FOX with Fe^{3+} to form a spectrophotometrically measurable blue-purple complex with an absorbance maximum at 560 nm. There are two widely used FOX assays: FOX 1 and FOX 2. FOX 1 assay is specifically used to determine low levels of hydrogen peroxide in aqueous media. On the other hand, FOX 2 assay is selectively used to determine lipid (PUFA) hydroperoxides in low concentrations despite the presence of high background levels of non-peroxidized fatty acids. The FOX assay is claimed to be convenient, inexpensive, and precise with broad applicability in the analysis of lipid (PUFA) peroxides in biological systems (37).

The extent of cellular membrane PUFA peroxidation is determined by the loss of PUFAs (both non-esterified and esterified) and the formation of lipid (PUFA) hydroperoxides. Alterations in the fatty acid composition and loss of PUFA in cellular membrane lipids (PLs) are determined by GC-MS after hydrolysis of membrane PLs and derivatizing the fatty acids into fatty acid methyl esters (39). By application of this method, the loss of PUFAs as a result of the peroxidative insult in different cellular membrane PL classes also can be determined (as changes in lipid phosphorus) as a reliable index of lipid peroxidation (39). Also, the changes in PUFA composition, loss of PLs, and nature and amount of membrane PUFA hydroperoxides have been determined as indices of lipid peroxidation in the biological membranes by gas chromatography (GC) and HPLC (39). The LOX-catalyzed synthesis of phosphatidylinositol peroxides has been characterized and determined (40).

The chemistry of oxidized lipids (PUFAs) is highly complex with diverse peroxidative products and poses analytical challenges encountered with tedious preparative methods (41). However, the application of mass spectrometry (MS) has been currently a sought-after analytical method that can distinguish and analyze up to 50 different lipid classes and a plethora of lipid species with precision and reproducibility (29). MS characterizes and determines the ionized lipid molecules according to their mass-to-charge ratio (m/z), and this can be applied to the MS analysis of the oxidatively modified lipids (42). The analytical advantage of MS is that the technique can be conveniently combined with the liquid chromatography (LC) method to avoid the rigorous chemical derivatization of the analytes as opposed to the GC-MS analysis wherein derivatization of PUFAs and oxidized PUFAs is a strict requirement to prior hydrolysis and release of the PUFAs for analysis or even the derivatization to produce the volatile molecules for the GC separation (42). Although electrospray ionization (ESI) and matrix-assisted laser desorption/ionization (MALDI) are known to be effective low energy ionization techniques used for the analysis of PLs, this method will not distinguish the positional distribution of the oxygenated groups without hydrolyzing the sample and analyzing the individual PUFAs (41). However, the ESI and MALDI techniques are used to simultaneously analyze multiple lipid classes giving rise to "shotgun lipidomics" (41), a direct infusion-based lipid analysis technique using the tandem, high mass accuracy, and multidimensional MS-based (MDMS-SL) methods (29). Structural properties of lipids can be obtained from the tandem MS-based methods where the molecule is analyzed by fragmenting ions. In this technique, a collision cell connects two analyzers, where in the first analyzer an ion is broken down by an inert gas in the collision cell to produce fragmented ions, which are then analyzed in the second analyzer (41). The quadrupole mass analyzer is the most common analyzer that only allows ions of specific m/z values to pass through the detector (41). This method is rapidly evolving, and the modern methods are equipped with a greater number of analyzers with enhanced fragmentation levels to elevate accuracy of the analysis (41). The MDMS-SL shotgun lipidomics method is capable of analyzing the unique molecular differences/properties among several classes of lipids even at low abundance with high precision and greater accuracy to analyze the oxidized PUFAs (e.g., LOX metabolites of PUFA), secondary peroxidative aldehyde products (e.g., reactive carbonyls such as hydroxyalkenals and 4-HNE), and oxidatively modified proteins in the biological systems undergoing lipid peroxidation (29). Therefore, the MS method of analysis of lipid peroxidation products (qualitative and quantitative) has emerged as a highly essential analytical technique in the field of oxidative lipidomics.

Electron paramagnetic spectroscopy (EPR spectroscopy) is a unique analytical technique useful for the analysis of lipid radicals. EPR spectroscopy uses a spin trap (diamagnetic compound) which reacts with a radical species to produce a spin adduct (stable free radical product) that is analyzed by EPR spectroscopy. The EPR spectrum of a radical exhibits a typical g value and hyperfine splitting pattern(s) for the characterization and determination of the radical species under investigation (43, 44). The g value signifies the orbital spin and angular momentum of an electron, whereas the hyperfine splitting patterns determine the distance of an electron from the nuclei and identify the specific nuclei and their abundance (43). EPR spin trapping is widely used for detection and analysis of oxygen radicals such as the hydroxyl and superoxide radicals, free radical metabolites, and most of the peroxidized lipid (PUFA) radicals. The EPR spin trap 5,5-dimethyl-pyrroline-1-oxide is widely used as it reacts with many radicals to form long-life spin adducts suitable for the EPR spectroscopy. However, to trap the carbon-centered radicals such as the lipid (PUFA) radicals, α-(4-pyridyl-1-oxide)-N-*tert*-butylnitrone (POBN) is selectively used, as the POBN–carbon-centered lipid radical adducts have longer life (hours to days) for the efficient EPR analysis of lipid radicals.

High-resolution NMR spectroscopy techniques such as ^{1}H-,^{13}C-, and ^{31}P-NMR spectroscopy are widely used in lipid analysis and lipidomics to characterize and determine the lipid (PUFA) peroxidation products (30, 45). The primary lipid oxidation products are analyzed by proton NMR spectroscopy by identifying/determining the levels of bis-allylic hydrogen atoms that arise from oxidation of

PUFAs. Proton NMR spectroscopy is advantageous for the analysis of lipid peroxidation products as no rigorous sample preparation and derivatization are necessary retaining the original integrity of the sample during the NMR analysis of the peroxidized lipids (36). In oxidized lipids, the ratios of aliphatic to olefinic protons and aliphatic to diallylmethylene protons increase (30). The PUFA hydroperoxides and the peroxidatively formed short-chain saturated and unsaturated aldehydes are characterized by typical resonances in the range of 10.5 to 8 ppm (45). Therefore, it is convincing that NMR spectroscopy appears to be a suitable method for the structural analysis of peroxidized PUFAs.

Lipid peroxidation causes extensive oxidative modifications of proteins that are both spatially and temporally controlled. As the oxidatively modified proteins may be trapped and localized in some cellular organelles and regions of the cell that are not as easily accessible, selective fluorescent dyes (e.g., boron-dipyrromethene [BODIPY: 4,4-difluoro-4-bora-3a,4a-diaza-s-indacene]) are combined with certain organelle-specific probes that facilitate the intracellular microscopic detection and localization of lipid peroxidation. BODIPY offers several advantages, such as its high sensitivity and narrow range of the excitation and emission wavelengths. The extent of lipid peroxidation and lipid radical formation can be visualized in the mitochondria by using the fluorogenic dyes that act as antioxidants scavenging the peroxy radicals (46).

Lipid (PUFA) peroxidation is also known to cause non-enzymatic and post-translational protein modifications in cells. These lipoperoxidative modifications of proteins are detected and analyzed. Most of these protein alterations are caused by the secondary products of lipid peroxidation (e.g., carbonyls), which can be analyzed and quantified by Western blotting and immunoassays (47). The covalent protein adducts formed by 4-HNE are analyzed and quantified by an enzyme-linked immunosorbent assay (ELISA) (48). Chemiluminescence methods are also applicable to study lipid peroxidation in biological systems as the singlet oxygen and triplet carbonyl compounds are chemiluminescent molecules arising from the peroxidized PUFAs (49, 50). The chemiluminescence assays offer greater sensitivity for detection as compared to spectrophotometry and fluorometry (49).

Several manufacturers and commercial vendors such as Abcam, Cayman Chemical, and Sigma-Aldrich sell premade kits to assay cellular and tissue lipid peroxidation by analyzing the levels of formation of products of lipid peroxidation including TBARS, MDA, lipid peroxides, hydroxyalkenals, and 4-HNE. Cell Biolabs supplies oxidized low-density lipoprotein (LDL), oxidized high-density lipoprotein (HDL), and IsoP ELISA kits, as well as specific HNE antibodies for the assay of lipid peroxidation in biological systems. The choice of the investigators to use commercial kits is on the rise as these kits are convenient to easily and quickly assay lipid peroxidation in biological systems. Oxidative lipidomics is a branch of lipidomics that analyzes structure, function, and quantification of oxidized lipids that involves LC-MS/MS analysis and data analysis, ultimately integrating the analytical data into a biological system or mechanism under study (51). Rapid developments in the areas of computer modeling and machine learning provide tools for the high throughput qualitative and quantitative analysis of lipids. Databases including LIPID MAPS, LipidHome, and SwissLipids are the most used databases for the MS analysis of lipids. The ALEX123 database provides fragmentation patterns and software for the automatic analysis of MS spectra. The Skyline proteomics software can be applied to complex lipid analysis (52). Bioinformatics tools offer advantages in the analysis and modeling of oxidized lipids, favoring the advancement of the discipline of oxidative lipidomics.

9.5 LIPIDS OF THE RESPIRATORY TRACT AND LUNGS

With modern advancements in the field of lipidomics, the diverse lipidomes of the respiratory tract and of the lungs have become less of a mystery in recent years. The resolution, or distinguishability between peaks, of MS technology has greatly improved over the last few decades. However, increases in resolution alone are often not enough to differentiate between certain lipid species. Although it may be possible to deduce the exact sum composition of a molecule through high-resolution mass spectrometry (HR-MS), there can exist dozens of isomers with the exact same composition. Further differentiation of molecules can be achieved through combining HR-MS with other methods of fragmentation, chromatography, or ion mobility (53), such as tandem mass spectrometry (MS/MS), LC-MS, or GC-MS.

The lipidome of the lung is diverse, but previous studies have demonstrated the existence of distinct enrichments of specific lipid classes within the major lung cell types, suggesting a level of cellular cooperation that supports critical lung function (54). Lung endothelial cells tend to have enrichments of sphingolipids and glycerophospholipids that have known roles in apoptosis and blood coagulation, such as ceramides, diacylglycerophosphoserines, plasmalogens, and diacylglycerophosphoethanolamines. Lung epithelial cells, specifically alveolar type II epithelial cells, are a major source for the lipids contained in pulmonary surfactant. Around 90% of surfactant consists of lipids by mass (the remaining 10% being protein), most of which are PLs. The most abundant PL, constituting approximately 40% of surfactant by mass, is a saturated phosphatidylcholine (PC) called dipalmitoyl phosphatidylcholine (DPPC) (55). Both mesenchymal cells and immune cells show enrichments of glycerolipids. Mesenchymal cells have high concentrations of short-chain triglycerides (TGs), primarily fully saturated and monounsaturated TGs. The fatty acids contained in these TGs are the main fatty acids found in surfactant lipids (54). Immune cells have high concentrations of long-chain TGs. Recent evidence suggests that these TGs are a potential source of fatty acids used as lipid mediators (much like the fatty acids of PLs), which may facilitate immune signaling (54).

9.6 PEROXIDATION AND OXIDATIVE MODIFICATION OF LIPIDS IN THE RESPIRATORY TRACT AND LUNGS: RELEVANCE TO STRESS AND DISEASE

Figure 9.8 highlights some of the pulmonary disorders and irritants that have been found to be associated with membrane PUFA peroxidation. Mechanisms of lipid peroxidation can be classified as

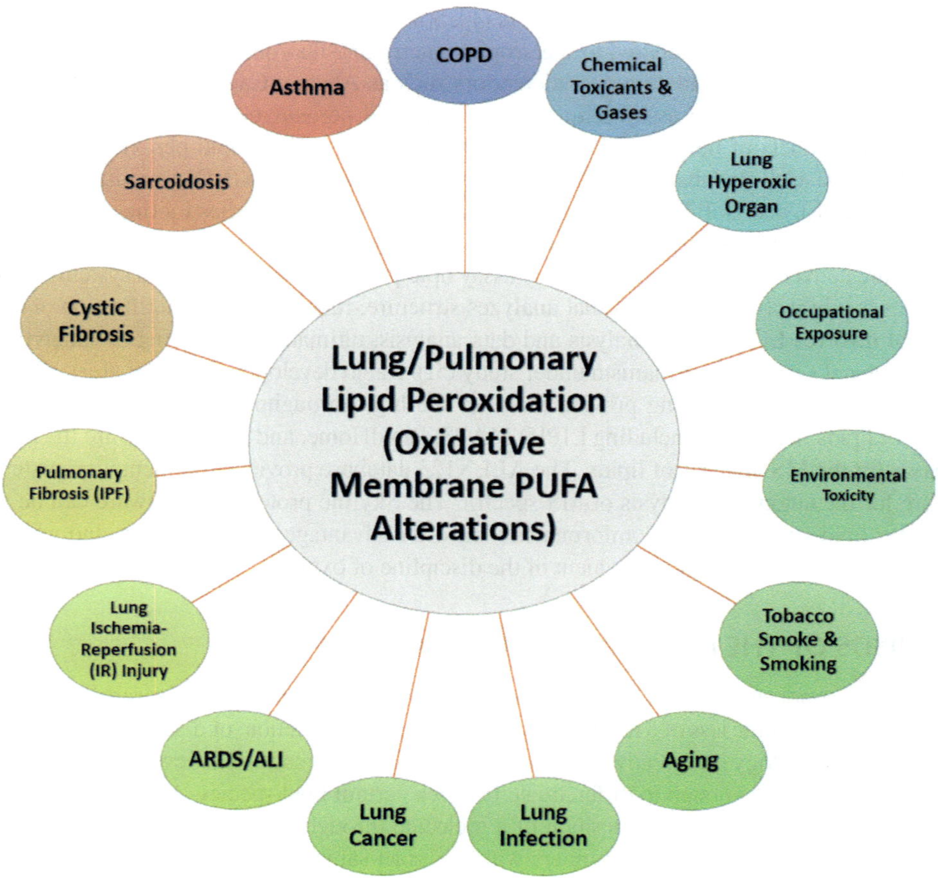

FIGURE 9.8 Association of membrane PUFA peroxidation and oxidative modifications with lung and pulmonary disorders and diseases.

either non-enzymatic and/or enzymatic. Enzymatic lipid peroxidation is catalyzed by the oxygenase enzymes including LOX, COX, CYP, human aldo-ketoreductases, and hydroxysteroid dehydrogenases (56). Omega-6 PUFAs, such as linoleic acid and arachidonic acid, can be acted upon by the LOX family of enzymes to generate 13-hydroperoxyoctadecadienoic acid and 15-hydroperoxyoctadecadienoic acid, respectively. These compounds serve as precursors to 4-HNE, which can be measured to indicate lipid peroxidation (57).

Arachidonic acid can be acted upon by COX to generate prostaglandin H2 (PGH_2). Thromboxane A2 synthase converts PGH_2 to thromboxane A_2, generating MDA as a side product (57). MDA is one of the most used biomarkers to indicate lipid peroxidation. Whereas PGs are formed in an enzymatic manner by COX, the generation of a similar class of compounds (IsoPs) is done non-enzymatically. IsoPs, specifically F_2-IsoPs, are also used as biomarkers for oxidative stress (58). Lipid peroxidation can be the result of toxicological or environmental effects. For example, PQ poisoning can lead to the production of excessive free radicals. PQ enters the cell and goes through the process of redox cycling, in which the PQ undergoes alternate reduction followed by reoxidation. The generated free radicals can extract hydrogen atoms from PUFAs (59).

Lipid peroxidation can also be brought on by infection and disease. In recent years, research surrounding the pathological promotion of lipid peroxidation within the lungs has largely been focused on a form of programmed cell death, ferroptosis. At its core, ferroptosis results from the accumulation of intracellular iron, which reduces H_2O_2 by a single electron to produce a hydroxyl radical in what is known as the Fenton reaction. The hydroxyl radicals go on to promote lipid peroxidation (60). There are several regulatory signaling pathways involved in ferroptosis. For example, glutathione peroxidase 4 (GPX4) uses glutathione to convert lipid hydroperoxides into nontoxic lipid alcohols. The eventual depletion of glutathione, and thus deactivation of GPX4, can lead to an accumulation of lipid ROS and ferroptosis. There is evidence to suggest a close relationship between ferroptosis and a multitude of lung diseases, including chronic obstructive pulmonary disease (COPD), asthma, cystic fibrosis, pulmonary fibrosis (PF), lung ischemia-reperfusion injury, acute lung injury (ALI)/acute respiratory distress syndrome (ARDS), lung cancer, and various pulmonary infections (60).

With so many stresses and diseases closely tied to lipid peroxidation, developing accurate ways to measure levels of lipid peroxidation within the lungs can lead to earlier diagnoses and treatments. One promising avenue is that of aldehyde detection. Aliphatic aldehydes are a common product of lipid peroxidation, which increase in production alongside the degree of fatty acid unsaturation (61). The measurement of these aldehydes can be performed quickly and noninvasively through the breath of patients suffering from lung disease (62).

9.7 SMOKING AND OXIDATIVE MODIFICATION OF LIPIDS IN THE RESPIRATORY TRACT AND LUNGS

Cigarette smoking is one of the leading causes of preventable death in the world and is closely tied to lung disease and severe damage. Exposure to cigarette smoking has been known to be associated with an increase in ROS within cells and a reduction in antioxidant glutathione (63). The lungs play a crucial role in being exposed to toxic chemicals emitted by smoking as they are the primary organs directly exposed to cigarette smoke. As a result, the lungs are extremely vulnerable to the toxic effects of oxidation and the production of free radicals. Cigarettes are composed of a plethora of harmful substances such as nicotine, an alkaloid compound. Nicotine is rapidly absorbed into the blood stream when individuals smoke a cigarette because the lungs come into direct contact with the chemicals. This process leads to nicotine having devastating effects primarily on the respiratory system and the lungs. Nicotine's presence in the lungs can alter typical functions and lead to the development of several respiratory conditions (64). From early research, nicotine has been associated with lipid peroxidation. In rats incubated with nicotine, there was an increase in lipid peroxidation level in pancreatic tissues (65), and Chinese hamster ovary cells that were incubated with nicotine also saw an increase in lipid peroxidation (66). Nicotine can promote oxidative stress in the lungs through several pathways; the ROS generated reacts with several biomolecules including lipids, which can ultimately lead to apoptosis (67).

CD44 is a receptor involved in clearing of lipid metabolism among other key processes; however, this receptor's levels were reduced in alveolar macrophages from smokers when compared with non-smokers. Chronic smoking can be associated with inflammation and lipid accumulation in alveolar macrophages, a key feature of foam cells where cells are enlarged due to a buildup of lipids. Several studies also corroborate that chronic smoking can lead to lipid peroxidation within the lungs (68). In one study, the delivery of oxidized lipids to rats closely mirrored the inflammation induced by cigarette smoking. The compound hOxLDL, an abundant source of oxidized lipids, was used. The compound was then administered intranasally to rats, and the results indicated effects that closely mimicked the pivotal cellular and molecular features in relation to inflammation onset by cigarette smoking. The delivery of hOxLDL also produced foamy macrophages similar to those produced through exposure to cigarette smoke. This study provides evidence that there is a close relationship between lipid peroxidation and chronic smoking (69). COPD can also be brought on by the oxidative stress caused by cigarette smoking (70). Another study performed on rats demonstrated that the direct administration of nicotine generated free radicals within tissue and significantly increased tissue injury and oxidative stress. By measuring MDA levels within the cells, a parallel pattern was displayed where MDA significantly increased in the lungs of the rats administered nicotine. High MDA levels indicate an abundance of lipid peroxidation and can be associated with the high levels of oxidative stress generated by nicotine (71).

9.8 ENVIRONMENTAL STRESS/TOXICITY AND LIPID PEROXIDATION IN THE RESPIRATORY TRACT AND LUNGS

9.8.1 METALS

Cadmium is a metal known to induce ROS and lipid peroxidation within a cell. In one study, the treatment of A549 cells with cadmium at concentrations ranging from 20 to 100 μM for 24 hours resulted in increased levels of lipid peroxidation. Lipid peroxidation levels were dose-dependent, so as the dose of cadmium increased, so did lipid peroxidation. Cells treated with 100 μM cadmium experienced the most lipid peroxidation by a significant amount. These results suggest that the heightened activity of lipid peroxidase when treated with cadmium may onset lipid peroxidation as well as membrane damage that can lead to cell death. MTT was also used as a method to measure cell death and confirmed the cytotoxic effects of cadmium when treating A549 cells (72).

Similarly, other metals such as aluminum and zinc are also known to induce lipid peroxidation. In one study, TBARS and NOx levels significantly increased when treated with aluminum and zinc compounds. Animals that were administered both Al_2O_3-nps and ZnO-nps displayed a tremendous increase in TBARS and NOx levels compared to the control groups (73). Chromium is another metal that can spike lipid peroxidation within cells, especially within the lungs. Studies have demonstrated that upon administration of chromium, the rate of lipid peroxidation in liver tissue increased significantly (74). In contrast, compounds can also alleviate the effects of oxidative stress. Curcumin is a compound that can do so and has the ability to protect from lipid peroxidation (75).

9.8.2 PARTICULATE MATTER

Airborne particulate matter (PM) is a known inducer of lipid peroxidation that can severely damage the body. In a study by Bin Luo, the lungs of rats were exposed to $PM_{2.5}$ and were then measured for TBARS levels. The results revealed that higher TBARS levels were found in rat lungs exposed to $PM_{2.5}$. This information suggests that ROS has a pivotal role in lung inflammation by $PM_{2.5}$ (76). Previous studies have demonstrated that exposure to PM_{10}, with a diameter of 10 μm or less, leads to a loss in function of antioxidant enzymes such as catalase, glutathione reductase (GR), SOD, and glutathione S-transferase in lung epithelial cells. Even when the researchers tried to use hydroxyl chelation, the damage caused by PM_{10} was irreversible. This study calls attention to the devastating effects of PM_{10}, especially on lung epithelial cells, because of its impairment of enzymatic activity and irreversible nature (77). The results obtained from one study revealed that exposure to PM_{10} resulted in a reshaping of lung lipids, specifically an increase in PL and cholesterol content, and caused oxidative stress that led to lipid peroxidation. These findings demonstrate that consistent exposure to PM_{10} in mice triggered lipid peroxidation induced by oxidative stress (78).

9.8.3 Toxic Gases

MDA levels have been shown to significantly increase in the lung tissue of piglets exposed to 50 ppm ammonia, demonstrating that lipid peroxidation within the lung tissue of these piglets increased significantly (79). Upon inhalation, ozone primarily interacts with cells and fluid lining the respiratory airways and interacts with components that are rich in lipids. Several studies have demonstrated ozone can lead to the formation of lipid peroxides. The process of ozone damaging the lungs induces oxidative stress, causes epithelial cell damage, and tarnishes antioxidant reservoirs (80). When ozone reacts with fatty acids, it results in the formation of stable lipid peroxides, which can then activate signaling cascades via oxidative stress pathways (81). In a 2010 study, it was observed that both chronic and acute exposure to ozone resulted in heightened levels of lipid peroxidation and superoxide accumulation in the hippocampus of rats (82). In a study analyzing lipid peroxidation after exposure to allophane derived from volcanic ash, TBARS levels increased with exposure to allophane with the highest production at concentrations of approximately 1000 ppm (83).

9.9 HOUSEHOLD, URBAN, AND INDUSTRIAL FACTORS AND OXIDATIVE MODIFICATION OF LIPIDS IN THE RESPIRATORY TRACT AND LUNGS

The combination of ethanol PUFAs has a severe damaging effect on cells that are partnered with a major increase in lipid peroxidation. Yeast cells specifically are sensitive to oxidative stress caused by PUFAs. The accumulation of ethanol during fermentation further exacerbates oxidative stress by inducing the overproduction of ROS (84). Noise is a common harmful factor in many working conditions causing severe damage to the auditory system. One study showed that the levels of MDA and oxidative stress enzymes (Cat and TAC) were significantly elevated in the group exposed to noise compared to the control group, implying that an increase in noise led to an increase in lipid peroxidation (85). A 2012 study revealed that workers, when compared to the control group, in a power plant exposed to noise displayed significantly greater serum levels of MDA, as well as elevated levels of GR and GPX, enzymes associated with the antioxidant defense system. The researchers suggested that the increased levels of MDA and antioxidant enzymes were a result of heightened production of free radicals induced by exposure to noise pollution (86). The findings suggest that noise pollution can be considered a risk factor for the increase in the generation of free radicals, including hydrogen peroxide, hydroxyl radicals, and superoxide. These free radicals can initiate lipid peroxidation, causing the breakdown of lipid molecules within the cell membrane, which leads to the release of MDA. The alteration of lipid peroxidation within cells in turn impacts the enzymatic antioxidant defense system (87).

Another common household item that can increase lipid peroxidation is sodium chloride, or table salt. The addition of salt to meat and seafood can promote lipid oxidation, a process that significantly contributes to quality deterioration in the food industry (88). Lipid oxidation in meat and seafood begins immediately after slaughter and continues throughout the postmortem aging process, as well as during processing and storage. Multiple factors impact the rate of lipid oxidation in meat. Pre-slaughter events, including stress and physical injuries, as well as post-slaughter factors such as pH, cold shortening, carcass temperature, and tenderization techniques, influence lipid oxidation. Additionally, processing methods play a key role in the oxidative stability of meat products (89), including the quality of raw meat; cooking or heating conditions; preparation methods (mincing or grinding); the use of additives such as salt, nitrite, antioxidants, and spices; packaging type; and temperature during distribution and storage.

It is widely recognized that salt has the tendency to accelerate lipid oxidation, leading to undesirable alterations in the flavor and color of meat, reducing their shelf life. However, in certain types of metals, lipid oxidation is preferred when a certain aroma or flavor is desired, such as in dry-cured meats and sausages (90). The induction of lipid oxidation was investigated in raw and cooked turkey and pork patties with 1% NaCl and a mixture of metal contaminants, including iron (0.118–1.995 ppm), manganese (0.100–1.700 ppm), and copper (0.005–0.019 ppm). Lipid peroxidation was measured by monitoring TBARS values during a frozen storage period of 9 months. The use of salt was found to have a pro-oxidant effect; however, no significant differences were observed among the patties containing differing concentrations of metal contaminants (91).

In contrast, several household spices are known to contain antioxidant properties. The incorporation of spice mixtures into high-fat food preparations can enhance antioxidant activity, causing a reduction in lipid peroxidation. Several studies have noted a decrease in lipid peroxidation when using spices, curcumin, and polyphenol sources (e.g., red wine powder, roasted coffee). The reduced lipid peroxidation in these food preparations has further been associated with beneficial effects, such as controlling for oxidative damage. Interestingly, another study also revealed that curcumin had the ability to decrease lipid peroxidation when cooking, but black pepper did not. However, when black pepper was added to turmeric root powder in a patty, MDA levels decreased when compared to the same amount of curcumin used in isolation to black pepper (92).

Bisphenol A (BPA) is another common chemical used extensively in the manufacturing of epoxy resins, polycarbonate plastics, and phenolic resins. This chemical is present in everyday household items, such as plastic baby bottles, food containers, paper products, and kitchen utensils (93). BPA can leach out from these items into water and food, especially when exposed to differing pH levels and/or elevated temperatures (94). Though dietary intake is the primary route of human exposure to BPA, the inhalation of dust and skin contact also contribute significantly to its absorption. Numerous studies have documented the presence of BPA in human urine, breast milk, serum, placental tissue, cord blood, and amniotic fluid (95, 96). Lipids and other macromolecules are the main targets of ROS induced by BPA. Treatment with high concentrations of BPA resulted in elevated levels of MDA (97).

9.10 AGROCHEMICAL EXPOSURE CAUSES OXIDATIVE MODIFICATION OF LIPIDS IN THE RESPIRATORY TRACT AND LUNGS

Many agrochemicals are composed of toxic compounds that are commonly used in fertilizers and other treatments used to promote agricultural growth and abundance (98). The unrestricted use of these chemicals has been linked to several issues, including the presence of insecticide residues in food and agricultural products, the emergence of insect resistance, and environmental contamination. The widespread use of these chemicals has detrimental effects on non-target species, including livestock, humans, and wildlife, which are consistently exposed to these pesticides (99). Several studies have been conducted on different pesticides and fertilizers that display a correlation between the exposure to these chemicals and an increase in lipid peroxidation.

Exposure to the chemical, bifenthrin, a type-I pyrethroid that finds extensive use in both agricultural and public health applications as an insecticide used against infestations, induces lipid peroxidation (100). Rats exposed to bifenthrin were studied and measured for MDA. The liver exhibited the highest level of oxidative stress, characterized by lipid peroxidation and changes in antioxidant enzyme activity, followed by the kidney and lung. These findings suggest that bifenthrin can induce significant oxidative stress throughout the body. Furthermore, the duration of exposure and the extent of oxidative stress have a positive correlation (101). Another toxic chemical that can induce damage to the human body at acute levels is PQ, an herbicide used primarily for weed and grass control. Previous studies have demonstrated that exposure to this chemical has been associated with lipid peroxidation as well as oxidative stress directed toward the medulla and adrenal gland cortex (102). In a study conducted on rats, a dose-dependent relation between lipid peroxidation induced by PQ was displayed. The levels of MDA, CnP, and 8-OhdG were significantly elevated in the liver and lungs of animals treated with PQ20 (20 mg/kg) and PQ30 (30 mg/kg) compared to the control groups (103).

The chemical fipronil, used as an insecticide to protect against pests and bugs, is another chemical that is known to cause damage to lipids within the body. In one study, quails were exposed to fipronil and were measured for lipid peroxide and nitric oxide levels. Exposure to fipronil led to increased levels of lipid peroxidation in the heart and lungs, indicating the accumulation of ROS and subsequent cellular damage. Fipronil can be seen to promote the generation of free radicals, leading to the peroxidation of PUFAs, degradation of PLs, and deterioration of cellular function (104). An additional harmful insecticide is malathion, used to control mosquitos and insects. A study on rats after exposure to malathion indicated a positive correlation between malathion levels and lipid peroxidation, measured through MDA. The findings of this study indicate that acute malathion poisoning leads to elevated levels of lipid peroxidation, disturbances in the antioxidant system, and an increase in apoptosis in lung tissue (105).

9.11 PULMONARY DISEASES AND OXIDATIVE MODIFICATION OF LIPIDS IN THE RESPIRATORY TRACT AND LUNG ASTHMA

It is well established that asthmatic patients are at an increased risk of oxidative stress due to a disrupted glutathione redox system. An early study of the role of ROS in bronchial asthma found no significant differences in reduced and oxidized glutathione (GSH, GSGG) levels between symptom-free asthmatic patients and controls. However, after GSH stability tests were performed by treating blood samples with acetylphenylhydrazine, concentrations of both GSH and GSGG were significantly reduced in asthmatic patients (106). Another study found a significant decrease of coenzyme Q_{10} (ubichinone or CoQ_{10}) in both the plasma and whole blood of patients with asthma compared to controls. CoQ_{10} is a vitamin-like substance with a strong antioxidative potential through the scavenging of free oxygen radicals (107). Traditionally, serum MDA levels have been used as an indicator of lipid peroxidation. Patients with bronchial asthma have serum MDA levels positively correlated with both the severity and the duration of their illness, with levels being highest at the time of an acute asthma attack and significantly decreasing with both treatment and symptom-free intervals (108). Paraoxonase-1 (PON1) is an enzyme shown to protect both LDL and HDL from oxidation by ROS generated by iron and copper ions. Serum PON1 levels are significantly reduced in asthmatic patients (109). The induced overexpression of PON1 in asthmatic mice has been shown to increase GSH production, suppress MDA activity, and reduce the incidence of inflammation (110). A recent study found significant decreases in the serum concentration of GPXs, catalase (CAT), Cu-Zn SOD, uric acid, and total antioxidant capacity (TAC) in patients with asthma compared to controls (111). As discussed earlier, GPX offers protection against lipid peroxidation by converting lipid hydroperoxides into lipid alcohols. Thus, the decreased levels of this enzyme in asthmatics leads to unchecked lipid peroxidation and, eventually, ferroptosis.

9.11.1 Chronic Obstructive Pulmonary Disease

Chronic obstructive pulmonary disease (COPD) is a progressive inflammation of the lung that affects the peripheral airways and lung parenchyma and is most often associated with the risk factor of cigarette smoking in high-income countries (112). As is the case with asthma, patients with COPD have increased levels of lipid peroxidation with disruptions to the antioxidant enzyme systems. For example, one study by Bukowska and colleagues (2015) focused on the oxidative stress and related damage to erythrocytes in patients with COPD. They found that patients with COPD had about a 30% increase in the lipid peroxidation products in erythrocyte membranes, an approximate 27% increase in SOD activity, and an approximate 17% decrease in GPX activity, but no difference in CAT activity when compared to controls (113). Although correlations between COPD and increased levels of lipid peroxidative markers found in blood and urine can often be observed, these may not necessarily be indicative of oxidative stress originating in the respiratory tract but merely evidence of systemic oxidative stress that could originate elsewhere. Markers found in exhaled air, exhaled breath condensate (EBC), induced sputum samples, or bronchoalveolar lavage fluid

(BALF) may serve as better evidence for lipid peroxidation occurring directly in the lungs. Ethane, a product of lipid peroxidation, can be measured in exhaled air using GC and has been shown to be elevated in those suffering from COPD, with significant decreases in patients undergoing treatment with steroids (114). Exhaled breath condensate can contain a variety of biomarkers correlated with COPD and lipid peroxidation. MDA levels in EBC, measured using HPLC, are increased in COPD patients and have even been found to be significantly higher than asthmatic patients (115). One of the products of non-enzymatic lipid peroxidation mentioned earlier, 8-isoprostane, can be measured in EBC by a specific enzyme immunoassay kit and are increased in cases of COPD (116, 117). A relatively simple ROS, hydrogen peroxide, can be measured in EBC spectrophotometrically and is increased in COPD cases and is correlated with severity of the disease (118). Immunostaining has been used to measure increased levels of 4-HNE in airway and alveolar epithelial cells, endothelial cells, and neutrophils in patients with COPD (70).

9.11.2 Acute Respiratory Distress Syndrome/Acute Lung Injury

Acute respiratory distress syndrome (ARDS) is a type of acute lung injury (ALI) characterized by inflammation of the lung's alveolar-capillary membrane, leading to an increase in permeability of this membrane and an exudate buildup within the alveoli that inactivates surfactant. The resulting collapse of alveoli leads to hypoxemia and pulmonary edema (119). Early studies linking ARDS and lipid peroxidation have found increased plasma concentrations of 4-HNE (120) and MDA (121) in patients with ARDS. More recently, GC-MS analysis of exhaled breath has proven to be a potential diagnostic or predictive tool for ARDS. Straight-chain aliphatic hydrocarbons found in the breath have been shown to be indicative of lipid peroxidation (122). Octane has specifically been shown to be predictive of ARDS (123). Like other lung diseases, progression of ALI/ARDS has been closely linked to ferroptosis (124). Treatments that target ferroptosis by focusing on the inhibition of iron metabolism and lipid peroxidation may be protective of ALI/ARDS (125).

9.11.3 Pulmonary Fibrosis

Interstitial lung disease (ILD) can refer to a variety of diseases that result in the irreversible scarring of lung tissue, also known as *pulmonary fibrosis*. Idiopathic pulmonary fibrosis (IPF), by definition, is progressive PF without a known etiology. Many of the lipid peroxidation biomarkers indicative of other lung diseases are also present in patients with IPF. Decreased glutathione levels in the epithelial lining fluid (ELF) (126) and BALF (127), increased MDA levels in serum (128), and increased levels of hydrogen peroxide and 8-isoprostane in the EBC (129) of patients with IPF are all noteworthy findings. In the BALF and cells of IPF patients, several ferroptosis-related genes have been found to be elevated (130, 131), suggesting ferroptosis plays an important role in the progression of IPF.

9.11.4 Sarcoidosis

Sarcoidosis is another ILD that is characterized by the development of granulomas in the lungs and lymph nodes and is often idiopathic in nature. Increased MDH levels and decreased SOD activity have been found in the serum of sarcoidosis patients (132), as well as elevated levels of 8-isoprostane in EBC (133) and carbonylated proteins in BALF (134).

9.11.5 Nasal Mucosa and Polyps

Nasal polyps (NPs) are benign growths of nasal mucosa that can be asymptomatic or associated with a variety of nasal problems including chronic rhinosinusitis, allergic rhinitis, and asthma (135). NPs have been shown to have higher levels of MDA compared to normal nasal mucosa tissue (136).

Patients with NPs also show higher levels of MDA and lower activity of SOD in plasma and erythrocytes (137). Significant decreases in the levels of GSH and the activities of antioxidant enzymes SOD, CAT, and GPX have been found in the nasal mucosa of mice with allergic rhinitis (138). Furthermore, there is some evidence linking the use of cannabis to disruptions of lipid composition and increased lipid peroxidation in the olfactory neuroepithelium (139). Odorant binding proteins (OBPs) found in nasal mucosa scavenge 4-HNE and preserve the viability of both epithelial cells and bacterial cells in vitro (140, 141), suggesting OBPs could potentially act as another biomarker for lipid peroxidation assessment. These studies connect lipid peroxidation within the nasal mucosa to local inflammations and diseases. However, there is a lack of attempts to draw a similar connection to lung-specific diseases. Such studies may be worth exploring because of the potential to develop noninvasive methods of detecting lung disease.

9.12 CONCLUSIONS

Choice of the method of analyzing lipid peroxidation (oxidative degradation of PUFA in membranes) is highly dependent upon the time/stage of analysis of lung lipids undergoing oxidative attack and the product of the oxidized PUFA of the cellular membrane PLs. Also, the method of analysis of those products of peroxidation/oxidation of PUFA of the cellular membrane is crucial. Therefore, one specific method to analyze/determine one specific product of the peroxidized/oxidized PUFA of the cellular membrane PLs may not yield satisfactory and dependable results to quantify/determine the involvement of lipid peroxidation process in a particular scenario of oxidative stress encountered by lung. Sophisticated and state-of-the-art methods do not always guarantee dependable, reproducible, and accurate results. Hence, the investigator initially should consider using appropriate, simple, and sensitive methods to analyze a few selected products of lipid peroxidation. The investigator should consider both temporal and spatial aspects for the analysis of peroxidized and oxidatively modified cellular membrane lipids because a plethora of products of membrane lipid peroxidation arise during oxidative insult in a cell and tissue. Also, the right standards should be used to quantify the products of membrane lipid peroxidation to confirm the right product formed and uniformly normalize the identified product in a given setting. Catching the right product of the lipid peroxidation pathway in a given scenario is critical, which dictates the choice of the appropriate method of analysis. Analytical sensitivity of the chosen method depends on the amount/kinetics of the analyte (product of lipid peroxidation) formed in the lung cell on a temporal scale, reactivity (metabolism), stability, and spatial localization.

9.13 ACKNOWLEDGMENTS

The authors acknowledge the encouragement and support of the Division of Pulmonary, Critical Care, and Sleep Medicine, Department of Medicine, and the Davis Heart & Lung Research of The Ohio State University Wexner Medical Center, Columbus, Ohio.

REFERENCES

1. McCord, J. M., and Fridovich, I. (1977) Superoxide dismutated: A history. In *Superoxide and Superoxide Dismutases* (Michelson, A., McCord, J. M., and Fridovich, I., eds.), pp. 1–11, Academic Press, New York
2. Samuelson, B., Ranwell, P., and Paolette, R. (1980) *Advances in Prostaglandin and Thromboxane Research*, Raven Press, New York
3. Bolton, J. R. (1972) *Biological Applications of Electron Spin Resonance* (Swartz, H. M., Bolton, J. R., and Borg, D. C., eds.), pp. 11–61, John Wiley & Sons Inc, New York
4. Pryor, W. A. (1971) Free radical pathology. *Chem Eng News Archive*. **49**, 34–51
5. Pryor, W. A. (1976) The role of free radical reactions in biological systems. In *Free Radicals in Biology* (Pryor, W. A., ed.), pp. 1–49, vol. I, Academic Press, New York

6. Bhagwat, A. S., and Sane, P. V. (1978) Evidence for the involvement of superoxide anions in the oxygenase reaction of ribulose-1,5-diphosphate carboxylase. *Biochem Biophys Res Commun.* **84**, 865–873
7. Fridovich, I. (1978) The biology of oxygen radicals. *Science (1979).* **201**, 875–880
8. Younes, M., and Weser, U. (1978) Involvement of superoxide in the catalytic cycle of diamine oxidase. *Biochim Biophys Acta (BBA)—Enzymol.* **526**, 644–647
9. Cohen, G., and Heikkila, R. E. (1974) The generation of hydrogen peroxide, superoxide radical, and hydroxyl radical by 6-hydroxydopamine, dialuric acid, and related cytotoxic agents. *J Biol Chem.* **249**, 2447–2452
10. Misra, H. P., and Fridovich, I. (1972) The generation of superoxide radical during the autoxidation of hemoglobin. *J Biol Chem.* **247**, 6960–6962
11. Flower, R. J. (2008) Biosynthesis of prostaglandins, pp. 123–142. https://doi.org/10.1002/9780470715413.ch8
12. Leibovitz, B. E., and Siegel, B. V. (1980) Aspects of free radical reactions in biological systems: Aging. *J Gerontol.* **35**, 45–56
13. Di Luzio, N. R. (1973) Antioxidants, lipid peroxidation and chemical-induced liver injury. *Fed Proc.* **32**, 1875–1881
14. Henderson, C. A., Metz, E. N., Balcerzak, S. P., and Sagone, A. L. (1978) Adriamycin and daunomycin generate reactive oxygen compounds in erythrocytes. *Blood.* **52**, 878–885
15. Jose, P. J., and Slater, T. F. (1972) Increased concentrations of malonaldehyde in the livers of rats treated with carbon tetrachloride. *Biochem J.* **128**, 141P
16. Köster-Albrecht, D., Kappus, H., and Remmer, H. (1978) Lipid peroxidation and cell damage in isolated hepatocytes due to bromotrichloromethane. *Toxicol Appl Pharmacol.* **46**, 499–505
17. Recknagel, R. O., Glende, E. A., and Hruszkewycz, A. M. (1977) Chemical mechanisms in carbon tetrachloride toxicity. In *Free Radicals in Biology*, pp. 97–132, Elsevier. https://doi.org/10.1016/B978-0-12-566503-2.50010-0
18. Pryor, W. A. (1982) Free radical biology: Xenobiotics, cancer, and aging. *Ann N Y Acad Sci.* **393**, 1–22
19. Pryor, W. A. (1980) Methods for detecting free radicals and free radical-mediated pathology in environmental toxicology. In *Molecular Basis of Environmental Toxicity* (Bhatnagar, R. S., ed.), pp. 3–36, Ann Arbor Science Publishers, Ann Arbor, MI
20. Slater, T. F. (1972) *Free Radical Mechanisms in Tissue Injury*, pp. 1–19, Pion Ltd, London
21. Mead, J. F. (1976) Free radical mechanisms of lipid damage and consequences for cellular membranes. In *Free Radicals in Biology*, pp. 51–68, Elsevier. https://doi.org/10.1016/B978-0-12-566501-8.50009-2
22. Dahle, L. K., Hill, E. G., and Holman, R. T. (1962) The thiobarbituric acid reaction and the autoxidations of polyunsaturated fatty acid methyl esters. *Arch Biochem Biophys.* **98**, 253–261
23. Tappel, A. L. (1973) Lipid peroxidation damage to cell components. *Fed Proc.* **32**, 1870–1874
24. Shimasaki, H., Nozawa, T., Privett, O. S., and Anderson, W. R. (1977) Detection of age-related fluorescent substances in rat tissues. *Arch Biochem Biophys.* **183**, 443–451
25. Dougherty, J. J., and Hoekstra, W. G. (1982) Effects of vitamin E and selenium on copper-induced lipid peroxidation in vivo and on acute copper toxicity. *Exp Biol Med.* **169**, 201–208
26. Fuchs, B. (2014) Mass spectrometry and inflammation—MS methods to study oxidation and enzyme-induced changes of phospholipids. *Anal Bioanal Chem.* **406**, 1291–1306
27. Gaschler, M. M., and Stockwell, B. R. (2017) Lipid peroxidation in cell death. *Biochem Biophys Res Commun.* **482**, 419–425
28. Davies, S. S., and Guo, L. (2014) Lipid peroxidation generates biologically active phospholipids including oxidatively N-modified phospholipids. *Chem Phys Lipids.* **181**, 1–33
29. Hu, C., Wang, M., and Han, X. (2017) Shotgun lipidomics in substantiating lipid peroxidation in redox biology: Methods and applications. *Redox Biol.* **12**, 946–955
30. Tsikas, D. (2017) Assessment of lipid peroxidation by measuring malondialdehyde (MDA) and relatives in biological samples: Analytical and biological challenges. *Anal Biochem.* **524**, 13–30
31. Gray, J. I. (1978) Measurement of lipid oxidation: A review. *J Am Oil Chem Soc.* **55**, 539–546
32. Kwon, C. W., Park, K.-M., Park, J. W., Lee, J., Choi, S. J., and Chang, P.-S. (2016) Rapid and sensitive determination of lipid oxidation using the reagent kit based on spectrophotometry (FOODLAB *fat* system). *J Chem.* **2016**, 1–6
33. Domijan, A.-M., Ralić, J., Radić Brkanac, S., Rumora, L., and Žanić-Grubišić, T. (2015) Quantification of malondialdehyde by HPLC-FL—application to various biological samples. *Biomed Chromatogr.* **29**, 41–46

34. Mendonça, R., Gning, O., Di Cesaré, C., Lachat, L., Bennett, N. C., Helfenstein, F., and Glauser, G. (2017) Sensitive and selective quantification of free and total malondialdehyde in plasma using UHPLC-HRMS. *J Lipid Res.* **58**, 1924–1931
35. Goldring, C., Casini, A. F., Maellaro, E., Del Bello, B., and Comporti, M. (1993) Determination of 4-hydroxynonenal by high-performance liquid chromatography with electrochemical detection. *Lipids.* **28**, 141–145
36. Tynkkynen, T., Mursu, J., Nurmi, T., Tuppurainen, K., Laatikainen, R., and Soininen, P. (2012) NMR protocol for determination of oxidation susceptibility of serum lipids and application of the protocol to a chocolate study. *Metabolomics.* **8**, 386–398
37. Orefice, I., Gerecht, A., d'Ippolito, G., Fontana, A., Ianora, A., and Romano, G. (2015) Determination of lipid hydroperoxides in marine diatoms by the FOX2 Assay. *Mar Drugs.* **13**, 5767–5783
38. Parinandi, N. L., Weis, B. K., and Schmid, H. H. O. (1988) Assay of cardiolipin peroxidation by high-performance liquid chromatography. *Chem Phys Lipids.* **49**, 215–220
39. Parinandi, N. L., Weis, B. K., Natarajan, V., and Schmid, H. H. O. (1990) Peroxidative modification of phospholipids in myocardial membranes. *Arch Biochem Biophys.* **280**, 45–52
40. O'Connor Butler, E. S., Mazerik, J. N., Cruff, J. P., Sherwani, S. I., Weis, B. K., Marsh, C. B., Raghavamenon, A. C., Uppu, R. M., Schmid, H. H. O., and Parinandi, N. L. (2010) Lipoxygenase-catalyzed phospholipid peroxidation: Preparation, purification, and characterization of phosphatidylinositol peroxides. *Methods Mol Biol.* **610**, pp. 387–401. https://doi.org/10.1007/978-1-60327-029-8_23
41. Spickett, C. M., and Pitt, A. R. (2015) Oxidative lipidomics coming of age: Advances in analysis of oxidized phospholipids in physiology and pathology. *Antioxid Redox Signal.* **22**, 1646–1666
42. Verrastro, I., Pasha, S., Jensen, K., Pitt, A., and Spickett, C. (2015) Mass spectrometry-based methods for identifying oxidized proteins in disease: Advances and challenges. *Biomolecules.* **5**, 378–411
43. Cui, L., Lahti, P. M., and Decker, E. A. (2017) Evaluating electron paramagnetic resonance (EPR) to measure lipid oxidation lag phase for shelf-life determination of oils. *J Am Oil Chem Soc.* **94**, 89–97
44. Venkataraman, S., Schafer, F. Q., and Buettner, G. R. (2004) Detection of lipid radicals using EPR. *Antioxid Redox Signal.* **6**, 631–638
45. Alexandri, E., Ahmed, R., Siddiqui, H., Choudhary, M., Tsiafoulis, C., and Gerothanassis, I. (2017) High resolution NMR spectroscopy as a structural and analytical tool for unsaturated lipids in solution. *Molecules.* **22**, 1663
46. Greene, L. E., Lincoln, R., and Cosa, G. (2018) Spatio-temporal monitoring of lipid peroxyl radicals in live cell studies combining fluorogenic antioxidants and fluorescence microscopy methods. *Free Radic Biol Med.* **128**, 124–136
47. Spickett, C. M. (2013) The lipid peroxidation product 4-hydroxy-2-nonenal: Advances in chemistry and analysis. *Redox Biol.* **1**, 145–152
48. Weber, D., Milkovic, L., Bennett, S. J., Griffiths, H. R., Zarkovic, N., and Grune, T. (2013) Measurement of HNE-protein adducts in human plasma and serum by ELISA—Comparison of two primary antibodies. *Redox Biol.* **1**, 226–233
49. Catalá, A. (2013) Five decades with polyunsaturated Fatty acids: Chemical synthesis, enzymatic formation, lipid peroxidation and its biological effects. *J Lipids.* **2013**, 1–19
50. Timmins, G. S., dos Santos, R. E., Whitwood, A. C., Catalani, L. H., Di Mascio, P., Gilbert, B. C., and Bechara, E. J. H. (1997) Lipid peroxidation-dependent chemiluminescence from the cyclization of alkylperoxyl radicals to dioxetane radical intermediates. *Chem Res Toxicol.* **10**, 1090–1096
51. Anthonymuthu, T. S., Kim-Campbell, N., and Bayır, H. (2017) Oxidative lipidomics: Applications in critical care. *Curr Opin Crit Care.* **23**, 251–256
52. Pauling, J. K., and Klipp, E. (2016) Computational lipidomics and lipid bioinformatics: Filling in the blanks. *J Integr Bioinform.* **13**, 34–51
53. Züllig, T., and Köfeler, H. C. (2021) High resolution mass spectrometry in lipidomics. *Mass Spectrom Rev.* **40**, 162–176
54. Kyle, J. E., Clair, G., Bandyopadhyay, G., Misra, R. S., Zink, E. M., Bloodsworth, K. J., Shukla, A. K., Du, Y., Lillis, J., Myers, J. R., Ashton, J., Bushnell, T., Cochran, M., Deutsch, G., Baker, E. S., Carson, J. P., Mariani, T. J., Xu, Y., Whitsett, J. A., Pryhuber, G., and Ansong, C. (2018) Cell type-resolved human lung lipidome reveals cellular cooperation in lung function. *Sci Rep.* **8**, 13455
55. Agudelo, C. W., Samaha, G., and Garcia-Arcos, I. (2020) Alveolar lipids in pulmonary disease. A review. *Lipids Health Dis.* https://doi.org/10.1186/s12944-020-01278-8

56. Hajeyah, A. A., Griffiths, W. J., Wang, Y., Finch, A. J., and O'Donnell, V. B. (2020) The biosynthesis of enzymatically oxidized lipids. *Front Endocrinol (Lausanne)*. https://doi.org/10.3389/fendo.2020.591819
57. Ayala, A., Muñoz, M. F., and Argüelles, S. (2014) Lipid peroxidation: Production, metabolism, and signaling mechanisms of malondialdehyde and 4-hydroxy-2-nonenal. *Oxid Med Cell Longev*. **2014**, 1–31
58. Vigor, C., Bertrand-Michel, J., Pinot, E., Oger, C., Vercauteren, J., Le Faouder, P., Galano, J.-M., Lee, J. C.-Y., and Durand, T. (2014) Non-enzymatic lipid oxidation products in biological systems: Assessment of the metabolites from polyunsaturated fatty acids. *J Chromatogr B*. **964**, 65–78
59. Liu, X., Yang, H., and Liu, Z. (2022) Signaling pathways involved in paraquat-induced pulmonary toxicity: Molecular mechanisms and potential therapeutic drugs. *Int Immunopharmacol*. **113**, 109301
60. Li, Y., Yang, Y., and Yang, Y. (2022) Multifaceted roles of ferroptosis in lung diseases. *Front Mol Biosci*. **9**, 919187
61. Yoshino, K., Sano, M., Fujita, M., and Tomita, I. (1991) Production of aliphatic aldehydes on peroxidation of various types of lipids. *Chem Pharm Bull (Tokyo)*. **39**, 1788–1791
62. Floss, M. A., Fink, T., Maurer, F., Volk, T., Kreuer, S., and Müller-Wirtz, L. M. (2022) Exhaled aldehydes as biomarkers for lung diseases: A narrative review. *Molecules*. **27**, 5258
63. Lugg, S. T., Scott, A., Parekh, D., Naidu, B., and Thickett, D. R. (2022) Cigarette smoke exposure and alveolar macrophages: Mechanisms for lung disease. *Thorax*. **77**, 94–101
64. El-Sokkary, G. H., Cuzzocrea, S., and Reiter, R. J. (2007) Effect of chronic nicotine administration on the rat lung and liver: Beneficial role of melatonin. *Toxicology*. **239**, 60–67
65. Wetscher, G. J., Bagchi, M., Bagchi, D., Perdikis, G., Hinder, P. R., Glaser, K., and Hinder, R. A. (1995) Free radical production in nicotine treated pancreatic tissue. *Free Radic Biol Med*. **18**, 877–882
66. Yildiz, D., Liu, Y.-S., Ercal, N., and Armstrong, D. W. (1999) Comparison of pure nicotine- and smokeless tobacco extract-induced toxicities and oxidative stress. *Arch Environ Contam Toxicol*. **37**, 434–439
67. Ames, B. N., Shigenaga, M. K., and Hagen, T. M. (1993) Oxidants, antioxidants, and the degenerative diseases of aging. *Proc Natl Acad Sci*. **90**, 7915–7922
68. Dong, Y., Arif, A. A., Guo, J., Ha, Z., Lee-Sayer, S. S. M., Poon, G. F. T., Dosanjh, M., Roskelley, C. D., Huan, T., and Johnson, P. (2020) CD44 loss disrupts lung lipid surfactant homeostasis and exacerbates oxidized lipid-induced lung inflammation. *Front Immunol*. https://doi.org/10.3389/fimmu.2020.00029
69. Morissette, M. C., Shen, P., Thayaparan, D., and Stämpfli, M. R. (2015) Disruption of pulmonary lipid homeostasis drives cigarette smoke-induced lung inflammation in mice. *Eur Respir J*. **46**, 1451–1460
70. Rahman, I., van Schadewijk, A. A. M., Crowther, A. J. L., Hiemstra, P. S., Stolk, J., MacNee, W., and De Boer, W. I. (2002) 4-hydroxy-2-nonenal, a specific lipid peroxidation product, is elevated in lungs of patients with chronic obstructive pulmonary disease. *Am J Respir Crit Care Med*. **166**, 490–495
71. Dhouib, H., Jallouli, M., Draief, M., Bouraoui, S., and El-Fazâa, S. (2015) Oxidative damage and histopathological changes in lung of rat chronically exposed to nicotine alone or associated to ethanol. *Pathol Biol*. **63**, 258–267
72. Kiran Kumar, K. M., Naveen Kumar, M., Patil, R. H., Nagesh, R., Hegde, S. M., Kavya, K., Babu, R. L., Ramesh, G. T., and Sharma, S. C. (2016) Cadmium induces oxidative stress and apoptosis in lung epithelial cells. *Toxicol Mech Methods*. **26**, 658–666
73. Yousef, M., Roychoudhury, S., Jafaar, K., Slama, P., Kesari, K., and Kamel, M. (2022) Aluminum oxide and zinc oxide induced nanotoxicity in rat brain, heart, and lung. *Physiol Res*. **71**, 677–694. https://doi.org/10.33549/physiolres.934831
74. Bagchi, D., Balmoori, J., Bagchi, M., Ye, X., Williams, C. B., and Stohs, S. J. (2002) Comparative effects of TCDD, endrin, naphthalene and chromium (VI) on oxidative stress and tissue damage in the liver and brain tissues of mice. *Toxicology*. **175**, 73–82
75. Chakraborty, R., Renu, K., Eladl, M. A., El-Sherbiny, M., Elsherbini, D. M. A., Mirza, A. K., Vellingiri, B., Iyer, M., Dey, A., and Valsala Gopalakrishnan, A. (2022) Mechanism of chromium-induced toxicity in lungs, liver, and kidney and their ameliorative agents. *Biomed Pharmacother*. **151**, 113119
76. Luo, B., Shi, H., Zhang, K., Wei, Q., Niu, J., Wang, J., Hammond, S. K., and Liu, S. (2019) Cold stress provokes lung injury in rats co-exposed to fine particulate matter and lipopolysaccharide. *Ecotoxicol Environ Saf*. **168**, 9–16
77. García-Cuellar, C. M., Chirino, Y. I., Morales-Bárcenas, R., Soto-Reyes, E., Quintana-Belmares, R., Santibáñez-Andrade, M., and Sánchez-Pérez, Y. (2020) Airborne particulate matter (PM10) inhibits apoptosis through PI3K/AKT/FoxO3a pathway in lung epithelial cells: The role of a second oxidant stimulus. *Int J Mol Sci*. **21**, 473

78. Rizzo, A. M., Corsetto, P. A., Farina, F., Montorfano, G., Pani, G., Battaglia, C., Sancini, G., and Palestini, P. (2014) Repeated intratracheal instillation of PM10 induces lipid reshaping in lung parenchyma and in extra-pulmonary tissues. *PLoS One*. **9**, e106855. https://doi.org/10.1371/journal.pone.0106855
79. Qin, W., Shen, L., Wang, Q., Gao, Y., She, M., Li, X., and Tan, Z. (2022) Chronic exposure to ammonia induces oxidative stress and enhanced glycolysis in lung of piglets. *Environ Toxicol*. **37**, 179–191
80. Bromberg, P. A. (2016) Mechanisms of the acute effects of inhaled ozone in humans. *Biochim Biophys Acta (BBA)—Gen Subj*. **1860**, 2771–2781
81. Cheng, W., Duncan, K. E., Ghio, A. J., Ward-Caviness, C., Karoly, E. D., Diaz-Sanchez, D., Conolly, R. B., and Devlin, R. B. (2018) Changes in metabolites present in lung-lining fluid following exposure of humans to ozone. *Toxicol Sci*. **163**, 430–439
82. Rivas-Arancibia, S., Guevara-Guzmán, R., López-Vidal, Y., Rodríguez-Martínez, E., Zanardo-Gomes, M., Angoa-Pérez, M., and Raisman-Vozari, R. (2010) Oxidative stress caused by ozone exposure induces loss of brain repair in the hippocampus of adult rats. *Toxicol Sci*. **113**, 187–197
83. Cervini-Silva, J., Antonio-Nieto-Camacho, Gomez-Vidales, V., Ramirez-Apan, M. T., Palacios, E., Montoya, A., Kaufhold, S., Abidin, Z., and Theng, B. K. G. (2014) Lipid peroxidation and cytotoxicity induced by respirable volcanic ash. *J Hazard Mater*. **274**, 237–246
84. Olmos-Orizaba, B. E., Arroyo-Peñaloza, J. S., Martínez-Alcántar, L., Montoya-Pérez, R., Flores-García, A., Rodríguez-Orozco, A. R., Calderón-Cortés, E., Saavedra-Molina, A., Campos-García, J., and Cortés-Rojo, C. (2022) Linolenic acid plus ethanol exacerbates cell death in Saccharomyces cerevisiae by promoting lipid peroxidation, cardiolipin loss, and necrosis. *Life*. **12**, 1052
85. Bagheri Hosseinabadi, M., Khanjani, N., Ebrahimi, M. H., Mirbadie, S. R., and Biganeh, J. (2019) The effects of industrial noise exposure on lipid peroxidation and antioxidant enzymes among workers. *Int Arch Occup Environ Health*. **92**, 1041–1046
86. Jubran, A. A.-S., and Almashhedy, L. (2012) Influence of noise pollution on oxidative stress parameters and acetylcholinesterase in gas electric stations. *J Babylon Univ Pure Appl Sci*. **22**
87. Del Rio, D., Stewart, A. J., and Pellegrini, N. (2005) A review of recent studies on malondialdehyde as toxic molecule and biological marker of oxidative stress. *Nutr Metab Cardiovasc Dis*. **15**, 316–328
88. Mariutti, L. R. B., and Bragagnolo, N. (2017) Influence of salt on lipid oxidation in meat and seafood products: A review. *Food Res Int*. **94**, 90–100
89. Erickson, M. C. (2002) Lipid oxidation of muscle foods. In *Food Lipids* (Akoh, C. C., and Min, D. B., eds.), pp. 365–413, CRC Press. https://doi.org/10.1201/9780203908815
90. Jin, G., He, L., Zhang, J., Yu, X., Wang, J., and Huang, F. (2012) Effects of temperature and NaCl percentage on lipid oxidation in pork muscle and exploration of the controlling method using response surface methodology (RSM). *Food Chem*. **131**, 817–825
91. Bess, K. N., Boler, D. D., Tavárez, M. A., Johnson, H. K., McKeith, F. K., Killefer, J., and Dilger, A. C. (2013) Texture, lipid oxidation and sensory characteristics of ground pork patties prepared with commercially available salts. *LWT—Food Sci Technol*. **50**, 408–413
92. Zhang, Y., Henning, S. M., Lee, R.-P., Huang, J., Zerlin, A., Li, Z., and Heber, D. (2015) Turmeric and black pepper spices decrease lipid peroxidation in meat patties during cooking. *Int J Food Sci Nutr*. **66**, 260–265
93. Staples, C. A., Dome, P. B., Klecka, G. M., Oblock, S. T., and Harris, L. R. (1998) A review of the environmental fate, effects, and exposures of bisphenol A. *Chemosphere*. **36**, 2149–2173
94. Nam, S.-H., Seo, Y.-M., and Kim, M.-G. (2010) Bisphenol A migration from polycarbonate baby bottle with repeated use. *Chemosphere*. **79**, 949–952
95. Kang, J.-H., Kondo, F., and Katayama, Y. (2006) Human exposure to bisphenol A. *Toxicology*. **226**, 79–89
96. Carwile, J. L., Luu, H. T., Bassett, L. S., Driscoll, D. A., Yuan, C., Chang, J. Y., Ye, X., Calafat, A. M., and Michels, K. B. (2009) Polycarbonate bottle use and urinary bisphenol A concentrations. *Environ Health Perspect*. **117**, 1368–1372
97. Huang, M., Liu, S., Fu, L., Jiang, X., and Yang, M. (2020) Bisphenol A and its analogues bisphenol S, bisphenol F and bisphenol AF induce oxidative stress and biomacromolecular damage in human granulosa KGN cells. *Chemosphere*. **253**, 126707
98. Rossetti, M. F., Stoker, C., and Ramos, J. G. (2020) Agrochemicals and neurogenesis. *Mol Cell Endocrinol*. **510**, 110820

99. Khan, A. M., Sultana, M., Raina, R., Dubey, N., and Dar, S. A. (2013) Effect of sub-acute toxicity of bifenthrin on antioxidant status and hematology after its oral exposure in goats. *Proc Natl Acad Sci India Sect B Biol Sci.* **83**, 545–549
100. Syed, F., Awasthi, K. K., Chandravanshi, L. P., Verma, R., Rajawat, N. K., Khanna, V. K., John, P. J., and Soni, I. (2018) Bifenthrin-induced neurotoxicity in rats: Involvement of oxidative stress. *Toxicol Res (Camb).* **7**, 48–58
101. Dar, M. A., Khan, A. M., Raina, R., Verma, P. K., and Wani, N. M. (2019) Effect of bifenthrin on oxidative stress parameters in the liver, kidneys, and lungs of rats. *Environ Sci Pollut Res.* **26**, 9365–9370
102. Rose, M. S., Crabtree, H. C., Fletcher, K., and Wyatt, I. (1974) Biochemical effects of diquat and paraquat. Disturbance of the control of corticosteroid synthesis in rat adrenal and subsequent effects on the control of liver glycogen utilization. *Biochem J.* **138**, 437–443
103. Novaes, R. D., Gonçalves, R. V., Cupertino, M. C., Santos, E. C., Bigonha, S. M., Fernandes, G. J. M., Maldonado, I. R. S. C., and Natali, A. J. (2016) Acute paraquat exposure determines dose-dependent oxidative injury of multiple organs and metabolic dysfunction in rats: Impact on exercise tolerance. *Int J Exp Pathol.* **97**, 114–124
104. Khalil, S. R., Mohammed, W. A., Zaglool, A. W., Elhady, W. M., Farag, M. R., and El Sayed, S. A. M. (2019) Inflammatory and oxidative injury is induced in cardiac and pulmonary tissue following fipronil exposure in Japanese quail: mRNA expression of the genes encoding interleukin 6, nuclear factor kappa B, and tumor necrosis factor-alpha. *Environ Pollut.* **251**, 564–572
105. Uysal, M., and Karaman, S. (2018) In vivo effects of intravenous lipid emulsion on lung tissue in an experimental model of acute malathion intoxication. *Toxicol Ind Health.* **34**, 110–118
106. Novák, Z., Németh, I., Gyurkovits, K., Varga, S. I., and Matkovics, B. (1991) Examination of the role of oxygen free radicals in bronchial asthma in childhood. *Clin Chim Acta.* https://doi.org/10.1016/0009-8981(91)90375-M
107. Gazdik, F., Gvozdjakova, A., Nadvornikova, R., Repicka, L., Jahnova, E., Kucharska, J., Pijak, M. R., and Gazdikova, K. (2002) Decreased levels of coenzyme Q10 in patients with bronchial asthma. *Allergy.* **57**, 811–814
108. Sharma, A., Bansal, S., and Nagpal, R. K. (2003) Lipid peroxidation in bronchial asthma. *Indian J Pediatr.* https://doi.org/10.1007/BF02724313
109. Ekmekci, O. B., Donma, O., Ekmekci, H., Yildirim, N., Uysal, O., Sardogan, E., Demirel, H., and Demir, T. (2006) Plasma paraoxonase activities, lipoprotein oxidation, and trace element interaction in asthmatic patients. *Biol Trace Elem Res.* https://doi.org/10.1385/BTER:111:1:41
110. Chen, W. Q., Xie, Z. Z., Wang, X., Zhao, J. H., Hu, Q., Chen, Y. H., Gao, W. Y., and Liu, Y. (2018) Influences of PON1 on airway inflammation and remodeling in bronchial asthma. *J Cell Biochem.* https://doi.org/10.1002/jcb.26242
111. Abboud, M. M., Al-Rawashde, F. A., and Al-Zayadneh, E. M. (2022) Alterations of serum and saliva oxidative markers in patients with bronchial asthma. *J Asthma.* https://doi.org/10.1080/02770903.2021.2008426
112. Barnes, P. J. (2022) Oxidative stress in chronic obstructive pulmonary disease. *Antioxidants.* **11**, 965
113. Bukowska, B., Sicińska, P., Pająk, A., Koceva-Chyla, A., Pietras, T., Pszczółkowska, A., Górski, P., and Koter-Michalak, M. (2015) Oxidative stress and damage to erythrocytes in patients with chronic obstructive pulmonary disease—changes in ATPase and acetylcholinesterase activity. *Biochem Cell Biol.* **93**, 574–580
114. Paredi, P., Kharitonov, S. A., Leak, D., Ward, S., Cramer, D., and Barnes, P. J. (2000) Exhaled ethane, a marker of lipid peroxidation, is elevated in chronic obstructive pulmonary disease. *Am J Respir Crit Care Med.* **162**, 369–373
115. Bartoli, M. L., Novelli, F., Costa, F., Malagrinò, L., Melosini, L., Bacci, E., Cianchetti, S., Dente, F. L., Di Franco, A., Vagaggini, B., and Paggiaro, P. L. (2011) Malondialdehyde in exhaled breath condensate as a marker of oxidative stress in different pulmonary diseases. *Mediators Inflamm.* **2011**, 1–7
116. Montuschi, P., Collins, J. V., Ciabattoni, G., Lazzeri, N., Corradi, M., Kharitonov, S. A., and Barnes, P. J. (2000) Exhaled 8-isoprostane as an in vivo biomarker of lung oxidative stress in patients with COPD and healthy smokers. *Am J Respir Crit Care Med.* **162**, 1175–1177
117. Biernacki, W. A. (2003) Increased leukotriene B4 and 8-isoprostane in exhaled breath condensate of patients with exacerbations of COPD. *Thorax.* **58**, 294–298
118. Dekhuijzen, P. N., Aben, K. K., Dekker, I., Aarts, L. P., Wielders, P. L., van Herwaarden, C. L., and Bast, A. (1996) Increased exhalation of hydrogen peroxide in patients with stable and unstable chronic obstructive pulmonary disease. *Am J Respir Crit Care Med.* **154**, 813–816

119. Mason, C., Dooley, N., and Griffiths, M. (2016) Acute respiratory distress syndrome. *Clin Med.* **16**, s66–s70
120. Quinlan, G. J., Lamb, N. J., Evans, T. W., and Gutteridge, J. M. C. (1996) Plasma fatty acid changes and increased lipid peroxidation in patients with adult respiratory distress syndrome. *Crit Care Med.* **24**, 241–246
121. Metnitz, P. G. H., Bartens, C., Fischer, M., Fridrich, P., Steltzer, H., and Druml, W. (1999) Antioxidant status in patients with acute respiratory distress syndrome. *Intensive Care Med.* **25**, 180–185
122. Frank Kneepkens, C. M., Lepage, G., and Roy, C. C. (1994) The potential of the hydrocarbon breath test as a measure of lipid peroxidation. *Free Radic Biol Med.* **17**, 127–160
123. Bos, L. D. J. (2018) Diagnosis of acute respiratory distress syndrome by exhaled breath analysis. *Ann Transl Med.* https://doi.org/10.21037/atm.2018.01.17
124. Qu, M., Zhang, H., Chen, Z., Sun, X., Zhu, S., Nan, K., Chen, W., and Miao, C. (2021) The role of ferroptosis in acute respiratory distress syndrome. *Front Med (Lausanne).* https://doi.org/10.3389/fmed.2021.651552
125. Yin, X., Zhu, G., Wang, Q., Fu, Y. D., Wang, J., and Xu, B. (2021) Ferroptosis, a new insight into acute lung injury. *Front Pharmacol.* https://doi.org/10.3389/fphar.2021.709538
126. Cantin, A. M., Hubbard, R. C., and Crystal, R. G. (1989) Glutathione deficiency in the epithelial lining fluid of the lower respiratory tract in idiopathic pulmonary fibrosis. *Am Rev Respir Dis.* **139**, 370–372
127. Rahman, I., Skwarska, E., Henry, M., Davis, M., O'Connor, C. M., FitzGerald, M. X., Greening, A., and MacNee, W. (1999) Systemic and pulmonary oxidative stress in idiopathic pulmonary fibrosis. *Free Radic Biol Med.* **27**, 60–68
128. Jack, C. I., Jackson, M. J., Johnston, I. D., and Hind, C. R. (1996) Serum indicators of free radical activity in idiopathic pulmonary fibrosis. *Am J Respir Crit Care Med.* **153**, 1918–1923
129. Psathakis, K., Mermigkis, D., Papatheodorou, G., Loukides, S., Panagou, P., Polychronopoulos, V., Siafakas, N. M., and Bouros, D. (2006) Exhaled markers of oxidative stress in idiopathic pulmonary fibrosis. *Eur J Clin Invest.* **36**, 362–367
130. He, J., Li, X., and Yu, M. (2022) Bioinformatics analysis identifies potential ferroptosis key genes in the pathogenesis of pulmonary fibrosis. *Front Genet.* https://doi.org/10.3389/fgene.2021.788417
131. Li, M., Wang, K., Zhang, Y., Fan, M., Li, A., Zhou, J., Yang, T., Shi, P., Li, D., Zhang, G., Chen, M., and Ren, H. (2021) Ferroptosis-related genes in bronchoalveolar lavage fluid serves as prognostic biomarkers for idiopathic pulmonary fibrosis. *Front Med (Lausanne).* https://doi.org/10.3389/fmed.2021.693959
132. Ivanišević, J., Kotur-Stevuljević, J., Stefanović, A., Jelić-Ivanović, Z., Spasić, S., Videnović-Ivanov, J., Vučinić-Mihailović, V., and Ilić, J. (2012) Dyslipidemia and oxidative stress in sarcoidosis patients. *Clin Biochem.* **45**, 677–682
133. Psathakis, K., Papatheodorou, G., Plataki, M., Panagou, P., Loukides, S., Siafakas, N. M., and Bouros, D. (2004) 8-isoprostane, a marker of oxidative stress, is increased in the expired breath condensate of patients with pulmonary sarcoidosis. *Chest.* **125**, 1005–1011
134. Bargagli, E., Rosi, E., Pistolesi, M., Lavorini, F., Voltolini, L., and Rottoli, P. (2017) Increased risk of atherosclerosis in patients with sarcoidosis. *Pathobiology.* **84**, 258–263
135. Ta, N. (2019) Will we ever cure nasal polyps? *Ann R Coll Surg Engl.* **101**, 35–39
136. Doğru, H., DeliBaş, N., Döner, F., Tüz, M., and Uygur, K. (2001) Free radical damage in nasal polyp tissue. *Otolaryngol Head Neck Surg.* **124**, 570–572
137. Karlidağ, T., Keles, E., İlhan, N., Yakclin, S., Kaygusuz, İ., and Yildiz, M. (2005) Roles of free radicals, nitric oxide, and scavenging enzymes in nasal polyp development. *Ann Otol Rhinol Laryngol.* **114**, 122–126
138. Li, X.-L., Zhou, A.-G., Zhang, L., and Chen, W.-J. (2011) Antioxidant status and immune activity of glycyrrhizin in allergic rhinitis mice. *Int J Mol Sci.* **12**, 905–916
139. Saladrigas-Manjón, S., Dučić, T., Galindo, L., Fernández-Avilés, C., Pérez, V., de la Torre, R., and Robledo, P. (2020) Effects of cannabis use on the protein and lipid profile of olfactory neuroepithelium cells from schizophrenia patients studied by synchrotron-based FTIR spectroscopy. *Biomolecules.* **10**, 329
140. Grolli, S., Merli, E., Conti, V., Scaltriti, E., and Ramoni, R. (2006) Odorant binding protein has the biochemical properties of a scavenger for 4-hydroxy-2-nonenal in mammalian nasal mucosa. *FEBS J.* **273**, 5131–5142
141. Macedo-Márquez, A., Vázquez-Acevedo, M., Ongay-Larios, L., Miranda-Astudillo, H., Hernández-Muñoz, R., González-Halphen, D., Grolli, S., and Ramoni, R. (2014) Overexpression of a monomeric form of the bovine odorant-binding protein protects *Escherichia coli* from chemical-induced oxidative stress. *Free Radic Res.* **48**, 814–822

Part III

Pathogenesis and Therapies of COVID-19

10 Renin-Angiotensin System and COVID-19

Exploring the Impact, Inhibition, and Treatment Strategies

Qing Lin and Hongpeng Jia

10.1 INTRODUCTION

The global outbreak of COVID-19, caused by the novel severe acute respiratory syndrome coronavirus 2 (SARS-CoV-2), has resulted in a significant global health crisis. Among the primary features of COVID-19 is lung inflammation and respiratory failure driven by an overactive immune response known as a cytokine storm. In recent research, the renin-angiotensin system (RAS), which plays a crucial role in regulating various physiological functions, has emerged as a key player in COVID-19 pathogenesis. Notably, the angiotensin-converting enzyme 2 (ACE2), a component of the RAS, serves as the functional receptor and entry point for SARS-CoV-2.

In this chapter, we will delve into the intricate relationship between the RAS and COVID-19, exploring the mechanisms of viral entry, the dysregulation of the RAS, and its impact on lung injury and inflammation. We will also examine the role of RAS inhibitors, such as angiotensin receptor blockers (ARBs) and angiotensin-converting enzyme inhibitors (ACEIs), in COVID-19 treatment and discuss the potential benefits and risks associated with their use. Furthermore, we will explore the regulation of ACE2 expression, the genetic variability of the ACE2 gene, and its implications for COVID-19 susceptibility and severity.

By understanding the complex interplay between the RAS and COVID-19, we can gain valuable insights into potential therapeutic strategies and personalized treatment approaches. This chapter aims to provide a comprehensive overview of the current understanding of the RAS in the context of COVID-19, shedding light on its implications for disease pathogenesis and treatment options.

10.2 THE ROLE OF THE RAS IN COVID-19

The global emergence of COVID-19, attributed to SARS-CoV-2, has posed significant public health and economic threats globally. Predominantly characterized by lung inflammation and respiratory failure, COVID-19 stems from a hyperactive immune response known as a cytokine storm (1). The RAS, which plays an instrumental role in the regulation of various functions in numerous tissues and organs including the cardiovascular system, kidneys, lungs, and liver, is vital for maintaining blood pressure, electrolyte balance, and homeostasis in inflammatory responses (2, 3). Reports indicate that RAS is implicated in the pathogenesis of COVID-19 and importantly, ACE2, a component of RAS, functions as the receptor and entry point for SARS-CoV-2 (4).

SARS-CoV-2 interacts with host tissues using its surface ectodomain spike glycoprotein, which binds to ACE2. This interaction is primed by the cellular serine protease TMPRSS2, thus enabling the virus's direct cellular entry via endocytosis (4, 5). Thereafter, the ACE2 on the cell surface membrane is cleaved by the metallopeptidase ADAM-17 (5). Consequently, SARS-CoV-2 leverages

DOI: 10.1201/9781003355243-13

the upregulation of ACE2 to avail more cellular targets for entry, thereby amplifying the infection (6). With the cleavage of SARS-CoV-2 spike proteins upon viral entry, ACE2 experiences endocytosis or shedding, causing a decrease in ACE2 expression and subsequent RAS imbalance (5). This imbalance culminates in heightened levels of angiotensin II (Ang II) and an overactivation of the Ang II type 1 receptor (AT1R) in lung cells, triggering and intensifying lung injury (7–9). The resultant RAS dysregulation could potentially lead to exacerbated inflammation (5) and long-term pulmonary fibrosis (10). The suppression of ACE2's protective and anti-inflammatory activities is perceived as a critical factor instigating the unchecked cytokine storm, a primary cause of COVID-19-related fatalities (11).

10.3 RAS INHIBITION AND COVID-19

Numerous studies have demonstrated the anti-inflammatory properties of angiotensin receptor 1 blockers (ARBs) (12). It has been reported that ARBs stimulate ACE2, Ang (1–7), and Mas expression, in parallel with the reduction of pro-inflammatory cytokines and the induction of IL-10, a key anti-inflammatory cytokine. The anti-inflammatory effects of these ARBs are associated with the downregulation of various proteins related to signaling pathways such as PI3K, phospho-Akt, phospho-p38 MAPK, phospho-JNK, phospho-ERK, and phospho-MAPK-2 (12, 13).

The infection site's ACE2 expression level determines the degree of viral entry and infectivity. Several studies have indicated a correlation between COVID-19 severity and older age, which is related to higher ACE2 gene expression in the airway and lungs (14, 15). Consequently, ARBs could potentially facilitate SARS-CoV-2 infection by increasing the virus's accessibility to the membrane-bound ACE2. ACEIs may also elevate the ACE2 expression level, as ACEIs lower the Ang II level, a substrate of the ACE2 enzyme (16, 17). Hence, both ACEIs and ARBs can be viewed as a double-edged sword (Figure 10.1), possibly increasing susceptibility to COVID-19 infection.

10.4 ANGIOTENSIN-CONVERTING ENZYME 2

ACE2, a more recent addition to the RAS, primarily functions as a robust negative regulator of RAS. It works to modulate RAS activity, preventing harmful consequences of excessive RAS activation. Significantly, ACE2 also serves as the main receptor for severe acute respiratory syndrome coronaviruses 1 and 2 (SARS-CoV and SARS-CoV-2), positioning it as a crucial research target in the face of the two pandemics that have transpired since the dawn of the 21st century.

Interestingly, ACE2, a zinc metallopeptidase, is encoded by a gene situated on the X chromosome Xp22. This gene, comprising 18 exons, is expressed in various cell types, from neurons to cells in the reproductive tracts. Despite its placement on the X chromosome, there are no reported differences in ACE2 gene expression between sexes under typical or disease conditions, indicating

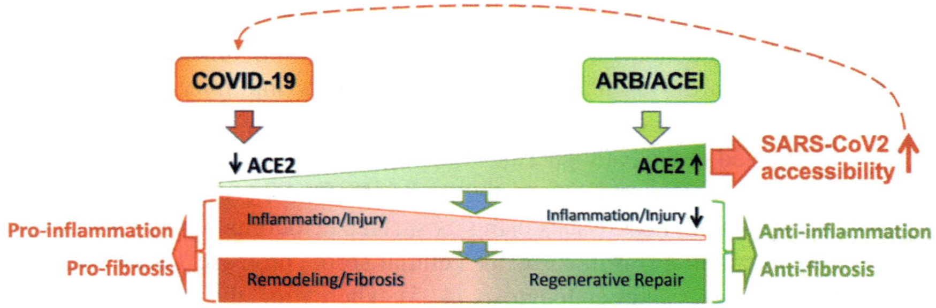

FIGURE 10.1 Schematic representation of the effects of RAS inhibition in COVID-19.

a mosaic pattern of allele expression. The broad expression of the ACE2 gene might partly account for why COVID-19 impacts a multitude of different organs, tissues, and cells.

As a monocarboxypeptidase, ACE2 chiefly exerts its regulatory influence on RAS by modulating the Ang II/AT1R and Ang (1–7)/Mas1 axis. Ang (1–7), an active metabolite of Ang II, is primarily produced by ACE2 cleaving one amino acid from the C terminal of Ang II (18). Therefore, maintaining equilibrium between Ang II/AT1R and Ang (1–7)/Mas 1R is crucial for ACE2 to fulfil its regulatory role in RAS. Although recent decades of research have identified more ACE2 substrates, including neocasomorphin, des-Arg9-bradykinin, β-casomorphin, Apelin13 and 36, and dynorphin A, the physiological and pathophysiological implications of these substrates regulated by ACE2 need further exploration.

In humans and animals, ACE2 predominantly exists in two forms: soluble and cell membrane–bound. The soluble form is produced through the cleavage of the cell membrane-bound form by A disintegrin and metalloprotease 17 (ADAM17), also known as tumor necrosis factor-alpha converting enzyme (TACE), near the boundary of the ectodomain and transmembrane domain (19, 20). Both forms can exhibit enzymatic activity and bind to the SARS-CoV and SARS-CoV-2 viruses, but only the membrane-bound ACE2 acts as the cognate receptor for these viruses. Yet, a study by Yeung et al. proposed that soluble ACE2 might aid SARS-CoV-2's cell entry via interaction with RAS-related proteins (21).

The discovery of delta ACE2 (dACE2) by Onabajo and colleagues in 2020 has added another layer of complexity to our understanding of ACE2 and its implications in human health (22). This novel isoform lacks 356 amino-terminal amino acids in comparison to the canonical ACE2, and interestingly, cannot bind to SARS-CoV-2 or function as a carboxypeptidase due to its N-terminal truncation. This suggests that it likely plays a distinct physiological role separate from the well-established functions of the ACE2 protein. As dACE2 is an interferon-stimulated gene, its expression might be important in immune responses. However, the precise biological role of this new form of ACE2 is still unclear and warrants further investigation.

10.5 ACE2 AND COVID-19

The outbreak of COVID-19, caused by the SARS-CoV-2 virus, has resulted in significant global health and economic consequences. One of the key features of COVID-19 is lung inflammation and respiratory failure due to an overactive immune response known as a cytokine storm. ACE2 acts as a functional receptor and entry point for SARS-CoV-2. The virus interacts with host tissues through its spike glycoprotein, which binds to ACE2. This complex is then primed by the serine protease TMPRSS2, allowing the virus to enter the host cell via endocytosis. ADAM-17 cleaves ACE2 on the cell membrane, providing additional targets for viral entry and enhancing infection. The shedding of ACE2 during viral entry leads to a decrease in ACE2 expression, disrupting the balance of the RAS. Dysregulation of the RAS can result in increased inflammation and potentially long-term pulmonary fibrosis. Loss of the protective and anti-inflammatory effects of ACE2 is considered a critical factor in the uncontrolled cytokine storm, which contributes to mortality in COVID-19 patients.

Clinical observations suggest that the level of ACE2 expression at the site of infection may determine viral infectivity. Advanced age, which is associated with higher ACE2 gene expression in the airway and lungs, has been linked to the severity of COVID-19. However, the relationship between ACE2 expression level and COVID-19 severity remains uncertain because of a lack of appropriate disease models. Additionally, the correlation between ACE2 expression and immune signature enrichment levels in patients may influence the clinical outcomes of SARS-CoV-2 infection. The precise factors within the host system that regulate ACE2 expression in COVID-19, particularly at the gene production level, are still unclear. Research is currently investigating the genetic variability of the ACE2 gene and the impact of specific polymorphic alleles on the course and outcome of COVID-19 (23). Functional polymorphisms can alter the structure of ACE2, affecting its interactions with the spike glycoprotein of SARS-CoV-2 (24–26). The

G8790A variation (rs2285666) is a key polymorphism that can increase ACE2 gene expression (27), whereas other variants such as rs5934250 can decrease ACE2 expression. Certain polymorphisms, like rs4646114 and rs4646115, are associated with increased susceptibility to viral infections, including COVID-19 (23), whereas others like rs536092258 and rs370596467 appear to confer a favorable response to infection (28). These ACE2 polymorphisms play a crucial role in the susceptibility, severity, and mortality of COVID-19. Furthermore, the frequency of specific ACE2 gene variants varies among populations, which may have implications for the epidemiology of COVID-19 in different regions.

10.6 ARB/ACEI TREATMENT FOR PATIENTS WITH COVID-19

ACEIs and ARBs are types of drugs commonly used to treat hypertension. In the context of the COVID-19 pandemic, there has been significant interest in understanding how these medications might impact the progression and severity of the disease.

Recent research seems to support the idea that ARB/ACEI treatment can improve outcomes for patients with hypertension who are infected by SARS-CoV-2. These individuals are particularly vulnerable to hyperinflammation and severe COVID-19 outcomes. Retrospective studies have demonstrated the safety of continuing ARB/ACEI treatments in COVID-19 patients (29, 30), especially those with heart failure, hypertension, or ischemic heart disease, as the disease tends to be more severe in these patient populations (31). Current evidence does not show any harmful effects from RAS inhibition in these patients, and discontinuing these medications may, in fact, be detrimental. ARBs and ACEIs help to convert angiotensin I (Ang I) to Ang (1–9) and Ang (1–7), respectively, both of which are known for their anti-inflammatory properties (32). There is a belief that ARB/ACEI treatment can reduce inflammation in COVID-19 pneumonia, potentially reducing mortality in patients without hypertension (33).

However, there are contrasting viewpoints and findings. Some studies suggest discontinuation of ARB/ACEI in COVID-19 may lead to a faster recovery (32). There are also observations suggesting long-term use of these medications is associated with severe renal dysfunction and acute kidney injury in severe COVID-19 patients (34). Other research has indicated that ARB/ACEI therapy is associated with increased in-hospital mortality, which suggests these medications may increase the risk of severe COVID-19 outcomes (35). In terms of ACE2 expression regulation, studies on experimental mice models have shown that administering captopril (an ACEI) and telmisartan (an ARB) decreased ACE2 protein expression in kidney-isolated membranes. This seems to support the view that the use of RAS blockers should be continued in patients at risk of contracting COVID-19. Interestingly, ACE2 expression in hypertensive patients and patients with type 1 or type 2 diabetes who are treated with ARB/ACEI is increased, which might make these patients more vulnerable.

The literature regarding the association of ARB/ACEI treatment with COVID-19 severity is complex. Whereas some studies reported that ACEI/ARB treatment does not influence the prognosis, including clinical severity, respiratory distress, and mortality, other in-hospital studies have found that use of ACEI/ARB is associated with a lower risk of mortality and critical illness in COVID-19 patients with hypertension. Some data have suggested that these treatments protect against intensive care unit admission and mortality, with chronic ACEI exposure associated with decreased odds of severe disease in COVID-19 patients. However, the decision to use ARB/ACEI should be individualized, considering the risk profile, indication for RAS inhibition, and the availability of alternative therapies and outpatient monitoring options.

Therefore, the relationship between ACEIs/ARBs and COVID-19 is multifaceted and requires further investigation. The variability in reported findings suggests that patient-specific factors, including baseline health status, comorbidities, and genetic factors, may significantly influence these outcomes. As such, personalized prescription of ACEI/ARBs for hypertensive COVID-19 patients may be the best approach to improve patient outcomes.

10.7 ARB/ACEI AND THE COMORBIDITIES/COMPLICATIONS OF COVID-19

Hypertension, diabetes, and coronary heart disease emerged as the primary comorbidities linked with COVID-19 fatalities during the early phase of the pandemic (36). An increase in ACE2 expression was observed in patients with hypertension or diabetes (type 1 or type 2) who were undergoing treatment with ARB/ACEI (28, 33). This sparked concerns that ARB/ACEI could make these patients more susceptible to the virus because of an enhanced entry and replication of the virus in type II alveolar cells (37).

However, for patients who were already on ARB/ACEI before contracting SARS-CoV-2, these medications could offer protection to the cardiovascular, renal, and pulmonary systems in instances of SARS-CoV-2–associated acute respiratory distress syndrome (ARDS) (32). This is attributed to the anti-inflammatory, anti-hypertensive, and anti-fibrotic properties of these medications (17, 38). Several studies have highlighted the safety and efficacy of ACEI/ARB drugs in patients with diabetes who contract COVID-19 (39). The safe use of ACE-I/ARBs has also been documented in patients with end-stage renal disease undergoing dialysis and infected with COVID-19 (40).

The literature further advocates the usage of the ACEI captopril in treating stone disease associated with cystinuria during the COVID-19 pandemic (41). The continuation of ACEI/ARB treatment has been linked with a reduced rate of gastrointestinal symptoms (e.g., diarrhea, vomiting, nausea, and abdominal pain), particularly in elderly patients and those with other comorbidities (42, 43).

Changes to the RAS can occur before the onset of clinically apparent disease in children (44). Children born prematurely with very low birth weight demonstrated higher plasma Ang II concentrations relative to lower Ang (1–7) when compared to full-term infants (45). ARB/ACEI therapy can shift the RAS toward the ACE2/Ang (1–7) pathway, potentially offering benefits for pediatric patients with COVID-19 (44).

However, risks remain associated with the usage of ARB/ACEI in COVID-19 patients. For instance, angioedema, an episodic side effect of ACE inhibitors, poses significant and distressing challenges.

10.8 COMPARISON OF RAS INHIBITORS AND THE COMBINATION OF THESE DRUGS FOR COVID-19 TREATMENT

Differences in the impacts of ACEIs and ARBs on COVID-19 have been examined in the scientific literature. Experiments on animals treated with ARB/ACEI observed more consistent upregulation of ACE2 with ARBs, whereas ACEIs had more variable effects on ACE2 modulation (46). In hypertensive patients, those treated with ARBs for a year displayed higher urinary ACE2 levels, but this was not the case with ACEI treatment (47). Among patients hospitalized with COVID-19, the use of ACEIs appeared to be linked with a lower risk of death, whereas ARB use might be connected with an increased risk (48).

However, in severe cases of COVID-19, high-dose ACEI usage was found to be detrimental in patients suffering from serious respiratory distress syndrome, leading to overactivation of bradykinin receptors (49). Suggestions have been made to switch patients from high-dose ACEIs to ARBs or other anti-hypertensive drugs in extreme cases of respiratory distress syndrome. This could potentially prevent the ACEI-triggered harmful pharmacological enhancement of severe bronchoalveolar bradykinin-related hyperinflammation in the lungs and the subsequent life-threatening virus-activated cytokine storm (49).

The advantages of ARBs over ACEIs in short-term treatment of hypertensive COVID-19 patients were further explored in studies by Zhao and colleagues (50). Compared to ACEIs, ARBs were found to inhibit non–ACE-mediated Ang II production and function under infection conditions, significantly reduce viscous sputum production and inflammatory responses, and potentially alleviate the lung fibrosis induced by mechanical ventilation in severe COVID-19 patients, thereby improving their prognosis (50). Interestingly, concurrent therapy with ACEIs was associated with an increased rate and severity of infection with the coronavirus HcoV-NL63, which shares the same host receptor as SARS-CoV-2 (51). This correlation was not found in infected patients on concurrent ARB therapy (51). These findings support the superiority of ARBs in treating COVID-19.

Studies comparing the efficacy of ACEIs/ARBs versus calcium channel blockers on the progression of hypertensive COVID-19 patients found no significant difference between the two groups (52). Research has also shown the potential benefits of combination treatment for hospitalized COVID-19 patients (53). Treatment with statins and ACEIs/ARBs significantly reduced mortality, with combination therapy resulting in a threefold reduction in the odds of hospital mortality (54). Moreover, ARB-based combination therapy was found to be safe and effective for patients with cardiovascular risk during the COVID-19 pandemic (54). These combination therapies include olmesartan (OLM)-based dual combinations such as OLM/hydrochlorothiazide, OLM/amlodipine, or ARB-based triple combination therapies such as OLM medoxomil/amlodipine/hydrochlorothiazide (54). Regardless, when it comes to therapies involving ACEIs/ARBs, individualized treatment decisions should be made based on the clinical condition and comorbidities of each COVID-19 patient (55).

REFERENCES

1. Jia, H., Neptune, E., and Cui, H. (2021) Targeting ACE2 for COVID-19 therapy: Opportunities and challenges. *Am J Respir Cell Mol Biol.* **64**, 416–425. https://doi.org/10.1165/rcmb.2020-0322PS
2. Pieruzzi, F., Abassi, Z. A., and Keiser, H. R. (1995) Expression of renin-angiotensin system components in the heart, kidneys, and lungs of rats with experimental heart failure. *Circulation.* **92**, 3105–3112. https://doi.org/10.1161/01.cir.92.10.3105
3. Balakumar, P., and Jagadeesh, G. (2014) A century old renin-angiotensin system still grows with endless possibilities: AT1 receptor signaling cascades in cardiovascular physiopathology. *Cell Signal.* **26**, 2147–2160. https://doi.org/10.1016/j.cellsig.2014.06.011
4. Hoffmann, M., Kleine-Weber, H., Schroeder, S., Kruger, N., Herrler, T., Erichsen, S., *et al.* (2020) SARS-CoV-2 cell entry depends on ACE2 and TMPRSS2 and is blocked by a clinically proven protease inhibitor. *Cell.* **181**, 271–280.e278. https://doi.org/10.1016/j.cell.2020.02.052
5. de la Cruz, A., Ashraf, S., Vittorio, T. J., and Bella, J. N. (2020) COVID-19 and renin-angiotensin system modulators: What do we know so far? *Expert Rev Cardiovasc Ther.* 1–6. https://doi.org/10.1080/14779072.2020.1814143
6. Ziegler, C. G. K., Allon, S. J., Nyquist, S. K., Mbano, I. M., Miao, V. N., Tzouanas, C. N., *et al.* (2020) SARS-CoV-2 receptor ACE2 is an interferon-stimulated gene in human airway epithelial cells and is detected in specific cell subsets across tissues. *Cell.* **181**, 1016–1035.e1019. https://doi.org/10.1016/j.cell.2020.04.035
7. Gong, J., Sun, Y., and Xie, L. (2021) ACEI/ARB drug therapy in COVID-19 patients: Yes or no? *J Transl Int Med.* **9**, 8–11. https://doi.org/10.2478/jtim-2021-0011
8. Imai, Y., Kuba, K., Rao, S., Huan, Y., Guo, F., Guan, B., *et al.* (2005) Angiotensin-converting enzyme 2 protects from severe acute lung failure. *Nature.* **436**, 112–116. https://doi.org/10.1038/nature03712
9. Kuba, K., Imai, Y., Ohto-Nakanishi, T., and Penninger, J. M. (2010) Trilogy of ACE2: A peptidase in the renin-angiotensin system, a SARS receptor, and a partner for amino acid transporters. *Pharmacol Ther.* **128**, 119–128. https://doi.org/10.1016/j.pharmthera.2010.06.003
10. McDonald, L. T. (2021) Healing after COVID-19: Are survivors at risk for pulmonary fibrosis? *Am J Physiol Lung Cell Mol Physiol.* **320**, L257–L265. https://doi.org/10.1152/ajplung.00238.2020
11. Zanza, C., Romenskaya, T., Manetti, A. C., Franceschi, F., La Russa, R., Bertozzi, G., *et al.* (2022) Cytokine storm in COVID-19: Immunopathogenesis and therapy. *Medicina (Kaunas).* **58**. https://doi.org/10.3390/medicina58020144
12. Jia, H. (2016) Pulmonary angiotensin-converting enzyme 2 (ACE2) and inflammatory lung disease. *Shock.* **46**, 239–248. https://doi.org/10.1097/SHK.0000000000000633
13. Simoes e Silva, A. C., Silveira, K. D., Ferreira, A. J., and Teixeira, M. M. (2013) ACE2, angiotensin-(1–7) and Mas receptor axis in inflammation and fibrosis. *Br J Pharmacol.* **169**, 477–492. https://doi.org/10.1111/bph.12159

14. Inde, Z., Croker, B. A., Yapp, C., Joshi, G. N., Spetz, J., Fraser, C., *et al.* (2021) Age-dependent regulation of SARS-CoV-2 cell entry genes and cell death programs correlates with COVID-19 severity. *Sci Adv.* **7**. https://doi.org/10.1126/sciadv.abf8609
15. Bunyavanich, S., Do, A., and Vicencio, A. (2020) Nasal gene expression of angiotensin-converting enzyme 2 in children and adults. *JAMA.* **323**, 2427–2429. https://doi.org/10.1001/jama.2020.8707
16. Kai, H., and Kai, M. (2020) Interactions of coronaviruses with ACE2, angiotensin II, and RAS inhibitors-lessons from available evidence and insights into COVID-19. *Hypertens Res.* **43**, 648–654. https://doi.org/10.1038/s41440-020-0455-8
17. Hajra, A., and Bandyopadhyay, D. (2020) COVID-19 and ACEI/ARB: Not associated? *Am J Hypertens.* **33**, 788. https://doi.org/10.1093/ajh/hpaa076
18. Turner, A. J., and Nalivaeva, N. N. (2022) Angiotensin-converting enzyme 2 (ACE2): Two decades of revelations and re-evaluation. *Peptides.* **151**, 170766. https://doi.org/10.1016/j.peptides.2022.170766
19. Lambert, D. W., Yarski, M., Warner, F. J., Thornhill, P., Parkin, E. T., Smith, A. I., *et al.* (2005) Tumor necrosis factor-alpha convertase (ADAM17) mediates regulated ectodomain shedding of the severe-acute respiratory syndrome-coronavirus (SARS-CoV) receptor, angiotensin-converting enzyme-2 (ACE2). *J Biol Chem.* **280**, 30113–30119. https://doi.org/10.1074/jbc.M505111200
20. Jia, H. P., Look, D. C., Tan, P., Shi, L., Hickey, M., Gakhar, L., *et al.* (2009) Ectodomain shedding of angiotensin converting enzyme 2 in human airway epithelia. *Am J Physiol Lung Cell Mol Physiol.* **297**, L84–L96. https://doi.org/10.1152/ajplung.00071.2009
21. Yeung, M. L., Teng, J. L. L., Jia, L., Zhang, C., Huang, C., Cai, J. P., *et al.* (2021) Soluble ACE2-mediated cell entry of SARS-CoV-2 via interaction with proteins related to the renin-angiotensin system. *Cell.* **184**, 2212–2228.e12. https://doi.org/10.1016/j.cell.2021.02.053.
22. Onabajo, O. O., Banday, A. R., Stanifer, M. L., Yan, W., Obajemu, A., Santer, D. M., *et al.* (2020) Interferons and viruses induce a novel truncated ACE2 isoform and not the full-length SARS-CoV-2 receptor. *Nat Genet.* **52**, 1283–1293. https://doi.org/10.1038/s41588-020-00731-9
23. Gintoni, I., Adamopoulou, M., and Yapijakis, C. (2022) The impact of ACE and ACE2 gene polymorphisms in pulmonary diseases including COVID-19. *In Vivo.* **36**, 13–29. https://doi.org/10.21873/invivo.12672
24. Benetti, E., Tita, R., Spiga, O., Ciolfi, A., Birolo, G., Bruselles, A., *et al.* (2020) ACE2 gene variants may underlie interindividual variability and susceptibility to COVID-19 in the Italian population. *Eur J Hum Genet.* **28**, 1602–1614. https://doi.org/10.1038/s41431-020-0691-z
25. Guo, X., Chen, Z., Xia, Y., Lin, W., and Li, H. (2020) Investigation of the genetic variation in ACE2 on the structural recognition by the novel coronavirus (SARS-CoV-2). *J Transl Med.* **18**, 321. https://doi.org/10.1186/s12967-020-02486-7
26. Pouladi, N., and Abdolahi, S. (2021) Investigating the ACE2 polymorphisms in COVID-19 susceptibility: An in silico analysis. *Mol Genet Genomic Med.* **9**, e1672. https://doi.org/10.1002/mgg3.1672
27. Mohlendick, B., Schonfelder, K., Breuckmann, K., Elsner, C., Babel, N., Balfanz, P., *et al.* (2021) ACE2 polymorphism and susceptibility for SARS-CoV-2 infection and severity of COVID-19. Pharmacogenet Genomics. **31**, 165–171. https://doi.org/10.1097/FPC.0000000000000436
28. Barash, A., Machluf, Y., Ariel, I., and Dekel, Y. (2020) The pursuit of COVID-19 biomarkers: Putting the spotlight on ACE2 and TMPRSS2 regulatory sequences. *Front Med (Lausanne).* **7**, 582793. https://doi.org/10.3389/fmed.2020.582793
29. Rubin, S. J. S., Falkson, S. R., Degner, N. R., and Blish, C. A. (2021) Safety of ACE-I and ARB medications in COVID-19: A retrospective cohort study of inpatients and outpatients in California. *J Clin Transl Sci.* **5**, e8. https://doi.org/10.1017/cts.2020.489
30. Kumar, S., Nikravesh, M., Chukwuemeka, U., Randazzo, M., Flores, P., Choday, P., *et al.* (2022) Safety of ACEi and ARB in COVID-19 management: A retrospective analysis. *Clin Cardiol.* https://doi.org/10.1002/clc.23836
31. Hu, J., Zhang, X., Zhang, X., Zhao, H., Lian, J., Hao, S., *et al.* (2020) COVID-19 is more severe in patients with hypertension; ACEI/ARB treatment does not influence clinical severity and outcome. *J Infect.* **81**, 979–997. https://doi.org/10.1016/j.jinf.2020.05.056
32. Rossi, G. P., Sanga, V., and Barton, M. (2020) Potential harmful effects of discontinuing ACE-inhibitors and ARBs in COVID-19 patients. *Elife.* **9**. https://doi.org/10.7554/eLife.57278
33. Khashkhusha, T. R., Chan, J. S. K., and Harky, A. (2020) ACEi and ARB with COVID-19. *J Card Surg.* **35**, 1388. https://doi.org/10.1111/jocs.14649

34. Oussalah, A., Gleye, S., Clerc Urmes, I., Laugel, E., Callet, J., Barbe, F., *et al.* (2020) Long-term ACE inhibitor/ARB use is associated with severe renal dysfunction and acute kidney injury in patients with severe COVID-19: Results from a referral center cohort in the northeast of France. *Clin Infect Dis.* **71**, 2447–2456. https://doi.org/10.1093/cid/ciaa677
35. Bauer, A., Schreinlechner, M., Sappler, N., Dolejsi, T., Tilg, H., Aulinger, B. A., *et al.* (2021) Discontinuation versus continuation of renin-angiotensin-system inhibitors in COVID-19 (ACEI-COVID): A prospective, parallel group, randomised, controlled, open-label trial. *Lancet Respir Med.* **9**, 863–872. https://doi.org/10.1016/S2213-2600(21)00214-9
36. Zhou, F., Yu, T., Du, R., Fan, G., Liu, Y., Liu, Z., *et al.* (2020) Clinical course and risk factors for mortality of adult inpatients with COVID-19 in Wuhan, China: A retrospective cohort study. *Lancet.* **395**, 1054–1062. https://doi.org/10.1016/S0140-6736(20)30566-3
37. Diaz, J. H. (2020) Hypothesis: Angiotensin-converting enzyme inhibitors and angiotensin receptor blockers may increase the risk of severe COVID-19. *J Travel Med.* **27**. https://doi.org/10.1093/jtm/taaa041
38. Schiffrin, E. L., Flack, J. M., Ito, S., Muntner, P., and Webb, R. C. (2020) Hypertension and COVID-19. *Am J Hypertens.* **33**, 373–374. https://doi.org/10.1093/ajh/hpaa057
39. Yang, Y., Liu, X., Liu, Y., Chen, Y., Tian, Y., Fu, X., *et al.* (2020) Efficacy and safety of ACEI/ARB drugs in patients with COVID-19 combined with diabetes mellitus: A protocol for systematic review and meta-analysis of randomized controlled trials. *Medicine (Baltimore).* **99**, e21723. https://doi.org/10.1097/MD.0000000000021723
40. Daoud, A., Ali, H., Rao, V., Rohan, V., and Soliman, K. (2021) ACE-inhibitor/angiotensin receptor blockers (ACE-I/ARBs) therapy in COVID-19 infected dialysis patients. *Ren Fail.* **43**, 1463–1464. https://doi.org/10.1080/0886022X.2021.1994419
41. Biyani, C. S., Palit, V., and Daga, S. (2020) The use of captopril-angiotensin converting enzyme (ACE) inhibitor for cystinuria during COVID-19 pandemic. *Urology.* **141**, 182–183. https://doi.org/10.1016/j.urology.2020.04.057
42. Tan, N. D., Qiu, Y., Xing, X. B., Ghosh, S., Chen, M. H., and Mao, R. (2020) Associations between angiotensin-converting enzyme inhibitors and angiotensin II receptor blocker use, gastrointestinal symptoms, and mortality among patients with COVID-19. *Gastroenterology.* **159**, 1170–1172.e1171. https://doi.org/10.1053/j.gastro.2020.05.034
43. Parigi, T. L., Vespa, E., and Pugliese, N. (2021) COVID-19, ACEI/ARBs, and gastrointestinal symptoms: The jury is still out on the association. *Gastroenterology.* **160**, 1896–1897. https://doi.org/10.1053/j.gastro.2020.06.095
44. South, A. M., Brady, T. M., and Flynn, J. T. (2020) ACE2 (angiotensin-converting enzyme 2), COVID-19, and ACE inhibitor and Ang II (angiotensin II) receptor blocker use during the pandemic: The pediatric. *Perspect Hypertens.* **76**, 16–22. https://doi.org/10.1161/HYPERTENSIONAHA.120.15291
45. South, A. M., Nixon, P. A., Chappell, M. C., Diz, D. I., Russell, G. B., Jensen, E. T., *et al.* (2018) Association between preterm birth and the renin-angiotensin system in adolescence: Influence of sex and obesity. *J Hypertens.* **36**, 2092–2101. https://doi.org/10.1097/HJH.0000000000001801
46. Kreutz, R., Algharably, E. A. E., Azizi, M., Dobrowolski, P., Guzik, T., Januszewicz, A., *et al.* (2020) Hypertension, the renin-angiotensin system, and the risk of lower respiratory tract infections and lung injury: Implications for COVID-19. *Cardiovasc Res.* **116**, 1688–1699. https://doi.org/10.1093/cvr/cvaa097
47. Furuhashi, M., Moniwa, N., Mita, T., Fuseya, T., Ishimura, S., Ohno, K., *et al.* (2015) Urinary angiotensin-converting enzyme 2 in hypertensive patients may be increased by olmesartan, an angiotensin II receptor blocker. *Am J Hypertens.* **28**, 15–21. https://doi.org/10.1093/ajh/hpu086
48. Chen, J., Mei, K., and Deng, C. (2021) Is there a difference in the effect between the ACEI and ARB on COVID-19? *Clin Cardiol.* **44**, 5–6. https://doi.org/10.1002/clc.23474
49. Szekacs, B., Varbiro, S., and Debreczeni, L. (2021) High-dose ACEi might be harmful in COVID-19 patients with serious respiratory distress syndrome by leading to excessive bradykinin receptor activation. *Physiol Int.* https://doi.org/10.1556/2060.2021.00007
50. Zhao, H. J., Li, Y., Wang, D. Y., and Yuan, H. T. (2021) ARB might be superior to ACEI for treatment of hypertensive COVID-19 patients. *J Cell Mol Med.* **25**, 11031–11034. https://doi.org/10.1111/jcmm.17051
51. Krvavac, A., Patel, T. P., Karle, E. M., Epstein, N. B., Reznikov, E. A., Gates, L. G., *et al.* (2020) Increased incidence, morbidity, and mortality in human coronavirus NL63 associated with ACE inhibitor therapy and implication in SARS-CoV-2 (COVID-19). *Mo Med.* **117**, 346–354. www.ncbi.nlm.nih.gov/pubmed/32848272

52. Liu, X., Liu, Y., Chen, K., Yan, S., Bai, X., Li, J., *et al.* (2021) Efficacy of ACEIs/ARBs vs CCBs on the progression of COVID-19 patients with hypertension in Wuhan: A hospital-based retrospective cohort study. *J Med Virol.* **93**, 854–862. https://doi.org/10.1002/jmv.26315
53. Byttebier, G., Belmans, L., Alexander, M., Saxberg, B. E. H., De Spiegeleer, B., De Spiegeleer, A., *et al.* (2021) Hospital mortality in COVID-19 patients in Belgium treated with statins, ACE inhibitors and/or ARBs. *Hum Vaccin Immunother.* **17**, 2841–2850. https://doi.org/10.1080/21645515.2021.1920271
54. Presta, V., Figliuzzi, I., Citoni, B., Gallo, G., Battistoni, A., Tocci, G., *et al.* (2021) ARB-based combination therapy for the clinical management of hypertension and hypertension-related comorbidities: A spotlight on their use in COVID-19 patients. *High Blood Press Cardiovasc Prev.* **28**, 255–262. https://doi.org/10.1007/s40292-021-00443-z
55. Sienko, J., Kotowski, M., Bogacz, A., Lechowicz, K., Drozdzal, S., Rosik, J., *et al.* (2020) COVID-19: The influence of ACE genotype and ACE-I and ARBs on the course of SARS-CoV-2 infection in elderly patients. *Clin Interv Aging.* **15**, 1231–1240. https://doi.org/10.2147/CIA.S261516

11 COVID-19 Therapeutics
Challenges and Opportunities

Travis B. Lear, Bill B. Chen, and Yuan Liu

11.1 INTRODUCTION

In the 3 years since the emergence of SARS-CoV-2, the COVID-19 pandemic has caused over 770 million cases and nearly 7 million deaths worldwide. Driven by infection of the beta coronavirus SARS-CoV-2, COVID-19 is a complicated disease, primarily manifesting in the lung but with systemic effects across the body. Most SARS-CoV-2 infections resolve after mild to moderate respiratory illness and recover without special treatment; however, some patients develop serious illness consisting of systemic inflammation and cytokine storm, which causes multiorgan damage and pulmonary distress known as acute respiratory distress syndrome (ARDS). This is similar to infectious cases with the SARS and MERS viruses; the extreme virulence is mostly due to dysfunctional immune responses and aggressive inflammatory responses (1–3). Extensive efforts from research groups in the pharmaceutical industry and translational academia have focused on the development of preventive and postinfection treatment strategies. These efforts have culminated with the successful and rapid development and deployment of several vaccines (4), two postinfection antiviral agents available on the market, and numerous potential interventions currently undergoing clinical trials. As part of the massive research effort to combat COVID-19, the many efforts in developing COVID-19 therapeutics have been extensively reviewed by several research groups (5–7). In this chapter, we hope to unify the many facets of anti–COVID-19 drug discovery into a cohesive summary and reflect on the continued efforts, challenges, and opportunities in developing antiviral therapeutics. A list of therapeutics is highlighted in Table 11.1, and the most updated and detailed treatment guidelines can be found at the NIH website COVID-19 Treatment Guidelines (https://www.covid19treatmentguidelines.nih.gov).

TABLE 11.1
COVID Therapuetics

Hospitalized and Requiring Conventional Oxygen	
1. Patients requiring minimal conventional oxygen	Remdesivir
2. Most patients	Remdesivir + dexamethasone
3. Patients receiving dexamethasone, having systemic inflammation and requiring increasing oxygen needs	Remdesivir + dexamethasone; add baricitinib or tocilizumab
Hospitalized and Requiring HFNC Oxygen or NIV	Remdesivir + dexamethasone; add baricitinib or tocilizumab
Hospitalized and Requiring MV or ECMO	Dexamethasone; add baricitinib or tocilizumab

Note: The potential additive, antagonistic, or synergistic effects and the safety of using certain combination therapies for the treatment of COVID-19 are unknown.

Baricitinib, Janus kinase inhibitor; dexamethasone, systematic corticosteroid; ECMO, extracorporeal membrane oxygenation; HFNC, high-flow nasal cannula; MV, mechanical ventilation; NIV, noninvasive ventilation; tocilizumab, anti–IL-6 receptor monoclonal antibody.

DOI: 10.1201/9781003355243-14

11.2 SARS-CoV-2 PATHOGENESIS

SARS-CoV-2 is an enveloped RNA virus belonging to the Coronaviridae family, with an exceptionally large genome, reaching almost 30 kb in length. From this genome, the SARS-CoV-2 proteome encodes four structural proteins (the spike, envelope, membrane, and nucleocapsid), 16 nonstructural proteins (NSP1–NSP16), and nine other accessory proteins (Figure 11.1).

Over the course of the viral infection, the viral surface spike protein interacts with the human angiotensin-converting enzyme 2 (ACE2) receptor on the host cell surface to enter the cell by either direct membrane fusion or by endosomal-mediated entry. An important cofactor in either entry mechanism is the role of human proteases such as TMPRSS2, furin, or endosomal cathepsins, which act to activate the spike protein for enhanced membrane tethering. Once internalized, the viral genome uses the host translational machinery to generate the viral proteome, which includes two massive polyproteins encoding NSP1 through NSP16. This pp1a/pp1b polypeptide is autocatalyzed by two viral proteases (PLpro and M^{pro}) to form individual, processed, and active viral proteins. Research continues to uncover varying functions and effects of each of the viral proteins, but key among them are the components of the RNA-dependent RNA polymerase (RdRp), which performs the vital replication function to generate a new viral genome. Viral components hijack the host organelle system to form replication organelles, in which the viral genome and key structural proteins are assembled into new virions. Finally, the nascent virion uses the endosomal egress system to traffic newly formed virus to the cell membrane for exocytosis. Researchers have denoted several points along the SARS-CoV-2 viral life cycle that can be targeted to blunt viral pathogenesis. These efforts will be detailed in this chapter.

11.3 CONCEPTUAL STRATEGY OF ANTIVIRAL EFFORTS

Extensive research efforts have yielded a wealth of information surrounding the life cycle and mechanism of SARS-CoV-2 infection, replication, and spread. Armed with this knowledge, drug discovery efforts diverged based on three distinct strategies: (1) the development of small-molecule and antibody agents that directly target viral components; (2) host-directed therapeutics that target essential human factors modulated by viral infection, including repurposing previously approved or the in-pipeline drugs for novel antiviral indications (8); and (3) modulating the human immune system to temper runaway inflammation and cytokine storm characteristic of later infection. Key in the first strategy is the inhibition of viral proteins essential for the viral life cycle. During the SARS-CoV-2 viral replication, several viral proteins have emerged as potential

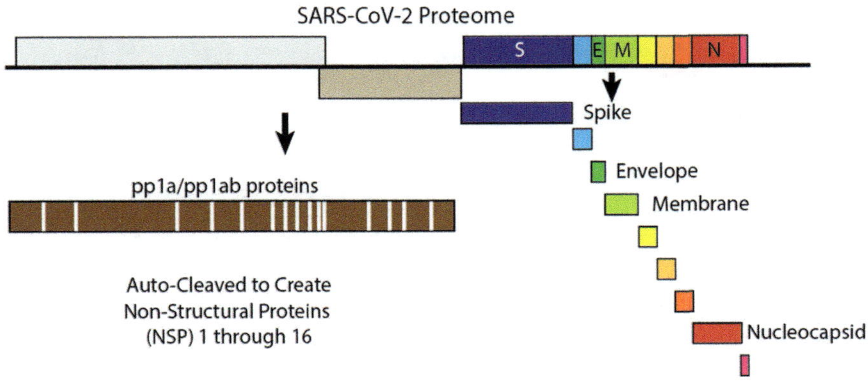

FIGURE 11.1 The SARS-CoV-2 proteome.

focuses of intervention to block different stages of the viral life cycle (Figure 11.2). These include agents to inhibit the action of the surface spike protein that mediates viral entry into the host cell, direct inhibitors of the viral main- and papain-like proteases to impair proteolytic processing, and several inhibitors of the multi-subunit polymerase and replication complex (comprising NSP12–NSP16) to impede viral message and structural assembly. A strength of the virus-directed approach is the high selectivity, especially when the targets lack human homologues. However, viral-directed strategies suffer the risk of drug resistance, as emerging variant sequences mutate the key drug-interacting domains (9). Research has noted that SARS-CoV-2 infection leads to the direct disruption and viral hijacking of over 300 human proteins, which we term "host factors." Viral manipulation of host factors, including key cellular proteins such as ACE2, are essential for completion of its viral life cycle (10). The second antiviral strategy centers the targeting of host factors essential to the viral life cycle (10, 11). This has included agents to inhibit ACE2 and the essential human protease cofactors such as TMPRSS2 and endosomal cathepsins. A benefit of this strategy lies in the potential for broad-spectrum antiviral activity and application against emerging variants and even other viruses. However, drawbacks to this strategy emerge in the limited selectivity of drugs to specific host factors and the potential for safety issues in off-targeting effects among noninfected tissues (12). The third strategy concentrates on the modulation of the human immune system to temper detrimental host responses to viral infection, such as "cytokine storm" and thrombosis (13). Early use of antiviral IFNs and corticosteroids in the later disease stage might prevent disease progression and expediate the recovery. These three strategical lines of thought are delineated as follows to highlight the achievements and obstacles in the COVID-19 therapy development to inspire novel tractable targeting strategy.

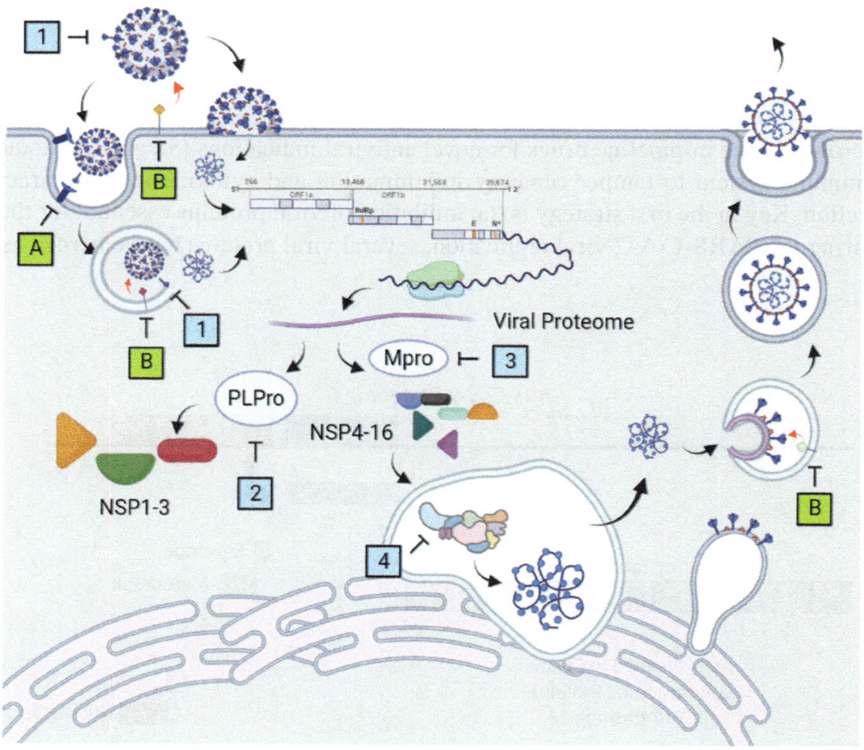

FIGURE 11.2 SARS-CoV-2 viral life cycle and therapeutic targets. Viral targets: (1) immunotherapy against spike protein; (2) PLpro inhibitors; (3) M^{pro} inhibitors; (4) RdRp inhibitors. Host factors: A, ACE2 inhibitors; B, protease (TMPRSS2, cathepsin, etc.) inhibitors.

11.4 DRUG REPURPOSING FOR COVID-19 TREATMENT

In parallel to the three-pronged targeting paradigms for antiviral drug discovery, rational design is an important consideration in drug development. Classically, the development process began de novo, using target-oriented or phenotypic screens of compound libraries to uncover a novel agent to serve as a structural basis for development. Due to many factors across business, including target validation, safety consideration, and regulatory compliance, the drug discovery process from hit compound to approval is long and fraught with a low success rate (14). To circumvent this, an alternative strategy lay in using previously developed agents for new purposes, including antiviral uses. This "repurposing" of drugs has recently emerged as an important strategic component of antiviral drug development to treat COVID-19. Drug repurposing involves the screening and validation of investigational or previously approved drugs for new antiviral indications. Importantly, a repurposed drug has usually undergone the rigorous processes to validate its safety in preclinical or even human trials as part of its original indication, as well as characterization of numerous pharmacological parameters (e.g., ADME, formulation, delivery). Collectively, the prior knowledge and completion of these essential steps accelerate a potential repurposed drug to overcome major hurdles in the validation and development of an antiviral agent. In the face of the rapid spread of COVID-19, the drug repurposing strategy was widely adopted to maximize the speed in producing an efficacious therapeutic. To initiate a drug repurposing project, researchers have several tools at their disposal to identify potential candidates, involving both computational and experimental approaches. Computational approaches leverage the massive amount of data accumulated, ranging from cell and tissue gene expression patterns, structure-activity relationship of specific chemical structures, to even electronic health records (EHRs) of patient cohorts (15). Additionally, accumulated cell viability studies from hundreds of compounds tested on immortalized human cancer cell lines are available for toxicological consideration (16). Collectively, this allows for the rational streamlining of a repurposing effort. Orthogonally, experimental efforts have been extraordinarily boosted by the instrumental and technological advancements essential for the high-throughput screening (HTS) campaigns. Following scalable assay design, such as a cell-based assay of viral entry, researchers can screen thousands of potential repurposing compounds for their antiviral efficacy. Beyond cell-based or biochemical means of hit identification and validation, in silico simulations of molecular docking based on chemical structure assist to predict potential ligand-target binding mechanism (17). In vitro, researchers use proteomic profiling via the purification of the purported target with the candidate compound coupled with mass spectrometry to validate drug binding partners. The other effective approach is to engage disease relevant model systems for phenotypic screen to identify biologically active compounds specific to targeted disease indication (18). Currently, several additional initiatives to identify candidate agents for COVID-19 treatment have been undertaken to test approved drugs, screen libraries of agents with confirmed safety profile, and evaluate novel therapies. The Scripps Research Institute has led a screen of a best-in-class drug repurposing library, ReFRAME, against high throughput and high content imaging infection assays. The ReFRAME library contains nearly 12,000 compounds approved by the FDA for other diseases or being extensively tested for human safety (19). However, the initial in vitro antiviral efficacy testing should be distinguished from a cellular toxicity to limit the failure of translation of in vitro antiviral activity into clinical benefit in COVID-19 treatment (20–22). In addition, mainly the agents demonstrating compelling antiviral effects using in vivo animal models should be considered for the clinical evaluation (22). Together, these efforts have demonstrated the advantages of drug repurposing as a fast-track drug discovery strategy given the needed speed for antiviral therapy development.

11.5 VIRAL PROTEIN TARGETS

11.5.1 The NSP5 M^{PRO}

Following viral internalization and hijacking of the host ribosome to produce the viral proteome, the massive polyprotein 1a/b is proteolytically cleaved by viral proteases to generate the individual *nonstructural* proteins essential for the viral life cycle. Among the two viral proteases, the Main protease (M^{pro}, also known as *nonstructural* protein 5 [NSP5]; or 3-chymotrypsin like protease [$3CL^{Pro}$]), plays the critical role of processing the latter portions of the pp1a/b protein, and has emerged as a primary strategic target for SARS-CoV-2 antiviral drug discovery. Indeed, one of the two currently FDA-approved small molecule therapies that target viral proteins contains the small molecule nirmatrelvir, functioning as an M^{pro} inhibitor. Several key factors support the rationale in targeting M^{pro} as an antiviral therapeutic strategy (5). Conceptually, cysteine proteases have been well characterized as druggable targets, and importantly, the sequence of the M^{pro} key active region and its cleavage specificity are distinct from human proteases, limiting the potential off-targeting issues. Further, historical efforts in targeting viral proteases such as HIV and HCV have successfully yielded multiple inhibitors in clinical use. Crystal structures of M^{pro} were reported rapidly at the onset of the pandemic, providing drug discovery efforts critically important data for modeling and compound binding studies (23, 24). Most importantly, tied to the concept of drug repurposing, compounds originally designed as therapeutics against the SARS-CoV M^{pro} during the 2003 outbreak (PF-00835231) demonstrated efficacy against SARS-CoV-2, and was evaluated in clinical trials as early as 2020 (NCT-04535167) (25–27). Owing to the extremely high sequence similarity (96%) in M^{pro} of both viruses, anti–SARS-CoV compound could be used as a starting point to further develop a SARS-CoV-2 M^{pro} inhibitory agent. This culminated with the development of the M^{pro} inhibitor nirmatrelvir, and its subsequent FDA EUA in the form of *Paxlovid*. Nirmatrelvir displays pan-antiviral activity against broad-spectrum human coronavirus, excellent off-target selectivity and safety profiles (28). However, the rapid metabolism of nirmatrelvir has necessitated co-administration with a CYP3A inhibitor, ritonavir, to ensure efficacious drug exposure. This co-treatment, though minimizing the metabolism of nirmatrelvir, also leads to major combination concerns with other drugs such as statins or blood thinners. Patients in need of Paxlovid treatment to avoid severe COVID-19 infection are proportionally more likely to be taking these contraindicated medications. Further, recent studies have noted that naturally occurring mutations in M^{pro} sequence among already circulating SARS-CoV-2 variants are experimentally unresponsive to nirmatrelvir treatment (5). Despite these drawbacks, M^{pro} remains a promising major target for SARS-CoV-2 antiviral development, with several more unique compounds in the development pipeline.

11.5.2 The NSP12 RdRp

After proteolytic processing by M^{pro}, the nonstructural viral proteins 8, 9, and 12 form a complex essential for the synthesis of new viral RNA genome, the RNA-dependent RNA polymerase (RdRp). Replication of the viral genome is a pivotal step in the viral life cycle, required for the generation of new virions for subsequent infection. Within this complex, NSP12 functions as the catalytic core of the polymerase. Stemming from this role, and the conservation of its sequence among differing SARS-CoV-2 variants and other coronaviruses, NSP12 has emerged as a major drug discovery target (29, 30). Akin to the previous successes in targeting viral proteases among other viruses, several inhibitors have been developed targeting the RdRps of HIV, HCV, influenzas A and B, and the respiratory syncytial virus (RSV) (31). One such compound, remdesivir, though originally designed for use against hepatitis C, demonstrated broad antiviral efficacy against several coronaviral species through its ribonucleotide analogue effect on RdRp (32, 33). As such, repurposing efforts of previous compounds for SARS-CoV-2 NSP12 targeting took place, prominently including remdesivir (34). Subsequent developments yielded the compound molnupiravir, although targeting RdRp, has a distinct mechanism compared to remdesivir (35). Remdesivir is metabolized into an active form that functions as a delayed chain extension terminator, stalling the RNA polymerase to prevent replication (36); however, molnupiravir's active metabolite instead mimics an RNA nucleoside and is directly incorporated into new synthesized viral RNA. Once incorporated, molnupiravir tautomers can be "read" as either cytidine or uridine, the incorrect reading thus leading to a mutated sequence in all subsequent genomic copies. Collectively, this causes an accumulated mutational burden that is eventually lethal to the viral life cycle (37). Along with *Paxlovid*, molnupiravir has received FDA approval for clinical use, however, molnupiravir targeting the RdRp has exhibited the ability to generate sublethal mutational burden resulting in still active, albeit mutated, virions. Some have posited that the effect of molnupiravir treatment and sublethal mutations could serve as an additional source of new viral mutation, potentially contributing to new viral variants. Additionally, although humans do not encode for an orthologous RdRp, chemical inhibitors have shown cross reactivity with human RNA polymerases, which could have unintended toxicity (38). Finally, beyond molnupiravir, other chemical agents targeting RdRp (favipiravir, sofosbuvir) had mixed results in clinical trial testing, resulting in no discernable benefit on patient hospitalization or mortality (39, 40), demonstrating the limits of drug repurposing as effective antiviral therapeutics. Despite this, targeting the RdRp, particularly with molnupiravir, shows clinical efficacy and may hold promise as a target for a pan-coronaviral therapeutic.

11.5.3 Papain-Like Protease of NSP3

Complementary to the Main protease M^{pro}, the SARS-CoV-2 nonstructural protein 3 (NSP3) or Papain-like protease (PLpro) aids in the processing of the polyprotein ppla/b from the N-terminal side. Following PLpro-mediated processing, the nonstructural proteins 1, 2, and 3 are produced, with multivariate effects on the replication process and modulating the host response. Whereas PLpro displays protease activity to aid in the viral protein maturation, PLpro's large protein sequence (>200 kDa) contains several domains with non-polyprotein processing activity. Notably, PLpro has been observed to manipulate post-translational modifications to human host proteins. Essential in the host antiviral signaling response is the modification of key host proteins with post-translational "tags" including the addition of ubiquitin (ubiquitination) and interferon-stimulated gene 15 (ISGylation). PLpro has been observed to directly remove ubiquitin from host proteins (deubiquitination) as well as ISG15 (deISGylation), with the net effect of dampening cellular antiviral cytokine signaling and suppressing the innate immune response (41, 42). Efforts to chemically target PLpro activity have focused on direct covalent inhibitors of the key catalytic domain, non-covalent inhibitors to sterically block catalytic activity, and allosteric inhibitors to impede PLpro function (43–45). Several candidate inhibitors that target the catalytic domain are peptidomimetics and are limited by the sequence affinity of PLpro for the motif LXGG. Intriguingly, allosterically targeted inhibitors have been reported to function by disrupting the Zn^{2+} binding pocket of PLpro, such as disulfiram (46). These efforts have led to decreased PLpro catalytic activity in removing ubiquitin and ISG15. Although demonstrating in vitro efficacy, the widespread use of Zn^{2+} in human protein functions may prove to be inescapable off-targeting toxicity of such an avenue. Further, efforts to inhibit PLpro's deubiquitinase-like activity (DUB) will likely have off-targeting effects on human DUBs and DUB-like enzymes due to structural similarity (47). Moreover, historic efforts to design inhibitors of human DUB enzymes for other diseases have not yet reached clinical approval, which may link to deeper fundamental issue in targeting the DUB-like activity of PLpro (48). Collectively, several studies note an antiviral effect in targeting and inhibiting PLpro; however, challenges remain in resolving specificity and selectivity in targeting. Emphasis on non-covalent active site inhibitors may hold the most promise moving forward.

11.5.4 The NSP3 Macrodomain

Within its massive protein sequence, the NSP3 protein harbors a functional domain with ADP-ribose catalysis activity. Termed the *macrodomain*, this 150-amino acid region is a domain conserved in all kingdoms of life and found with numerous viruses (49). Macrodomains consist of a globular structure involving a mixed $\alpha/\beta/\alpha$ sandwich fold that contains a conserved cleft region with affinity for ADP-ribose, leading to the enzymatic removal of ADP-ribose units from proteins (49–51). Importantly, host ADP-ribosyltransferases have been noted to modify viral proteins with

ADP-ribose transfer to impede viral protein activity (52). The catalytic activity of viral *macrodomain* functions to oppose host ADP-ribosylation as a means to evade antiviral immune efforts. Macrodomain activity has been shown to be tied to virulence, as inactivating mutations within this domain lead to decreased virulence (53). As such, chemical inhibition of the NSP3 macrodomain may be an effective antiviral strategy with the screening results from early discovery stage suggesting the antiviral efficacy of inhibiting NSP3 macrodomain (51, 54–56).

11.5.5 The NSP13 Helicase

As part of the viral life cycle, coronaviruses use an unusually large RNA-synthesizing and RNA-processing enzyme suite to mediate the expression, replication, and quality control of their large RNA genome. As described above, the RdRp complex is the central component of genome synthesis, and an attractive therapeutic target. Alternatively, the associate nonstructural proteins required for RdRp fidelity may constitute additional targets for antiviral drug development. The nonstructural protein 13 (NSP13) is a cognate to the RdRp replication complex, functioning to unwind the template RNA from 5′ to 3′ (57, 58). Additionally, NSP13 mediates proofreading and template switching, collectively being an essential part of the replication process (59, 60). Intriguingly, NSP13 is highly conserved among coronaviral family members, and studies have shown that loss of NSP13 impairs viral replication (61, 62). Given its vital role for viral function and evolutionary conservation, targeting NSP13 may have pan-coronaviral activity, along with the antiviral efficacy indicated in the early screening efforts, there remain no clinically advanced agents (63). Further, an imposing hurdle lay in the similarity between coronaviral helicase active sites, and human helicases, such DDX, requiring the vigilant monitoring of off-target side effects (64).

11.5.6 The NSP14 and NSP16 Methyltransferases

The SARS-CoV-2 proteome encodes two methyltransferase enzymes, NSP14, and NSP16 that function to modify the viral genome to evade the host immune response. During the antiviral host response, pattern recognition receptors (PRR) such as RIG-I sense viral nucleic acids and initiate an immune response. To avoid this detection, the viral methyltransferases modify the viral genome to form an RNA cap. Specifically, NSP14, methylates guanosine bases within the viral RNA, followed by the action of NSP16, along with NSP10, to catalyze the 2′-O-methylation of ribose backbone (65). This modification effectively blunts a major cellular antiviral signaling pathway (66), suggesting that inhibition of these methyltransferases may be an avenue to augment antiviral signaling. Recent efforts have uncovered the underlying enzymatic mechanism of methyltransferases (67, 68); and several groups have reported potential agents to target NSP14 activity (69, 70), albeit with limited antiviral effects in vitro. Of note, in addition to its methyltransferase activity, NSP14 exhibits proofreading activity through a 3′-5′ exoribonuclease domain, in conjugation with NSP10 (71, 72). The NSP14-mediated proofreading activity is vital for effective replication, leading to the consideration of NSP14 as a potential antiviral drugging target. Although some agents have been identified as potential hits, similarities to human exonucleases (e.g., DEDD family) present the potential liability of off-targeting (73). Likewise, NSP16 has structural similarities to human methyltransferases, such as CMTR1. Experimental inhibition of NSP16, while blunting viral replication, has affected host proteins. Importantly, other studies have noted that inhibition of host methyltransferase may have neurological and nephrological off-targeting effects, which may limit the utility in direct NSP16 inhibition (74, 75).

11.5.7 The NSP15 Endonuclease

In addition to the immune evasive measures of NSP14 and NSP16, the NSP15 is a viral endonuclease that functions to dampen host immune responses. NSP15 specifically targets and cleaves uridine-rich regions of viral RNA, which otherwise would be signaling molecules sensed by cellular PRRs to initiate an innate immune response (76–79). The catalytic domain of NSP15 belongs to the EndoU family (80), and is conserved among several Coronaviridae family members (81). Importantly, NSP15 shares no close homology with human proteins, and its conservation among different viruses makes it an intriguing target with potential for pan-antiviral efficacy. Current discovery efforts have yielded several hits as potential NSP15 inhibitors, but development remains nascent (82). However, inhibitors of orthologous endonucleases in influenza have been clinically used in the form of baloxacir marboxil, suggesting the potential for anti-coronaviral endonuclease development (83, 84).

Directly targeting viral proteins to block viral replication cycle might have high selectivity and safety benefits when the viral target has no human homologues. However, therapies targeting viral proteins can rapidly lose efficacy in the evolving variant strains. The monoclonal antibodies that target the SARS-CoV-2 spike protein have shown clinical benefits in treating SARS-CoV-2 infection during the early stage of pandemic. However, the in vitro neutralization activities of those anti–SARS-CoV-2 monoclonal antibodies against the current variants and subvariants (e.g., Omicron XBB and Omicron XBB.1.5) drastically reduced, which leads these antibodies unlikely to be active in clinical use. Thus, host-focused targeting strategy regulating human proteins involved in the viral life cycle can serve as a supplemental therapy, particularly for broad-spectrum antiviral usage.

11.6 HOST PROTEIN TARGETS

The alternative host targeting strategy to block the viral life cycle lies in focusing on human factors that are either hijacked for viral use or dysregulated to promote virulence. SARS-CoV-2 relies on more than 300 host factors to complete its viral life cycle including cell entry and viral replication (85). Conceptually, the strategy targeting human host factors has increased risk of non-selectivity and safety concerns, as target inhibition may have deleterious and pleiotropic effects in noninfected tissues. However, a fundamental advantage lay in circumventing the selective mutational pressure and chance for resistance that comes with a viral-directed inhibitor strategy. Indeed, efforts in targeting essential host factors tied to other viruses have shown success, such as an antibody targeting human T-cell CD4 to impair the viral entry of HIV strains (86). Ultimately, the promise of host-directed antiviral therapeutic strategy lay in the potential of developing agents with broad-spectrum efficacy that retain new and emerging variants (11). In this section, we will discuss the several key human host factors hijacked and modulated by SARS-CoV-2 and the drug discovery efforts centering them.

11.6.1 Receptors for SARS-CoV-2 Host Cell Entry

One of the first host factors identified for SARS-CoV-2 infectivity is the primary and essential entry receptor, angiotensin-converting enzyme 2 (ACE2). Early efforts to uncover the SARS-CoV-2 entry receptor were informed based on homology with other human coronaviruses, as SARS-CoV and HcoV-NL63 both share ACE2 as a primary entry mechanism (87). During an entry event, the viral spike protein on the envelope surface is recognized by the ACE2 extracellular peptidase domain on the host cell, leading to either endosomal-mediated entry or direct cell surface fusion for viral entry (88). Because of the high level of ACE2 receptors in the host lung, the respiratory tract is the primary entry route for SARS-CoV-2 (89). Although research has suggested that SARS-CoV-2 may use alternative host receptor proteins to facilitate viral entry, ACE2 remains the most observed and best characterized mechanism (90). From this, drug development efforts from the pandemic onset have attempted to target ACE2 to prevent viral entry. Development of mimetic peptide agents of the N-terminal ACE2 region and soluble recombinant ACE2 protein have been tested in an attempt to compete out spike protein interaction with cell surface ACE2 (91, 92). Additionally, small molecule ligands of ACE2 such as SB27041 and SB27047 have been tested to prevent viral entry, putatively working by changing the confirmational orientation of ACE2 to not recognize the spike protein (93). Potentially this strategy can circumvent mutations within the spike protein and still retain efficacy. However, these efforts have shown little effect, including the lack of clinical improvement of mortality with treatment of recombinant ACE2 (APN01, NCT04335136). Further, direct ACE2 catalytic inhibitors have failed to prevent infection in vitro, which has helped to uncouple ACE2 enzymatic activity from the mechanism of viral entry.

11.6.2 Host Proteases Regulate SARS-CoV-2 Viral Life Cycle

In addition to the central role of ACE2 as the cellular receptor for SARS-CoV-2 viral entry, the surface spike protein takes advantage of several human proteases to enhance entry ability. This is accomplished by the proteolytic cleaving of the spike protein to produce a processed form that is required for membrane anchoring (94). The spike protein has been noted to be cleaved by host proteases such as furin, transmembrane protease serine 2 (TMPRSS2), vesicular cathepsins, and membrane ADAM enzymes (95–97). Importantly, the cofactor role of human proteases is associated with both the rapid viral entry of direct membrane fusion and the slower endosomal-mediated internalization (98, 99). As such, these human protease cofactors for viral entry have been targeted to prevent viral entry. Two small molecule inhibitors of the human TMPRSS2, nafamostat mesylate and camostat mesylate, have been developed and shown to have efficacy in reducing in vitro infection in both airway cell culture and primary epithelial cells (95). Moreover, nafamostat treatment to a transgenic mouse model of SARS-CoV-2 infection (K18-hACE2) reduced viral levels and improved mortality (100). However, these successes have failed to translate to clinical improvements, as several double-blinded, randomized control trials failed to show a clinical benefit of either camostat or nafamostat against COVID-19 infection (NCT04657497, NCT04321096, NCT04623021, NCT04473053). Inhibitors against other human proteases implicated in viral processing and entry, such as cathepsin L and furin, have shown in vitro promise, but these strategies may not be sufficient because of the potential compensation of multiple avenues of viral entry. Moreover, combination treatment to fully block all pathways faces the drawback of severe toxicity with a multiple protease inhibitor regimen. As an orthogonal approach, augmenting cellular mechanisms to promote TMPRSS2 inhibition may be beneficial, such as through increasing the degradation of TMPRSS2 through the ubiquitin proteasome system (101).

11.6.3 Modulating the Autophagy-Lysosomal Pathway

A major surveillance mechanism for cellular homeostasis maintenance is through the autophagy-lysosomal pathway. This pathway facilitates the elimination and recycling of numerous cellular components, including individual proteins up to entire organelles. Along a carefully orchestrated signaling cascade, cellular components are engulfed in a double-membraned autophagosome, which reaches its terminus when fused to the lysosome (102). As an organelle attracting extensive research interests, the lysosome is a sequestered cellular region of acidic pH and numerous proteases, in which it functions to break down cellular components for reuse and recycling (103). Beyond its roles in protein degradation, recent research has uncovered roles for the lysosome in nutrient sensing, metabolic flux, modulating of gene expression, and an immunoprotective role through actions such as xenophagy and antigen presentation (104). Xenophagy serves as a direct antiviral host defense strategy, facilitating the capture and lysosomal degradation of exogenous threats (105). Additionally, viral component processing by the lysosome can be used as antigen presentation mechanisms to augment the activity of immune cells (106). However, the autophagy-lysosomal system can be hijacked by viral infection, including de-acidification as an exit route for newly generated virions (107). Efforts to target the lysosomal system have met with mixed results. Early in the COVID-19 pandemic, lysosome modulators such as chloroquine were observed to have efficacy in vitro but failed to translate to clinical efficacy. Other research efforts have sought to promote lysosomal biogenesis during infection to stimulate pathogen clearance (108). Because of the complexity of the autophagy-lysosomal system, there is a wealth of targets that remain unexplored as potential antiviral avenues.

11.6.4 Novel Host Protein Targets

Beyond the known host factors that are influenced by SARS-CoV-2 infection, drug discovery efforts have used high-throughput screening approaches to uncover new potential targets. Multiple groups have carried out genetic screens using whole-genome CRISPR-based knockout approaches followed by viral infection, with the endpoint of cell viability postinfection (109–116). This approach hypothesizes that viral treatment will cause cytopathy, and that genes overrepresented among the survivor pool of infected cells relative to control cells may function as pro-viral factors. Upon the loss of these host factors, viral infectivity is diminished, leading to lessened cytopathy. These screens have tested various coronaviruses and infection conditions, generating several hits such as RAB7A, TMEM41B, PI3K type3, and the lysosomal protein TMEM106B as essential pro-viral host factors. However, important drawbacks of this screen design result from using cellular viability as the primary outcome, as genes related to later stages of the viral life cycle such as assembly and release are biased against. Beyond genomic screens for SARS-CoV-2 related host factors, several other candidate human proteins have been explored as potential targets for antiviral therapeutics. These have included the farnesoid X receptor, caspase-6, eukaryotic translation elongation factor 1A, bromodomain-containing protein 2, CD147, and transmembrane protein 16F (117–122). Each of these efforts remains preliminary, however, suggesting that host protein targets have provided a means to avoid antiviral drug resistance. Notably, drug development targeting host factors needs to be closely monitoring the safety performance.

11.7 HOST IMMUNOLOGICAL TARGETS

Beyond viral-centered targets for drug inhibition or even host proteins that are hijacked by the viral invasion, the appropriate host immune response is a strategic target to ameliorate viral infection and promote recovery. In many cases, acute SARS-CoV-2 infection in the lung resolves without extensive inflammatory response, however a subset of cases progress to a hyperinflammatory state characterized by systemic immune activation and cytokine storm. Some cases progress into ARDS, in which runaway inflammatory activation leads to diffuse alveolar damage, fluid buildup, and even respiratory failure. Paradoxically, these catastrophic effects often occur after host tissues have mostly contained the viral infection; many patients with SARS-CoV-2 ARDS have already cleared the virus from their system. In these situations, efforts to temper the immune system response and inflammatory signaling are crucial to improve patient outcomes.

11.7.1 Immunotherapy

An important consideration in dissecting the systemic inflammatory response during SARS-CoV-2 is the heterogenous response observed among patients. The course of viral infection, disease progression in the lung, and extent of immune activation rely on a complex set of variables not only among distinct types and doses of viral infection, but also the inherent differences among the infected patients. Moreover, viral infection has been observed to yield a variety of pathophysiological consequences, including hyperinflammation, inactive lymphocytic activity, tissue fibrosis, and dysfunction in thrombopoiesis. Collectively, SARS-CoV-2 infection causes a broadly dysregulated immune response with repercussions beyond the lung. From this, several clinical trials have sought to approach COVID-19 therapy by using immunotherapy.

11.7.2 Virus-Targeted Immunotherapies

With the goal of avoiding the complications associated with late-stage immune activation, several drug development strategies have sought to combat viral infection early on, through the use of immunotherapy strategy. Historically, infectious disease management has used nascent immunotherapy techniques, such as the use of convalescent plasma administration from survivors to give to newly infected patients beginning in the 1930s (123). In the early stages of COVID-19, the use of high-titer convalescent plasma has improved patient outcomes successfully (124–126). Development of immunotherapy directed at SARS-CoV-2 surface proteins has progressed, including targeting of the spike protein. Monoclonal derived anti-spike agents, such as bamlanivimab and etesevimab treated together, have demonstrated success in reducing SARS-CoV-2 viral load among patients with mild to moderate COVID-19; however, these effects were lost in the context of solely bamlanivimab treatment (127). This lack of efficacy has led to the revocation of the EUA for *sole use* bamlanivimab treatment in 2021. Importantly, viral targeted immunotherapies rely on viral antigenicity, the emergence of new SARS-CoV-2 variants and continuing mutations within the spike protein leaves the earlier viral-targeted immunotherapies unlikely to be active in clinical use.

11.7.3 Host-Targeted Immunotherapies

To dampen runaway host immune activation, immunotherapy development has targeted key activators of host inflammatory signaling pathways. COVID-19 patient samples have displayed elevated markers of systemic inflammation, including the accumulation of pro-inflammatory cytokines (e.g., interleukin [IL]-1, IL-6, IL-8 and tumor necrosis factor) as well as other markers of damage, such as D-dimer, ferritin, and C-reactive protein) (128, 129). A cohort study of 201 COVID-19 patients with pneumonia noted that factors such as age, neutrophilia, organ and coagulation dysfunction were significant risk factors of ARDS progression and mortality. Early interventions to modulate this host immune activity have used both broad and targeted anti-inflammatory avenues. During the early onset of COVID-19, corticosteroids were deployed to combat deleterious inflammation and ARDS phenotypes. Preliminary studies using the corticosteroid methylprednisolone treatment was associated with reduced mortality among patients that developed ARDS with COVID-19 infection (130). However, a larger randomized control trial (NCT04381936) testing the effect of dexamethasone among 2104 COVID-19 patients led to lower mortality, but only of a sub-cohort that received supplemental respiratory support with mechanical ventilation of oxygen (131). Importantly, there was no benefit of dexamethasone treatment to patients that received no respiratory support. A potential explanation for this discrepancy lay in the pleiotropic, whole-body effects of corticosteroid treatment, that may be beneficial in more severe COVID-19 cases with diffuse pathophysiological manifestations. A crucial factor in the decision of corticosteroid administration is the correct identification of the key patient subset that would benefit. Efforts to characterize the immunological signatures and subsets of at-risk and severe COVID-19 patients would greatly inform treatment paradigms.

11.7.4 Cytokine Blockade

In contrast to systematic corticosteroid treatment options, targeted anti-inflammatory interventions have been developed to identify discrete agents of immune regulation. An aberrantly elevated inflammatory response consistently observed among patient samples, animals, and cell models of SARS-CoV-2 infection, characterized by high levels of cytokines and low levels of type I and II interferons (132). During severe infection, late stage pathophysiology entails low soluble oxygen levels in the blood and shortness of breath with the potential to progress to ARDS (133). Key in this late stage of infection is the involvement of the overabundant pro-inflammatory cytokines and their signaling pathways, driving the detrimental immune activation and cytokine storm. Notable upregulated cytokines include interleukin-6 (IL-6), IL-1β, IL-18, and GM-CSF (granulocyte-macrophage colony–stimulating factor); together, the activity of these cytokines results in advanced lung injury and respiratory failure (134–137). Of note, SARS-CoV-2 infection stimulates IL-1 and IL-6 production to a much greater extent than other coronaviral family member infection, potentially explaining the severity of distinct disease (132). Additionally, SARS-CoV-2 infection drives cellular signaling pathways promoting pro-inflammatory cytokine signaling. Viral infection stimulates the assembly and activation of the NLRP3 inflammasome, leading to IL-1β and IL-18 production. Importantly, IL-1β and IL-18 levels have been negatively correlated with COVID-19 patient outcomes (138). From this, efforts have focused on neutralizing the activity of host cytokines to blunt the runaway inflammation and immune response. The therapeutic antibody targeting IL-1β, canakinumab, has been tested in COVID-19 patients noted to have a high inflammatory response; however, beneficial effects of IL-1β neutralization were only seen in a sub-cohort of patients who also underwent mechanical ventilation. In contrast, efforts to target IL-6 have been more successful. Patients were treated with tocilizumab, a recombinantly derived humanized anti–IL-6 receptor-antibody fusion protein that functions to inhibit the binding of IL-6 to receptors both localized on cell membranes, and soluble receptor complexes, subsequently binding IL-6 signaling. In this trial, tocilizumab treatment to COVID-19 patients led to improved survival and other clinical outcomes, including patients without respiratory support or co-treatment with corticosteroids (NCT04381936) (139). These effects were replicated in additional studies, including treatment with another anti–IL-6 agent, sarilumab; IL-6 inhibition resulted in improved survival and outcomes in the REMAP-CAP trail (NCT02735707) (140). Beyond IL-6 treatment, the targeting of other cytokines has been explored, such as GM-CSF to blunt neutrophil activation. Better understanding of the underlying cytokine profile of individual patients will be a critical advantage to optimize specific cytokine inhibition as part of COVID-19 therapy.

11.7.5 Tyrosine Kinase Inhibitor

As a complement to targeting the circulating pro-inflammatory cytokines, agents have been tested to target the key cellular receptors transducing the cytokine's inflammatory signal. Previously studied for several other inflammatory diseases, tyrosine kinase inhibitors (TKIs) have undergone extensive optimization to establish a clinical safety profile (141). These agents have been tested for use against COVID-19 infection. A double-blind randomized control trial with the Janus tyrosine kinase (JAK) inhibitor baricitinib for COVID-19 patients resulted in reduced mortality and faster recovery (142, 143). These studies spurred the recent FDA authorization of baricitinib for clinical use under an EUA. Additional TKI agents have been tested for use against SARS-CoV-2 infection, including inhibitors of PI3-kinases, Bruton's tyrosine kinases, and even mechanistic target of rapamycin (mTOR) (144). Inhibition of cytokine signaling at the level of the cellular receptor or signaling transducer may be an alternative way to blunt inflammatory signaling in severe COVID-19 disease conditions.

11.7.6 Immune-Thrombotic Modulator

Beyond the manifestations of respiratory tract viral infection and lung injury, COVID-19 infection can lead to dysregulated immune responses that affect the vasculature. SARS-CoV-2 infection has been observed to dysregulate immunothrombosis in host vessels, in which myeloid cells activate platelets to form thrombi or clots, in both large and small vessels. Research has suggested that thrombi formation in small vessels, a process termed microthrombosis, may be a contributing factor to ARDS progression (145). Patients with COVID-19 display arterial and venous thrombi even in the context of thromboprophylaxis treatment, and thrombotic events are associated with poor prognosis (146). As such, the dose of thromboprophylactic agents may be increased for COVID-19 patients to assist the removal of deleterious thrombi (147). Further study into the measures available to alleviate vascular injury is needed.

11.7.7 Antiviral Defense Stimulation: Interferons

Contrary to pro-inflammatory cytokines discussed above, antiviral associated interferons (IFNs) have been observed to be downregulated during SARS-CoV-2 infection. Efforts to reintroduce these agents have been tested as an antiviral strategy, specifically type I interferons, and have noted some success (148). COVID-19 patients were treated with an inhaled form of IFNβ-1a (SNG001) for 14 days, and the treated group demonstrated an accelerated recovery from infection (149). IFN agents have also been used in complex; the triple combination of IFNβ-1b, lopinavir-ritonavir, and ribavirin was noted to aid patients with mild to moderate COVID-19, improving symptoms and reducing hospital stay length (150). An important note is the crosstalk that co-treatment with corticosteroids may have, as combination treatment may have unintended effects on IFN signaling. In addition to Type I interferons, efforts to manipulate Type II interferons have been tested. Many at-risk patient groups have impaired immune responses and suffer extensive viral load upon SARS-CoV-2 infection. Treatment with interferon-gamma (IFNγ), a type II interferon, has been noted to strongly activate the innate antiviral response; subcutaneous treatment of IFNγ proved beneficial to critical ill COVID-19 patients with dysfunctional immune responses, leading to lower viral load and improved outcomes (151).

11.8 PERSPECTIVES FOR FUTURE DRUG DISCOVERY EFFORTS

11.8.1 Models for SARS-CoV-2 Drug Screening

To maximize the antiviral drug discovery efforts, consideration of the model system is paramount. Among the extensive efforts and attempts to develop an anti–SARS-CoV-2 agent, in vitro efficacy of many agents failed to translate to the benefits in animal models or in the clinical trials. When developing a model system for anti–SARS-CoV-2 agent screening, a key consideration is ensuring the physiological relevance of the model to the actual biology. This includes care to recapitulate the viral life cycle and the viral interactions with host proteins. Several tools are invaluable models to test the efficacy of a putative antiviral agent. Assays such as the plaque reduction neutralization test (PRNT), are preferred tests to measure the levels of neutralizing antibodies in subject serum and plasma. Additionally, new technological developments reduce assay time and increase sample throughput in tests such as the microneutralization assay (MNA), and pseudotyped-virus neutralization assay (PNA). Various model systems are established and optimized to test SARS-CoV-2 infection and inhibition, including monolayer cells, patient tissue explants, and extensive animal models ranging from rodents to nonhuman primates (NHPs) (152, 153).

11.8.2 Cell Culture Models

Cell-based models of SARS-CoV-2 infection are key to understanding the mechanism and kinetics of viral replication and toxicity, and the efficacy of antiviral agents. Conceptually, cell culture models can be organized into two tiers: tier 1, with the potential for higher throughput translation, and tier 2, lower throughput more suitable to measure biological efficacy readout. Within tier 1 models, monolayer cell culture systems, such Vero E6 and Calu-3, have been extensively characterized for their response to viral infection and scaled for uses in multiplexed assays. When used in combination with an HTS system, tier 1 models can be screened for multiple conditions of infection and compound dosing arrangement with varying dosage, infection time, and measurements of viral replication kinetics. Tier 2 systems use more physiologically relevant cell models but at the expense of throughput. For example, systems using primary-derived human airway epithelial cells (e.g., nasal or bronchial) retain their differentiated characteristics and the high endogenous expression of the key SARS-CoV-2 human receptors, ACE2 and TMPRSS2. These models have been effectively used as a screening platform with direct translational relevance (154, 155). Additionally, primary human airway epithelial (HAE) cells can be cultured in an air-liquid interface (ALI), which closely resembles the physiological context of the host environment, and can be used to test infection and agent effects on cellular ultrastructure and immunological signatures (156, 157). This system can be modified to generate "polarized" HAE cells in ALI culture for extended treatment periods to measure chronic exposure (158, 159). These models are ideal to evaluate drug efficacy in viral entry and replication, host innate immune response as a biologically relevant pre-clinical model for emerging variant strains. The choice of cellular model system has important consequences for the downstream development process. Repurposing screening efforts using Vero E6 and Calu3 as model systems of efficacy yielded chloroquine and ivermectin as putative antiviral agents, yet subsequent testing revealed no clinical benefit for treatment of COVID-19 patients (160, 161). These cases are difficult lessons as to the importance of model choice when initiating a drug discovery screening campaign.

11.8.3 Animal Models

During drug development, preclinical models of disease, in particular animal models, are essential to investigate viral pathogenesis, vaccine protection, and postexposure therapies. The development of effective small animal and NHP models of SARS-CoV-2 infection has been its own massive undertaking, and is described in detail at the NIH COVID-19 Open Data Portal and has been thoroughly reviewed by several groups (162, 163). A major challenge early in the COVID-19 was the unsuitability of the mouse model for SARS-CoV-2 infection because of the lack of recognition by the murine ACE2 receptor, thus impairing viral cellular entry. Because the early preclinical models relied on the ferret and hamster with their inherent susceptibility to SARS-CoV-2, this observation in 2020 led to their rapid deployment as model species (164–167). Despite this, logistical issues and cost limited the large-scale application of ferrets and hamsters as primary models, necessitating the development of a competent mouse model. To accomplish this, researchers repurposed models previously developed for testing SARS-CoV, such as the K18-hACE2 transgenic mouse model expressing human ACE2 and demonstrated its susceptibility to SARS-CoV-2 infection (168). In contrast, research efforts also adapted the SARS-CoV-2 virus for recognition by mouse cellular machinery, by generating mouse-adapted viral strains (SARS-CoV-2 MA) (169). In addition, emerging SARS-CoV-2 variants containing mutations in spike (such as N501Y-carrying variants) have resulted in increased interaction with murine ACE2, potentially extending the capability to infect additional rodent species. Testing of the 501Y-carrying variant led to the infection of rodents and aged animals, although at overall lower viral loads, representing a physiologically relevant model on pathogenesis, immunology, age, and antivirals (170). Validation of COVID-19 vaccines in preclinical studies of NHPs have proved remarkably predictive of the outcome of clinical efficacy studies in humans, and the observed immunological signatures of NHP are similar to the human correlates of protection necessary for vaccine efficacy (171–174). Together these have established NHP as a valuable research model for therapeutic testing, however, their usage in small molecule drug discovery is limited, particularly because of a global NHP supply shortage.

11.9 OUTLOOK

The unprecedented global pandemic of COVID-19 has been met with an also unprecedented response of drug discovery research across academia and industry to generate antiviral agents. Within less than 2 years, exceptional collaboration across biopharmaceutical companies together with open science consortia, rapid implementation of clinical trial networks, along with streamlined regulatory review bodies has led to several antiviral agents being discovered, clinically evaluated and approved. However, with the emerging virus variants, COVID rebound, and long COVID, we still need to deepen our understanding of COVID-19 pathogenesis and drugging target selection to facilitate the antiviral drug treatment of post-COVID conditions. Within virus-focused strategy, efforts continue in the goal of systematically exploring, screening, and targeting each of the relevant members of the SARS-CoV-2 proteome to uncover the new viral targets in the context of host response and explore the combination therapies. In parallel, host factor–centered strategies continue in the profiling and understanding of key features of host response to SARS-CoV-2 infection at the structural genomic, transcriptomic, proteomic, and metabolomic levels. These efforts help to interrogate the evolutionally conserved biological pathways and identify key protein drivers of host immune response against pan-coronaviruses. Systematically developing chemical agent libraries targeting both viral and host components will further assist the validation of the drugging strategy and may prevent a future endemic from turning into a pandemic.

ABBREVIATIONS

ACE2	angiotensin-converting enzyme 2
ADME	absorption, distribution, metabolism, excretion
ADP	adenosine diphosphate
ALI	air-liquid interface
ARDS	acute respiratory distress syndrome
COVID-19	coronavirus disease
CRISPR	clustered regularly interspaced short palindromic repeats
CRP	C-reactive protein
CYP	cytochrome P450
DUB	deubiquitinase
ECMO	extracorporeal membrane oxygenation
EHR	electronic health record
GM–CSF	granulocyte-macrophage colony–stimulating factor
HAE	human airway epithelium
hCoV	human coronavirus
HFNC	high-flow nasal cannula
HTS	high-throughput screening
IFN	interferon
IL	interleukin
IMV	invasive mechanical ventilation
ISG	interferon-stimulated gene
JAK	Janus tyrosine kinase
K18-hACE2	human keratin 18 promoter-driven human angiotensin-converting enzyme 2 transgenic mouse
MNA	microneutralization assay
M^{pro}	main protease
mTOR	mammalian target of rapamycin
MV	mechanical ventilation
NHP	nonhuman primate
NIV	noninvasive ventilation
NSP	nonstructural protein
PI3K	phosphatidylinositol-3-kinase
pLpro	papain-like protease
PNA	pseudotyped virus neutralization assay
PRNT	plaque reduction neutralization test
PRR	pattern recognition receptors
RBD	receptor-binding domain
RCT	randomized controlled trial
RdRp	RNA-dependent RNA polymerase
RSV	respiratory syncytial virus

SARS-CoV-2	severe acute respiratory syndrome coronavirus 2
MA SARS-CoV-2	mouse-adapted viral strain
TKI	tyrosine kinase inhibitor
TMPRSS2	transmembrane protease serine 2

REFERENCES

1. Diamond, M. S., and Kanneganti, T.-D. (2022) Innate immunity: The first line of defense against SARS-CoV-2. *Nat Immunol.* **23**, 165–176
2. Tay, M. Z., Poh, C. M., Rénia, L., MacAry, P. A., and Ng, L. F. P. (2020) The trinity of COVID-19: Immunity, inflammation and intervention. *Nat Rev Immunol.* **20**, 363–374
3. Merad, M., Blish, C. A., Sallusto, F., and Iwasaki, A. (2022) The immunology and immunopathology of COVID-19. *Science.* **375**, 1122–1127
4. Chaudhary, N., Weissman, D., and Whitehead, K. A. (2021) mRNA vaccines for infectious diseases: Principles, delivery and clinical translation. *Nat Rev Drug Discov.* **20**, 817–838
5. von Delft, A., Hall, M. D., Kwong, A. D., Purcell, L. A., Saikatendu, K. S., Schmitz, U., Tallarico, J. A., and Lee, A. A. (2023) Accelerating antiviral drug discovery: Lessons from COVID-19. *Nat Rev Drug Discov.* **22**, 585–603. https://doi.org/10.1038/s41573-023-00692-8
6. Toussi, S. S., Hammond, J. L., Gerstenberger, B. S., and Anderson, A. S. (2023) Therapeutics for COVID-19. *Nat Microbiol.* **8**, 771–786
7. Li, G., Hilgenfeld, R., Whitley, R., and De Clercq, E. (2023) Therapeutic strategies for COVID-19: Progress and lessons learned. *Nat Rev Drug Discov.* **22**, 449–475
8. Meganck, R. M., and Baric, R. S. (2021) Developing therapeutic approaches for twenty-first-century emerging infectious viral diseases. *Nat Med.* **27**, 401–410
9. De Clercq, E., and Li, G. (2016) Approved antiviral drugs over the past 50 years. *Clin Microbiol Rev.* **29**, 695–747
10. Baggen, J., Vanstreels, E., Jansen, S., and Daelemans, D. (2021) Cellular host factors for SARS-CoV-2 infection. *Nat Microbiol.* **6**, 1219–1232
11. Kaufmann, S. H. E., Dorhoi, A., Hotchkiss, R. S., and Bartenschlager, R. (2018) Host-directed therapies for bacterial and viral infections. *Nat Rev Drug Discov.* **17**, 35–56
12. Muñoz-Fontela, C., and Rafael, D. (2022) *Antiviral Discovery for Highly Pathogenic Emerging Viruses*, Royal Society of Chemistry, Cambridge
13. Gu, S. X., Tyagi, T., Jain, K., Gu, V. W., Lee, S. H., Hwa, J. M., Kwan, J. M., Krause, D. S., Lee, A. I., Halene, S., Martin, K. A., Chun, H. J., and Hwa, J. (2021) Thrombocytopathy and endotheliopathy: Crucial contributors to COVID-19 thromboinflammation. *Nat Rev Cardiol.* **18**, 194–209
14. Pushpakom, S., Iorio, F., Eyers, P. A., Escott, K. J., Hopper, S., Wells, A., Doig, A., Guilliams, T., Latimer, J., McNamee, C., Norris, A., Sanseau, P., Cavalla, D., and Pirmohamed, M. (2019) Drug repurposing: Progress, challenges and recommendations. *Nat Rev Drug Discov.* **18**, 41–58
15. Hurle, M. R., Yang, L., Xie, Q., Rajpal, D. K., Sanseau, P., and Agarwal, P. (2013) Computational drug repositioning: From data to therapeutics. *Clin Pharmacol Ther.* **93**, 335–341
16. Huang, Y. H., and Vakoc, C. R. (2016) A biomarker harvest from one thousand cancer cell lines. *Cell.* **166**, 536–537
17. Kitchen, D. B., Decornez, H., Furr, J. R., and Bajorath, J. (2004) Docking and scoring in virtual screening for drug discovery: Methods and applications. *Nat Rev Drug Discov.* **3**, 935–949
18. Moffat, J. G., Vincent, F., Lee, J. A., Eder, J., and Prunotto, M. (2017) Opportunities and challenges in phenotypic drug discovery: An industry perspective. *Nat Rev Drug Discov.* **16**, 531–543
19. Bakowski, M. A., Beutler, N., Wolff, K. C., Kirkpatrick, M. G., Chen, E., Nguyen, T. H., Riva, L., Shaabani, N., Parren, M., Ricketts, J., Gupta, A. K., Pan, K., *et al.* (2021) Drug repurposing screens identify chemical entities for the development of COVID-19 interventions. *Nat Commun.* **12**, 3309
20. Self, W. H., Semler, M. W., Leither, L. M., Casey, J. D., Angus, D. C., Brower, R. G., Chang, S. Y., Collins, S. P., Eppensteiner, J. C., Filbin, M. R., Files, D. C., Gibbs, K. W., Ginde, A. A., Gong, M. N., Harrell, F. E., Jr., *et al.* (2020) Effect of hydroxychloroquine on clinical status at 14 days in hospitalized patients with COVID-19: A randomized clinical trial. *Jama.* **324**, 2165–2176

21. WHO Solidarity Trial Consortium, Pan, H., Peto, R., Henao-Restrepo, A. M., Preziosi, M. P., Sathiyamoorthy, V., Abdool Karim, Q., Alejandria, M. M., Hernández García, C., Kieny, M. P., Malekzadeh, R., Murthy, S., Reddy, K. S., Roses Periago, M., Abi Hanna, P., Ader, F., *et al.* (2021). Repurposed antiviral drugs for Covid-19 – interim WHO solidarity trial results. *N Engl J Med.* **384**, 497–511. https://doi.org/ 10.1056/NEJMoa2023184
22. Watashi, K. (2021) Identifying and repurposing antiviral drugs against severe acute respiratory syndrome coronavirus 2 with in silico and in vitro approaches. *Biochem Biophys Res Commun.* **538**, 137–144
23. Unoh, Y., Uehara, S., Nakahara, K., Nobori, H., Yamatsu, Y., Yamamoto, S., Maruyama, Y., Taoda, Y., Kasamatsu, K., Suto, T., Kouki, K., Nakahashi, A., Kawashima, S., Sanaki, T., Toba, S., Uemura, K., Mizutare, T., Ando, S., Sasaki, M., Orba, Y., Sawa, H., Sato, A., Sato, T., Kato, T., and Tachibana, Y. (2022) Discovery of S-217622, a noncovalent oral SARS-CoV-2 3CL protease inhibitor clinical candidate for treating COVID-19. *J Med Chem.* **65**, 6499–6512
24. Consortium, T. C. M., Achdout, H., Aimon, A., Alonzi, D. S., Arbon, R., Bar-David, E., Barr, H., Ben-Shmuel, A., Bennett, J., Bilenko, V. A., Bilenko, V. A., Boby, M. L., Borden, B., Boulet, P., Bowman, G. R., Brun, J., Brwewitz, L., *et al.* (2023) Open science discovery of potent non-covalent SARS-CoV-2 main protease inhibitors. *bioRxiv*, 2020.2010.2029.339317
25. Yang, H., and Yang, J. (2021) A review of the latest research on M(pro) targeting SARS-COV inhibitors. *RSC Med Chem.* **12**, 1026–1036
26. Boras, B., Jones, R. M., Anson, B. J., Arenson, D., Aschenbrenner, L., Bakowski, M. A., Beutler, N., Binder, J., Chen, E., Eng, H., Hammond, H., Hammond, J., Haupt, R. E., Hoffman, R., Kadar, E. P., Kania, R., Kimoto, E., *et al.* (2021) Preclinical characterization of an intravenous coronavirus 3CL protease inhibitor for the potential treatment of COVID19. *Nat Commun.* **12**, 6055
27. Hoffman, R. L., Kania, R. S., Brothers, M. A., Davies, J. F., Ferre, R. A., Gajiwala, K. S., He, M., Hogan, R. J., Kozminski, K., Li, L. Y., Lockner, J. W., Lou, J., Marra, M. T., Mitchell, L. J., Jr., Murray, B. W., Nieman, J. A., Noell, S., Planken, S. P., Rowe, T., Ryan, K., Smith, G. J., Solowiej, J. E., Steppan, C. M., and Taggart, B. (2020) Discovery of ketone-based covalent inhibitors of coronavirus 3CL proteases for the potential therapeutic treatment of COVID-19. *J Med Chem.* **63**, 12725–12747
28. Owen, D. R., Allerton, C. M. N., Anderson, A. S., Aschenbrenner, L., Avery, M., Berritt, S., Boras, B., Cardin, R. D., Carlo, A., Coffman, K. J., Dantonio, A., Di, L., Eng, H., Ferre, R., Gajiwala, K. S., Gibson, S. A., *et al.* (2021) An oral SARS-CoV-2 M(pro) inhibitor clinical candidate for the treatment of COVID-19. *Science.* **374**, 1586–1593
29. Gao, Y., Yan, L., Huang, Y., Liu, F., Zhao, Y., Cao, L., Wang, T., Sun, Q., Ming, Z., Zhang, L., Ge, J., Zheng, L., Zhang, Y., Wang, H., Zhu, Y., Zhu, C., Hu, T., Hua, T., Zhang, B., Yang, X., Li, J., Yang, H., Liu, Z., Xu, W., Guddat, L. W., Wang, Q., Lou, Z., and Rao, Z. (2020) Structure of the RNA-dependent RNA polymerase from COVID-19 virus. *Science.* **368**, 779–782
30. Yazdani, S., De Maio, N., Ding, Y., Shahani, V., Goldman, N., and Schapira, M. (2021) Genetic variability of the SARS-CoV-2 pocketome. *J Proteome Res.* **20**, 4212–4215
31. Tian, L., Qiang, T., Liang, C., Ren, X., Jia, M., Zhang, J., Li, J., Wan, M., YuWen, X., Li, H., Cao, W., and Liu, H. (2021) RNA-dependent RNA polymerase (RdRp) inhibitors: The current landscape and repurposing for the COVID-19 pandemic. *Eur J Med Chem.* **213**, 113201
32. Sheahan, T. P., Sims, A. C., Graham, R. L., Menachery, V. D., Gralinski, L. E., Case, J. B., Leist, S. R., Pyrc, K., Feng, J. Y., Trantcheva, I., Bannister, R., Park, Y., Babusis, D., Clarke, M. O., Mackman, R. L., Spahn, J. E., Palmiotti, C. A., Siegel, D., Ray, A. S., Cihlar, T., Jordan, R., Denison, M. R., and Baric, R. S. (2017) Broad-spectrum antiviral GS-5734 inhibits both epidemic and zoonotic coronaviruses. *Sci Transl Med.* **9**
33. Tchesnokov, E. P., Feng, J. Y., Porter, D. P., and Götte, M. (2019) Mechanism of inhibition of Ebola virus RNA-dependent RNA polymerase by remdesivir. *Viruses.* **11**
34. Beigel, J. H., Tomashek, K. M., Dodd, L. E., Mehta, A. K., Zingman, B. S., Kalil, A. C., Hohmann, E., Chu, H. Y., Luetkemeyer, A., Kline, S., Lopez de Castilla, D., Finberg, R. W., Dierberg, K., Tapson, V., *et al.* (2020) Remdesivir for the treatment of Covid-19—Final report. *N Engl J Med.* **383**, 1813–1826

35. Jayk Bernal, A., Gomes da Silva, M. M., Musungaie, D. B., Kovalchuk, E., Gonzalez, A., Delos Reyes, V., Martín-Quirós, A., Caraco, Y., Williams-Diaz, A., Brown, M. L., Du, J., Pedley, A., Assaid, C., Strizki, J., Grobler, J. A., Shamsuddin, H. H., Tipping, R., Wan, H., Paschke, A., Butterton, J. R., Johnson, M. G., and De Anda, C. (2022) Molnupiravir for oral treatment of Covid-19 in nonhospitalized patients. *N Engl J Med.* **386**, 509–520
36. Yin, W., Mao, C., Luan, X., Shen, D. D., Shen, Q., Su, H., Wang, X., Zhou, F., Zhao, W., Gao, M., Chang, S., Xie, Y. C., Tian, G., Jiang, H. W., Tao, S. C., Shen, J., Jiang, Y., Jiang, H., Xu, Y., Zhang, S., Zhang, Y., and Xu, H. E. (2020) Structural basis for inhibition of the RNA-dependent RNA polymerase from SARS-CoV-2 by remdesivir. *Science.* **368**, 1499–1504
37. Kabinger, F., Stiller, C., Schmitzová, J., Dienemann, C., Kokic, G., Hillen, H. S., Höbartner, C., and Cramer, P. (2021) Mechanism of molnupiravir-induced SARS-CoV-2 mutagenesis. *Nat Struct Mol Biol.* **28**, 740–746
38. Arnold, J. J., Sharma, S. D., Feng, J. Y., Ray, A. S., Smidansky, E. D., Kireeva, M. L., Cho, A., Perry, J., Vela, J. E., Park, Y., Xu, Y., Tian, Y., Babusis, D., Barauskus, O., Peterson, B. R., Gnatt, A., Kashlev, M., Zhong, W., and Cameron, C. E. (2012) Sensitivity of mitochondrial transcription and resistance of RNA polymerase II dependent nuclear transcription to antiviral ribonucleosides. *pLoS Pathog.* **8**, e1003030
39. Vegivinti, C. T. R., Evanson, K. W., Lyons, H., Akosman, I., Barrett, A., Hardy, N., Kane, B., Keesari, P. R., Pulakurthi, Y. S., Sheffels, E., Balasubramanian, P., Chibbar, R., Chittajallu, S., Cowie, K., Karon, J., Siegel, L., Tarchand, R., Zinn, C., Gupta, N., Kallmes, K. M., Saravu, K., and Touchette, J. (2022) Efficacy of antiviral therapies for COVID-19: A systematic review of randomized controlled trials. *BMC Infect Dis.* **22**, 107
40. Brimacombe, K. R., Zhao, T., Eastman, R. T., Hu, X., Wang, K., Backus, M., Baljinnyam, B., Chen C. Z., Chen, L., Eicher, T., Ferrer, M., Fu, Y., Gorshkov, K., Guo, H., Hanson, Q. M., Itkin, Z., Kales, S. C., Klumpp-Thomas, C., Lee, E. M., Michael, S., Mierzwa, T., Patt, A., Pradhan, M., Renn, A., Shinn, P., Shrimp, J. H., Viraktamath, A., Wilson, K. M., Xu, M., Zakharov, A. V., Zhu, W., Zheng, W., Simeonov, A., Mathé, E. A., Lo, D. C., Hall, M. D., and Shen, M. (2020) An OpenData portal to share COVID-19 drug repurposing data in real time. *bioRxiv* [Preprint]. https://doi.org/10.1101/2020.06.04.135046
41. Barretto, N., Jukneliene, D., Ratia, K., Chen, Z., Mesecar, A. D., and Baker, S. C. (2005) The papain-like protease of severe acute respiratory syndrome coronavirus has deubiquitinating activity. *J Virol.* **79**, 15189–15198
42. Shin, D., Mukherjee, R., Grewe, D., Bojkova, D., Baek, K., Bhattacharya, A., Schulz, L., Widera, M., Mehdipour, A. R., Tascher, G., Geurink, P. P., Wilhelm, A., van der Heden van Noort, G. J., Ovaa, H., Müller, S., Knobeloch, K. P., Rajalingam, K., Schulman, B. A., Cinatl, J., Hummer, G., Ciesek, S., and Dikic, I. (2020) Papain-like protease regulates SARS-CoV-2 viral spread and innate immunity. *Nature.* **587**, 657–662
43. Rut, W., Lv, Z., Zmudzinski, M., Patchett, S., Nayak, D., Snipas, S. J., El Oualid, F., Huang, T. T., Bekes, M., Drag, M., and Olsen, S. K. (2020) Activity profiling and crystal structures of inhibitor-bound SARS-CoV-2 papain-like protease: A framework for anti-COVID-19 drug design. *Sci Adv.* **6**
44. Shen, Z., Ratia, K., Cooper, L., Kong, D., Lee, H., Kwon, Y., Li, Y., Alqarni, S., Huang, F., Dubrovskyi, O., Rong, L., Thatcher, G. R. J., and Xiong, R. (2022) Design of SARS-CoV-2 PLpro inhibitors for COVID-19 antiviral therapy leveraging binding cooperativity. *J Med Chem.* **65**, 2940–2955
45. Yuan, S., Gao, X., Tang, K., Cai, J. P., Hu, M., Luo, P., Wen, L., Ye, Z. W., Luo, C., Tsang, J. O., Chan, C. C., Huang, Y., Cao, J., Liang, R., Qin, Z., Qin, B., Yin, F., Chu, H., Jin, D. Y., Sun, R., Chan, J. F., Cui, S., and Yuen, K. Y. (2022) Targeting papain-like protease for broad-spectrum coronavirus inhibition. *Protein Cell.* **13**, 940–953
46. Sargsyan, K., Lin, C. C., Chen, T., Grauffel, C., Chen, Y. P., Yang, W. Z., Yuan, H. S., and Lim, C. (2020) Multi-targeting of functional cysteines in multiple conserved SARS-CoV-2 domains by clinically safe Zn-ejectors. *Chem Sci.* **11**, 9904–9909
47. Ratia, K., Saikatendu, K. S., Santarsiero, B. D., Barretto, N., Baker, S. C., Stevens, R. C., and Mesecar, A. D. (2006) Severe acute respiratory syndrome coronavirus papain-like protease: Structure of a viral deubiquitinating enzyme. *Proc Natl Acad Sci U S A.* **103**, 5717–5722

48. Harrigan, J. A., Jacq, X., Martin, N. M., and Jackson, S. P. (2018) Deubiquitylating enzymes and drug discovery: Emerging opportunities. *Nat Rev Drug Discov*. **17**, 57–78
49. Fehr, A. R., Jankevicius, G., Ahel, I., and Perlman, S. (2018) Viral macrodomains: Unique mediators of viral replication and pathogenesis. *Trends Microbiol*. **26**, 598–610
50. Karras, G. I., Kustatscher, G., Buhecha, H. R., Allen, M. D., Pugieux, C., Sait, F., Bycroft, M., and Ladurner, A. G. (2005) The macro domain is an ADP-ribose binding module. *EMBO J*. **24**, 1911–1920
51. Schuller, M., Correy, G. J., Gahbauer, S., Fearon, D., Wu, T., Díaz, R. E., Young, I. D., Carvalho Martins, L., Smith, D. H., Schulze-Gahmen, U., Owens, T. W., Deshpande, I., Merz, G. E., Thwin, A. C., Biel, J. T., Peters, J. K., *et al*. (2021) Fragment binding to the Nsp3 macrodomain of SARS-CoV-2 identified through crystallographic screening and computational docking. *Sci Adv*. **7**
52. Du, Q., Miao, Y., He, W., and Zheng, H. (2023) ADP-ribosylation in antiviral innate immune response. *Pathogens*. **12**
53. Fehr, A. R., Channappanavar, R., Jankevicius, G., Fett, C., Zhao, J., Athmer, J., Meyerholz, D. K., Ahel, I., and Perlman, S. (2016) The conserved coronavirus macrodomain promotes virulence and suppresses the innate immune response during severe acute respiratory syndrome coronavirus infection. *mBio*. **7**
54. Sherrill, L. M., Joya, E. E., Walker, A., Roy, A., Alhammad, Y. M., Atobatele, M., Wazir, S., Abbas, G., Keane, P., Zhuo, J., Leung, A. K. L., Johnson, D. K., Lehtiö, L., Fehr, A. R., and Ferraris, D. (2022) Design, synthesis and evaluation of inhibitors of the SARS-CoV-2 nsp3 macrodomain. *Bioorg Med Chem*. **67**, 116788
55. Gahbauer, S., Correy, G. J., Schuller, M., Ferla, M. P., Doruk, Y. U., Rachman, M., Wu, T., Diolaiti, M., Wang, S., Neitz, R. J., Fearon, D., Radchenko, D. S., Moroz, Y. S., Irwin, J. J., Renslo, A. R., Taylor, J. C., Gestwicki, J. E., von Delft, F., Ashworth, A., Ahel, I., Shoichet, B. K., and Fraser, J. S. (2023) Iterative computational design and crystallographic screening identifies potent inhibitors targeting the Nsp3 macrodomain of SARS-CoV-2. *Proc Natl Acad Sci U S A*. **120**, e2212931120
56. Roy, A., Alhammad, Y. M., McDonald, P., Johnson, D. K., Zhuo, J., Wazir, S., Ferraris, D., Lehtiö, L., Leung, A. K. L., and Fehr, A. R. (2022) Discovery of compounds that inhibit SARS-CoV-2 Mac1-ADP-ribose binding by high-throughput screening. *Antiviral Res*. **203**, 105344
57. Tanner, J. A., Watt, R. M., Chai, Y. B., Lu, L. Y., Lin, M. C., Peiris, J. S., Poon, L. L., Kung, H. F., and Huang, J. D. (2003) The severe acute respiratory syndrome (SARS) coronavirus NTPase/helicase belongs to a distinct class of 5′ to 3′ viral helicases. *J Biol Chem*. **278**, 39578–39582
58. Malone, B., Urakova, N., Snijder, E. J., and Campbell, E. A. (2022) Structures and functions of coronavirus replication—transcription complexes and their relevance for SARS-CoV-2 drug design. *Nat Rev Mol Cell Biol*. **23**, 21–39
59. Chen, J., Malone, B., Llewellyn, E., Grasso, M., Shelton, P. M. M., Olinares, P. D. B., Maruthi, K., Eng, E. T., Vatandaslar, H., Chait, B. T., Kapoor, T. M., Darst, S. A., and Campbell, E. A. (2020) Structural basis for helicase-polymerase coupling in the SARS-CoV-2 replication-transcription complex. *Cell*. **182**, 1560–1573.e1513
60. Malone, B., Chen, J., Wang, Q., Llewellyn, E., Choi, Y. J., Olinares, P. D. B., Cao, X., Hernandez, C., Eng, E. T., Chait, B. T., Shaw, D. E., Landick, R., Darst, S. A., and Campbell, E. A. (2021) Structural basis for backtracking by the SARS-CoV-2 replication-transcription complex. *Proc Natl Acad Sci U S A*. **118**
61. Yan, L., Zhang, Y., Ge, J., Zheng, L., Gao, Y., Wang, T., Jia, Z., Wang, H., Huang, Y., Li, M., Wang, Q., Rao, Z., and Lou, Z. (2020) Architecture of a SARS-CoV-2 mini replication and transcription complex. *Nat Commun*. **11**, 5874
62. White, M. A., Lin, W., and Cheng, X. (2020) Discovery of COVID-19 inhibitors targeting the SARS-CoV-2 Nsp13 helicase. *J Phys Chem Lett*. **11**, 9144–9151
63. Newman, J. A., Douangamath, A., Yadzani, S., Yosaatmadja, Y., Aimon, A., Brandão-Neto, J., Dunnett, L., Gorrie-Stone, T., Skyner, R., Fearon, D., Schapira, M., von Delft, F., and Gileadi, O. (2021) Structure, mechanism and crystallographic fragment screening of the SARS-CoV-2 NSP13 helicase. *Nat Commun*. **12**, 4848
64. Squeglia, F., Romano, M., Ruggiero, A., Maga, G., and Berisio, R. (2020) Host DDX helicases as possible SARS-CoV-2 proviral factors: A structural overview of their hijacking through multiple viral proteins. *Front Chem*. **8**, 602162

65. Krafcikova, P., Silhan, J., Nencka, R., and Boura, E. (2020) Structural analysis of the SARS-CoV-2 methyltransferase complex involved in RNA cap creation bound to sinefungin. *Nat Commun.* **11**, 3717
66. Daffis, S., Szretter, K. J., Schriewer, J., Li, J., Youn, S., Errett, J., Lin, T. Y., Schneller, S., Zust, R., Dong, H., Thiel, V., Sen, G. C., Fensterl, V., Klimstra, W. B., Pierson, T. C., Buller, R. M., Gale, M., Jr., Shi, P. Y., and Diamond, M. S. (2010) 2′-O methylation of the viral mRNA cap evades host restriction by IFIT family members. *Nature.* **468**, 452–456
67. Rosas-Lemus, M., Minasov, G., Shuvalova, L., Inniss, N. L., Kiryukhina, O., Brunzelle, J., and Satchell, K. J. F. (2020) High-resolution structures of the SARS-CoV-2 2′-O-methyltransferase reveal strategies for structure-based inhibitor design. *Sci Signal.* **13**
68. Wilamowski, M., Sherrell, D. A., Minasov, G., Kim, Y., Shuvalova, L., Lavens, A., Chard, R., Maltseva, N., Jedrzejczak, R., Rosas-Lemus, M., Saint, N., Foster, I. T., Michalska, K., Satchell, K. J. F., and Joachimiak, A. (2021) 2′-O methylation of RNA cap in SARS-CoV-2 captured by serial crystallography. *Proc Natl Acad Sci U S A.* **118**
69. Ahmed-Belkacem, R., Hausdorff, M., Delpal, A., Sutto-Ortiz, P., Colmant, A. M. G., Touret, F., Ogando, N. S., Snijder, E. J., Canard, B., Coutard, B., Vasseur, J. J., Decroly, E., and Debart, F. (2022) Potent inhibition of SARS-CoV-2 nsp14 N7-methyltransferase by sulfonamide-based bisubstrate analogues. *J Med Chem.* **65**, 6231–6249
70. Otava, T., Šála, M., Li, F., Fanfrlík, J., Devkota, K., Perveen, S., Chau, I., Pakarian, P., Hobza, P., Vedadi, M., Boura, E., and Nencka, R. (2021) The structure-based design of SARS-CoV-2 nsp14 methyltransferase ligands yields nanomolar inhibitors. *ACS Infect Dis.* **7**, 2214–2220
71. Liu, C., Shi, W., Becker, S. T., Schatz, D. G., Liu, B., and Yang, Y. (2021) Structural basis of mismatch recognition by a SARS-CoV-2 proofreading enzyme. *Science.* **373**, 1142–1146
72. Case, J. B., Li, Y., Elliott, R., Lu, X., Graepel, K. W., Sexton, N. R., Smith, E. C., Weiss, S. R., and Denison, M. R. (2018) Murine hepatitis virus nsp14 exoribonuclease activity is required for resistance to innate immunity. *J Virol.* **92**
73. Baddock, H. T., Brolih, S., Yosaatmadja, Y., Ratnaweera, M., Bielinski, M., Swift, L. P., Cruz-Migoni, A., Fan, H., Keown, J. R., Walker, A. P., Morris, G. M., Grimes, J. M., Fodor, E., Schofield, C. J., Gileadi, O., and McHugh, P. J. (2022) Characterization of the SARS-CoV-2 ExoN (nsp14ExoN-nsp10) complex: Implications for its role in viral genome stability and inhibitor identification. *Nucleic Acids Res.* **50**, 1484–1500
74. Bergant, V., Yamada, S., Grass, V., Tsukamoto, Y., Lavacca, T., Krey, K., Mühlhofer, M. T., Wittmann, S., Ensser, A., Herrmann, A., Vom Hemdt, A., Tomita, Y., Matsuyama, S., Hirokawa, T., Huang, Y., Piras, A., Jakwerth, C. A., Oelsner, M., Thieme, S., Graf, A., Krebs, S., Blum, H., Kümmerer, B. M., Stukalov, A., Schmidt-Weber, C. B., Igarashi, M., Gramberg, T., Pichlmair, A., and Kato, H. (2022) Attenuation of SARS-CoV-2 replication and associated inflammation by concomitant targeting of viral and host cap 2′-O-ribose methyltransferases. *EMBO J.* **41**, e111608
75. Lhuissier, E., Aury-Landas, J., Bouet, V., Bazille, C., Repesse, Y., Freret, T., Boumédiene, K., and Baugé, C. (2018) Evaluation of the impact of S-adenosylmethionine-dependent methyltransferase inhibitor, 3-deazaneplanocin A, on tissue injury and cognitive function in mice. *Oncotarget.* **9**, 20698–20708
76. Ulferts, R., and Ziebuhr, J. (2011) Nidovirus ribonucleases: Structures and functions in viral replication. *RNA Biol.* **8**, 295–304
77. Kindler, E., Gil-Cruz, C., Spanier, J., Li, Y., Wilhelm, J., Rabouw, H. H., Züst, R., Hwang, M., V'Kovski, P., Stalder, H., Marti, S., Habjan, M., Cervantes-Barragan, L., Elliot, R., Karl, N., Gaughan, C., van Kuppeveld, F. J., Silverman, R. H., Keller, M., Ludewig, B., Bergmann, C. C., Ziebuhr, J., Weiss, S. R., Kalinke, U., and Thiel, V. (2017) Early endonuclease-mediated evasion of RNA sensing ensures efficient coronavirus replication. *pLoS Pathog.* **13**, e1006195
78. Deng, X., Hackbart, M., Mettelman, R. C., O'Brien, A., Mielech, A. M., Yi, G., Kao, C. C., and Baker, S. C. (2017) Coronavirus nonstructural protein 15 mediates evasion of dsRNA sensors and limits apoptosis in macrophages. *Proc Natl Acad Sci U S A.* **114**, e4251–e4260
79. Hackbart, M., Deng, X., and Baker, S. C. (2020) Coronavirus endoribonuclease targets viral polyuridine sequences to evade activating host sensors. *Proc Natl Acad Sci U S A.* **117**, 8094–8103

80. Kim, Y., Jedrzejczak, R., Maltseva, N. I., Wilamowski, M., Endres, M., Godzik, A., Michalska, K., and Joachimiak, A. (2020) Crystal structure of Nsp15 endoribonuclease NendoU from SARS-CoV-2. *Protein Sci.* **29**, 1596–1605
81. Ivanov, K. A., Hertzig, T., Rozanov, M., Bayer, S., Thiel, V., Gorbalenya, A. E., and Ziebuhr, J. (2004) Major genetic marker of nidoviruses encodes a replicative endoribonuclease. *Proc Natl Acad Sci U S A.* **101**, 12694–12699
82. Saramago, M., Costa, V. G., Souza, C. S., Bárria, C., Domingues, S., Viegas, S. C., Lousa, D., Soares, C. M., Arraiano, C. M., and Matos, R. G. (2022) The nsp15 nuclease as a good target to combat SARS-CoV-2: Mechanism of action and its inactivation with FDA-approved drugs. *Microorganisms.* **10**, 342
83. Noshi, T., Kitano, M., Taniguchi, K., Yamamoto, A., Omoto, S., Baba, K., Hashimoto, T., Ishida, K., Kushima, Y., Hattori, K., Kawai, M., Yoshida, R., Kobayashi, M., Yoshinaga, T., Sato, A., Okamatsu, M., Sakoda, Y., Kida, H., Shishido, T., and Naito, A. (2018) In vitro characterization of baloxavir acid, a first-in-class cap-dependent endonuclease inhibitor of the influenza virus polymerase PA subunit. *Antiviral Res.* **160**, 109–117
84. Hayden, F. G., Sugaya, N., Hirotsu, N., Lee, N., de Jong, M. D., Hurt, A. C., Ishida, T., Sekino, H., Yamada, K., Portsmouth, S., Kawaguchi, K., Shishido, T., Arai, M., Tsuchiya, K., Uehara, T., and Watanabe, A. (2018) Baloxavir marboxil for uncomplicated influenza in adults and adolescents. *N Engl J Med.* **379**, 913–923
85. Baggen, J., Vanstreels, E., Jansen, S., and Daelemans, D. (2021) Cellular host factors for SARS-CoV-2 infection. *Nat Microbiol.* **6**, 1219–1232
86. Le Hingrat, Q., Collin, G., Bachelard, A., Ghosn, J., Chalal, S., Pacanowski, J., Peytavin, G., Weinheimer, S., Marsolais, C., Damond, F., Matheron, S., Charpentier, C., and Descamps, D. (2022) Ibalizumab shows in-vitro activity against group A and group B HIV-2 clinical isolates. *Aids.* **36**, 1055–1060
87. Millet, J. K., Jaimes, J. A., and Whittaker, G. R. (2021) Molecular diversity of coronavirus host cell entry receptors. *FEMS Microbiol Rev.* **45**
88. Yan, R., Zhang, Y., Li, Y., Xia, L., Guo, Y., and Zhou, Q. (2020) Structural basis for the recognition of SARS-CoV-2 by full-length human ACE2. *Science.* **367**, 1444–1448
89. Harrison, A. G., Lin, T., and Wang, P. (2020) Mechanisms of SARS-CoV-2 transmission and pathogenesis. *Trends Immunol.* **41**, 1100–1115
90. Lim, S., Zhang, M., and Chang, T. L. (2022) ACE2-independent alternative receptors for SARS-CoV-2. *Viruses.* **14**
91. Karoyan, P., Vieillard, V., Gómez-Morales, L., Odile, E., Guihot, A., Luyt, C. E., Denis, A., Grondin, P., and Lequin, O. (2021) Human ACE2 peptide-mimics block SARS-CoV-2 pulmonary cells infection. *Commun Biol.* **4**, 197
92. Monteil, V., Eaton, B., Postnikova, E., Murphy, M., Braunsfeld, B., Crozier, I., Kricek, F., Niederhöfer, J., Schwarzböck, A., Breid, H., Devignot, S., Klingström, J., Thålin, C., Kellner, M. J., Christ, W., Havervall, S., Mereiter, S., Knapp, S., *et al.* (2022) Clinical grade ACE2 as a universal agent to block SARS-CoV-2 variants. *EMBO Mol Med.* **14**, e15230
93. Shin, Y.-H., Jeong, K., Lee, J., Lee, H. J., Yim, J., Kim, J., Kim, S., and Park, S. B. (2022) Inhibition of ACE2-spike interaction by an ACE2 binder suppresses SARS-CoV-2 entry. *Angew Chem Int Ed.* **61**, e202115695
94. Jackson, C. B., Farzan, M., Chen, B., and Choe, H. (2022) Mechanisms of SARS-CoV-2 entry into cells. *Nat Rev Mol Cell Biol.* **23**, 3–20
95. Hoffmann, M., Kleine-Weber, H., Schroeder, S., Krüger, N., Herrler, T., Erichsen, S., Schiergens, T. S., Herrler, G., Wu, N. H., Nitsche, A., Müller, M. A., Drosten, C., and Pöhlmann, S. (2020) SARS-CoV-2 cell entry depends on ACE2 and TMPRSS2 and is blocked by a clinically proven protease inhibitor. *Cell.* **181**, 271–280.e278
96. Koch, J., Uckeley, Z. M., Doldan, P., Stanifer, M., Boulant, S., and Lozach, P. Y. (2021) TMPRSS2 expression dictates the entry route used by SARS-CoV-2 to infect host cells. *EMBO J.* **40**, e107821
97. Jocher, G., Grass, V., Tschirner, S. K., Riepler, L., Breimann, S., Kaya, T., Oelsner, M., Hamad, M. S., Hofmann, L. I., Blobel, C. P., Schmidt-Weber, C. B., Gokce, O., Jakwerth, C. A., Trimpert, J., Kimpel, J., Pichlmair, A., and Lichtenthaler, S. F. (2022) ADAM10 and ADAM17 promote SARS-CoV-2 cell entry and spike protein-mediated lung cell fusion. *EMBO Rep.* **23**, e54305
98. Peacock, T. P., Goldhill, D. H., Zhou, J., Baillon, L., Frise, R., Swann, O. C., Kugathasan, R., Penn, R., Brown, J. C., Sanchez-David, R. Y., Braga, L., Williamson, M. K., Hassard, J. A., Staller, E., Hanley, B., Osborn, M., Giacca, M., Davidson, A. D., Matthews, D. A., and Barclay, W. S. (2021) The furin cleavage site in the SARS-CoV-2 spike protein is required for transmission in ferrets. *Nat Microbiol.* **6**, 899–909

99. Laporte, M., Raeymaekers, V., Van Berwaer, R., Vandeput, J., Marchand-Casas, I., Thibaut, H. J., Van Looveren, D., Martens, K., Hoffmann, M., Maes, P., Pöhlmann, S., Naesens, L., and Stevaert, A. (2021) The SARS-CoV-2 and other human coronavirus spike proteins are fine-tuned towards temperature and proteases of the human airways. *pLoS Pathog*. **17**, e1009500
100. Li, K., Meyerholz, D. K., Bartlett, J. A., and McCray, P. B., Jr. (2021) The TMPRSS2 inhibitor nafamostat reduces SARS-CoV-2 pulmonary infection in mouse models of COVID-19. *mBio*. **12**, e0097021
101. Chen, Y., Lear, T. B., Evankovich, J. W., Larsen, M. B., Lin, B., Alfaras, I., Kennerdell, J. R., Salminen, L., Camarco, D. P., Lockwood, K. C., Tuncer, F., Liu, J., Myerburg, M. M., McDyer, J. F., Liu, Y., Finkel, T., and Chen, B. B. (2021) A high-throughput screen for TMPRSS2 expression identifies FDA-approved compounds that can limit SARS-CoV-2 entry. *Nat Commun*. **12**, 3907
102. Aman, Y., Schmauck-Medina, T., Hansen, M., Morimoto, R. I., Simon, A. K., Bjedov, I., Palikaras, K., Simonsen, A., Johansen, T., Tavernarakis, N., Rubinsztein, D. C., Partridge, L., Kroemer, G., Labbadia, J., and Fang, E. F. (2021) Autophagy in healthy aging and disease. *Nat Aging*. **1**, 634–650
103. Yim, W. W., and Mizushima, N. (2020) Lysosome biology in autophagy. *Cell Discov*. **6**, 6
104. Yim, W. W.-Y., and Mizushima, N. (2020) Lysosome biology in autophagy. *Cell Discov*. https://doi.org/10.1038/s41421-020-0141-7
105. Sharma, V., Verma, S., Seranova, E., Sarkar, S., and Kumar, D. (2018) Selective autophagy and xenophagy in infection and disease. *Front Cell Dev Biol*. **6**, 147
106. Choi, Y., Bowman, J. W., and Jung, J. U. (2018) Autophagy during viral infection—a double-edged sword. *Nat Rev Microbiol*. **16**, 341–354
107. Ghosh, S., Dellibovi-Ragheb, T. A., Kerviel, A., Pak, E., Qiu, Q., Fisher, M., Takvorian, P. M., Bleck, C., Hsu, V. W., Fehr, A. R., Perlman, S., Achar, S. R., Straus, M. R., Whittaker, G. R., de Haan, C. A. M., Kehrl, J., Altan-Bonnet, G., and Altan-Bonnet, N. (2020) β-Coronaviruses use lysosomes for egress instead of the biosynthetic secretory pathway. *Cell*. **183**, 1520–1535.e1514
108. Liu, Y., Lear, T., Larsen, M., Lin, B., Cao, Q., Alfaras, I., Kennerdell, J., Salminen, L., Camarco, D., Lockwood, K., Ma, J., Liu, J., Tan, J., Myerburg, M., Chen, Y., Croix, C. S., Sekine, Y., Evankovich, J., Finkel, T., and Chen, B. (2021) Modulation of lysosomal function as a therapeutic approach for coronaviral infections. *Res Sq* [Preprint]. rs.3.rs-419305. https://doi.org/10.21203/rs.3.rs-419305/v1
109. Wei, J., Alfajaro, M. M., DeWeirdt, P. C., Hanna, R. E., Lu-Culligan, W. J., Cai, W. L., Strine, M. S., Zhang, S.-M., Graziano, V. R., Schmitz, C. O., Chen, J. S., Mankowski, M. C., Filler, R. B., Ravindra, N. G., Gasque, V., de Miguel, F. J., Patil, A., Chen, H., Oguntuyo, K. Y., Abriola, L., Surovtseva, Y. V., Orchard, R. C., Lee, B., Lindenbach, B. D., Politi, K., van Dijk, D., Kadoch, C., Simon, M. D., Yan, Q., Doench, J. G., and Wilen, C. B. (2021) Genome-wide CRISPR screens reveal host factors critical for SARS-CoV-2 infection. *Cell*. **184**, 76–91.e13
110. Schneider, W. M., Luna, J. M., Hoffmann, H. H., Sánchez-Rivera, F. J., Leal, A. A., Ashbrook, A. W., Le Pen, J., Ricardo-Lax, I., Michailidis, E., Peace, A., Stenzel, A. F., Lowe, S. W., MacDonald, M. R., Rice, C. M., and Poirier, J. T. (2021) Genome-scale identification of SARS-CoV-2 and pan-coronavirus host factor networks. *Cell*. **184**, 120–132.e114
111. Hoffmann, H. H., Sánchez-Rivera, F. J., Schneider, W. M., Luna, J. M., Soto-Feliciano, Y. M., Ashbrook, A. W., Le Pen, J., Leal, A. A., Ricardo-Lax, I., Michailidis, E., Hao, Y., Stenzel, A. F., Peace, A., Zuber, J., Allis, C. D., Lowe, S. W., MacDonald, M. R., Poirier, J. T., and Rice, C. M. (2021) Functional interrogation of a SARS-CoV-2 host protein interactome identifies unique and shared coronavirus host factors. *Cell Host Microbe*. **29**, 267–280.e265
112. Baggen, J., Persoons, L., Vanstreels, E., Jansen, S., Van Looveren, D., Boeckx, B., Geudens, V., De Man, J., Jochmans, D., Wauters, J., Wauters, E., Vanaudenaerde, B. M., Lambrechts, D., Neyts, J., Dallmeier, K., Thibaut, H. J., Jacquemyn, M., Maes, P., and Daelemans, D. (2021) Genome-wide CRISPR screening identifies TMEM106B as a proviral host factor for SARS-CoV-2. *Nat Genet*. **53**, 435–444
113. Zhu, Y., Feng, F., Hu, G., Wang, Y., Yu, Y., Zhu, Y., Xu, W., Cai, X., Sun, Z., Han, W., Ye, R., Qu, D., Ding, Q., Huang, X., Chen, H., Xu, W., Xie, Y., Cai, Q., Yuan, Z., and Zhang, R. (2021) A genome-wide CRISPR screen identifies host factors that regulate SARS-CoV-2 entry. *Nat Commun*. **12**, 961
114. Daniloski, Z., Jordan, T. X., Wessels, H. H., Hoagland, D. A., Kasela, S., Legut, M., Maniatis, S., Mimitou, E. P., Lu, L., Geller, E., Danziger, O., Rosenberg, B. R., Phatnani, H., Smibert, P., Lappalainen, T., tenOever, B. R., and Sanjana, N. E. (2021) Identification of required host factors for SARS-CoV-2 infection in human cells. *Cell*. **184**, 92–105.e116

115. Wang, R., Simoneau, C. R., Kulsuptrakul, J., Bouhaddou, M., Travisano, K. A., Hayashi, J. M., Carlson-Stevermer, J., Zengel, J. R., Richards, C. M., Fozouni, P., Oki, J., Rodriguez, L., Joehnk, B., Walcott, K., Holden, K., Sil, A., Carette, J. E., Krogan, N. J., Ott, M., and Puschnik, A. S. (2021) Genetic screens identify host factors for SARS-CoV-2 and common cold coronaviruses. *Cell.* **184**, 106–119.e114
116. Flynn, R. A., Belk, J. A., Qi, Y., Yasumoto, Y., Wei, J., Alfajaro, M. M., Shi, Q., Mumbach, M. R., Limaye, A., DeWeirdt, P. C., Schmitz, C. O., Parker, K. R., Woo, E., Chang, H. Y., Horvath, T. L., Carette, J. E., Bertozzi, C. R., Wilen, C. B., and Satpathy, A. T. (2021) Discovery and functional interrogation of SARS-CoV-2 RNA-host protein interactions. *Cell.* **184**, 2394–2411.e2316
117. Brevini, T., Maes, M., Webb, G. J., John, B. V., Fuchs, C. D., Buescher, G., Wang, L., Griffiths, C., Brown, M. L., Scott, W. E., Pereyra-Gerber, P., Gelson, W. T. H., Brown, S., Dillon, S., Muraro, D., Sharp, J., Neary, M., Box, H., Tatham, L., Stewart, J., *et al.* (2023) FXR inhibition may protect from SARS-CoV-2 infection by reducing ACE2. *Nature.* **615**, 134–142
118. Samelson, A. J., Tran, Q. D., Robinot, R., Carrau, L., Rezelj, V. V., Kain, A. M., Chen, M., Ramadoss, G. N., Guo, X., Lim, S. A., Lui, I., Nuñez, J. K., Rockwood, S. J., Wang, J., Liu, N., Carlson-Stevermer, J., Oki, J., Maures, T., Holden, K., Weissman, J. S., Wells, J. A., Conklin, B. R., TenOever, B. R., Chakrabarti, L. A., Vignuzzi, M., Tian, R., and Kampmann, M. (2022) BRD2 inhibition blocks SARS-CoV-2 infection by reducing transcription of the host cell receptor ACE2. *Nat Cell Biol.* **24**, 24–34
119. Chu, H., Hou, Y., Yang, D., Wen, L., Shuai, H., Yoon, C., Shi, J., Chai, Y., Yuen, T. T., Hu, B., Li, C., Zhao, X., Wang, Y., Huang, X., Lee, K. S., Luo, C., Cai, J. P., Poon, V. K., Chan, C. C., Zhang, A. J., Yuan, S., Sit, K. Y., Foo, D. C., Au, W. K., Wong, K. K., Zhou, J., Kok, K. H., Jin, D. Y., Chan, J. F., and Yuen, K. Y. (2022) Coronaviruses exploit a host cysteine-aspartic protease for replication. *Nature.* **609**, 785–792
120. Geng, J., Chen, L., Yuan, Y., Wang, K., Wang, Y., Qin, C., Wu, G., Chen, R., Zhang, Z., Wei, D., Du, P., Zhang, J., Lin, P., Zhang, K., Deng, Y., Xu, K., Liu, J., Sun, X., Guo, T., Yang, X., Wu, J., Jiang, J., Li, L., Zhang, K., Wang, Z., Zhang, J., Yan, Q., Zhu, H., Zheng, Z., Miao, J., Fu, X., Yang, F., Chen, X., Tang, H., Zhang, Y., Shi, Y., Zhu, Y., Pei, Z., Huo, F., Liang, X., Wang, Y., Wang, Q., Xie, W., Li, Y., Shi, M., Bian, H., Zhu, P., and Chen, Z. N. (2021) CD147 antibody specifically and effectively inhibits infection and cytokine storm of SARS-CoV-2 and its variants delta, alpha, beta, and gamma. *Signal Transduct Target Ther.* **6**, 347
121. White, K. M., Rosales, R., Yildiz, S., Kehrer, T., Miorin, L., Moreno, E., Jangra, S., Uccellini, M. B., Rathnasinghe, R., Coughlan, L., Martinez-Romero, C., Batra, J., Rojc, A., Bouhaddou, M., Fabius, J. M., Obernier, K., Dejosez, M., Guillén, M. J., Losada, A., Avilés, P., Schotsaert, M., Zwaka, T., Vignuzzi, M., Shokat, K. M., Krogan, N. J., and García-Sastre, A. (2021) Plitidepsin has potent preclinical efficacy against SARS-CoV-2 by targeting the host protein eEF1A. *Science.* **371**, 926–931
122. Braga, L., Ali, H., Secco, I., Chiavacci, E., Neves, G., Goldhill, D., Penn, R., Jimenez-Guardeño, J. M., Ortega-Prieto, A. M., Bussani, R., Cannatà, A., Rizzari, G., Collesi, C., Schneider, E., Arosio, D., Shah, A. M., Barclay, W. S., Malim, M. H., Burrone, J., and Giacca, M. (2021) Drugs that inhibit TMEM16 proteins block SARS-CoV-2 spike-induced syncytia. *Nature.* **594**, 88–93
123. Zinsser, H., and Batchelder, A. P. (1930) Studies on Mexican typhus fever. I. *J Exp Med.* **51**, 847–858
124. O'Donnell, M. R., Grinsztejn, B., Cummings, M. J., Justman, J. E., Lamb, M. R., Eckhardt, C. M., Philip, N. M., Cheung, Y. K., Gupta, V., João, E., Pilotto, J. H., Diniz, M. P., Cardoso, S. W., Abrams, D., Rajagopalan, K. N., Borden, S. E., Wolf, A., Sidi, L. C., Vizzoni, A., *et al.* (2021) A randomized double-blind controlled trial of convalescent plasma in adults with severe COVID-19. *J Clin Invest.* **131**
125. Chai, K. L., Valk, S. J., Piechotta, V., Kimber, C., Monsef, I., Doree, C., Wood, E. M., Lamikanra, A. A., Roberts, D. J., McQuilten, Z., So-Osman, C., Estcourt, L. J., and Skoetz, N. (2020) Convalescent plasma or hyperimmune immunoglobulin for people with COVID-19: A living systematic review. *Cochrane Database Syst Rev.* **10**, CD013600. https://doi.org/10.1002/14651858.CD013600.pub2
126. Abani, O., Abbas, A., Abbas, F., Abbas, M., Abbasi, S., Abbass, H., Abbott, A., Abdallah, N., Abdelaziz, A., and Abdelfattah, M. (2021) Convalescent plasma in patients admitted to hospital with COVID-19 (RECOVERY): A randomised controlled, open-label, platform trial. *Lancet.* **397**, 2049–2059
127. Gottlieb, R. L., Nirula, A., Chen, P., Boscia, J., Heller, B., Morris, J., Huhn, G., Cardona, J., Mocherla, B., Stosor, V., Shawa, I., Kumar, P., Adams, A. C., Van Naarden, J., Custer, K. L., Durante, M., Oakley, G., Schade, A. E., Holzer, T. R., Ebert, P. J., Higgs, R. E., Kallewaard, N. L., Sabo, J., Patel, D. R., Klekotka, P., Shen, L., and Skovronsky, D. M. (2021) Effect of bamlanivimab as monotherapy or in combination with etesevimab on viral load in patients with mild to moderate COVID-19: A randomized clinical trial. *JAMA.* **325**, 632–644

128. Chen, N., Zhou, M., Dong, X., Qu, J., Gong, F., Han, Y., Qiu, Y., Wang, J., Liu, Y., Wei, Y., Xia, J., Yu, T., Zhang, X., and Zhang, L. (2020) Epidemiological and clinical characteristics of 99 cases of 2019 novel coronavirus pneumonia in Wuhan, China: A descriptive study. *Lancet.* **395**, 507–513
129. Del Valle, D. M., Kim-Schulze, S., Huang, H. H., Beckmann, N. D., Nirenberg, S., Wang, B., Lavin, Y., Swartz, T. H., Madduri, D., Stock, A., Marron, T. U., Xie, H., Patel, M., Tuballes, K., Van Oekelen, O., Rahman, A., Kovatch, P., Aberg, J. A., Schadt, E., Jagannath, S., Mazumdar, M., Charney, A. W., Firpo-Betancourt, A., Mendu, D. R., Jhang, J., Reich, D., Sigel, K., Cordon-Cardo, C., Feldmann, M., Parekh, S., Merad, M., and Gnjatic, S. (2020) An inflammatory cytokine signature predicts COVID-19 severity and survival. *Nat Med.* **26**, 1636–1643
130. Wu, C., Chen, X., Cai, Y., Xia, J. a., Zhou, X., Xu, S., Huang, H., Zhang, L., Zhou, X., Du, C., Zhang, Y., Song, J., Wang, S., Chao, Y., Yang, Z., Xu, J., Zhou, X., Chen, D., Xiong, W., Xu, L., Zhou, F., Jiang, J., Bai, C., Zheng, J., and Song, Y. (2020) Risk factors associated with acute respiratory distress syndrome and death in patients with coronavirus disease 2019 pneumonia in Wuhan, China. *JAMA Int Med.* **180**, 934–943
131. Horby, P., Lim, W. S., Emberson, J. R., Mafham, M., Bell, J. L., Linsell, L., Staplin, N., Brightling, C., Ustianowski, A., Elmahi, E., Prudon, B., Green, C., Felton, T., Chadwick, D., Rege, K., Fegan, C., Chappell, L. C., Faust, S. N., Jaki, T., Jeffery, K., Montgomery, A., Rowan, K., Juszczak, E., Baillie, J. K., Haynes, R., and Landray, M. J. (2021) Dexamethasone in hospitalized patients with Covid-19. *N Engl J Med.* **384**, 693–704
132. Blanco-Melo, D., Nilsson-Payant, B. E., Liu, W. C., Uhl, S., Hoagland, D., Møller, R., Jordan, T. X., Oishi, K., Panis, M., Sachs, D., Wang, T. T., Schwartz, R. E., Lim, J. K., Albrecht, R. A., and tenOever, B. R. (2020) Imbalanced host response to SARS-CoV-2 drives development of COVID-19. *Cell.* **181**, 1036–1045.e1039
133. Siddiqi, H. K., and Mehra, M. R. (2020) COVID-19 illness in native and immunosuppressed states: A clinical—therapeutic staging proposal. *J Heart Lung Transplant.* **39**, 405–407
134. Huang, C., Wang, Y., Li, X., Ren, L., Zhao, J., Hu, Y., Zhang, L., Fan, G., Xu, J., Gu, X., Cheng, Z., Yu, T., Xia, J., Wei, Y., Wu, W., Xie, X., Yin, W., Li, H., Liu, M., Xiao, Y., Gao, H., Guo, L., Xie, J., Wang, G., Jiang, R., Gao, Z., Jin, Q., Wang, J., and Cao, B. (2020) Clinical features of patients infected with 2019 novel coronavirus in Wuhan, China. *Lancet.* **395**, 497–506
135. Bonaventura, A., Vecchié, A., Wang, T. S., Lee, E., Cremer, P. C., Carey, B., Rajendram, P., Hudock, K. M., Korbee, L., Van Tassell, B. W., Dagna, L., and Abbate, A. (2020) Targeting GM-CSF in COVID-19 pneumonia: Rationale and strategies. *Front Immunol.* **11**, 1625
136. Fajgenbaum, D. C., and June, C. H. (2020) Cytokine storm. *N Engl J Med.* **383**, 2255–2273
137. Mehta, P., McAuley, D. F., Brown, M., Sanchez, E., Tattersall, R. S., and Manson, J. J. (2020) COVID-19: Consider cytokine storm syndromes and immunosuppression. *Lancet.* **395**, 1033–1034
138. Rodrigues, T. S., de Sá, K. S. G., Ishimoto, A. Y., Becerra, A., Oliveira, S., Almeida, L., Gonçalves, A. V., Perucello, D. B., Andrade, W. A., Castro, R., Veras, F. P., Toller-Kawahisa, J. E., Nascimento, D. C., de Lima, M. H. F., *et al.* (2020) Inflammasomes are activated in response to SARS-CoV-2 infection and are associated with COVID-19 severity in patients. *J Exp Med.* **218**
139. RECOVERY Collaborative Group. (2021). Tocilizumab in patients admitted to hospital with COVID-19 (RECOVERY): A randomised, controlled, open-label, platform trial. *Lancet.* **397**, 1637–1645. https://doi.org/10.1016/S0140-6736(21)00676-0
140. REMAP-CAP Investigators, Gordon, A. C., Mouncey, P. R., Al-Beidh, F., Rowan, K. M., Nichol, A. D., Arabi, Y. M., Annane, D., Beane, A., van Bentum-Puijk, W., Berry, L. R., Bhimani, Z., Bonten, M. J. M., Bradbury, C. A., Brunkhorst, F. M., Buzgau, A., Cheng, A. C., Detry, M. A., Duffy, E. J., Estcourt, L. J., *et al.* (2021) Interleukin-6 receptor antagonists in critically Ill patients with Covid-19. *N Engl J Med.* **384**, 1491–1502. https://doi.org/ 10.1056/NEJMoa2100433
141. Ferguson, F. M., and Gray, N. S. (2018) Kinase inhibitors: The road ahead. *Nat Rev Drug Discov.* **17**, 353–377
142. Marconi, V. C., Ramanan, A. V., de Bono, S., Kartman, C. E., Krishnan, V., Liao, R., Piruzeli, M. L. B., Goldman, J. D., Alatorre-Alexander, J., and de Cassia Pellegrini, R. (2021) Efficacy and safety of baricitinib for the treatment of hospitalised adults with COVID-19 (COV-BARRIER): A randomised, double-blind, parallel-group, placebo-controlled phase 3 trial. *Lancet Respir Med.* **9**, 1407–1418

143. Kalil, A. C., Patterson, T. F., Mehta, A. K., Tomashek, K. M., Wolfe, C. R., Ghazaryan, V., Marconi, V. C., Ruiz-Palacios, G. M., Hsieh, L., Kline, S., Tapson, V., Iovine, N. M., Jain, M. K., Sweeney, D. A., El Sahly, H. M., Branche, A. R., Regalado Pineda, J., Lye, D. C., Sandkovsky, U., Luetkemeyer, A. F., Cohen, S. H., Finberg, R. W., *et al.* (2021) Baricitinib plus remdesivir for hospitalized adults with Covid-19. *N Engl J Med.* **384**, 795–807
144. Jacobs, C. F., Eldering, E., and Kater, A. P. (2021) Kinase inhibitors developed for treatment of hematologic malignancies: Implications for immune modulation in COVID-19. *Blood Adv.* **5**, 913–925
145. Bonaventura, A., Vecchié, A., Dagna, L., Martinod, K., Dixon, D. L., Van Tassell, B. W., Dentali, F., Montecucco, F., Massberg, S., Levi, M., and Abbate, A. (2021) Endothelial dysfunction and immunothrombosis as key pathogenic mechanisms in COVID-19. *Nat Rev Immunol.* **21**, 319–329
146. Muñoz-Rivas, N., Abad-Motos, A., Mestre-Gómez, B., Sierra-Hidalgo, F., Cortina-Camarero, C., Lorente-Ramos, R. M., Torres-Rubio, P., Arranz-García, P., Franco-Moreno, A. I., Gómez-Mariscal, E., Mauleón-Fernández, C., Alonso-García, S., Rogado, J., Saez-Vaquero, T., Such-Diaz, A., Ryan, P., Moya-Mateo, E., Martín-Navarro, J. A., Hernández-Rivas, J. A., Torres-Macho, J., and Churruca, J. (2021) Systemic thrombosis in a large cohort of COVID-19 patients despite thromboprophylaxis: A retrospective study. *Thromb Res.* **199**, 132–142
147. Talasaz, A. H., Sadeghipour, P., Kakavand, H., Aghakouchakzadeh, M., Kordzadeh-Kermani, E., Van Tassell, B. W., Gheymati, A., Ariannejad, H., Hosseini, S. H., Jamalkhani, S., Sholzberg, M., Monreal, M., Jimenez, D., Piazza, G., Parikh, S. A., Kirtane, A. J., Eikelboom, J. W., Connors, J. M., Hunt, B. J., Konstantinides, S. V., Cushman, M., Weitz, J. I., Stone, G. W., Krumholz, H. M., Lip, G. Y. H., Goldhaber, S. Z., and Bikdeli, B. (2021) Recent randomized trials of antithrombotic therapy for patients with COVID-19: JACC state-of-the-art review. *J Am Coll Cardiol.* **77**, 1903–1921
148. Cinatl, J., Morgenstern, B., Bauer, G., Chandra, P., Rabenau, H., and Doerr, H. W. (2003) Treatment of SARS with human interferons. *Lancet.* **362**, 293–294
149. Monk, P. D., Marsden, R. J., Tear, V. J., Brookes, J., Batten, T. N., Mankowski, M., Gabbay, F. J., Davies, D. E., Holgate, S. T., Ho, L. P., Clark, T., Djukanovic, R., and Wilkinson, T. M. A. (2021) Safety and efficacy of inhaled nebulised interferon beta-1a (SNG001) for treatment of SARS-CoV-2 infection: A randomised, double-blind, placebo-controlled, phase 2 trial. *Lancet Respir Med.* **9**, 196–206
150. Hung, I. F.-N., Lung, K.-C., Tso, E. Y.-K., Liu, R., Chung, T. W.-H., Chu, M.-Y., Ng, Y.-Y., Lo, J., Chan, J., Tam, A. R., Shum, H.-P., Chan, V., Wu, A. K.-L., Sin, K.-M., Leung, W.-S., Law, W.-L., Lung, D. C., Sin, S., Yeung, P., *et al.* (2020) Triple combination of interferon beta-1b, lopinavir—ritonavir, and ribavirin in the treatment of patients admitted to hospital with COVID-19: An open-label, randomised, phase 2 trial. *Lancet.* **395**, 1695–1704
151. van Laarhoven, A., Kurver, L., Overheul, G. J., Kooistra, E. J., Abdo, W. F., van Crevel, R., Duivenvoorden, R., Kox, M., Ten Oever, J., Schouten, J., van de Veerdonk, F. L., van der Hoeven, H., Rahamat-Langendoen, J., van Rij, R. P., Pickkers, P., and Netea, M. G. (2021) Interferon gamma immunotherapy in five critically ill COVID-19 patients with impaired cellular immunity: A case series. *Med.* **2**, 1163–1170.e1162
152. Bestion, E., Halfon, P., Mezouar, S., and Mège, J.-L. (2022) Cell and animal models for SARS-CoV-2 research. *Viruses.* **14**, 1507
153. Pandamooz, S., Jurek, B., Meinung, C. P., Baharvand, Z., Sahebi Shahem-Abadi, A., Haerteis, S., Miyan, J. A., Downing, J., Dianatpour, M., Borhani-Haghighi, A., and Salehi, M. S. (2022) Experimental models of SARS-CoV-2 infection: Possible platforms to study COVID-19 pathogenesis and potential treatments. *Annu Rev Pharmacol Toxicol.* **62**, 25–53
154. Lamers, M. M., van der Vaart, J., Knoops, K., Riesebosch, S., Breugem, T. I., Mykytyn, A. Z., Beumer, J., Schipper, D., Bezstarosti, K., Koopman, C. D., Groen, N., Ravelli, R. B. G., Duimel, H. Q., Demmers, J. A. A., Verjans, G., Koopmans, M. P. G., Muraro, M. J., Peters, P. J., Clevers, H., and Haagmans, B. L. (2021) An organoid-derived bronchioalveolar model for SARS-CoV-2 infection of human alveolar type II-like cells. *EMBO J.* **40**, e105912
155. Sungnak, W., Huang, N., Bécavin, C., Berg, M., Queen, R., Litvinukova, M., Talavera-López, C., Maatz, H., Reichart, D., Sampaziotis, F., Worlock, K. B., Yoshida, M., and Barnes, J. L. (2020) SARS-CoV-2 entry factors are highly expressed in nasal epithelial cells together with innate immune genes. *Nat Med.* **26**, 681–687
156. Sheahan, T. P., Sims, A. C., Zhou, S., Graham, R. L., Pruijssers, A. J., Agostini, M. L., Leist, S. R., Schäfer, A., Dinnon, K. H., 3rd, Stevens, L. J., Chappell, J. D., Lu, X., Hughes, T. M., George, A. S., Hill, C. S., Montgomery, S. A., Brown, A. J., Bluemling, G. R., Natchus, M. G., Saindane, M., Kolykhalov, A.

A., Painter, G., Harcourt, J., Tamin, A., Thornburg, N. J., Swanstrom, R., Denison, M. R., and Baric, R. S. (2020) An orally bioavailable broad-spectrum antiviral inhibits SARS-CoV-2 in human airway epithelial cell cultures and multiple coronaviruses in mice. *Sci Transl Med.* **12**

157. Pizzorno, A., Padey, B., Julien, T., Trouillet-Assant, S., Traversier, A., Errazuriz-Cerda, E., Fouret, J., Dubois, J., Gaymard, A., Lescure, F. X., Dulière, V., Brun, P., Constant, S., Poissy, J., Lina, B., Yazdanpanah, Y., Terrier, O., and Rosa-Calatrava, M. (2020) Characterization and treatment of SARS-CoV-2 in nasal and bronchial human airway epithelia. *Cell Rep Med.* **1**, 100059
158. Hao, S., Ning, K., Kuz, C. A., Vorhies, K., Yan, Z., and Qiu, J. (2020) Long-term modeling of SARS-CoV-2 infection of in vitro cultured polarized human airway epithelium. *mBio.* **11**
159. Liu, X., Wu, Y., and Rong, L. (2020) Conditionally reprogrammed human normal airway epithelial cells at ALI: A physiological model for emerging viruses. *Virol Sin.* **35**, 280–289
160. Wang, M., Cao, R., Zhang, L., Yang, X., Liu, J., Xu, M., Shi, Z., Hu, Z., Zhong, W., and Xiao, G. (2020) Remdesivir and chloroquine effectively inhibit the recently emerged novel coronavirus (2019-nCoV) in vitro. *Cell Res.* **30**, 269–271
161. Hoffmann, M., Mösbauer, K., Hofmann-Winkler, H., Kaul, A., Kleine-Weber, H., Krüger, N., Gassen, N. C., Müller, M. A., Drosten, C., and Pöhlmann, S. (2020) Chloroquine does not inhibit infection of human lung cells with SARS-CoV-2. *Nature.* **585**, 588–590
162. Chu, H., Chan, J. F., and Yuen, K. Y. (2022) Animal models in SARS-CoV-2 research. *Nat Methods.* **19**, 392–394
163. de Vries, R. D., Rockx, B., Haagmans, B. L., Herfst, S., Koopmans, M. P., and de Swart, R. L. (2021) Animal models of SARS-CoV-2 transmission. *Curr Opin Virol.* **50**, 8–16
164. Shi, J., Wen, Z., Zhong, G., Yang, H., Wang, C., Huang, B., Liu, R., He, X., Shuai, L., Sun, Z., Zhao, Y., Liu, P., Liang, L., Cui, P., Wang, J., Zhang, X., Guan, Y., Tan, W., Wu, G., Chen, H., and Bu, Z. (2020) Susceptibility of ferrets, cats, dogs, and other domesticated animals to SARS-coronavirus 2. *Science.* **368**, 1016–1020
165. Ryan, K. A., Bewley, K. R., Fotheringham, S. A., Slack, G. S., Brown, P., Hall, Y., Wand, N. I., Marriott, A. C., Cavell, B. E., Tree, J. A., Allen, L., Aram, M. J., Bean, T. J., Brunt, E., Buttigieg, K. R., Carter, D. P., Cobb, R., Coombes, N. S., Findlay-Wilson, S. J., Godwin, K. J., *et al.* (2021) Dose-dependent response to infection with SARS-CoV-2 in the ferret model and evidence of protective immunity. *Nat Commun.* **12**, 81
166. Imai, M., Iwatsuki-Horimoto, K., Hatta, M., Loeber, S., Halfmann, P. J., Nakajima, N., Watanabe, T., Ujie, M., Takahashi, K., Ito, M., Yamada, S., Fan, S., Chiba, S., Kuroda, M., Guan, L., Takada, K., Armbrust, T., *et al.* (2020) Syrian hamsters as a small animal model for SARS-CoV-2 infection and countermeasure development. *Proc Natl Acad Sci U S A.* **117**, 16587–16595
167. Chan, J. F., Zhang, A. J., Yuan, S., Poon, V. K., Chan, C. C., Lee, A. C., Chan, W. M., Fan, Z., Tsoi, H. W., Wen, L., Liang, R., Cao, J., Chen, Y., Tang, K., Luo, C., Cai, J. P., Kok, K. H., Chu, H., Chan, K. H., Sridhar, S., Chen, Z., Chen, H., To, K. K., and Yuen, K. Y. (2020) Simulation of the clinical and pathological manifestations of coronavirus disease 2019 (COVID-19) in a golden Syrian hamster model: Implications for disease pathogenesis and transmissibility. *Clin Infect Dis.* **71**, 2428–2446
168. Bao, L., Deng, W., Huang, B., Gao, H., Liu, J., Ren, L., Wei, Q., Yu, P., Xu, Y., Qi, F., Qu, Y., Li, F., Lv, Q., Wang, W., Xue, J., Gong, S., Liu, M., Wang, G., Wang, S., Song, Z., Zhao, L., Liu, P., Zhao, L., Ye, F., Wang, H., Zhou, W., *et al.* (2020) The pathogenicity of SARS-CoV-2 in hACE2 transgenic mice. *Nature.* **583**, 830–833
169. Dinnon, K. H., Leist, S. R., Schäfer, A., Edwards, C. E., Martinez, D. R., Montgomery, S. A., West, A., Yount, B. L., Hou, Y. J., Adams, L. E., Gully, K. L., Brown, A. J., Huang, E., Bryant, M. D., Choong, I. C., Glenn, J. S., Gralinski, L. E., Sheahan, T. P., and Baric, R. S. (2020) A mouse-adapted model of SARS-CoV-2 to test COVID-19 countermeasures. *Nature.* **586**, 560–566
170. Shuai, H., Chan, J. F., Yuen, T. T., Yoon, C., Hu, J. C., Wen, L., Hu, B., Yang, D., Wang, Y., Hou, Y., Huang, X., Chai, Y., Chan, C. C., Poon, V. K., Lu, L., Zhang, R. Q., Chan, W. M., Ip, J. D., Chu, A. W., Hu, Y. F., Cai, J. P., Chan, K. H., Zhou, J., Sridhar, S., Zhang, B. Z., Yuan, S., Zhang, A. J., Huang, J. D., To, K. K., Yuen, K. Y., and Chu, H. (2021) Emerging SARS-CoV-2 variants expand species tropism to murines. *eBioMedicine.* **73**, 103643

171. Mercado, N. B., Zahn, R., Wegmann, F., Loos, C., Chandrashekar, A., Yu, J., Liu, J., Peter, L., McMahan, K., Tostanoski, L. H., He, X., Martinez, D. R., Rutten, L., Bos, R., van Manen, D., Vellinga, J., Custers, J., Langedijk, J. P., Kwaks, T., *et al.* (2020) Single-shot Ad26 vaccine protects against SARS-CoV-2 in rhesus macaques. *Nature*. **586**, 583–588
172. Corbett, K. S., Flynn, B., Foulds, K. E., Francica, J. R., Boyoglu-Barnum, S., Werner, A. P., Flach, B., O'Connell, S., Bock, K. W., Minai, M., Nagata, B. M., Andersen, H., Martinez, D. R., Noe, A. T., Douek, N., Donaldson, M. M., Nji, N. N., *et al.* (2020) Evaluation of the mRNA-1273 vaccine against SARS-CoV-2 in nonhuman primates. *N Engl J Med*. **383**, 1544–1555
173. Vogel, A. B., Kanevsky, I., Che, Y., Swanson, K. A., Muik, A., Vormehr, M., Kranz, L. M., Walzer, K. C., Hein, S., Güler, A., Loschko, J., Maddur, M. S., Ota-Setlik, A., Tompkins, K., Cole, J., Lui, B. G., Ziegenhals, T., Plaschke, A., Eisel, D., *et al.* (2021) BNT162b vaccines protect rhesus macaques from SARS-CoV-2. *Nature*. **592**, 283–289
174. Muñoz-Fontela, C., Widerspick, L., Albrecht, R. A., Beer, M., Carroll, M. W., de Wit, E., Diamond, M. S., Dowling, W. E., Funnell, S. G. P., García-Sastre, A., Gerhards, N. M., de Jong, R., Munster, V. J., Neyts, J., Perlman, S., Reed, D. S., Richt, J. A., Riveros-Balta, X., Roy, C. J., Salguero, F. J., Schotsaert, M., Schwartz, L. M., Seder, R. A., Segalés, J., Vasan, S. S., Henao-Restrepo, A. M., and Barouch, D. H. (2022) Advances and gaps in SARS-CoV-2 infection models. *pLoS Pathog*. **18**, e1010161

12 Clinical Manifestations and Pathogenesis of COVID-19 Pulmonary Disease

Moemen Eltobgy, Finny Johns, and Joseph S. Bednash

12.1 INTRODUCTION

Coronavirus disease 2019 (COVID-19) is an acute viral lung infection caused by severe acute respiratory syndrome coronavirus 2 (SARS-CoV-2). In early 2019, COVID-19 emerged in Wuhan, China, and evolved into a global pandemic by early 2020 (1). As of June 2023, the COVID-19 pandemic is responsible for more than 6.9 million deaths worldwide, with around 1.1 million deaths alone in the United States (2). Due to far-reaching implications for health, wellness, the global economy, social life, and political systems, much of the global scientific infrastructure shifted focus to study and better understand this newly emerged pathogen. The COVID-19 scientific literature rapidly expanded in a very short time. Yet, our understanding of pathobiology continues to evolve because of the relative novelty of SARS-CoV-2, and research concerning long-term sequelae of COVID-19 will take many years to fully mature.

SARS-CoV-2, the causative agent of COVID-19 (3), is a respiratory virus of the betacoronavirus (β-CoV) genera (4), and the species, severe acute respiratory syndrome-related coronavirus (SARS-CoV), which includes severe acute respiratory syndrome (SARS)-CoV that caused an outbreak in the early 2000s (5). SARS-CoV-2 first emerged in 2019; since then, thousands of variants have emerged, which are grouped by different schemas depending on the governing body (6). Dominant variants include the alpha, beta, delta, gamma, and omicron strains. β-CoV is a group of RNA viruses distinguished by its namesake "corona," or crown, which is a "fringe" of projections on the viral surface, as viewed by electron microscopy (7). This fringe is composed of viral spike glycoproteins, present on the surface of the viral particles (8). Similar to other β-CoV, SARS-CoV-2 enters host cells facilitated by the viral spike proteins interactions with a host surface protein, angiotensin-converting enzyme 2 (ACE2) (9). ACE2 is a component of the renin-angiotensin system that regulates systemic vascular tone. ACE2 is required to establish SARS-CoV-2 viral infection in host cells (9, 10). Coronaviruses can also be transmitted through cell-to-cell contact (11, 12), which contributes to viral spread within host tissue. ACE2 enhances but is not required for cell-to-cell transmission (13). ACE2 has significant tissue tropism, which impacts disease patterns in humans (14).

Here we will discuss the pathogenesis of COVID-19 with focus on pulmonary manifestations of the disease. In-depth discussion of virology or pathobiology of disease specifically affecting other organ systems is beyond the scope of this review. We will begin with discussion of the clinical manifestations of SARS-CoV-2 infection to frame our review of the pathobiology that leads to disease in humans. Finally, we will review literature regarding therapeutics that target dysregulated biology in COVID-19.

12.2 COVID-19 CLINICAL MANIFESTATIONS

COVID-19 manifests as an acute viral lung infection transmitted from human to human through respiratory droplets and aerosols (15, 16), containing SARS-CoV-2 virus, formed from coughing, sneezing, speaking, and exhaling (17–19). Studies estimate that one-third of patients may be

DOI: 10.1201/9781003355243-15

asymptomatic (20). COVID-19 symptoms are characteristic of a viral pneumonia, including fever, dyspnea, cough, headache, fatigue, malaise, anorexia, and myalgias (21–24). Other common symptoms less characteristic of common viral pneumonia include loss of taste and smell (25) and gastrointestinal symptoms, such as nausea, vomiting, and diarrhea (26, 27). Whereas most patients develop mild to moderate disease, severe COVID-19 commonly leads to respiratory failure, acute lung injury, and acute respiratory distress syndrome (ARDS) (28, 29). Other complications include venous thromboembolism, viral myocarditis, cardiac arrhythmia, sepsis, and multiorgan dysfunction (30, 31). Older adults and comorbid diseases, such as COPD, asthma, diabetes mellitus, and hypertension, are associated with increased incidence of severe COVID-19 disease (26, 32–34). Here we will discuss common clinical manifestations of COVID-19 before considering pathophysiology that leads to patterns of disease in COVID-19.

12.2.1 Viral Pneumonia

The most common clinical presentation of COVID-19 is a viral pneumonia that affects both the upper and lower respiratory tract (35). Studies suggest that initial infection is established in the upper respiratory tract, evidenced by a gradient of infection, highest in the upper respiratory tract and lower in the distal airways in the early stages of infection (36). Viral load in the upper airways is highest during the first week of infection (37, 38). There is a median delay of 4–5 days between virus exposure and development of symptoms (39). Diagnosis is based on clinical signs and symptoms, chest radiography, and positive testing for SARS-CoV-2 mRNA or antigen on upper respiratory tract sampling. The most common radiology findings are ground glass opacities, detected by chest x-ray or computerized tomography (CT) (40). Chest CT is a reliable tool to detect pathology in COVID-19 but is not routinely recommended for diagnosis (41). The most common finding on CT is ground glass diffuse opacities. However, focal consolidation, nodules, lobar infiltrate, or multi-focal infiltrates may be seen in more severe disease (33, 42–46). Chest ultrasound may be of use as a rapid, bedside tool to assess for pleural effusions and superficial pulmonary structures (43, 47). Abnormal lab findings commonly observed in COVID-19 positive patients include lymphopenia, leukopenia, elevated C-reactive protein (CRP), ferritin, and interleukin (IL)-6 with suggestion that lab abnormalities correlate with disease severity. Less common lab abnormalities include elevated liver enzymes and elevated D-dimer (33).

12.2.2 Respiratory Failure and ARDS

Most patients infected with COVID-19 manifest mild and self-limited disease. However, some patients progress to severe pneumonia, respiratory failure, and/or acute respiratory distress syndrome (ARDS) (48). Severe COVID-19 most commonly results in hypoxemic respiratory failure, necessitating delivery of supplemental oxygen for organ support. As disease progresses, patients may require advanced oxygen delivery devices, such as heated high flow oxygen, and noninvasive or invasive mechanical ventilation. Severe COVID-19 may be complicated by ARDS, which is characterized by non-cardiogenic pulmonary edema manifest as bilateral opacities on chest imaging with clinical hypoxemia. In ARDS, diffuse lung inflammation enacts dysfunction of barrier defenses in the alveolar epithelium and breakdown of the alveolar-capillary membrane. Influx of edema fluid into the alveolar space leads to disrupted pulmonary perfusion, hypoxemia, and decreased lung compliance, which is the ability of the lungs to stretch and expand, which may necessitate the need for ventilatory support (43). Some studies have noticed a unique feature of ARDS in COVID-19, which is hypoxemia out of proportion to impairments in lung mechanics (43, 49, 50). Unlike many other bacterial or viral pneumonias, respiratory failure is thought to be a principal driver of mortality in COVID-19 (51, 52). The time to severe disease is variable, with studies reporting respiratory failure within 8–15 days of initial infection (53, 54). Older patients (over 65 years of age) and those with chronic illness, such as COPD, asthma, diabetes mellitus, and chronic hypertension, are at increased risk of severe COVID-19 and respiratory failure (26, 32–34), characterized by increased risk of mechanical ventilation, prolonged duration of hospitalization, and higher mortality rates (54, 55).

12.2.3 Sepsis and Multiorgan Failure

Beyond local pulmonary disease, patients with severe COVID-19 may develop sepsis and septic shock with resultant multiorgan dysfunction. The Third International Consensus Definitions for Sepsis and Septic Shock (Sepsis-3) defines sepsis as a life-threatening organ dysfunction because of dysregulated host responses to infection (56). A systematic review and meta-analysis showed that the majority (77.9%) of critically ill hospitalized adults with COVID-19 (patients in the intensive care unit) met Sepsis-3 criteria (57). Although primary SARS-CoV-2 infection can lead to sepsis and critical illness, these patients are at risk for secondary infections. The presence of co-infection (such as viral co-infection) or secondary infections (such as bacterial or fungal) can further worsen disease course (58, 59). Among critically ill patients with COVID-19, pulmonary dysfunction and ARDS are the most common manifestations (87.5%), followed by septic shock, with less common prevalence of other organ dysfunction such as kidney failure, coagulopathy, and liver failure (57).

12.2.4 Thromboembolic Disease

Thromboembolic disease is a condition where blood clots form in the blood vessels, either impacting the local vessel or detaching to become lodged in a distant vessel. Common examples are deep venous thrombosis (DVT) of the legs, pulmonary embolism (PE), myocardial infarction, and cerebrovascular accident (CVA). Among hospitalized and critically ill patients with COVID-19, thrombosis is a common complication and may be the first presenting manifestation of SARS-CoV-2 infection (60, 61). While activation of coagulation pathways is a known complication of severe infection and sepsis, COVID-19 is notable as an especially thrombo-inflammatory condition with resultant thromboembolic events leading to the complication termed *COVID-19-associated coagulopathy* (CAC) (62, 63). CAC is notable from other causes of coagulopathy and sepsis-induced coagulopathy, characterized by increased D-dimer and fibrinogen levels often without initial changes in prothrombin time and platelet count (63). This coagulopathy is the result of the interplay of multiple factors including direct viral endothelial injury and immune system activation. Thrombosis in the setting of COVID-19 manifests as venous thromboembolism, including pulmonary thrombosis and DVT, or arterial thrombosis, affecting the brain causing acute ischemic stroke or the peripheral arteries causing arterial thrombosis (64). Alternatively, increased incidence of bleeding is seen in severe COVID-19, underscoring the perturbation in coagulation systems (65). An autopsy series of COVID-19 patients demonstrated diffuse microthrombi primarily in the lungs and rarely at extra-pulmonary sites (66). Incidence of pulmonary microthrombi in COVID-19 cases was found to be nine times higher compared to the H1N1 Influenza virus (67). Despite this increased incidence, clinical guidelines for COVID-19 recommend imaging for PE only when suspected and not as a screening practice (68). In more severe cases, elevated coagulation factors, specifically factor VIII, and von Willebrand factor may be detected (61). Elevated D-dimer is associated with poor prognosis in COVID-19 (69). Thromboembolism management does not differ from non-COVID thrombosis and recommends use of standard anticoagulants, such as low-molecular-weight heparin (70). The ANTICOVID randomized control trial showed that among patients with hypoxemic pneumonia and no initial macrovascular thrombotic pathology, high-dose prophylactic anticoagulation was associated with less incidence of thrombosis and major bleeding compared to standard-dose or therapeutic-dose anticoagulants (71).

12.2.5 Other Organ System Manifestations

12.2.5.1 Cardiovascular Manifestations

Cardiovascular complications may represent the most significant extra-pulmonary organ manifestation of COVID-19. Potential complications include myocardial infarction, myocarditis, and heart failure. One study reported that 71% of recently recovered COVID-19 patients showed elevations in high-sensitivity troponin T, a biomarker for cardiac damage, regardless of clinical signs and symptoms of myocardial infarction (72). Another study showed a significantly positive correlation between troponin levels and risk of clinical myocardial infarction (73). In a study of 100 recently recovered COVID-19 patients, 78% showed signs of cardiac involvement by cardiac magnetic resonance imaging (MRI), and 60% showed evidence of myocardial inflammation (72).

12.2.5.2 Renal Manifestations

Renal injury and renal failure, specifically proximal tubule dysfunction and glomerular injury, are complications linked to COVID-19 (74, 75). Most cases of COVID-19 complicated by renal failure are attributable to pre-renal disease secondary to hypotension and sepsis in critically ill patients. However, immune activation may lead to inflammatory renal disease, specifically collapsing glomerulonephritis in COVID-19-infected males of African ancestry with a predisposition to the APOL-1 genotype (75).

12.2.5.3 Neurological Manifestations

Neurological manifestations of COVID-19 include, but are not limited to, myopathy, delirium, encephalitis, and fatigue (75, 76). Furthermore, one observational study reported that cryptogenic stroke (strokes that lack a clear cause) was significantly associated with COVID-19-positive patients when compared with a COVID-19-negative control group (77).

12.2.5.4 Endocrine Manifestations

COVID-19 is suspected to play a role in thyroid dysfunction, with one study reporting a 20% prevalence of thyrotoxicosis (abnormally elevated circulating thyroid hormones) and a 5% prevalence of hypothyroidism among hospitalized COVID-19-positive patients (75, 78). Another endocrine manifestation that is thought to be associated with COVID-19 is diabetes mellitus, with one study showing that COVID-19-positive patients have an increased risk of developing diabetes up to a year after infection (79).

12.2.6 Post-Acute Sequelae of COVID-19

Long COVID, or post-acute sequelae of COVID-19 (PASC) (80), is defined by the US Centers for Disease Control and Prevention (CDC) as persistent, new, or recurring symptoms of COVID-19 experienced by patients 4 or more weeks after the infection. By contrast, the World Health Organization (WHO) defined long COVID as symptoms occurring 3 months after symptomatic SARS-CoV-2 infection with a duration of more than 2 months (81). Most common symptoms include fatigue, difficulty with memory, and concentration disturbance (82). There are multiple hypotheses explaining the pathophysiology of PASC, including persistence of viral infection, persistent tissue damage, dysbiosis, or autoimmune response with persistent innate immunity activation and presence of autoantibodies (81, 83, 84). Inflammatory cytokines such as IL-6, TNF-α, and IL-1β are found to be elevated in these reports (85).

12.3 PULMONARY HOST RESPONSE TO COVID-19

Pathogenesis of COVID-19 is multifaceted. Our understanding of pathogenesis is evolving. As a viral lung infection, COVID-19 shares common mechanisms of disease with other viral respiratory infections. Specifically, SARS-CoV-2 may cause direct cellular and tissue lung injury. Further, host innate and adaptive immune responses are activated in response to infection to fight off the pathogen but can result in bystander tissue injury, which, when significant, results in acute lung injury and acute respiratory distress syndrome. Here, we highlight important causative mechanisms of host defense in COVID-19 to discuss mechanisms common to viral pneumonia or unique in SARS-CoV-2 infection.

12.3.1 Innate Immunity

The innate immune system represents a non-specific first line of defense against pathogens, including the SARS-CoV-2 virus (86). Innate immunity is induced by stimulation of pattern recognition receptors (PRR) with pathogen-associated molecular patterns (PAMPs), which are non-specific conserved molecular patterns. In COVID-19, PAMPs such as viral RNA and viral proteins interact with PRRs to induce inflammation. This process is exacerbated by resultant release of damage-associated molecular patterns from affected host cells (87–89). PRRs are present on the entire epithelial lining of the respiratory tracts, alveoli (90–92), alveolar macrophages, neutrophils, and recruited monocytes (93) and detect COVID-19 PAMPs, such as viral RNA or viral fragments (94, 95). Innate immune activation results in a subsequent adaptive immune response as well (96). A unique feature of the COVID-19 pandemic was the novelty of the SARS-CoV-2 virus. As the virus descended upon a virus-naïve population without adaptive immunity to SARS-CoV-2, innate immune responses were the primary drivers of host defense with resultant bystander inflammatory injury through multiple mechanisms.

12.3.1.1 Viral Entry and Cytopathic Damage

In addition to viral RNA, the SARS-CoV-2 virion is composed of multiple proteins, including spike (S), envelope (E), membrane (M), 16 nonstructural proteins (NSPs), and nine accessory proteins (97). Viral entry is mediated through the binding of the S protein to the ACE2 receptor on host cells (98), facilitated by proteolytic cleavage of the S protein by TMPRSS2, a serine protease (99). Viral entry and viral replication represent multiple opportunities for innate immune activation through direct cytopathic effects, exposure of PRRs to PAMPs, and cell death with subsequent release of DAMPs (86, 100). In cultures of human airway epithelial cells, SARS-CoV-2 demonstrates high efficiency of viral replication and is known to infect both ciliated and secretory cells, whereas other Coronaviridae and respiratory viruses display preferences for either ciliated or secretory cells (101). Macrophages may also be directly infected by SARS-CoV-2 virus and contribute to the hyperinflammation in COVID-19 through activation of intracellular inflammatory mediators such as the inflammasome with resultant signaling with the potent cytokine IL-1β and the highly inflammatory cell death pathway, pyroptosis (102).

12.3.1.2 Toll-Like Receptors

Toll-like receptors (TLRs) are membrane-spanning receptors that function as PRRs and are central to innate immune function (87, 103). Thirteen different TLRs have been described (103). Among them, six are important for interaction with viral PAMPs, including TLR3, TLR7, RLT8, and TLR9 for the viral nucleic acids, and TLR2 and TLR4 for viral glycoproteins (87). SARS-CoV-2 is classified as a single-stranded RNA virus (97); single-stranded RNA is detected within endosomes primarily by TLR7 and TLR8, which are highly expressed by myeloid dendritic cells and macrophages (81, 104). However, during the replication cycle of SARS-CoV-2, double-stranded RNA is also present and recognized by TLR3 (105) expressed by myeloid cells and macrophages (106). TLR activation results in recruitment of downstream adaptor molecules, such as MyD88 innate immune signal transduction adaptor for all TLRs and TIR domain-containing adapter molecule 1 (TRIF) for TLR3. Signaling by MyD88 and TRIF converge to activate transcription factors, specifically nuclear factor-kappa-B (NF-κB) and interferon (IFN)-regulatory factor (IRF). Activation of NF-κB increases transcription of multiple important cytokines and chemokines, including tumor necrosis factor (TNF), IL-6, and IL-1β, and some other innate immune sensors such as the inflammasome constituent NACHT, LRR, and PYD domain-containing protein 3 (NLRP3) (107–110), all known to be upregulated in COVID-19. Further, these cytokines and NLRP3 are implicated in inflammatory lung injury that contributes to the pathogenesis of ARDS. IRF results in transcription of multiple interferons, including type I (mostly α and β) and type III (λ1/2/3) IFNs. TLR pathway expression, including TLR1, TLR2, TLR4, TLR5, TLR8, TLR9, MyD88 and TRIF adaptor molecules was found to be overexpressed in patients with severe COVID-19 compared to healthy controls (111, 112). Conversely, TLR3 signaling has been suggested to be protective during COVID-19 infection (113–115). TLR3 mutations and decreased TLR3 expression in peripheral blood cells have been correlated with severe disease in COVID-19 (116). Viral protein E has been implicated to interact with TLR2 with subsequent proinflammatory cytokine production (111), and TLR2 knockout mice demonstrated decreased IL6, among other cytokines, upon exposure to protein E (111, 117). Taken together, these findings suggest a central role for TLRs to activate the innate immune response to SARS-CoV-2 to drive the cytokine response vital for host defense, which may also contribute to the inflammatory lung injury characteristic of COVID-19.

12.3.1.3 RIG-I–Like Receptors

Additional PRRs that function in COVID-19 are the RIG-I-like receptors (RLRs), intracellular PRRs that detect viral RNA (100, 118). Two primary RLRs implicated to function in COVID-19 host defense are antiviral innate immune response receptor RIG-I and the interferon-induced helicase C domain-containing protein 1 (IFIH1 or MDA5) (119–121). These RLGs stimulate type I and III IFN gene transcription through activation of TNF receptor-associated factor (TRAF)3, TANK-binding kinase (TBK)1 and IκB kinase (IKK) to induce phosphorylation of IRF3 (108, 118). In cell models, MDA5 silencing or knockout resulted in reduced expression of type I IFNs with exacerbation of SARS-CoV-2 infection (119, 120), whereas results have been mixed on the role of RIG-I in the control of SARS-CoV-2 viral infection (119–121).

12.3.1.4 Inflammasome Activation

Nod-like receptors (NLRs) are PRRs that sense a variety of intracellular PAMPs and DAMPs and are activated by SARS-CoV-2 infection (86). Specifically NLRP3, an inflammasome constituent and canonical inflammasome activator, is vital for appropriate innate immune responses in COVID-19. NLRP3 disruption impairs cytokine responses and host defense in SARS-CoV-2 infection (113, 122–126). The NLRP3 inflammasome enacts secretion of the potent inflammatory cytokines IL-1β and IL-18 through cleavage of their precursors pro-IL-1β and pro-IL-18 and activation of pyroptosis or inflammatory cell death (127, 128). Elevated plasma levels of IL-1β and IL-18 were associated with increased COVID-19 severity and mortality (129, 130).

12.3.1.5 Cytokine Storm

SARS-CoV-2 viral RNA and proteins interact with multiple PRRs to induce a robust innate immune response. As infection evolves, cytopathic effects, cell death, and immune activation can perpetuate immune activation as additional virus and many DAMPs are released secondary to cellular damage. The end result is a robust and potentially excessive cytokine response, including multiple interferons and other proinflammatory cytokines. Indeed, studies have demonstrated a robust cytokine response in COVID-19 (131). The involved cytokines, including IL-1β, IL-6, TNF, IL-12, IFN-β, IFN-γ, and IL-17, are significantly upregulated in patients with COVID-19 and are important for host defense and elimination of the SARS-CoV-2 virus (23, 112, 132). Excessive cytokine release is termed *cytokine storm*, a life-threatening condition of cell death and inflammatory tissue damage (133). Cytokine storm is a systemic hyperinflammatory condition that leads to hemodynamic instability and multiorgan failure. SARS-CoV-2, being highly inflammatory, is demonstrated to activate multiple cell death pathways, including pyroptosis (134), apoptosis (135, 136), and necroptosis (137, 138). A cell death pathway termed PANoptosis, a convergence of pyroptosis, apoptosis, and necroptosis pathways leads to cell death in COVID-19 (139), likely because of excessive inflammatory signaling in SARS-CoV-2 infection. In PANoptosis, the synergistic effect of TNF and IFN-γ results in signal transducer and activator of transcription (STAT)1 and IRF1 activation with subsequent caspase 8–mediated cell death (139, 140). Additional cytokines are released during inflammatory cell death, which can enact more cell death in a dangerous positive feedback loop (133, 139). In severe COVID-19 infection, cytokine storm results in sepsis, septic shock, or inflammatory organ damage, including ARDS (67, 86). Severe COVID-19 is distinct from other causes of cytokine storm because of the presence of lymphopenia (141), thought to be a result of lymphocyte cell death in lymph nodes and the spleen (142) induced by TNF and IFN-γ (139, 143). Lymphopenia may lead to blunting of the adaptive immune response with poor disease outcomes (144).

12.3.1.6 Innate Immunity Evasion Strategies

IFN responses are important for antiviral response to reduce viral loads to limit non-specific acute inflammation. SARS-CoV-2 also evolved multiple mechanisms to evade innate immune responses leading to reduced levels of type I and III IFNs in the lung and peripheral blood (104, 112, 145, 146) through multiple mechanisms along the interferon signaling pathway (86, 100). SARS-CoV-2 non-structural proteins Nsp1 and Nsp14 inhibit cellular protein translation that dampens production of proteins along the IFN signaling pathway (147–150). MDA5, a cytoplasmic viral nucleic acid sensor that activates type I IFNs and other cytokines, is inhibited by SARS-CoV-2 papain-like protease (PLpro) that inhibits MDA5 activation by de-ISGylation (151, 152). Downstream from interferon sensors, IFN gene transcription is reliant on multiple transcription factors. SARS-CoV-2 inhibits IRF3 nuclear translocation by SARS-CoV-2 proteins Nsp12 and ORF6, and NF-κB translocation is blocked by Nsp13 (148, 149, 153, 154). IFNs signal through interferon receptors and JAK/STAT second messengers. SARS-CoV-2 ORF3a decreases JAK2 by activation of ubiquitination and increases SOCS1, which is a negative regulator of JAK2, leading to inhibition of IFN signal propagation (155). STAT1 and STAT2 signaling requires phosphorylation, which is inhibited by SARS-CoV-2 proteins Nsp1, Nsp6, Nsp13, Nsp14, ORF3a, ORF7a, and ORF7b (147, 156). Taken together, SARS-CoV-2 has evolved strategies to inhibit interferon signaling in response to viral infection.

12.3.2 Adaptive Immunity

The adaptive immune response is equally important in COVID-19 pathogenesis and determines clinical course during SARS-CoV-2 infection. Adaptive immunity differs from innate immune responses in its specificity to particular pathogens. As a novel virus, the majority of the population had no adaptive immunity to SARS-CoV-2 early in the COVID-19 pandemic, which allowed the innate immune system to dominate. An important feature of adaptive immunity is immunologic memory, which allows for swifter elimination of pathogens with repeated insults, often with less acute inflammation and bystander tissue injury. Here we will review humoral and cellular-based adaptive immune responses.

12.3.2.1 Humoral Response

Humoral immunity is mediated by secreted macromolecules, including antibodies, complement, and antimicrobial peptides. B cells produce multiple immunoglobulins (Ig), including IgD, IgM, IgA, IgG, and IgE, which have important antiviral properties (157). IgM is a pentamer that is important for opsonization that tags pathogens for elimination by phagocytic cells (158). IgM levels increase within 1 week of SARS-CoV-2 infection or roughly 2 days after symptom onset (159). Elevated IgM, specific to SARS-CoV-2, indicate recent SARS-CoV-2 infection or immunization (158). In previously unexposed patients, IgG secretion is noted around 2 weeks post–SARS-CoV-2 infection, later than IgM. IgG antibodies are characterized by long-term immune memory, high-affinity binding to antigens, complement activation, and antibody-dependent cellular cytotoxicity (ADCC) (157, 160). In COVID-19, IgG1 and IgG3 are the dominant IgG isoforms (161). In respiratory infections, IgA antibodies play an important role in mucous membrane host defense. In the respiratory tract, IgA functions as a first line of defense by binding and neutralizing pathogens, preventing host cell entry. IgA antibodies are usually detectable shortly after IgM (157). However, one study demonstrated IgA detection before IgM in COVID-19 (162). Neutralizing antibodies against

SARS-CoV-2 remain elevated for around 3 months postinfection. After 3 months, patients may have increased susceptibility for reinfection (157). Antibodies produced in response to other respiratory coronaviruses may play a role in the immune response against COVID-19 and shape the course of the disease (163, 164). The antibody response is important for activation of cellular immunity and may impact the extent of T cell activation (165, 166). In severe COVID-19, some studies suggest a delay in neutralizing antibody production against the spike protein contributes to disease severity (167–169). In addition, $CD4^+$ T cell depletion seen in severe COVID-19 is associated with defective lymph node germinal centers, which are essential for B cell differentiation and subsequent antibody production (142).

12.3.2.2 Cellular Response

The cellular immune response is vital for the control of intracellular pathogens, including viral invaders. Antigen-presenting cells (APCs), such as monocytes, macrophages, and dendritic cells, express ACE2, which functions as a viral receptor, facilitating direct infection of these APCs by SARS-CoV-2 virus. By contrast, lymphocytes do not express ACE2 (170). APCs subsequently present antigens and epitopes to T cells. Of note, B cells may participate in SARS-CoV-2 antigen presentation as well. Acquired cellular responses are mediated by T cells that recognize antigens and epitopes from individual pathogens (157). In COVID-19, T cell responses are essential for viral clearance and are associated with protection from virus and severe disease. Indeed, early and effective cytotoxic $CD8^+$ T cell responses with associated with viral clearance (171) and levels of $CD8^+$ T cell activation helps distinguish mild and severe COVID-19 disease (172). In severe COVID-19, patients display severe lymphopenia in the peripheral blood (173), which impairs adaptive immune responses. Lymphopenia occurs through multiple cell death pathways, including a cell-autonomous death and activation of apoptosis, evidenced by high caspase levels (174) and PANoptosis, seen in cytokine storm (131). Lymphopenia results in loss of both $CD4^+$ (helper) and $CD8^+$ T cells and disrupts functioning of $CD16^+$ cells (131, 174). Lymphopenia and resultant impaired cellular adaptive immunity has been implicated in COVID-19 pathogenesis and correlates with severe disease (157, 174, 175). Cellular immunity in COVID-19 is an important determinant of vaccine response as memory T cells may be activated well over a decade after initial infection (176).

12.4 IMPLICATIONS FOR CLINICAL MANAGEMENT

12.4.1 COVID-19 and Preexisting Lung Disease

Fundamentally, COVID-19 begins as an acute viral lung infection. Multiple studies have assessed the risk of COVID-19 incidence and severity in patients with preexisting lung disease, including asthma, COPD, and structural lung disease. Asthma is a chronic respiratory condition characterized by intermittent or persistent inflammation and narrowing of the airways. Patients experience wheezing and dyspnea and may exhibit obstructive lung physiology on pulmonary function testing. Initial reports suggested a higher incidence of COVID-19 among asthmatic patients (26, 32–34, 177). However, a meta-analysis of 51 studies demonstrated a lower risk of developing COVID-19 in patients with asthma compared to unobstructed patients, with no statistically significant difference between groups regarding hospitalization, intensive care unit admission, and ventilator use (178). Further studies examined asthma phenotypes, specifically type 2 asthma (allergic and eosinophilic) and non–type 2 (neutrophilic and paucigranulocytic) and suggested disparate incidence of severe COVID-19 among the two groups (179, 180). Immune responses in type 2 asthma are thought to be driven by cytokines from $CD4^+$ T cells, predominantly IL-4, IL-5, and IL-13; IL-13 may decrease ACE2 expression in the epithelial cells, which could result in decreased viral entry and infectivity (181–183). INF-induced expression of ACE2 may be blunted by type 2 cytokines and IgE cross-linking (184, 185). Conversely, non–type 2 asthma is associated with increased risk of severe COVID-19 (179). Studies suggest non–type 2 asthma may have higher ACE2 expression (186) and a predominantly pro-inflammatory response in COVID-19 with release of IL-1β, IL-8, IL-6, and IL-17, and potential cytokine storm (180). Among patients with chronic obstructive pulmonary disease (COPD), incidence of COVID-19 is not reported to be higher (179). However, COPD patients are more susceptible to severe COVID-19 disease and hospitalization (187, 188). Impaired lung functions present in COPD reduces compensatory reserve for the pathological effects of COVID-19 viral pneumonia and potential respiratory complications (189, 190). COPD patients may exhibit a decreased type I INF responses with inefficient antiviral immune responses (191) and increased TLR2 and TLR4 expression, which may lead to ACE2 upregulation and more severe innate immune activation (192–194). One particularly at risk population includes patients with underlying bronchiectasis. These patients, in contrast to other chronic lung diseases, have a higher incidence of COVID-19 infection (195) and higher risk of severe disease, likely because of structural lung abnormalities and impaired immune responses (179, 196).

12.4.2 COVID-19 Treatment Strategies

Host defense in SARS-CoV-2 infection is multilayered and complex. Following viral entry, innate immune responses are activated to combat the virus but put the host at risk for bystander tissue damage and inflammatory lung and systemic injury. Antibody and cellular adaptive immune responses specifically target the virus with less collateral damage. These adaptive immune responses were lacking in an unexposed population early in the pandemic. Further, SARS-CoV-2 has multiple mechanisms to evade host defenses, and lymphocytopenia seen in COVID-19 leads to ineffective adaptive immune responses. Treatment strategies to limit morbidity and mortality in COVID-19 include (1) directly targeting the virus to limit infection and replication, (2) limiting inflammation and inflammatory tissue damage, and (3) augmenting adaptive immune responses. Timing of intervention is critical to success in each strategy, which will be discussed below.

12.4.2.1 Antiviral Agents

Like many viruses, the viral life cycle of SARS-CoV-2 is dependent upon both viral and host factors. When considering antiviral agents, therapeutic strategies have emerged targeting both viral proteins (197–200) and essential host proteins for SARS-CoV-2 entry and replication (201, 202). SARS-CoV-2 entry occurs as a result of the interaction between the viral S protein and the host ACE2 (203), and several agents, including neutralizing antibodies, peptides, and small molecules have been suggested to SARS-CoV-2 and ACE2 interaction and prevent viral entry. Monoclonal antibodies, bebtelovimab, sotrovimab, regdanvimab, and cilgavimab, against SARS-CoV-2 protein S have been investigated (200). However, SARS-CoV-2 protein S is a highly mutable protein rendering these monoclonal antibodies inefficient or inactive against the emerging new variants of the virus, such as omicron variants (204–206). Moreover, the clinical value of these monoclonal antibodies is restricted to the early stages of COVID-19 (207) before the host produces endogenous antibodies (208). In addition to monoclonal antibodies, anti-spike small molecules such as clofazimine inhibit protein S interaction with the ACE2 viral receptor but are still under investigation (200, 209).

Another SARS-CoV-2 viral protein, papain-like protease (PLpro) is a cysteine protease that functions to release the viral proteins NSP1, NSP2, and NSP3 (152). Inhibition of PLpro both interferes with viral protein maturation and enhances the innate response against the virus as PLpro functions in SARS-CoV-2 immunity evasion (152, 210). The Main protease (M^{pro}) is a SARS-CoV2 protease whose inhibition leads to the failure of maturation of key viral proteins, such as NSP12 and NSP13 (200, 211). Numerous M^{pro} inhibitors have been proposed, including nirmatrelvir and ensitrelvir (212). Nirmatrelvir, the most promising M^{pro} inhibitor (200), has shown clinical efficiency in COVID-19 treatment, and when administered with ritonavir it has demonstrated reduced COVID-19 hospitalization and mortality rate in adult patients in the phase II/III EPIC-HR study (213). SARS-CoV-2 replication and transcription enzymes have been considered as antiviral targets. NSP12 encodes SARS-CoV2 RNA-dependent RNA polymerase (RdRp), a guanylyltransferase that caps the 5′ end of viral RNA (214). RdRp can be inhibited using several repurposed inhibitors (215), such as nucleoside analogs (remdesivir, molnupiravir, favipiravir, and bemnifosbuvir) and non-nucleoside analogs (suramin) (200, 216). In particular, remdesivir (GS-5734) has shown clinical benefit in certain patients with COVID-19 and has been adopted for widespread use in outpatients and certain hospitalized patients (217). Molnupiravir, another nucleoside analog, induces lethal GC mutations in the viral RNA (218, 219), with studies reporting decreased hospitalization and mortality rates in COVID-19 patients (220). Of note, other studies failed support the clinical utility of remdesivir (221, 222) or molnupiravir (223) in certain populations.

SARS-CoV-2 uses host proteins and other factors for its viral life cycle (202). Targeting these host factors proposes favorable broad-spectrum efficiency against the emerging viral variants at the risk of undesired off-target effects as the targets are host in origin (199, 201). ACE2 targeting and host cellular proteases are examples of such targets. ACE2 viral receptor interacts with the protein S of SARS-CoV-2 for viral entry (203). Cellular proteases, such as transmembrane protease serine 2 (TMPRSS2) and cathepsin, enact proteolytic processing of the protein S to facilitate viral entry (99, 224, 225). Antiviral agents against ACE2 and cellular proteases are still under investigation and include peptide fragments (APN01) mimicking ACE2 (226) and camostat mesylate and nafamostat mesylate (protease inhibitors) (99, 227).

12.4.2.2 Immunomodulatory Agents

Immune system modulators function by targeting host proteins involved in the dysfunctional, exaggerated, and harmful immune response to viral infection implicated in the pathogenesis of cytokine storm, ARDS, multiorgan failure, and thrombotic complications (81, 86, 228).

Corticosteroids, such as dexamethasone, hydrocortisone, prednisolone, and methylprednisolone, are well known for their anti-inflammatory and immunosuppressive actions (229–231). Dexamethasone was shown to have clinical benefit for hospitalized patients with COVID-19 requiring oxygen or respiratory support (232, 233). Of note, systemic steroids carry a risk of adverse effects, especially when used in high doses for prolonged time, including osteoporosis, hyperglycemia, adrenal suppression (234). Initiation of steroids is reserved until later in disease course as steroids early in disease course may worsen outcomes because of their immunosuppressive function leading to increased viral replication (235).

Janus kinases are receptor-associated tyrosine kinases of the JAK-STAT signaling pathway, involved in the response to extracellular cytokines, interferons, and growth factors (236). Multiple studies have demonstrated the clinical efficacy of the JAK inhibitor, barcitinib (237–239), which gained US FDA approval in 2022 for treatment of COVID-19 in inpatients who require respiratory support (200). Other JAK inhibitors, such as fostamatinib (240), nezulcitinib (241), ruxolitinib (242), and pacritinib (243), have been evaluated in clinical trials but have not advanced through the clinical pipeline because of inadequate statistical power or inability to meet study endpoints.

As excessive cytokine release plays a key role in COVID-19 pathogenesis, cytokine antagonists represent potential therapeutics. Specifically, excessive circulating IL-6 has been implicated in severe COVID-19 pathogenesis (81, 86, 228). An IL-6 receptor antagonist, tocilizumab, has demonstrated favorable clinical outcomes when used in hospitalized hypoxemic COVID-19 patients with systemic inflammation, evidenced by elevated IL-6 (244). IL-1 receptor antagonist, anakinra, has been investigated in hospitalized COVID-19 patients and showed clinical efficacy of all-cause mortality and hospitalization reduction in the SAVE-MORE trial (245) and no added benefit in other studies, such as REMAP-CAP (NCT02735707).

Type I IFNs are vital to the immune response against SARS-CoV-2. Further SARS-CoV-2 inhibition of type I IFNs is a viral immune evasion mechanism (246, 247). Studies have investigated and demonstrated clinical value of exogenous IFNβ-1b (248) and Peg-IFNλ-1a (249) in COVID-19 inpatients or outpatients, respectively. Conversely, IFNβ-1a showed no clinical benefit in COVID-19 (250) with increased adverse events, such as development of respiratory failure (251).

12.4.2.3 Boosting Adaptive Immunity

The production of effective mRNA vaccines against SARS-CoV-2 infection has been the most impactful weapon in restraining the pandemic. Unlike traditional vaccines, which contain either weakened, non-live, or inactivated pathogens, mRNA vaccines deliver packages of synthetic mRNA encoding viral proteins. These strands of mRNA are protected from degradation by a lipid nanoparticle envelope. Cells uptake the mRNA followed by translation to its corresponding protein. The proteins then cause an immune response, which leads to the development of immunologic memory. The mRNA vaccine offers advantages compared with more conventional vaccines (252). The importance of the S protein to viral attachment and entry into cells prompted focused efforts to develop mRNA vaccines targeting S protein (252). The foundational research to develop mRNA vaccines spans several decades and was implemented on a massive scale against the COVID-19 pandemic with impressive results (253). Among others, Pfizer's BNT162b2 vaccine and Moderna's mRNA-1723 vaccine were 95% and 94.5% successful in the prevention of COVID-19, respectively (252). SARS-CoV-2 mRNA have been modified as the pandemic evolves to account for ongoing mutation in the S protein, but the targeting strategy remains the same.

Passive immunization refers to the transfer of antibodies from an exogenous source to fight infection. Monoclonal antibodies (mABs) are lab-designed antibodies that bind to a specific epitope of an antigen. mABs can neutralize viral targets and thus are of therapeutic value. However, mABs are difficult to produce and not as quickly deployed as convalescent plasma therapy (254). convalescent plasma therapy (CPT) refers to the transferring of the plasma of a recovering COVID-19 patient to an infected patient. Convalescent plasma contains polyclonal neutralizing antibodies that facilitate the removal of viral load from the bloodstream (255, 256). However, a recent meta-analysis reported mixed results for the efficacy of CPT, stating that although CPT decreased the risk of hospitalization, there was no strong association between CPT and survival times (257). As the pandemic has evolved, the efficacy of CPT is restricted to patient without preexisting immunity obtained from prior infection or vaccination.

12.5 CONCLUSIONS

COVID-19 caused by SARS-CoV-2 leads to viral pneumonia and portends risk for numerous pulmonary and systemic complications. Various elements of viral pathology and host defense coalesce to inform the clinical course of patients infected with the SARS-CoV-2 virus. Our understanding of the mechanisms that contribute to pathobiology continues to evolve. Multiple therapeutic strategies have been designed and tested to combat viral entry, replication, and resultant injurious inflammation that leads to clinical disease. Efforts are ongoing as the scientific community continues to combat the COVID-19 pandemic and prepare for future SARS-CoV-2 strains and the potential emergence of novel respiratory viruses.

REFERENCES

1. Zhu, N., Zhang, D., Wang, W., Li, X., Yang, B., Song, J., *et al.* (2020) A novel coronavirus from patients with pneumonia in China, 2019. *N Engl J Med.* **382**, 727–733. https://doi.org/10.1056/NEJMoa2001017
2. [dataset] *Who COVID-19 Dashboard.* Geneva: World Health Organization, 2020. https://covid19.who.int/ (Accessed June 8, 2023)
3. Zhu, N., Zhang, D., Wang, W., Li, X., Yang, B., Song, J., *et al.* (2020) A novel coronavirus from patients with pneumonia in China, 2019. *N Engl J Med.* **382**, 727–733. https://doi.org/10.1056/NEJMoa2001017
4. Chan, J. F., Kok, K. H., Zhu, Z., Chu, H., To, K. K., Yuan, S., *et al.* (2020) Genomic characterization of the 2019 novel human-pathogenic coronavirus isolated from a patient with atypical pneumonia after visiting Wuhan. *Emerg Microbes Infect.* **9**, 221–236. https://doi.org/10.1080/22221751.2020.1719902
5. (2020) The species Severe acute respiratory syndrome-related coronavirus: Classifying 2019-nCoV and naming it SARS-CoV-2. *Nat Microbiol.* **5**, 536–544. https://doi.org/10.1038/s41564-020-0695-z

6. Rambaut, A., Holmes, E. C., O'Toole, Á., Hill, V., McCrone, J. T., Ruis, C., *et al.* (2020) A dynamic nomenclature proposal for SARS-CoV-2 lineages to assist genomic epidemiology. *Nat Microbiol.* **5**, 1403–1407. https://doi.org/10.1038/s41564-020-0770-5
7. (1968) Virology: Coronaviruses. *Nature.* **220**(5168), 650. https://doi.org/10.1038/220650b0
8. Sturman, L. S., and Holmes, K. V. (1983) The molecular biology of coronaviruses. *Adv Virus Res.* **28**, 35–112. https://doi.org/10.1016/s0065-3527(08)60721-6
9. Hoffmann, M., Kleine-Weber, H., Schroeder, S., Kruger, N., Herrler, T., Erichsen, S., *et al.* (2020) SARS-CoV-2 cell entry depends on ACE2 and TMPRSS2 and is blocked by a clinically proven protease inhibitor. *Cell.* **181**, 271–280.e278. https://doi.org/10.1016/j.cell.2020.02.052
10. Shang, J., Wan, Y., Luo, C., Ye, G., Geng, Q., Auerbach, A., *et al.* (2020) Cell entry mechanisms of SARS-CoV-2. *Proc Natl Acad Sci U S A.* **117**, 11727–11734. https://doi.org/10.1073/pnas.2003138117
11. Sattentau, Q. (2008) Avoiding the void: Cell-to-cell spread of human viruses. *Nat Rev Microbiol.* **6**, 815–826. https://doi.org/10.1038/nrmicro1972
12. Mothes, W., Sherer, N. M., Jin, J., and Zhong, P. (2010) Virus cell-to-cell transmission. *J Virol.* **84**, 8360–8368. https://doi.org/10.1128/JVI.00443-10
13. Zeng, C., Evans, J. P., King, T., Zheng, Y. M., Oltz, E. M., Whelan, S. P. J., *et al.* (2022) SARS-CoV-2 spreads through cell-to-cell transmission. *Proc Natl Acad Sci U S A.* **119**. https://doi.org/10.1073/pnas.2111400119
14. Hikmet, F., Méar, L., Edvinsson, Å., Micke, P., Uhlén, M., and Lindskog, C. (2020) The protein expression profile of ACE2 in human tissues. *Mol Syst Biol.* **16**, e9610. https://doi.org/10.15252/msb.20209610
15. Li, J. Y., You, Z., Wang, Q., Zhou, Z. J., Qiu, Y., Luo, R., *et al.* (2020) The epidemic of 2019-novel-coronavirus (2019-nCoV) pneumonia and insights for emerging infectious diseases in the future. *Microbes Infect.* **22**, 80–85. https://doi.org/10.1016/j.micinf.2020.02.002
16. Lewis, D. (2020) Mounting evidence suggests coronavirus is airborne—but health advice has not caught up. *Nature.* **583**, 510–513. https://doi.org/10.1038/d41586-020-02058-1
17. Stadnytskyi, V., Bax, C. E., Bax, A., and Anfinrud, P. (2020) The airborne lifetime of small speech droplets and their potential importance in SARS-CoV-2 transmission. *Proc Natl Acad Sci U S A.* **117**, 11875–11877. https://doi.org/10.1073/pnas.2006874117
18. Anfinrud, P., Stadnytskyi, V., Bax, C. E., and Bax, A. (2020) Visualizing speech-generated oral fluid droplets with laser light scattering. *N Engl J Med.* **382**, 2061–2063. https://doi.org/10.1056/NEJMc2007800
19. Klompas, M., Baker, M. A., and Rhee, C. (2020) Airborne transmission of SARS-CoV-2: Theoretical considerations and available evidence. *Jama.* **324**, 441–442. https://doi.org/10.1001/jama.2020.12458
20. Oran, D. P., and Topol, E. J. (2021) The proportion of SARS-CoV-2 infections that are asymptomatic: A systematic review. *Ann Intern Med.* **174**, 655–662. https://doi.org/10.7326/m20-6976
21. Islam, M. A., Kundu, S., Alam, S. S., Hossan, T., Kamal, M. A., and Hassan, R. (2021) Prevalence and characteristics of fever in adult and paediatric patients with coronavirus disease 2019 (COVID-19): A systematic review and meta-analysis of 17515 patients. *PLoS ONE.* **16**, e0249788. https://doi.org/10.1371/journal.pone.0249788
22. Islam, M. A., Alam, S. S., Kundu, S., Hossan, T., Kamal, M. A., and Cavestro, C. (2020) Prevalence of headache in patients with coronavirus disease 2019 (COVID-19): A systematic review and meta-analysis of 14,275 patients. *Front Neurol.* **11**, 562634. https://doi.org/10.3389/fneur.2020.562634
23. Huang, C., Wang, Y., Li, X., Ren, L., Zhao, J., Hu, Y., *et al.* (2020) Clinical features of patients infected with 2019 novel coronavirus in Wuhan, China. *Lancet.* **395**, 497–506. https://doi.org/10.1016/s0140-6736(20)30183-5
24. Richardson, S., Hirsch, J. S., Narasimhan, M., Crawford, J. M., McGinn, T., Davidson, K. W., *et al.* (2020) Presenting characteristics, comorbidities, and outcomes among 5700 patients hospitalized with COVID-19 in the New York City area. *Jama.* **323**, 2052–2059. https://doi.org/10.1001/jama.2020.6775
25. Grant, M. C., Geoghegan, L., Arbyn, M., Mohammed, Z., McGuinness, L., Clarke, E. L., *et al.* (2020) The prevalence of symptoms in 24,410 adults infected by the novel coronavirus (SARS-CoV-2; COVID-19): A systematic review and meta-analysis of 148 studies from 9 countries. *PLoS ONE.* **15**, e0234765. https://doi.org/10.1371/journal.pone.0234765
26. Wang, D., Hu, B., Hu, C., Zhu, F., Liu, X., Zhang, J., *et al.* (2020) Clinical characteristics of 138 hospitalized patients with 2019 novel coronavirus-infected pneumonia in Wuhan, China. *JAMA.* **323**, 1061–1069. https://doi.org/10.1001/jama.2020.1585
27. Zhang, J. J., Dong, X., Cao, Y. Y., Yuan, Y. D., Yang, Y. B., Yan, Y. Q., *et al.* (2020) Clinical characteristics of 140 patients infected with SARS-CoV-2 in Wuhan, China. *Allergy.* **75**, 1730–1741. https://doi.org/10.1111/all.14238

28. Murthy, S., Gomersall, C. D., and Fowler, R. A. (2020) Care for critically Ill patients with COVID-19. *Jama.* **323**, 1499–1500. https://doi.org/10.1001/jama.2020.3633
29. Heymann, D. L., and Shindo, N. (2020) COVID-19: What is next for public health? *Lancet.* **395**, 542–545. https://doi.org/10.1016/s0140-6736(20)30374-3
30. Long, B., Brady, W. J., Koyfman, A., and Gottlieb, M. (2020) Cardiovascular complications in COVID-19. *Am J Emerg Med.* **38**, 1504–1507. https://doi.org/10.1016/j.ajem.2020.04.048
31. Lindner, D., Fitzek, A., Bräuninger, H., Aleshcheva, G., Edler, C., Meissner, K., *et al.* (2020) Association of cardiac infection with SARS-CoV-2 in confirmed COVID-19 autopsy cases. *JAMA Cardiol.* **5**, 1281–1285. https://doi.org/10.1001/jamacardio.2020.3551
32. COVID-19 treatment guidelines panel. Coronavirus disease 2019 (COVID-19) treatment guidelines. *National Institutes of Health.* www.covid19treatmentguidelines.nih.gov/ (Accessed April 30, 2023)
33. Guan, W.-J., Ni, Z.-Y., Hu, Y., Liang, W.-H., Ou, C.-Q., He, J.-X., *et al.* (2020) Clinical characteristics of coronavirus disease 2019 in China. *N Engl J Med.* **382**, 1708–1720. https://doi.org/10.1056/NEJMoa2002032
34. Wu, C., Chen, X., Cai, Y., Xia, J. A., Zhou, X., Xu, S., *et al.* (2020) Risk factors associated with acute respiratory distress syndrome and death in patients with coronavirus disease 2019 pneumonia in Wuhan, China. *JAMA Int Med.* **180**, 934–943. https://doi.org/10.1001/jamainternmed.2020.0994
35. Harrison, A. G., Lin, T., and Wang, P. (2020) Mechanisms of SARS-CoV-2 transmission and pathogenesis. *Trends Immunol.* **41**, 1100–1115. https://doi.org/10.1016/j.it.2020.10.004
36. Hou, Y. J., Okuda, K., Edwards, C. E., Martinez, D. R., Asakura, T., Dinnon, K. H., *et al.* (2020) SARS-CoV-2 reverse genetics reveals a variable infection gradient in the respiratory. *Tract Cell.* **182**, 429–446. e414. https://doi.org/10.1016/j.cell.2020.05.042
37. Wölfel, R., Corman, V. M., Guggemos, W., Seilmaier, M., Zange, S., Müller, M. A., *et al.* (2020) Virological assessment of hospitalized patients with COVID-2019. *Nature.* **581**, 465–469. https://doi.org/10.1038/s41586-020-2196-x
38. To, K. K., Tsang, O. T., Leung, W. S., Tam, A. R., Wu, T. C., Lung, D. C., *et al.* (2020) Temporal profiles of viral load in posterior oropharyngeal saliva samples and serum antibody responses during infection by SARS-CoV-2: An observational cohort study *Lancet Infect Dis.* **20**, 565–574. https://doi.org/10.1016/s1473-3099(20)30196-1
39. Gandhi, R. T., Lynch, J. B., and Del Rio, C. (2020) Mild or moderate Covid-19. *N Engl J Med.* **383**, 1757–1766. https://doi.org/10.1056/NEJMcp2009249
40. Li, Y., and Xia, L. (2020) Coronavirus disease 2019 (COVID-19): Role of chest CT in diagnosis and management. *AJR Am J Roentgenol.* **214**, 1280–1286. https://doi.org/10.2214/ajr.20.22954
41. Fang, Y. C., Zhang, H. Q., Xie, J. C., Lin, M. J., Ying, L. J., Pang, P. P., *et al.* (2020) Sensitivity of chest CT for COVID-19: Comparison to RT-PCR. *Radiology.* **296**, E115–E117. https://doi.org/10.1148/radiol.2020200432
42. Li, X. W., Geng, M. M., Peng, Y. Z., Meng, L. S., and Lu, S. M. (2020) Molecular immune pathogenesis and diagnosis of COVID-19. *J Pharm Anal.* **10**, 102–108. https://doi.org/10.1016/j.jpha.2020.03.001
43. Montenegro, F., Unigarro, L., Paredes, G., Moya, T., Romero, A., Torres, L., *et al.* (2021) Acute respiratory distress syndrome (ARDS) caused by the novel coronavirus disease (COVID-19): A practical comprehensive literature review. *Expert Rev Respir Med.* **15**, 183–195. https://doi.org/10.1080/17476348.2020.1820329
44. Zhang, R., Ouyang, H., Fu, L., Wang, S., Han, J., Huang, K., *et al.* (2020) CT features of SARS-CoV-2 pneumonia according to clinical presentation: A retrospective analysis of 120 consecutive patients from Wuhan city. *Eur Radiol.* **30**, 4417–4426. https://doi.org/10.1007/s00330-020-06854-1
45. Phua, J., Weng, L., Ling, L., Egi, M., Lim, C.-M., Divatia, J. V., *et al.* (2020) Intensive care management of coronavirus disease 2019 (COVID-19): Challenges and recommendations. *Lancet Respir Med.* **8**, 506–517. https://doi.org/10.1016/S2213-2600(20)30161-2
46. Shi, H., Han, X., Jiang, N., Cao, Y., Alwalid, O., Gu, J., *et al.* (2020) Radiological findings from 81 patients with COVID-19 pneumonia in Wuhan, China: A descriptive study. *Lancet Infect Dis.* **20**, 425–434. https://doi.org/10.1016/S1473-3099(20)30086-4
47. Tusman, G., Acosta, C. M., and Costantini, M. (2016) Ultrasonography for the assessment of lung recruitment maneuvers. *Crit Ultrasound J.* **8**, 8. https://doi.org/10.1186/s13089-016-0045-9
48. Yang, X., Yu, Y., Xu, J., Shu, H., Xia, J., Liu, H., *et al.* (2020) Clinical course and outcomes of critically ill patients with SARS-CoV-2 pneumonia in Wuhan, China: A single-centered, retrospective, observational study. *Lancet Respir Med.* **8**, 475–481. https://doi.org/10.1016/s2213-2600(20)30079-5

49. Song, Y. G., and Shin, H.-S. (2020) COVID-19, a clinical syndrome manifesting as hypersensitivity pneumonitis. *Infect Chemother*. **52**, 110–112. https://doi.org/10.3947/ic.2020.52.1.110. www.ncbi.nlm.nih.gov/pmc/articles/PMC7113449/pdf/ic-52-110.pdf
50. Chen, J., Lau, Y. F., Lamirande, E. W., Paddock, C. D., Bartlett, J. H., Zaki, S. R., *et al.* (2010) Cellular immune responses to severe acute respiratory syndrome coronavirus (SARS-CoV) infection in senescent BALB/c mice: $CD4^+$ T cells are important in control of SARS-CoV infection. *J Virol*. **84**, 1289–1301. https://doi.org/10.1128/JVI.01281-09
51. Baud, D., Qi, X., Nielsen-Saines, K., Musso, D., Pomar, L., and Favre, G. (2020) Real estimates of mortality following COVID-19 infection. *Lancet Infect Dis*. **20**, 773. https://doi.org/10.1016/s1473-3099(20)30195-x
52. Gattinoni, L., Coppola, S., Cressoni, M., Busana, M., Rossi, S., and Chiumello, D. (2020) COVID-19 does not lead to a "typical" acute respiratory distress syndrome. *Am J Respir Crit Care Med*. **201**, 1299–1300. https://doi.org/10.1164/rccm.202003-0817LE
53. Zhou, F., Yu, T., Du, R., Fan, G., Liu, Y., Liu, Z., *et al.* (2020) Clinical course and risk factors for mortality of adult inpatients with COVID-19 in Wuhan, China: A retrospective cohort study. *Lancet*. **395**, 1054–1062. https://doi.org/10.1016/S0140-6736(20)30566-3
54. Denson, J. L., Gillet, A. S., Zu, Y., Brown, M., Pham, T., Yoshida, Y., *et al.* (2021) Metabolic syndrome and acute respiratory distress syndrome in hospitalized patients with COVID-19. *JAMA Netw Open*. **4**, e2140568–e2140568. https://doi.org/10.1001/jamanetworkopen.2021.40568
55. Domecq, J. P., Lal, A., Sheldrick, C. R., Kumar, V. K., Boman, K., Bolesta, S., *et al.* (2021) Outcomes of patients with coronavirus disease 2019 receiving organ support therapies: The international viral infection and respiratory illness universal study registry. *Crit Care Med*. **49**, 437–448. https://doi.org/10.1097/ccm.0000000000004879
56. Singer, M., Deutschman, C. S., Seymour, C. W., Shankar-Hari, M., Annane, D., Bauer, M., *et al.* (2016) The third international consensus definitions for sepsis and septic shock (Sepsis-3). *Jama*. **315**, 801–810. https://doi.org/10.1001/jama.2016.0287
57. Karakike, E., Giamarellos-Bourboulis, E. J., Kyprianou, M., Fleischmann-Struzek, C., Pletz, M. W., Netea, M. G., *et al.* (2021) Coronavirus disease 2019 as cause of viral sepsis: A systematic review and meta-analysis. *Crit Care Med*. **49**, 2042–2057. https://doi.org/10.1097/ccm.0000000000005195
58. Koçak Tufan, Z., Kayaaslan, B., and Mer, M. (2021) COVID-19 and sepsis. *Turk J Med Sci*. **51**, 3301–3311. https://doi.org/10.3906/sag-2108-239
59. Patil, M., Singh, S., Henderson, J., and Krishnamurthy, P. (2021) Mechanisms of COVID-19-induced cardiovascular disease: Is sepsis or exosome the missing link? *J Cell Physiol*. **236**, 3366–3382. https://doi.org/10.1002/jcp.30109
60. Klok, F. A., Kruip, M. J. H. A., van der Meer, N. J. M., Arbous, M. S., Gommers, D., Kant, K. M., *et al.* (2020) Confirmation of the high cumulative incidence of thrombotic complications in critically ill ICU patients with COVID-19: An updated analysis. *Thromb Res*. **191**, 148–150 https://doi.org/10.1016/j.thromres.2020.04.041
61. Thomas, M. R., and Scully, M. (2022) Clinical features of thrombosis and bleeding in COVID-19. *Blood*. **140**, 184–195. https://doi.org/10.1182/blood.2021012247
62. Becker, R. C., and Ortel, T. L. (2023) Anticoagulation in patients with COVID-19 pneumonia—what is the optimal intensity? *JAMA Int Med*. https://doi.org/10.1001/jamainternmed.2023.0625
63. Iba, T., Levy, J. H., Connors, J. M., Warkentin, T. E., Thachil, J., and Levi, M. (2020) The unique characteristics of COVID-19 coagulopathy. *Crit Care*. **24**, 360. https://doi.org/10.1186/s13054-020-03077-0
64. Helms, J., Tacquard, C., Severac, F., Leonard-Lorant, I., Ohana, M., Delabranche, X., *et al.* (2020) High risk of thrombosis in patients with severe SARS-CoV-2 infection: A multicenter prospective cohort study. *Intensive Care Med*. **46**, 1089–1098. https://doi.org/10.1007/s00134-020-06062-x
65. Dorgalaleh, A. (2020) Bleeding and bleeding risk in COVID-19. *Semin Thromb Hemost*. **46**, 815–818. https://doi.org/10.1055/s-0040-1713434
66. Buja, L. M., Wolf, D. A., Zhao, B., Akkanti, B., McDonald, M., Lelenwa, L., *et al.* (2020) The emerging spectrum of cardiopulmonary pathology of the coronavirus disease 2019 (COVID-19): Report of 3 autopsies from Houston, Texas, and review of autopsy findings from other United States cities. *Cardiovasc Pathol*. **48**, 107233. https://doi.org/10.1016/j.carpath.2020.107233
67. Ackermann, M., Verleden, S. E., Kuehnel, M., Haverich, A., Welte, T., Laenger, F., *et al.* (2020) Pulmonary vascular endothelialitis, thrombosis, and angiogenesis in Covid-19. *N Engl J Med*. **383**, 120–128. https://doi.org/10.1056/NEJMoa2015432

68. Moores, L. K., Tritschler, T., Brosnahan, S., Carrier, M., Collen, J. F., Doerschug, K., *et al.* (2020) Prevention, diagnosis, and treatment of VTE in patients with coronavirus disease 2019: CHEST guideline and expert panel report. *Chest.* **158**, 1143–1163. https://doi.org/10.1016/j.chest.2020.05.559
69. Bikdeli, B., Madhavan, M. V., Jimenez, D., Chuich, T., Dreyfus, I., Driggin, E., *et al.* (2020) COVID-19 and thrombotic or thromboembolic disease: Implications for prevention, antithrombotic therapy, and follow-up: JACC state-of-the-art review. *J Am Coll Cardiol.* **75**, 2950–2973. https://doi.org/10.1016/j.jacc.2020.04.031
70. Hirsh, J., Bauer, K. A., Donati, M. B., Gould, M., Samama, M. M., and Weitz, J. I. (2008) Parenteral anticoagulants: American college of chest physicians evidence-based clinical practice guidelines (8th edition). *Chest.* **133**, 141S–159S. https://doi.org/10.1378/chest.08-0689
71. Labbé, V., Contou, D., Heming, N., Megarbane, B., Razazi, K., Boissier, F., *et al.* (2023) Effects of standard-dose prophylactic, high-dose prophylactic, and therapeutic anticoagulation in patients with hypoxemic COVID-19 pneumonia: The ANTICOVID randomized clinical trial. *JAMA Int Med.* https://doi.org/10.1001/jamainternmed.2023.0456
72. Puntmann, V. O., Carerj, M. L., Wieters, I., Fahim, M., Arendt, C., Hoffmann, J., *et al.* (2020) Outcomes of cardiovascular magnetic resonance imaging in patients recently recovered from coronavirus disease 2019 (COVID-19). *JAMA Cardiol.* **5**, 1265–1273. https://doi.org/10.1001/jamacardio.2020.3557
73. Lala, A., Johnson, K. W., Januzzi, J. L., Russak, A. J., Paranjpe, I., Richter, F., *et al.* (2020) Prevalence and impact of myocardial injury in patients hospitalized with COVID-19 infection. *J Am Coll Cardiol.* **76**, 533–546. https://doi.org/10.1016/j.jacc.2020.06.007
74. Werion, A., Belkhir, L., Perrot, M., Schmit, G., Aydin, S., Chen, Z., *et al.* (2020) SARS-CoV-2 causes a specific dysfunction of the kidney proximal tubule. *Kidney Int.* **98**, 1296–1307. https://doi.org/10.1016/j.kint.2020.07.019
75. Ramos-Casals, M., Brito-Zerón, P., and Mariette, X. (2021) Systemic and organ-specific immune-related manifestations of COVID-19. *Nat Rev Rheumatol.* **17**, 315–332. https://doi.org/10.1038/s41584-021-00608-z
76. Pirzadeh, M., and Prescott, H. C. (2023) Impact of COVID-19 on nonpulmonary critical illness: Prevalence, clinical manifestations, management, and outcomes. *Clin Chest Med.* **44**, 249–262. https://doi.org/10.1016/j.ccm.2022.11.011
77. Dhamoon, M. S., Thaler, A., Gururangan, K., Kohli, A., Sisniega, D., Wheelwright, D., *et al.* (2021) Acute cerebrovascular events with COVID-19 infection. *Stroke.* **52**, 48–56. https://doi.org/10.1161/strokeaha.120.031668
78. Lania, A., Sandri, M. T., Cellini, M., Mirani, M., Lavezzi, E., and Mazziotti, G. (2020) Thyrotoxicosis in patients with COVID-19: The THYRCOV study. *Eur J Endocrinol.* **183**, 381–387. https://doi.org/10.1530/eje-20-0335
79. Xie, Y., and Al-Aly, Z. (2022) Risks and burdens of incident diabetes in long COVID: A cohort study. *Lancet Diabetes Endocrinol.* **10**, 311–321. https://doi.org/10.1016/s2213-8587(22)00044-4
80. Kenny, G., Townsend, L., Savinelli, S., and Mallon, P. W. G. (2023) Long COVID: Clinical characteristics, proposed pathogenesis and potential therapeutic targets. *Front Mol Biosci.* **10**, 1157651. https://doi.org/10.3389/fmolb.2023.1157651
81. Merad, M., Blish, C. A., Sallusto, F., and Iwasaki, A. (2022) The immunology and immunopathology of COVID-19. *Science.* **375**, 1122–1127. https://doi.org/10.1126/science.abm8108
82. Bliddal, S., Banasik, K., Pedersen, O. B., Nissen, J., Cantwell, L., Schwinn, M., *et al.* (2021) Acute and persistent symptoms in non-hospitalized PCR-confirmed COVID-19 patients. *Sci Rep.* **11**, 13153. https://doi.org/10.1038/s41598-021-92045-x
83. Massey, D., Berrent, D., Akrami, A., Assaf, G., Davis, H., Harris, K., *et al.* (2021) Change in symptoms and immune response in people with post-acute sequelae of SARS-Cov-2 infection (PASC) after SARS-Cov-2 vaccination. *medRxiv.* 2021.2007.2021.21260391. https://doi.org/10.1101/2021.07.21.21260391
84. Stein, S. R., Ramelli, S. C., Grazioli, A., Chung, J.-Y., Singh, M., Yinda, C. K., *et al.* (2022) SARS-CoV-2 infection and persistence in the human body and brain at autopsy. *Nature.* **612**, 758–763. https://doi.org/10.1038/s41586-022-05542-y
85. Peluso, M. J., Lu, S., Tang, A. F., Durstenfeld, M. S., Ho, H.-E., Goldberg, S. A., *et al.* (2021) Markers of immune activation and inflammation in individuals with post-acute sequelae of SARS-CoV-2 infection. *medRxiv.* 2021.2007.2009.21260287. https://doi.org/10.1101/2021.07.09.21260287
86. Diamond, M. S., and Kanneganti, T.-D. (2022) Innate immunity: The first line of defense against SARS-CoV-2. *Nat Immunol.* **23**, 165–176. https://doi.org/10.1038/s41590-021-01091-0

87. Akira, S., Uematsu, S., and Takeuchi, O. (2006) Pathogen recognition and innate immunity. *Cell*. **124**, 783–801. https://doi.org/10.1016/j.cell.2006.02.015
88. Herwald, H., and Egesten, A. (2016) On PAMPs and DAMPs. *J Innate Immun*. **8**, 427–428. https://doi.org/10.1159/000448437
89. Brubaker, S. W., Bonham, K. S., Zanoni, I., and Kagan, J. C. (2015) Innate immune pattern recognition: A cell biological perspective. *Ann Rev Immunol*. **33**, 257–290. https://doi.org/10.1146/annurev-immunol-032414-112240
90. Golebski, K., Hoepel, W., van Egmond, D., de Groot, E. J., Amatngalim, G. D., Beekman, J. M., *et al.* (2019) FcγRIII stimulation breaks the tolerance of human nasal epithelial cells to bacteria through cross-talk with TLR4. *Mucosal Immunol*. **12**, 425–433. https://doi.org/10.1038/s41385-018-0129-x
91. van Tongeren, J., Röschmann, K. I. L., Reinartz, S. M., Luiten, S., Fokkens, W. J., de Jong, E. C., *et al.* (2015) Expression profiling and functional analysis of toll-like receptors in primary healthy human nasal epithelial cells shows no correlation and a refractory LPS response. *Clin Transl Allergy*. **5**, 42. https://doi.org/10.1186/s13601-015-0086-3
92. Yamamoto, K., Ferrari, J. D., Cao, Y., Ramirez, M. I., Jones, M. R., Quinton, L. J., *et al.* (2012) Type I alveolar epithelial cells mount innate immune responses during pneumococcal pneumonia. *J Immunol*. **189**, 2450–2459. https://doi.org/10.4049/jimmunol.1200634
93. Gasteiger, G., Osualdo, A., Schubert, D. A., Weber, A., Bruscia, E. M., *et al.* (2016) Cellular innate immunity: An old game with new players. *J Innate Immun*. **9**, 111–125. https://doi.org/10.1159/000453397
94. Kanneganti, T. D. (2020) Intracellular innate immune receptors: Life inside the cell. *Immunol Rev*. **297**, 5–12. https://doi.org/10.1111/imr.12912
95. Janeway, C. A., Jr. (1989) Approaching the asymptote? Evolution and revolution in immunology. *Cold Spring Harb Symp Quant Biol*. **54**(Pt 1), 1–13. https://doi.org/10.1101/sqb.1989.054.01.003
96. Sonnenberg, G. F., and Hepworth, M. R. (2019) Functional interactions between innate lymphoid cells and adaptive immunity. *Nat Rev Immunol*. **19**, 599–613. https://doi.org/10.1038/s41577-019-0194-8
97. V'Kovski, P., Kratzel, A., Steiner, S., Stalder, H., and Thiel, V. (2021) Coronavirus biology and replication: Implications for SARS-CoV-2. *Nat Rev Microbiol*. **19**, 155–170. https://doi.org/10.1038/s41579-020-00468-6
98. Zhou, P., Yang, X. L., Wang, X. G., Hu, B., Zhang, L., Zhang, W., *et al.* (2020) A pneumonia outbreak associated with a new coronavirus of probable bat origin. *Nature*. **579**, 270–273. https://doi.org/10.1038/s41586-020-2012-7
99. Hoffmann, M., Kleine-Weber, H., Schroeder, S., Krüger, N., Herrler, T., Erichsen, S., *et al.* (2020) SARS-CoV-2 cell entry depends on ACE2 and TMPRSS2 and is blocked by a clinically proven protease inhibitor. *Cell*. **181**, 271–280.e278. https://doi.org/10.1016/j.cell.2020.02.052
100. Madden, E. A., and Diamond, M. S. (2022) Host cell-intrinsic innate immune recognition of SARS-CoV-2. *Curr Opin Virol*. **52**, 30–38. https://doi.org/10.1016/j.coviro.2021.11.002
101. Zhu, N., Wang, W., Liu, Z., Liang, C., Wang, W., Ye, F., *et al.* (2020) Morphogenesis and cytopathic effect of SARS-CoV-2 infection in human airway epithelial cells. *Nat Commun*. **11**, 3910. https://doi.org/10.1038/s41467-020-17796-z
102. Sefik, E., Qu, R., Junqueira, C., Kaffe, E., Mirza, H., Zhao, J., *et al.* (2022) Inflammasome activation in infected macrophages drives COVID-19 pathology. *Nature*. **606**, 585–593. https://doi.org/10.1038/s41586-022-04802-1
103. Li, D., and Wu, M. (2021) Pattern recognition receptors in health and diseases. *Signal Transduct Target Ther*. **6**, 291. https://doi.org/10.1038/s41392-021-00687-0
104. Kasuga, Y., Zhu, B., Jang, K. J., and Yoo, J. S. (2021) Innate immune sensing of coronavirus and viral evasion strategies. *Exp Mol Med*. **53**, 723–736. https://doi.org/10.1038/s12276-021-00602-1
105. Mdkhana, B., Saheb Sharif-Askari, N., Ramakrishnan, R. K., Goel, S., Hamid, Q., and Halwani, R. (2021) Nucleic acid-sensing pathways during SARS-CoV-2 infection: Expectations versus reality. *J Inflamm Res*. **14**, 199–216. https://doi.org/10.2147/jir.S277716
106. Liu, G., and Zhao, Y. (2007) Toll-like receptors and immune regulation: Their direct and indirect modulation on regulatory CD4+ CD25+ T cells. *Immunology*. **122**, 149–156. https://doi.org/10.1111/j.1365-2567.2007.02651.x
107. Carty, M., Guy, C., and Bowie, A. G. (2021) Detection of viral infections by innate immunity. *Biochem Pharmacol*. **183**, 114316 https://doi.org/10.1016/j.bcp.2020.114316
108. Amor, S., Fernández Blanco, L., and Baker, D. (2020) Innate immunity during SARS-CoV-2: Evasion strategies and activation trigger hypoxia and vascular damage. *Clin Exp Immunol*. **202**, 193–209. https://doi.org/10.1111/cei.13523

109. Liu, Q., Chi, S., Dmytruk, K., Dmytruk, O., and Tan, S. (2022) Coronaviral infection and interferon response: The virus-host arms race and COVID-19. *Viruses*. **14**, 1349. www.mdpi.com/1999-4915/14/7/1349
110. Akira, S., and Takeda, K. (2004) Toll-like receptor signalling. *Nat Rev Immunol*. **4**, 499–511. https://doi.org/10.1038/nri1391
111. Zheng, M., Karki, R., Williams, E. P., Yang, D., Fitzpatrick, E., Vogel, P., *et al.* (2021) TLR2 senses the SARS-CoV-2 envelope protein to produce inflammatory cytokines. *Nat Immunol*. **22**, 829–838. https://doi.org/10.1038/s41590-021-00937-x
112. Hadjadj, J., Yatim, N., Barnabei, L., Corneau, A., Boussier, J., Smith, N., *et al.* (2020) Impaired type I interferon activity and inflammatory responses in severe COVID-19 patients. *Science*. **369**, 718-+. https://doi.org/10.1126/science.abc6027
113. Lee, S., Channappanavar, R., and Kanneganti, T. D. (2020) Coronaviruses: Innate immunity, inflammasome activation, inflammatory cell death, and cytokines. *Trends Immunol*. **41**, 1083–1099. https://doi.org/10.1016/j.it.2020.10.005
114. Zhao, J., Wohlford-Lenane, C., Zhao, J., Fleming, E., Lane, T. E., McCray, P. B., *et al.* (2012) Intranasal treatment with poly(I·C) protects aged mice from lethal respiratory virus infections. *J Virol*. **86**, 11416–11424. https://doi.org/10.1128/JVI.01410-12
115. Totura, A. L., Whitmore, A., Agnihothram, S., Schäfer, A., Katze, M. G., Heise, M. T., *et al.* (2015) Toll-like receptor 3 signaling via TRIF contributes to a protective innate immune response to severe acute respiratory syndrome coronavirus infection. *mBio*. **6**, e00638–00615. https://doi.org/10.1128/mBio.00638-15
116. Menezes, M. C. S., Veiga, A. D. M., Martins de Lima, T., Kunimi Kubo Ariga, S., Vieira Barbeiro, H., de Lucena Moreira, C., *et al.* (2021) Lower peripheral blood Toll-like receptor 3 expression is associated with an unfavorable outcome in severe COVID-19 patients. *Sci Rep*. **11**, 15223. https://doi.org/10.1038/s41598-021-94624-4
117. Potapov, I., Kanneganti, T. D., and Del Sol, A. (2022) Fostering experimental and computational synergy to modulate hyperinflammation. *Trends Immunol*. **43**, 4–7. https://doi.org/10.1016/j.it.2021.11.004
118. Loo, Y. M., and Gale, M., Jr. (2011) Immune signaling by RIG-I-like receptors. *Immunity*. **34**, 680–692. https://doi.org/10.1016/j.immuni.2011.05.003
119. Yin, X., Riva, L., Pu, Y., Martin-Sancho, L., Kanamune, J., Yamamoto, Y., *et al.* (2021) MDA5 governs the innate immune response to SARS-CoV-2 in lung epithelial cells. *Cell Rep*. **34**, 108628. https://doi.org/10.1016/j.celrep.2020.108628
120. Rebendenne, A., Valadão, A. L. C., Tauziet, M., Maarifi, G., Bonaventure, B., McKellar, J., *et al.* (2021) SARS-CoV-2 triggers an MDA-5-dependent interferon response which is unable to control replication in lung epithelial cells. *J Virol*. **95**. https://doi.org/10.1128/jvi.02415-20
121. Thorne, L. G., Reuschl, A. K., Zuliani-Alvarez, L., Whelan, M. V. X., Turner, J., Noursadeghi, M., *et al.* (2021) SARS-CoV-2 sensing by RIG-I and MDA5 links epithelial infection to macrophage inflammation. *EMBO J*. **40**, e107826. https://doi.org/10.15252/embj.2021107826
122. Zheng, M., Williams, E. P., Malireddi, R. K. S., Karki, R., Banoth, B., Burton, A., *et al.* (2020) Impaired NLRP3 inflammasome activation/pyroptosis leads to robust inflammatory cell death via caspase-8/RIPK3 during coronavirus infection. *J Biol Chem*. **295**, 14040–14052. https://doi.org/10.1074/jbc.RA120.015036
123. Siu, K. L., Yuen, K. S., Castaño-Rodriguez, C., Ye, Z. W., Yeung, M. L., Fung, S. Y., *et al.* (2019) Severe acute respiratory syndrome coronavirus ORF3a protein activates the NLRP3 inflammasome by promoting TRAF3-dependent ubiquitination of ASC. *FASEB J*. **33**, 8865–8877. https://doi.org/10.1096/fj.201802418R
124. Shi, C. S., Nabar, N. R., Huang, N. N., and Kehrl, J. H. (2019) SARS-coronavirus open reading frame-8b triggers intracellular stress pathways and activates NLRP3 inflammasomes. *Cell Death Discov*. **5**, 101. https://doi.org/10.1038/s41420-019-0181-7
125. Nieto-Torres, J. L., Verdiá-Báguena, C., Jimenez-Guardeño, J. M., Regla-Nava, J. A., Castaño-Rodriguez, C., Fernandez-Delgado, R., *et al.* (2015) Severe acute respiratory syndrome coronavirus E protein transports calcium ions and activates the NLRP3 inflammasome. *Virology*. **485**, 330–339. https://doi.org/10.1016/j.virol.2015.08.010
126. Campbell, G. R., To, R. K., Hanna, J., and Spector, S. A. (2021) SARS-CoV-2, SARS-CoV-1, and HIV-1 derived ssRNA sequences activate the NLRP3 inflammasome in human macrophages through a non-classical pathway. *iScience*. **24**, 102295. https://doi.org/10.1016/j.isci.2021.102295

127. Christgen, S., and Kanneganti, T. D. (2020) Inflammasomes and the fine line between defense and disease. *Curr Opin Immunol.* **62**, 39–44. https://doi.org/10.1016/j.coi.2019.11.007
128. Kayagaki, N., Kornfeld, O. S., Lee, B. L., Stowe, I. B., O'Rourke, K., Li, Q., *et al.* (2021) NINJ1 mediates plasma membrane rupture during lytic cell death. *Nature.* **591**, 131–136. https://doi.org/10.1038/s41586-021-03218-7
129. Qin, C., Zhou, L., Hu, Z., Zhang, S., Yang, S., Tao, Y., *et al.* (2020) Dysregulation of immune response in patients with coronavirus 2019 (COVID-19) in Wuhan, China. *Clin Infect Dis.* **71**, 762–768. https://doi.org/10.1093/cid/ciaa248
130. Laing, A. G., Lorenc, A., Del Molino Del Barrio, I., Das, A., Fish, M., Monin, L., *et al.* (2020) A dynamic COVID-19 immune signature includes associations with poor prognosis. *Nat Med.* **26**, 1623–1635. https://doi.org/10.1038/s41591-020-1038-6
131. Schultze, J. L., and Aschenbrenner, A. C. (2021) COVID-19 and the human innate immune system. *Cell.* **184**, 1671–1692. https://doi.org/10.1016/j.cell.2021.02.029
132. Lucas, C., Wong, P., Klein, J., Castro, T. B. R., Silva, J., Sundaram, M., *et al.* (2020) Longitudinal analyses reveal immunological misfiring in severe COVID-19. *Nature.* **584**, 463-+. https://doi.org/10.1038/s41586-020-2588-y
133. Karki, R., and Kanneganti, T. D. (2021) The "cytokine storm": Molecular mechanisms and therapeutic prospects. *Trends Immunol.* **42**, 681–705. https://doi.org/10.1016/j.it.2021.06.001
134. Ferreira, A. C., Soares, V. C., de Azevedo-Quintanilha, I. G., Dias, S., Fintelman-Rodrigues, N., Sacramento, C. Q., *et al.* (2021) SARS-CoV-2 engages inflammasome and pyroptosis in human primary monocytes. *Cell Death Discov.* **7**, 43. https://doi.org/10.1038/s41420-021-00428-w
135. Yang, Y., Wu, Y., Meng, X., Wang, Z., Younis, M., Liu, Y., *et al.* (2022) SARS-CoV-2 membrane protein causes the mitochondrial apoptosis and pulmonary edema via targeting BOK. *Cell Death Differ.* **29**, 1395–1408. https://doi.org/10.1038/s41418-022-00928-x
136. Li, F., Li, J., Wang, P. H., Yang, N., Huang, J., Ou, J., *et al.* (2021) SARS-CoV-2 spike promotes inflammation and apoptosis through autophagy by ROS-suppressed PI3K/AKT/mTOR signaling. *Biochim Biophys Acta Mol Basis Dis.* **1867**, 166260. https://doi.org/10.1016/j.bbadis.2021.166260
137. Santos, L. D., Antunes, K. H., Cassão, G., Gonçalves, J. I., Abbadi, B. L., Bizarro, C. V., *et al.* (2023) SARS-CoV-2 immune complex triggers human monocyte necroptosis. *Int Immunopharmacol.* **117**, 109954. https://doi.org/10.1016/j.intimp.2023.109954
138. Bednash, J. S., Kagan, V. E., Englert, J. A., Farkas, D., Tyurina, Y. Y., Tyurin, V. A., *et al.* (2022) Syrian hamsters as a model of lung injury with SARS-CoV-2 infection: Pathologic, physiologic, and detailed molecular profiling. *Transl Res.* **240**, 1–16. https://doi.org/10.1016/j.trsl.2021.10.007
139. Karki, R., Sharma, B. R., Tuladhar, S., Williams, E. P., Zalduondo, L., Samir, P., *et al.* (2021) Synergism of TNF-α and IFN-γ triggers inflammatory cell death, tissue damage, and mortality in SARS-CoV-2 infection and cytokine shock syndromes. *Cell.* **184**, 149–168.e117. https://doi.org/10.1016/j.cell.2020.11.025
140. Malireddi, R. K. S., Karki, R., Sundaram, B., Kancharana, B., Lee, S., Samir, P., *et al.* (2021) Inflammatory cell death, PANoptosis, mediated by cytokines in diverse cancer lineages inhibits tumor growth. *Immunohorizons.* **5**, 568–580. https://doi.org/10.4049/immunohorizons.2100059
141. Fajgenbaum, D. C., and June, C. H. (2020) Cytokine storm. *N Engl J Med.* **383**, 2255–2273. https://doi.org/10.1056/NEJMra2026131
142. Kaneko, N., Kuo, H. H., Boucau, J., Farmer, J. R., Allard-Chamard, H., Mahajan, V. S., *et al.* (2020) Loss of Bcl-6-expressing T follicular helper cells and germinal centers in COVID-19. *Cell.* **183**, 143–157.e113. https://doi.org/10.1016/j.cell.2020.08.025
143. Bhattacharjee, S., and Banerjee, M. (2020) Immune thrombocytopenia secondary to COVID-19: A systematic review. *SN Compr Clin Med.* **2**, 2048–2058. https://doi.org/10.1007/s42399-020-00521-8
144. Zohar, T., Loos, C., Fischinger, S., Atyeo, C., Wang, C., Slein, M. D., *et al.* (2020) Compromised humoral functional evolution tracks with SARS-CoV-2 mortality. *Cell.* **183**, 1508–1519.e1512. https://doi.org/10.1016/j.cell.2020.10.052
145. Blanco-Melo, D., Nilsson-Payant, B. E., Liu, W. C., Uhl, S., Hoagland, D., Moller, R., *et al.* (2020) Imbalanced host response to SARS-CoV-2 drives development of COVID-19. *Cell.* **181**, 1036-+. https://doi.org/10.1016/j.cell.2020.04.026
146. Galani, I. E., Rovina, N., Lampropoulou, V., Triantafyllia, V., Manioudaki, M., Pavlos, E., *et al.* (2021) Untuned antiviral immunity in COVID-19 revealed by temporal type I/III interferon patterns and flu comparison. *Nat Immunol.* **22**, 32–40. https://doi.org/10.1038/s41590-020-00840-x

147. Xia, H., Cao, Z., Xie, X., Zhang, X., Chen, J. Y., Wang, H., *et al.* (2020) Evasion of type I interferon by SARS-CoV-2. *Cell Rep.* **33**, 108234. https://doi.org/10.1016/j.celrep.2020.108234
148. Lei, X., Dong, X., Ma, R., Wang, W., Xiao, X., Tian, Z., *et al.* (2020) Activation and evasion of type I interferon responses by SARS-CoV-2. *Nat Commun.* **11**, 3810. https://doi.org/10.1038/s41467-020-17665-9
149. Vazquez, C., Swanson, S. E., Negatu, S. G., Dittmar, M., Miller, J., Ramage, H. R., *et al.* (2021) SARS-CoV-2 viral proteins NSP1 and NSP13 inhibit interferon activation through distinct mechanisms. *PLoS ONE.* **16**, e0253089. https://doi.org/10.1371/journal.pone.0253089
150. Hsu, J. C., Laurent-Rolle, M., Pawlak, J. B., Wilen, C. B., and Cresswell, P. (2021) Translational shutdown and evasion of the innate immune response by SARS-CoV-2 NSP14 protein. *Proc Natl Acad Sci U S A.* **118**. https://doi.org/10.1073/pnas.2101161118
151. Munnur, D., Teo, Q., Eggermont, D., Lee, H. H. Y., Thery, F., Ho, J., *et al.* (2021) Altered ISGylation drives aberrant macrophage-dependent immune responses during SARS-CoV-2 infection. *Nat Immunol.* **22**, 1416–1427. https://doi.org/10.1038/s41590-021-01035-8
152. Shin, D., Mukherjee, R., Grewe, D., Bojkova, D., Baek, K., Bhattacharya, A., *et al.* (2020) Papain-like protease regulates SARS-CoV-2 viral spread and innate immunity. *Nature.* **587**, 657–662. https://doi.org/10.1038/s41586-020-2601-5
153. Wang, W., Zhou, Z., Xiao, X., Tian, Z., Dong, X., Wang, C., *et al.* (2021) SARS-CoV-2 nsp12 attenuates type I interferon production by inhibiting IRF3 nuclear translocation. *Cell Mol Immunol.* **18**, 945–953. https://doi.org/10.1038/s41423-020-00619-y
154. Kimura, I., Konno, Y., Uriu, K., Hopfensperger, K., Sauter, D., Nakagawa, S., *et al.* (2021) Sarbecovirus ORF6 proteins hamper induction of interferon signaling. *Cell Rep.* **34**, 108916. https://doi.org/10.1016/j.celrep.2021.108916
155. Wang, R., Yang, X., Chang, M., Xue, Z., Wang, W., Bai, L., *et al.* (2021) ORF3a protein of severe acute respiratory syndrome coronavirus 2 inhibits interferon-activated Janus kinase/signal transducer and activator of transcription signaling via elevating suppressor of cytokine signaling 1. *Front Microbiol.* **12**, 752597. https://doi.org/10.3389/fmicb.2021.752597
156. Hayn, M., Hirschenberger, M., Koepke, L., Nchioua, R., Straub, J. H., Klute, S., *et al.* (2021) Systematic functional analysis of SARS-CoV-2 proteins uncovers viral innate immune antagonists and remaining vulnerabilities. *Cell Rep.* **35**, 109126. https://doi.org/10.1016/j.celrep.2021.109126
157. Li, Q., Wang, Y., Sun, Q., Knopf, J., Herrmann, M., Lin, L., *et al.* (2022) Immune response in COVID-19: What is next? *Cell Death Differ.* **29**, 1107–1122. https://doi.org/10.1038/s41418-022-01015-x
158. Denning, D. W., Kilcoyne, A., and Ucer, C. (2020) Non-infectious status indicated by detectable IgG antibody to SARS-CoV-2. *Br Dent J.* **229**, 521–524. https://doi.org/10.1038/s41415-020-2228-9
159. Long, Q. X., Liu, B. Z., Deng, H. J., Wu, G. C., Deng, K., Chen, Y. K., *et al.* (2020) Antibody responses to SARS-CoV-2 in patients with COVID-19. *Nat Med.* **26**, 845–848. https://doi.org/10.1038/s41591-020-0897-1
160. Nakano, Y., Kurano, M., Morita, Y., Shimura, T., Yokoyama, R., Qian, C., *et al.* (2021) Time course of the sensitivity and specificity of anti-SARS-CoV-2 IgM and IgG antibodies for symptomatic COVID-19 in Japan. *Sci Rep.* **11**, 2776. https://doi.org/10.1038/s41598-021-82428-5
161. Moura, A. D., da Costa, H. H. M., Correa, V. A., de, S. L. A. K., Lindoso, J. A. L., De Gaspari, E., *et al.* (2021) Assessment of avidity related to IgG subclasses in SARS-CoV-2 Brazilian infected patients. *Sci Rep.* **11**, 17642. https://doi.org/10.1038/s41598-021-95045-z
162. Sterlin, D., Mathian, A., Miyara, M., Mohr, A., Anna, F., Claër, L., *et al.* (2020) IgA dominates the early neutralizing antibody response to SARS-CoV-2. *Sci Transl Med.* **13**, eabd2223. https://doi.org/10.1126/scitranslmed.abd2223
163. Ng, K. W., Faulkner, N., Cornish, G. H., Rosa, A., Harvey, R., Hussain, S., *et al.* (2020) Preexisting and de novo humoral immunity to SARS-CoV-2 in humans. *Science.* **370**, 1339-+. https://doi.org/10.1126/science.abe1107
164. Sette, A., and Crotty, S. (2020) Pre-existing immunity to SARS-CoV-2: The knowns and unknowns. *Nat Rev Immunol.* **20**, 457–458. https://doi.org/10.1038/s41577-020-0389-z
165. Zheng, H. Y., Zhang, M., Yang, C. X., Zhang, N., Wang, X. C., Yang, X. P., *et al.* (2020) Elevated exhaustion levels and reduced functional diversity of T cells in peripheral blood may predict severe progression in COVID-19 patients. *Cell Mol Immunol.* **17**, 541–543. https://doi.org/10.1038/s41423-020-0401-3
166. Rydyznski Moderbacher, C., Ramirez, S. I., Dan, J. M., Grifoni, A., Hastie, K. M., Weiskopf, D., *et al.* (2020) Antigen-specific adaptive immunity to SARS-CoV-2 in acute COVID-19 and associations with age and disease severity. *Cell.* **183**, 996–1012.e1019. https://doi.org/10.1016/j.cell.2020.09.038

167. Nielsen, S. C. A., Yang, F., Jackson, K. J. L., Hoh, R. A., Röltgen, K., Jean, G. H., *et al.* (2020) Human B cell clonal expansion and convergent antibody responses to SARS-CoV-2. *Cell Host Microbe.* **28**, 516–525.e515. https://doi.org/10.1016/j.chom.2020.09.002
168. Lucas, C., Klein, J., Sundaram, M. E., Liu, F., Wong, P., Silva, J., *et al.* (2021) Delayed production of neutralizing antibodies correlates with fatal COVID-19. *Nat Med.* **27**, 1178–1186. https://doi.org/10.1038/s41591-021-01355-0
169. Li, K., Huang, B., Wu, M., Zhong, A., Li, L., Cai, Y., *et al.* (2020) Dynamic changes in anti-SARS-CoV-2 antibodies during SARS-CoV-2 infection and recovery from COVID-19. *Nat Commun.* **11**, 6044. https://doi.org/10.1038/s41467-020-19943-y
170. Zhou, L., Niu, Z., Jiang, X., Zhang, Z., Zheng, Y., Wang, Z., *et al.* (2020) SARS-CoV-2 targets by the pscRNA profiling of ACE2, TMPRSS2 and furin proteases. *iScience.* **23**, 101744. https://doi.org/10.1016/j.isci.2020.101744
171. Notarbartolo, S., Ranzani, V., Bandera, A., Gruarin, P., Bevilacqua, V., Putignano, A. R., *et al.* (2021) Integrated longitudinal immunophenotypic, transcriptional and repertoire analyses delineate immune responses in COVID-19 patients. *Sci Immunol.* **6**. https://doi.org/10.1126/sciimmunol.abg5021
172. Bergamaschi, L., Mescia, F., Turner, L., Hanson, A. L., Kotagiri, P., Dunmore, B. J., *et al.* (2021) Longitudinal analysis reveals that delayed bystander CD8+ T cell activation and early immune pathology distinguish severe COVID-19 from mild disease. *Immunity.* **54**, 1257–1275.e1258. https://doi.org/10.1016/j.immuni.2021.05.010
173. Wiersinga, W. J., Rhodes, A., Cheng, A. C., Peacock, S. J., and Prescott, H. C. (2020) Pathophysiology, transmission, diagnosis, and treatment of coronavirus disease 2019 (COVID-19): A review. *JAMA.* **324**, 782–793. https://doi.org/10.1001/jama.2020.12839
174. André, S., Picard, M., Cezar, R., Roux-Dalvai, F., Alleaume-Butaux, A., Soundaramourty, C., *et al.* (2022) T cell apoptosis characterizes severe Covid-19 disease. *Cell Death Differ.* **29**, 1486–1499. https://doi.org/10.1038/s41418-022-00936-x
175. Georg, P., Astaburuaga-García, R., Bonaguro, L., Brumhard, S., Michalick, L., Lippert, L. J., *et al.* (2022) Complement activation induces excessive T cell cytotoxicity in severe COVID-19. *Cell.* **185**, 493–512.e425. https://doi.org/10.1016/j.cell.2021.12.040
176. Le Bert, N., Tan, A. T., Kunasegaran, K., Tham, C. Y. L., Hafezi, M., Chia, A., *et al.* (2020) SARS-CoV-2-specific T cell immunity in cases of COVID-19 and SARS, and uninfected controls. *Nature.* **584**, 457–462. https://doi.org/10.1038/s41586-020-2550-z
177. Izquierdo, J. L., Almonacid, C., González, Y., Del Rio-Bermudez, C., Ancochea, J., Cárdenas, R., *et al.* (2021) The impact of COVID-19 on patients with asthma. *Eur Respir J.* **57**. https://doi.org/10.1183/13993003.03142-2020
178. Sunjaya, A. P., Allida, S. M., Di Tanna, G. L., and Jenkins, C. R. (2022) Asthma and COVID-19 risk: A systematic review and meta-analysis. *Eur Respir J.* **59**. https://doi.org/10.1183/13993003.01209-2021
179. Calver, J. F., Fabbri, L., May, J., and Jenkins, R. G. (2023) COVID-19 in patients with chronic lung disease. *Clin Chest Med.* **44**, 385–393. https://doi.org/10.1016/j.ccm.2022.11.013
180. Lombardi, C., Gani, F., Berti, A., Comberiati, P., Peroni, D., and Cottini, M. (2021) Asthma and COVID-19: A dangerous liaison? *Asthma Res Pract.* **7**, 9. https://doi.org/10.1186/s40733-021-00075-z
181. Jackson, D. J., Busse, W. W., Bacharier, L. B., Kattan, M., O'Connor, G. T., Wood, R. A., *et al.* (2020) Association of respiratory allergy, asthma, and expression of the SARS-CoV-2 receptor ACE2. *J Allergy Clin Immunol.* **146**, 203–206.e203. https://doi.org/10.1016/j.jaci.2020.04.009
182. Kimura, H., Francisco, D., Conway, M., Martinez, F. D., Vercelli, D., Polverino, F., *et al.* (2020) Type 2 inflammation modulates ACE2 and TMPRSS2 in airway epithelial cells. *J Allergy Clin Immunol.* **146**, 80–88.e88. https://doi.org/10.1016/j.jaci.2020.05.004
183. Sajuthi, S. P., DeFord, P., Li, Y., Jackson, N. D., Montgomery, M. T., Everman, J. L., *et al.* (2020) Type 2 and interferon inflammation regulate SARS-CoV-2 entry factor expression in the airway epithelium. *Nat Commun.* **11**, 5139. https://doi.org/10.1038/s41467-020-18781-2
184. Contoli, M., Ito, K., Padovani, A., Poletti, D., Marku, B., Edwards, M. R., *et al.* (2015) Th2 cytokines impair innate immune responses to rhinovirus in respiratory epithelial cells. *Allergy.* **70**, 910–920. https://doi.org/10.1111/all.12627
185. Gill, M. A., Bajwa, G., George, T. A., Dong, C. C., Dougherty, II, Jiang, N., *et al.* (2010) Counterregulation between the FcepsilonRI pathway and antiviral responses in human plasmacytoid dendritic cells. *J Immunol.* **184**, 5999–6006. https://doi.org/10.4049/jimmunol.0901194

186. Camiolo, M., Gauthier, M., Kaminski, N., Ray, A., and Wenzel, S. E. (2020) Expression of SARS-CoV-2 receptor ACE2 and coincident host response signature varies by asthma inflammatory phenotype. *J Allergy Clin Immunol.* **146**, 315–324.e317. https://doi.org/10.1016/j.jaci.2020.05.051
187. Lacedonia, D., Scioscia, G., Santomasi, C., Fuso, P., Carpagnano, G. E., Portacci, A., *et al.* (2021) Impact of smoking, COPD and comorbidities on the mortality of COVID-19 patients. *Sci Rep.* **11**, 19251. https://doi.org/10.1038/s41598-021-98749-4
188. Zhao, Q., Meng, M., Kumar, R., Wu, Y., Huang, J., Lian, N., *et al.* (2020) The impact of COPD and smoking history on the severity of COVID-19: A systemic review and meta-analysis. *J Med Virol.* **92**, 1915–1921. https://doi.org/10.1002/jmv.25889
189. Hu, W., Dong, M., Xiong, M., Zhao, D., Zhao, Y., Wang, M., *et al.* (2020) Clinical courses and outcomes of patients with chronic obstructive pulmonary disease during the COVID-19 epidemic in Hubei, China. *Int J Chron Obstruct Pulmon Dis.* **15**, 2237–2248. https://doi.org/10.2147/copd.S265004
190. Thomas, M., Price, O. J., and Hull, J. H. (2021) Pulmonary function and COVID-19. *Curr Opin Physiol.* **21**, 29–35. https://doi.org/10.1016/j.cophys.2021.03.005
191. García-Valero, J., Olloquequi, J., Montes, J. F., Rodríguez, E., Martín-Satué, M., Texidó, L., *et al.* (2019) Deficient pulmonary IFN-β expression in COPD patients. *PLoS One.* **14**, e0217803. https://doi.org/10.1371/journal.pone.0217803
192. Guo-Parke, H., Linden, D., Weldon, S., Kidney, J. C., and Taggart, C. C. (2022) Deciphering respiratory-virus-associated interferon signaling in COPD airway epithelium. *Medicina (Kaunas).* **58**. https://doi.org/10.3390/medicina58010121
193. Hansbro, P. M., Haw, T. J., Starkey, M. R., and Miyake, K. (2017) Toll-like receptors in COPD. *Eur Respir J.* **49**. https://doi.org/10.1183/13993003.00739-2017
194. Ziegler, C. G. K., Allon, S. J., Nyquist, S. K., Mbano, I. M., Miao, V. N., Tzouanas, C. N., *et al.* (2020) SARS-CoV-2 receptor ACE2 Is an interferon-stimulated gene in human airway epithelial cells and is detected in specific cell subsets across tissues. *Cell.* **181**, 1016–1035.e1019. https://doi.org/10.1016/j.cell.2020.04.035
195. Choi, H., Lee, H., Lee, S. K., Yang, B., Chung, S. J., Yeo, Y., *et al.* (2021) Impact of bronchiectasis on susceptibility to and severity of COVID-19: A nationwide cohort study. *Ther Adv Respir Dis.* **15**, 1753466621995043. https://doi.org/10.1177/1753466621995043
196. Crichton, M. L., Shoemark, A., and Chalmers, J. D. (2021) The impact of the COVID-19 pandemic on exacerbations and symptoms in bronchiectasis: A prospective study. *Am J Respir Crit Care Med.* **204**, 857–859. https://doi.org/10.1164/rccm.202105-1137LE
197. Meganck, R. M., and Baric, R. S. (2021) Developing therapeutic approaches for twenty-first-century emerging infectious viral diseases. *Nat Med.* **27**, 401–410. https://doi.org/10.1038/s41591-021-01282-0
198. De Clercq, E., and Li, G. (2016) Approved antiviral drugs over the past 50 years. *Clin Microbiol Rev.* **29**, 695–747. https://doi.org/10.1128/cmr.00102-15
199. Li, G., and De Clercq, E. (2021) Overview of antiviral drug discovery and development: Viral versus host targets. In *Antiviral Discovery for Highly Pathogenic Emerging Viruses* (Muñoz-Fontela, C., and Delgado, R., eds.), The Royal Society of Chemistry, Cambridge
200. Li, G., Hilgenfeld, R., Whitley, R., and De Clercq, E. (2023) Therapeutic strategies for COVID-19: Progress and lessons learned. *Nat Rev Drug Discov.* **22**, 449–475. https://doi.org/10.1038/s41573-023-00672-y
201. Kaufmann, S. H. E., Dorhoi, A., Hotchkiss, R. S., and Bartenschlager, R. (2018) Host-directed therapies for bacterial and viral infections. *Nat Rev Drug Discov.* **17**, 35–56. https://doi.org/10.1038/nrd.2017.162
202. Baggen, J., Vanstreels, E., Jansen, S., and Daelemans, D. (2021) Cellular host factors for SARS-CoV-2 infection. *Nat Microbiol.* **6**, 1219–1232. https://doi.org/10.1038/s41564-021-00958-0
203. Zhao, P., Praissman, J. L., Grant, O. C., Cai, Y., Xiao, T., Rosenbalm, K. E., *et al.* (2020) Virus-receptor interactions of glycosylated SARS-CoV-2 spike and human ACE2 receptor. *Cell Host Microbe.* **28**, 586–601.e586. https://doi.org/10.1016/j.chom.2020.08.004
204. Cox, M., Peacock, T. P., Harvey, W. T., Hughes, J., Wright, D. W., Willett, B. J., *et al.* (2023) SARS-CoV-2 variant evasion of monoclonal antibodies based on in vitro studies. *Nat Rev Microbiol.* **21**, 112–124. https://doi.org/10.1038/s41579-022-00809-7
205. Liu, L., Iketani, S., Guo, Y., Chan, J. F., Wang, M., Liu, L., *et al.* (2022) Striking antibody evasion manifested by the Omicron variant of SARS-CoV-2. *Nature.* **602**, 676–681. https://doi.org/10.1038/s41586-021-04388-0

206. Cao, Y., Wang, J., Jian, F., Xiao, T., Song, W., Yisimayi, A., *et al.* (2022) Omicron escapes the majority of existing SARS-CoV-2 neutralizing antibodies. *Nature.* **602**, 657–663. https://doi.org/10.1038/s41586-021-04385-3
207. RECOVERY Collaborative Group. (2022) Casirivimab and imdevimab in patients admitted to hospital with COVID-19 (RECOVERY): A randomised, controlled, open-label, platform trial. *Lancet.* **399**, 665–676 https://doi.org/10.1016/S0140-6736(22)00163-5
208. ACTIV-3/TICO Bamlanivimab Study Group, Lundgren, J. D., Grund, B., Barkauskas, C. E., Holland, T. L., Gottlieb, R. L., *et al.* (2022) Responses to a neutralizing monoclonal antibody for hospitalized patients with COVID-19 according to baseline antibody and antigen levels. *Ann Intern Med.* **175**, 234–243. https://doi.org/10.7326/m21-3507%m34928698
209. Yuan, S., Yin, X., Meng, X., Chan, J. F.-W., Ye, Z.-W., Riva, L., *et al.* (2021) Clofazimine broadly inhibits coronaviruses including SARS-CoV-2. *Nature.* **593**, 418–423. https://doi.org/10.1038/s41586-021-03431-4
210. Tan, H., Hu, Y., Jadhav, P., Tan, B., and Wang, J. (2022) Progress and challenges in targeting the SARS-CoV-2 papain-like protease. *J Med Chem.* **65**, 7561–7580. https://doi.org/10.1021/acs.jmedchem.2c00303
211. Gao, K., Wang, R., Chen, J., Tepe, J. J., Huang, F., and Wei, G. W. (2021) Perspectives on SARS-CoV-2 main protease inhibitors. *J Med Chem.* **64**, 16922–16955. https://doi.org/10.1021/acs.jmedchem.1c00409
212. Noske, G. D., de Souza Silva, E., de Godoy, M. O., Dolci, I., Fernandes, R. S., Guido, R. V. C., *et al.* (2023) Structural basis of nirmatrelvir and ensitrelvir activity against naturally occurring polymorphisms of the SARS-CoV-2 main protease. *J Biol Chem.* **299**, 103004. https://doi.org/10.1016/j.jbc.2023.103004
213. Hammond, J., Leister-Tebbe, H., Gardner, A., Abreu, P., Bao, W., Wisemandle, W., *et al.* (2022) Oral nirmatrelvir for high-risk, nonhospitalized adults with Covid-19. *N Engl J Med.* **386**, 1397–1408. https://doi.org/10.1056/NEJMoa2118542
214. Walker, A. P., Fan, H., Keown, J. R., Knight, M. L., Grimes, J. M., and Fodor, E. (2021) The SARS-CoV-2 RNA polymerase is a viral RNA capping enzyme. *Nucleic Acids Res.* **49**, 13019–13030. https://doi.org/10.1093/nar/gkab1160
215. Li, G., Wang, Y., and De Clercq, E. (2022) Approved HIV reverse transcriptase inhibitors in the past decade. *Acta Pharm Sin B.* **12**, 1567–1590. https://doi.org/10.1016/j.apsb.2021.11.009
216. Aleissa, M. M., Silverman, E. A., Paredes Acosta, L. M., Nutt, C. T., Richterman, A., and Marty, F. M. (2020) New perspectives on antimicrobial agents: Remdesivir treatment for COVID-19. *Antimicrob Agents Chemother.* **65**. https://doi.org/10.1128/aac.01814-20
217. Qaseem, A., Yost, J., Etxeandia-Ikobaltzeta, I., Abraham, G. M., Jokela, J. A., Forciea, M. A., *et al.* (2021) Should remdesivir be used for the treatment of patients with COVID-19? Rapid, living practice points from the American College of Physicians (Version 2). *Ann Intern Med.* **174**, 673–679. https://doi.org/10.7326/m20-8101
218. Janion, C., and Glickman, B. W. (1980) N4-hydroxycytidine: A mutagen specific for AT to GC transitions. *Mutat Res.* **72**, 43–47. https://doi.org/10.1016/0027-5107(80)90218-3
219. Kabinger, F., Stiller, C., Schmitzová, J., Dienemann, C., Kokic, G., Hillen, H. S., *et al.* (2021) Mechanism of molnupiravir-induced SARS-CoV-2 mutagenesis. *Nat Struct Mol Biol.* **28**, 740–746. https://doi.org/10.1038/s41594-021-00651-0
220. Jayk Bernal, A., Gomes da Silva, M. M., Musungaie, D. B., Kovalchuk, E., Gonzalez, A., Delos Reyes, V., *et al.* (2022) Molnupiravir for oral treatment of Covid-19 in nonhospitalized patients. *N Engl J Med.* **386**, 509–520. https://doi.org/10.1056/NEJMoa2116044
221. WHO Solidarity Trial Consortium. (2022) Remdesivir and three other drugs for hospitalised patients with COVID-19: Final results of the WHO Solidarity randomised trial and updated meta-analyses. *Lancet.* **399**, 1941–1953. https://doi.org/10.1016/s0140-6736(22)00519-0
222. Ader, F., Bouscambert-Duchamp, M., Hites, M., Peiffer-Smadja, N., Poissy, J., Belhadi, D., *et al.* (2022) Remdesivir plus standard of care versus standard of care alone for the treatment of patients admitted to hospital with COVID-19 (DisCoVeRy): A phase 3, randomised, controlled, open-label trial. *Lancet Infect Dis.* **22**, 209–221. https://doi.org/10.1016/s1473-3099(21)00485–0
223. Yip, T. C., Lui, G. C., Lai, M. S., Wong, V. W., Tse, Y. K., Ma, B. H., *et al.* (2023) Impact of the use of oral antiviral agents on the risk of hospitalization in community coronavirus disease 2019 patients (COVID-19). *Clin Infect Dis.* **76**, e26–e33. https://doi.org/10.1093/cid/ciac687
224. Koch, J., Uckeley, Z. M., Doldan, P., Stanifer, M., Boulant, S., and Lozach, P. Y. (2021) TMPRSS2 expression dictates the entry route used by SARS-CoV-2 to infect host cells. *EMBO J.* **40**, e107821. https://doi.org/10.15252/embj.2021107821

225. Jocher, G., Grass, V., Tschirner, S. K., Riepler, L., Breimann, S., Kaya, T., *et al.* (2022) ADAM10 and ADAM17 promote SARS-CoV-2 cell entry and spike protein-mediated lung cell fusion. *EMBO Rep.* **23**, e54305. https://doi.org/10.15252/embr.202154305
226. Monteil, V., Eaton, B., Postnikova, E., Murphy, M., Braunsfeld, B., Crozier, I., *et al.* (2022) Clinical grade ACE2 as a universal agent to block SARS-CoV-2 variants. *EMBO Mol Med.* **14**, e15230. https://doi.org/10.15252/emmm.202115230
227. Li, K., Meyerholz, D. K., Bartlett, J. A., and McCray, P. B., Jr. (2021) The TMPRSS2 inhibitor nafamostat reduces SARS-CoV-2 pulmonary infection in mouse models of COVID-19. *mBio.* **12**, e0097021. https://doi.org/10.1128/mBio.00970-21
228. Tay, M. Z., Poh, C. M., Rénia, L., MacAry, P. A., and Ng, L. F. P. (2020) The trinity of COVID-19: Immunity, inflammation and intervention. *Nat Rev Immunol.* **20**, 363–374. https://doi.org/10.1038/s41577-020-0311-8
229. Cain, D. W., and Cidlowski, J. A. (2017) Immune regulation by glucocorticoids. *Nat Rev Immunol.* **17**, 233–247. https://doi.org/10.1038/nri.2017.1
230. Sinha, S., Rosin, N. L., Arora, R., Labit, E., Jaffer, A., Cao, L., *et al.* (2022) Dexamethasone modulates immature neutrophils and interferon programming in severe COVID-19. *Nat Med.* **28**, 201–211. https://doi.org/10.1038/s41591-021-01576-3
231. Hardy, R. S., Raza, K., and Cooper, M. S. (2020) Therapeutic glucocorticoids: Mechanisms of actions in rheumatic diseases. *Nat Rev Rheumatol.* **16**, 133–144. https://doi.org/10.1038/s41584-020-0371-y
232. Bhimraj, A., Morgan, R. L., Shumaker, A. H., Baden, L., Cheng, V. C. C., Edwards, K. M. *et al.* (2022) Infectious Diseases Society of America guidelines on the treatment and management of patients with COVID-19. *Clin Infect Dis.* https://doi.org/10.1093/cid/ciac724
233. RECOVERY Collaborative Group, Horby, P., Lim, W. S., Emberson, J. R., Mafham, M., Bell, J. L., *et al.* (2020) Dexamethasone in hospitalized patients with Covid-19. *N Engl J Med.* **384**, 693–704. https://doi.org/10.1056/NEJMoa2021436
234. Liu, D., Ahmet, A., Ward, L., Krishnamoorthy, P., Mandelcorn, E. D., Leigh, R., *et al.* (2013) A practical guide to the monitoring and management of the complications of systemic corticosteroid therapy. *Allergy Asthma Clin Immunol.* **9**, 30. https://doi.org/10.1186/1710-1492-9-30
235. Matthay, M. A., and Thompson, B. T. (2020) Dexamethasone in hospitalised patients with COVID-19: Addressing uncertainties. *Lancet Respir Med.* **8**, 1170–1172. https://doi.org/10.1016/s2213-2600(20)30503-8
236. Philips, R. L., Wang, Y., Cheon, H., Kanno, Y., Gadina, M., Sartorelli, V., *et al.* (2022) The JAK-STAT pathway at 30: Much learned, much more to do. *Cell.* **185**, 3857–3876. https://doi.org/10.1016/j.cell.2022.09.023
237. RECOVERY Collaborative Group. (2022) Baricitinib in patients admitted to hospital with COVID-19 (RECOVERY): A randomised, controlled, open-label, platform trial and updated meta-analysis. *Lancet.* **400**, 359–368 https://doi.org/10.1016/S0140-6736(22)01109-6
238. Marconi, V. C., Ramanan, A. V., de Bono, S., Kartman, C. E., Krishnan, V., Liao, R., *et al.* (2021) Efficacy and safety of baricitinib for the treatment of hospitalised adults with COVID-19 (COV-BARRIER): A randomised, double-blind, parallel-group, placebo-controlled phase 3 trial. *Lancet Respir Med.* **9**, 1407–1418. https://doi.org/10.1016/s2213-2600(21)00331-3
239. Kalil, A. C., Patterson, T. F., Mehta, A. K., Tomashek, K. M., Wolfe, C. R., Ghazaryan, V., *et al.* (2021) Baricitinib plus remdesivir for hospitalized adults with Covid-19. *N Engl J Med.* **384**, 795–807. https://doi.org/10.1056/NEJMoa2031994
240. Strich, J. R., Tian, X., Samour, M., King, C. S., Shlobin, O., Reger, R., *et al.* (2022) Fostamatinib for the treatment of hospitalized adults with coronavirus disease 2019: A randomized trial. *Clin Infect Dis.* **75**, e491–e498. https://doi.org/10.1093/cid/ciab732
241. Singh, D., Bogus, M., Moskalenko, V., Lord, R., Moran, E. J., Crater, G. D., *et al.* (2021) A phase 2 multiple ascending dose study of the inhaled pan-JAK inhibitor nezulcitinib (TD-0903) in severe COVID-19. *Eur Respir J.* **58**. https://doi.org/10.1183/13993003.00673-2021
242. Rein, L., Calero, K., Shah, R., Ojielo, C., Hudock, K. M., Lodhi, S., *et al.* (2022) Randomized phase 3 trial of ruxolitinib for COVID-19-associated acute respiratory distress syndrome. *Crit Care Med.* **50**, 1701–1713. https://doi.org/10.1097/ccm.0000000000005682
243. Cafardi, J., Miller, C., Terebelo, H., Tewell, C., Benzaquen, S., Park, D., *et al.* (2022) Efficacy and safety of pacritinib vs placebo for patients with severe COVID-19: A phase 2 randomized clinical trial. *JAMA Netw Open.* **5**, e2242918. https://doi.org/10.1001/jamanetworkopen.2022.42918

244. RECOVERY Collaborative Group. (2021) Tocilizumab in patients admitted to hospital with COVID-19 (RECOVERY): A randomised, controlled, open-label, platform trial. *Lancet.* **397**, 1637–1645. https://doi.org/10.1016/S0140-6736(21)00676-0
245. Kyriazopoulou, E., Poulakou, G., Milionis, H., Metallidis, S., Adamis, G., Tsiakos, K., *et al.* (2021) Early treatment of COVID-19 with anakinra guided by soluble urokinase plasminogen receptor plasma levels: A double-blind, randomized controlled phase 3 trial. *Nat Med.* **27**, 1752–1760. https://doi.org/10.1038/s41591-021-01499-z
246. McNab, F., Mayer-Barber, K., Sher, A., Wack, A., and O'Garra, A. (2015) Type I interferons in infectious disease. *Nat Rev Immunol.* **15**, 87–103. https://doi.org/10.1038/nri3787
247. Hoagland, D. A., Møller, R., Uhl, S. A., Oishi, K., Frere, J., Golynker, I., *et al.* (2021) Leveraging the antiviral type I interferon system as a first line of defense against SARS-CoV-2 pathogenicity. *Immunity.* **54**, 557–570.e555. https://doi.org/10.1016/j.immuni.2021.01.017
248. Tam, A. R., Zhang, R. R., Lung, K. C., Liu, R., Leung, K. Y., Liu, D., *et al.* (2023) Early treatment of high-risk hospitalized coronavirus disease 2019 (COVID-19) patients with a combination of interferon beta-1b and remdesivir: A phase 2 open-label randomized controlled trial. *Clin Infect Dis.* **76**, e216–e226. https://doi.org/10.1093/cid/ciac523
249. Reis, G., Moreira Silva, E. A. S., Medeiros Silva, D. C., Thabane, L., Campos, V. H. S., Ferreira, T. S., *et al.* (2023) Early treatment with pegylated interferon lambda for Covid-19. *N Engl J Med.* **388**, 518–528. https://doi.org/10.1056/NEJMoa2209760
250. WHO Solidarity Trial Consortium, Pan, H., Peto, R., Henao-Restrepo, A., Preziosi, M., Sathiyamoorthy, V., *et al.* (2020) Repurposed antiviral drugs for Covid-19—Interim WHO solidarity trial results. *N Engl J Med.* **384**, 497–511. https://doi.org/10.1056/NEJMoa2023184
251. Kalil, A. C., Mehta, A. K., Patterson, T. F., Erdmann, N., Gomez, C. A., Jain, M. K., *et al.* (2021) Efficacy of interferon beta-1a plus remdesivir compared with remdesivir alone in hospitalised adults with COVID-19: A double-bind, randomised, placebo-controlled, phase 3 trial. *Lancet Respir Med.* **9**, 1365–1376. https://doi.org/10.1016/s2213-2600(21)00384-2
252. Teo, S. P. (2022) Review of COVID-19 mRNA vaccines: BNT162b2 and mRNA-1273. *J Pharm Pract.* **35**, 947–951. https://doi.org/10.1177/08971900211009650
253. Verbeke, R., Lentacker, I., De Smedt, S. C., and Dewitte, H. (2021) The dawn of mRNA vaccines: The COVID-19 case. *J Control Release.* **333**, 511–520. https://doi.org/10.1016/j.jconrel.2021.03.043
254. Taylor, P. C., Adams, A. C., Hufford, M. M., de la Torre, I., Winthrop, K., and Gottlieb, R. L. (2021) Neutralizing monoclonal antibodies for treatment of COVID-19. *Nat Rev Immunol.* **21**, 382–393. https://doi.org/10.1038/s41577-021-00542-x
255. Focosi, D., Anderson, A. O., Tang, J. W., and Tuccori, M. (2020) Convalescent plasma therapy for COVID-19: State of the art. *Clin Microbiol Rev.* **33**. https://doi.org/10.1128/cmr.00072-20
256. Cheng, Y., Wong, R., Soo, Y. O., Wong, W. S., Lee, C. K., Ng, M. H., *et al.* (2005) Use of convalescent plasma therapy in SARS patients in Hong Kong. *Eur J Clin Microbiol Infect Dis.* **24**, 44–46. https://doi.org/10.1007/s10096-004-1271-9
257. Filippatos, C., Ntanasis-Stathopoulos, I., Sekeri, K., Ntanasis-Stathopoulos, A., Gavriatopoulou, M., Psaltopoulou, T., *et al.* (2023) Convalescent plasma therapy for COVID-19: A systematic review and meta-analysis on randomized controlled trials. *Viruses.* **15**. https://doi.org/10.3390/v15030765

Part IV

Traditional Chinese Medicine in Lung Diseases

13 Pulmonary Diseases and Traditional Chinese Medicine

Baochang Zhao, Xiaohong Shu, and Hong Zhang

13.1 INTRODUCTION

Traditional Chinese medicine (TCM) is a brilliant component of Chinese culture with more than 3000 years of history. The ancient medicine books *Huangdi Neijing* (黄帝内经, Huangdi's internal classic), *Shennong Ben Cao Jing* (神农本草经, Shennong's herbal classic), and *Shanghan Lun* (伤寒论, Treatise on febrile disease) have clearly described the symptoms of various respiratory diseases, as well as approaches for prevention and treatment. TCM has a unique theoretical system based on the dialectical viewpoint of ancient Chinese philosophical thought, that is the guiding principles of holism and syndrome differentiation and treatment.

The traditional Chinese herbs are an especially important component of whole TCM. The specific parts of plants (roots, rhizomes, stems, flowers, fruits, seeds, plant products, extracts, and metabolites) are widely applied in treating respiratory disorders and alleviating symptoms such as cough, cold, flu, bronchitis, asthma, respiratory tract infections, as well as weakness, gastrointestinal disorders, and even alleviating anxiety. The plant-based components account for one-quarter of the most commonly used medicinal compounds (1).

The Chinese patent medicines, formulas, or prescriptions are all mixtures of various kinds of materials and have the function of treating both symptoms and root causes. The herbs in each prescription have their specific pharmacological roles in treating patients. According to the theory of TCM, the roles of these herbs in each formula, for example, can be divided into Jun (君, monarch), Chen (臣, minister), Zuo (佐, assistant), and Shi (使, guide) (2). The monarch herbs are the principal herbs with the main therapeutic activity. The minister herbs are the secondary principal components of the formula for enhancing or assisting the effect of the monarch herb. The assistant and guide herbs in the formula are for treating accompanying symptoms, enhancing the delivery of herbal ingredients, and/or controlling the toxicity of the primary components. Therefore, TCM also possess the characteristics of multiple components, multiple targets, and multiple pathways for treating patients against diseases.

With the modernization of TCM research, hundreds of Chinese herbal medicines for treating respiratory diseases have revealed the molecular mechanisms of action. Various bioactive compounds or metabolites of these herbs include flavonoids, saponins, alkaloids, phenols, terpenes, and others. Therefore, ancient Chinese medicine has been further developed, and it shows more powerful vitality when combined with Western medicine in an organic manner. In this chapter, we will introduce some common Chinese herbal remedies for treating respiratory diseases, their effective chemical compositions, and possible mechanisms of action according to modern biological theories.

13.2 INFECTION OF VIRUSES

The epidemic diseases caused by respiratory viruses, such as severe acute respiratory syndrome coronavirus (SARS-CoV)-1 and SARS-CoV-2, influenza A viruses, and respiratory syncytial virus, often pose a great harm to society. In China, there are enriched experiences and effective prescriptions that have existed for more than 2000 years for controlling and treating epidemic infectious diseases. The theory of prevention and treatment of "pestilence" (fatal epidemic diseases) has not

DOI: 10.1201/9781003355243-17

only recommended many prescriptions but also provided principles of epidemic disease prevention, which have been followed by TCM practitioners until now.

According to TCM theory, there are two main strategies in antiviral treatment: dispelling the pathogenic factors and strengthening the body. That means on one hand to directly eliminate or inhibit virus, and on the other hand to regulate immunity, control inflammation, as well as repair tissues and protect the host. In this section, we will only discuss the coronavirus infection.

13.2.1 Treatment of SARS-CoV-2 Infection with TCM in China

Since the end of 2019, the outbreak of SARS-CoV-2 infection (named COVID-19) had become a global health emergency on a pandemic scale for more than 3 years. In order to defeat the COVID-19 disease, the Chinese National Health Commission & National Administration of Traditional Chinese Medicine officially issued ten versions of *Diagnosis and Treatment Protocol for COVID-19 Patients (Trial)*. The ninth version (3), released on March 15, 2022, includes a combination strategy of TCM and Western medicine in preventing and treating COVID-19. In the TCM section, there are 27 Chinese patent medicines, formulas, and recommended prescriptions with more than 100 herbs involved. In the protocol, several acupuncture points are also recommended, including Hegu (合谷), Houxi (后溪), Yinlingquan (阴陵泉), Taixi (太溪), Feishu (肺俞), and Pishu (脾俞), Neiguan (内关), Kongzui (孔最), Quchi (曲池), Qihai (气海), Zhongwan (中脘), Dazhui (大椎), Lieque (列缺), Taichong (太冲), Guanyuan (关元), and Baihui (百会), Zusanli (足三里). In addition, separated moxibustion is also recommended to select Dazhui (大椎), Feishu (肺俞), Pishu (脾俞), and Kongzui (孔最). The therapeutic strategies of the protocol cover the entire processes of the disease with different degrees of severity.

The protocol recommended many formulas for different degrees of cases, such as Huoxiang Zhengqi (藿香正气) capsule and Shufeng Jiedu (疏风解毒) capsules for medical observation cases; Jinhua Qinggan (金花清感) granules and Lianhua Qingwen (连花清瘟) capsules for both medical observation and mild cases; Qingfei Paidu (清肺排毒) decoction for mild, moderate, and severe cases; lung-diffusing and toxin-resolving (散肺解毒) formula, lung-diffusing, dryness moistening, and toxin-resolving (散肺润燥解毒) formula for moderate cases; dampness-removing and toxin-resolving (祛湿解毒) formula, Huashi Baidu (化湿败毒) granule, and Xiyanping (喜炎平) injection for severe cases; Xuebijing (血必净) injection, Reduning (热毒宁) injection, Tanreqing (痰热清) injection and Xingnaojing (醒脑静) injection for both severe cases and critically ill cases; Suhexiang (苏合香) pill or Angong Niuhuang (安宫牛黄) pill, Shenfu (参附) injection, Shengmai (生脉) injection, Shenmai (参麦) injection for critically ill cases and the cases on mechanical ventilation with abdominal distention or constipation; as well as some recommended prescriptions for cases in convalescent period.

Through the screening in clinical practice, “three medicines and three formulas” is considered to have the most obvious curative effects. The “three medicines” include Jinhua Qinggan granules, Lianhua Qingwen capsules, and Xuebijing injection. The “three formulas” are Qingfei Paidu decoction, Huashi Baidu formula, and Xuanfei Baidu formula. All the medicines, except Xuebijing injection, are modified on the basis of a famous TCM named Ma Xing Shi Gan (麻杏石甘, ephedrae herba, armeniacae semen, gypsum fibrosum, glycyrrhizae radix) decoction from Zhang Zongjing’s *Treatise on Cold Damage Diseases* published in the late Eastern Han Dynasty (4, 5).

A plethora of clinical evidence indicates that TCM is beneficial for treating COVID-19 in (1) relieving the typical symptoms of fever, cough, fatigue, dry throat, sore throat, sputum production, shortness of breath, myalgia, and diarrhea, reducing the time to symptom recovery; (2) improving the lung features including lung inflammatory absorption, CT imaging, lung function, and related laboratory index; (3) shortening the duration of positive viral nucleic acid, and the progression to severe cases; and (4) protecting against multiorgan injury. Clinical practice has demonstrated that the strategy of combining TCM and Western medicine in the treatment of COVID-19 can achieve satisfactory results compared with the treatment of TCM or Western medicine alone (4).

13.2.2 The Active Ingredients of TCM in the Treatment of COVID-19

13.2.2.1 The Herbs and Their Active Ingredients of TCM

In addition to constantly adjusting and developing recipes for effectively treating COVID-19 patients, many scientists have also focused on discovering the effective herbs and their active ingredients. According to the analytic results of numerous prescriptions in treating COVID-19, the following herbs are most frequently used to suppress the virus: glycyrrhizae radix, armeniacae semen, ephedrae herba, poria, forsythiae fructus, scutellariae radix, pinellinae rhizoma praeparatum, pogostemonis herba, lonicerae japonicae, bupleuri radix, citri reticulatae pericarpium, atractylodis macrocephalae, rheum palmatum, magnoliae officinalis, isatidis radix, tsaoko fructus, mint, zingiberis rhizoma recens, dryopteridis crassirhizomatis rhizoma, coptidis rhizoma, rosae laevigatae fructus, propolis, linderae radix, green aucklandiae radix, polygoni cuspidati rhizoma et radix, sargassum, luffae fructus retinervus, and granati pericarpium, as well as mineral-based gypsum fibrosum (6, 7).

The active ingredients of hundreds of herbs have been identified. The ingredients that are most effective in treatment of COVID-19 include aescin, andrographolide, apigenin, astragaloside IV, baicalin, bavachinin, betulinic acid, blancoxanthone, celastrol, corylifol A, curcumin, emodin, ginkgetin, glycyrrhizin, α-hederin, hesperidin, iguesterin, isobavachalcone, jubanine G, kaempferol, lignan, luteolin, lycorine, matrine, myricetin, naringenin, neobavaisoflavone, nigellidine, patchouli alcohol, psoralidin, quercetin, resveratrol, saikosaponins, shikonin, silvestrol, tanshinone, and tingenone (8, 9).

There is a wide variety of sources of these ingredients, but primarily they can be any parts of a plant. Each plant always contains several active ingredients. A single ingredient may have several pharmacological functions and can be found in several plants.

13.2.2.2 The Mechanisms of TCM Ingredients Acting on COVID-19

The treatment of COVID-19 is largely inspired by the successful experience of SARS treatment in 2002–2003. This is because (1) SARS-CoV-2 and SARS-CoV-1 have more than 80% genetic homology, and (2) the critical targets of a spike, 3-chymotripsin-like protease (3CLpro), papain-like protease (PLpro), and RNA dependent RNA polymerase (RdRp) have 76%, 96%, 83%, and 96% sequence similarity, respectively (10). Similarly, SARS-CoV-2 virus invades into respiratory epithelial cells of host by identifying the angiotensin-converting enzyme 2 (ACE2) as the main entry point.

The key step for SARS-CoV-2 host invasion is the interaction between the virus spike glycoprotein (SGp) and host ACE2, medicated by serine protease transmembrane protease serine 2 (TMPRSS2) (10). Once the virus enters into host cells, RdRp plays a pivotal role in the replication and transcription of SARS-CoV-2. SARS-CoV-2 also encodes two proteases: 3CLpro, the viral main protease (M^{pro}), and PLpro for proteolytic processing during viral maturation. 3CLpro cleaves polyprotein pp1a and pp1ab for releasing several functional polypeptides, which establish a favorable environment for viral RNA synthesis. PLpro is required for SARS-CoV-2 replication and promotes the dysregulation of signaling cascades in the infected host cells. Nsp14, a nonstructural protein, is a functional enzyme involved in replication, fidelity, and mRNA capping (6, 11). Accordingly, the functional importance of the proteins RdRp, ACE2, SGp, 3CLpro and PLpro in the viral life cycle makes them potential targets for drug development against SARS-CoV-2 infection. On the other hand, after endocytosis, SARS-CoV-2 infection causes PAK1 (serine/threonine p21 protein–activated kinase-1) upregulation, inducing lung inflammation, pulmonary fibrosis, and other critical mortality factors. Increases in PAK1 levels also suppress the adaptive immune response, facilitating viral replication. Chemokines and activated pro-inflammatory cytokines, such as IL-6, IL-1, and TNF-α, can lead to the development of the disease with rapid respiratory impairment and pulmonary failure (12).

In short, the targets of TCM and its ingredients are focused on ACE2/TMPRSS2, 3CLpro, PLpro, SGp, RdRp, and N-terminal 3′-5′ exonuclease (Nsp14) for antivirus in addition to some key points for anti-inflammatory and immune regulation. Interestingly, each herb ingredient always has several target points that it attacks; conversely, each target can also be acted on by several different herbal ingredients. For example, rheum palmatum, glycyrrhizae radix, pogostemonis herba, and armeniacae semen are the most often used herbs in the "three medicines and three formulas." Their active ingredients, including emodin, glycyrrhizin, quercetin, patchoulic alcohol, and kaempferol, have high affinity for binding their target PLpro, 3CL pro, ACE2, RdRp, and SGp (4, 7) (Figure 13.1).

Naringenin, a possible candidate against SARS-CoV-2 infection and the pathogenesis of COVID-19, is present in a wide variety of fruits and vegetables, especially grapefruits, tangerines, oranges, and tomatoes. Naringenin is of great interest because of its numerous beneficial activities, such as analgesic, antioxidant, hypolipidemic, hepatoprotective, elastase inhibiting, anti-inflammatory, anti-mutagenic, anti-tumor, and antimicrobial effects. Importantly, naringenin, as a natural flavonoid, is a potential inhibitor of SARS-CoV-2 infection-related proteins, such as SGp, TMPRSS2, ACE2, M^{pro}, PLpro, RdRp, and endo-lysosomal two-pore channels (TPCs), as well as anti-inflammatory and antioxidant effects (13) (Figure 13.2).

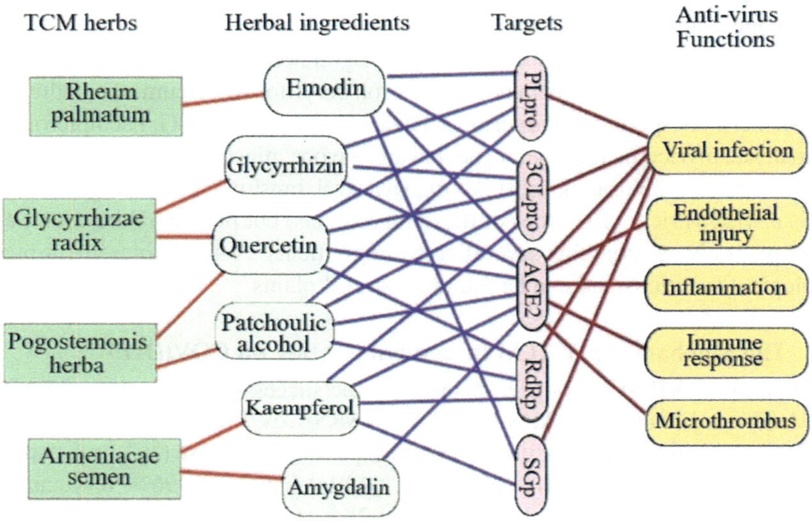

FIGURE 13.1 Network of some representative herbs and their major bioactive ingredients and functions in treating COVID-19.

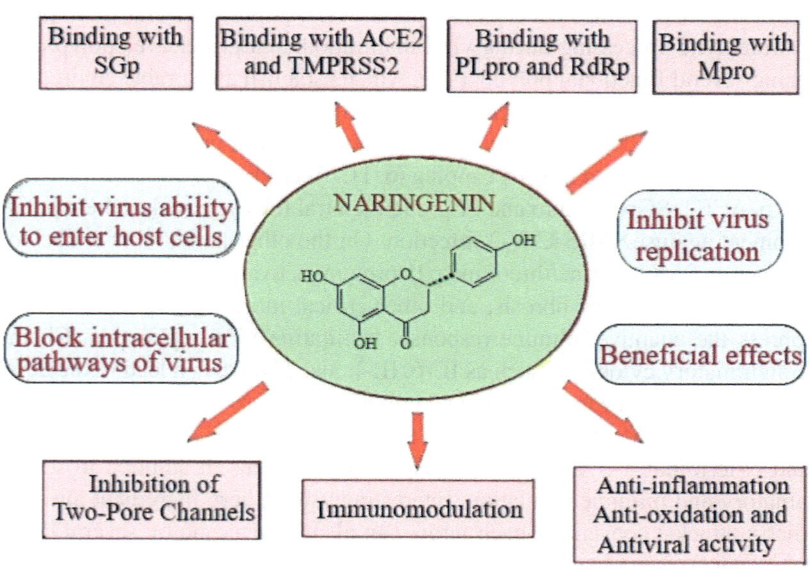

FIGURE 13.2 The possible targets and beneficial effects of naringenin against COVID-19.

Nevertheless, there is still a long way to go in the development of effective and precise drugs for the prevention and treatment of virus infections. The mechanisms of action, the characteristics of mutations of viruses, and the routes of transmission of virus infection also need to be clarified.

13.3 ACUTE LUNG INJURY AND ACUTE RESPIRATORY DISTRESS SYNDROME

Acute respiratory distress syndrome (ARDS) is a kind of acute lung injury (ALI) with diffuse inflammation and acute respiratory failure with a series pathophysiological changes. The treatment of ALI/ARDS is still a worldwide problem. Western medicine is mostly focused on supportive treatment, which emphasizes lung-protective ventilation and fluid management. There is a lack of effective and specific pharmacotherapy being identified. Although Western medicine could improve the treatment of ALI/ARDS, there has been no substantial decrease in the fatality rate of ALI/ARDS patients. TCM, as an adjuvant therapy, is widely used in clinical practice.

13.3.1 TCM Prescriptions in Treating ALI/ARDS

In the disease list of TCM, there is no appellation of ARDS. According to TCM theory, based on the clinical characteristics of ARDS patients, such as dyspnea, shallow and rapid breathing, severe suffocation, and chest constriction, ARDS belongs to "asthma," caused by lung failure, lung Qi (气, vital energy) inversion, or Qi exchange disorder in lung and kidney (14). TCM has good potential not only in preventing and treating ARDS, but also in reducing the dosages of Western medicines and subsequently limiting adverse drug reactions.

A review article (14) analyzed 16 studies and found eight formulas were often used. They are Da Chengqi decoction (大承气汤), modified Xuanbai Chengqi decoction (宣白承气汤), Ephedra aconite asarum decoction (麻黄附子细辛汤), Qingfei decoction (清肺汤), Baihu Chengqi decoction (白虎承气汤), Tingli Dazao Xiefei decoction (葶苈大枣泻肺汤), Xiao Qinglong decoction (小青龙汤), and Fusu mixture (复苏合剂). These formulas could reduce the mortality rate of patients, improve the effective rate of clinical treatment, reduce the average mechanical ventilation time, increase the levels of PaO_2, PaO_2/FiO_2, and reduce the level of inflammatory factors. Besides the conventional therapy, Chinese medicine injection (CMI) is also wildly used in ALI/ARDS treatment, such as Tanreqing (痰热清) injection, Shengmai (生脉) injection, Shenfu (参附) injection, Danshen (Salviae miltiorrhizae radix) injection, Reduning (热毒宁) injection, and Xuebijing (血必净) injection. However, the curative effect is controversial, and current evidence cannot be determined (15).

13.3.2 The Pharmacological Functions of TCM in Treating ALI/ARDS

Dysregulated inflammation, coagulation, and oxidative stress all play central roles in ARDS pathology. Many researchers have focused their attention on the role of TCM in modulating inflammatory imbalances, abnormal coagulation, antioxidation processes, and immunomodulation in ALI/ARDS, as well as the molecular mechanisms behind its effects.

13.3.2.1 Signaling Pathways and Receptors Related to ALI/ARDS

Accumulating evidence indicated that many signaling pathways are related to alveolar injury and repair of ALI/ARDS, including NF-κB, JAK/STAT, MAPK, mTOR, Notch, TLR4, HMGB1/RAGE, and macrophage activation (16–18). Under the condition of ALI/ARDS, the injured alveolar epithelial cells, injured pulmonary capillary endothelial cells, and activated and infiltrated immune cells all release various cytokines. The constantly activated inflammatory cells can form a vicious cycle, ultimately resulting in a cytokine storm. These activated intracellular signaling pathways play vital roles in this process. Among these, NF-κB signaling, as a prototypical proinflammatory signaling pathway, plays an important role in promoting inflammatory and oxidative storms as marked by promoting the expression of cytokines IL-1, IL-2, IL-6, and TNF-α, chemokines IL-8 and MIP-1α, adhesion molecules, and COX-2. Moreover, NF-κB, as a redox-sensitive transcription factor, is an important rendezvous with a broad spectrum of inflammatory signaling pathways to weave a complicated network.

Macrophages in lung tissue play a critical role in the inflammatory response to ALI/ARDS. In response to stimulation by foreign pathogens, the alveolar macrophages can be activated to form M1 macrophages by polarization, which subsequently release various inflammatory cytokines and chemokines to initiate a cascade of amplified inflammatory responses and mediate lung tissue injury. STAT3, NF-κB, MAPK, mTOR, and Notch can enhance the action of M1 macrophages (19, 20). The imbalance of Treg/Th17 in ALI/ARDS can increase the production of IL-2, IL-6, IL-10, IL-17, and TGF-β1. Inhibition of STAT3 activation can restore the balance (19, 21). In addition, alveolar macrophage pyroptosis may be involved in NLRP3 inflammasome participation and the development of ALI; p38-MAPK and TLR4/MyD88 may promote lung inflammation through formation of NLRP3 inflammasome and alveolar macrophage pyroptosis. Moreover, during the pyroptosis process, high-mobility group box 1 (HMGB1) is released into extracellular matrix to aggravate the inflammation of ALI (22). Nrf2/HO-1 pathway can activate the expression of HO-1, which in turn both inhibits proinflammatory cytokines activity and activates the activity of IL-10. PI3K/Akt signaling can inhibit NF-κB related inflammatory signals and exert anti-inflammatory and antioxidative stress properties. The Notch signaling pathway can also promote dendritic cells to produce IL-10 for inhibition of ALI/ARDS (17). In addition, upregulation of HO-1 by activation of PPARγ could inhibit the HMGB1-receptor of the advanced glycation end product (RAGE) signaling pathway and ameliorate the development of ALI/ARDS (23). The signaling pathways involved in ALI/ARDS are briefly illustrated in Figure 13.3.

13.3.2.2 The Molecular Mechanisms of Some Representative Formulas and Ingredients

Xuanbai Chengqi decoction (XBCQ), as a representative TCM prescription in "Systematized Identification of 'Warm Diseases,'" written by Wu Jutong in the Qing Dynasty, has been widely used to treat a variety of common respiratory diseases in China. XBCQ contains only four constituents, mineral-based gypsum fibrosum and the herbs rheum palmatum, armeniacae semen, and pericarpium trichosanthis. In the LPS-induced ALI of rats, XBCQ inhibits the PI3K/AKT/mTOR/HIF-1α/VEGF signaling pathway to decrease TNF-α, IL-6, and IL-1β production, resulting in reducing the injuries of ALI, such as exaggerated inflammatory response and pulmonary edema (24).

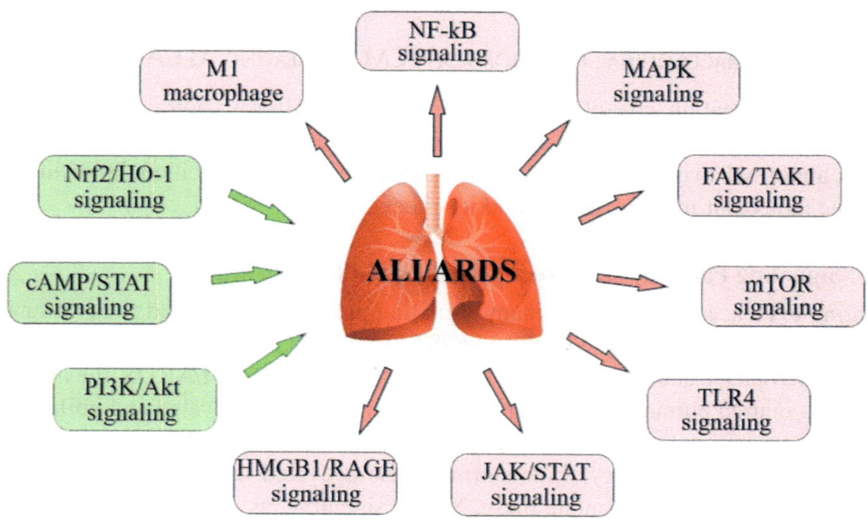

FIGURE 13.3 ALI/ARDS-related signaling pathways.

Qingfei Litan decoction (QFLT, 清肺理痰汤), included in the *Wind-Heat in Lung Syndrome TCM Diagnosis and Treatment Plan* by the TCM State Administration, has been effectively used to clinically treat ALI/ARDS for years. The combination of QFLT and Western medicine is more effective in the treatment of pneumonia cough than Western medicine alone (25). QFLT consists of gypsum fibrosum and nine herbs: herba nervilia plicatae, pericarpium trichosanthis, scutellariae baicalensis, fritillaria thunbergii, houttuynia cordata, phragmites rhizoma, armeniacae semen, platycodonis radix, and glycyrrhizae radix. Acacetin, baicalein, luteolin, liquiritigenin, wogonin, and isorhamnetin may be the important anti-inflammatory and antioxidative ingredients of QFLT against ALI/ARDS. Acacetin (25, 26), from platycodonis radix, decreases iNOS and COX-2 expression and increases HO-1 expression and SODs activity. Luteolin (27, 28), also from platycodonis radix, inhibits NF-κB activity and production of ROS, elevates the cellular GSH level, and activates the Akt/Nrf2 pathway. Baicalein (25, 27, 29), from scutellariae baicalensis, inhibits the NF-κB, MAPK, and JAK/STAT3 signaling pathways. Liquiritigenin (25, 30) from glycyrrhizae radix inhibits NF-κB activity in macrophages and decreases the production of iNOS and proinflammatory cytokines. Wogonin (5, 25, 27), from scutellariae baicalensis, armeniacae semen, platycodonis radix, and glycyrrhizae radix, can suppress IL-10 production in B cells via inhibiting the STAT3 and ERK signaling pathway. Isorhamnetin (25), from glycyrrhizae radix, blocks the MAPK and NF-κB signaling pathway, inhibiting the release of inflammatory mediators (TNF-α, IL-1β, IL-6, iNOS, and COX2) and MDA, as well as increasing SOD levels in LPS-induced ALI mice. Thus QLFT implements its therapeutic effects on ALI/ARDS mainly by anti-inflammation and antioxidation via downregulation of TLR4/NF-κB signaling pathway and upregulation of Nrf2 signaling pathway (Figure 13.4).

13.4 ASTHMA

Asthma is an allergic heterogeneous disease with a chronic airway inflammatory response in which eosinophils, mast cells, T cells, and an array of cytokines are involved. Controlling airway inflammation has become the consensus of scientists for asthma treatment. The recent treatment of asthma in clinic is mainly based on asthma calming, anti-inflammation, and bronchodilatation by using glucocorticoids, leukotriene receptor antagonists, and β2-agonists, which can only temporarily control symptoms but cannot effectively control disease recurrence. In addition, long-term use of these drugs may have certain side effects. It is estimated that up to 30% of adults and 60% of children in the United States are using complementary and alternative medicine (CAM) for asthma management (31). Due to the special theory of TCM and thousands of years of clinical practice history, the main pathogenesis of asthma is an accumulation of phlegm in lungs, classified as Tan Re Yong Fei

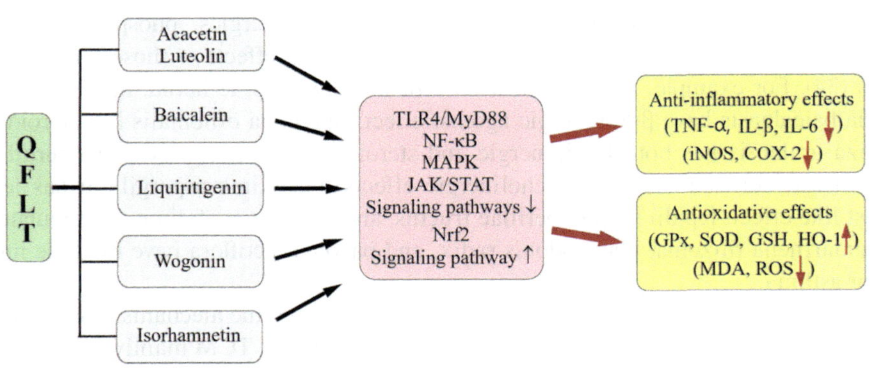

FIGURE 13.4 The proposal signaling pathways of anti-inflammatory and antioxidative mechanisms of treatment with QFLT against ALI/ARDS.

(痰热壅肺, phlegm-heat blocking the lungs) syndrome, combined with external influences, diet, emotions, and fatigue (32). TCM offers the principle of Yin (阴) rehydration and Gu Yang Qu Feng (固阳祛风, strengthening the lungs and dispelling wind), which can strengthen immune function and assist asthma patients in their recovery from the condition (33). In addition, the TCM prescriptions can also be flexibly formulated according to the patients' physical condition to achieve personalized drug use with little toxicity and side effects. Therefore, Chinese herbal medicine becomes widely used as CAM for add-on treatment of bronchial asthma.

13.4.1 The Commonly Used TCM Formulas and Active Ingredients Involved

There are variety of commonly used TCM formulas for effectively treatment of bronchial asthma including Ma Xing Shi Gan decoction, Xiaoqinglong decoction, Shegan Mahuang (射干麻黄, belamcandae rhizoma-ephedrae herba) decoction, Xiao Chaihu decoction (小柴胡汤), Suzi Jiangqi decoction (苏子降气汤), Zhichuan (治喘) formula, Liujunzi (六君子) decoction, San Ao (三拗) decoction, Fuling Xingren Gancao (茯苓杏仁甘草, poria-armeniacae semen-glycyrrhizae radix) decoction, Shaoyao Gancao (芍药甘草, paeonia lactiflora-glycyrrhizae radix) decoction, San Zi Yang Qin (三子养亲) decoction, Ban Xia Hou Po (半夏厚朴, pinellia ternata-magnoliae officinalis) decoction, Gui Zhi Jia Hou Po Xing Ren (桂枝加厚朴杏仁, cinnamomi ramulus plus magnoliae officinalis-armeniacae semen) decoction, Ling Gui Shu Gan (苓桂术甘, poria-cinnamomi ramulus-atractylodis macrocephalae-glycyrrhizae radix) decoction, Bu-Shen-Yi-Qi (补肾益气) formula, fructus schisandrae (五味子) syrup, Fang-xiao (防哮) formula, Jin Gui Shen Qi (金匮肾气) pill, Mahuang (麻黄, ephedrae herba) decoction, Pheretimaas pergillum (地龙) decoction, Soufeng Yuchuan (搜风愈喘) decoction, Wu-Hu (五虎) decoction, and Yu-Ping-Feng (玉屏风) powder (32–34).

A review article (34) rigorously screened 128 studies and identified the 26 most commonly used herbs in TCM prescriptions primarily for asthma treatment. They are glycyrrhiza uralensis, armeniacae semen, pinellia ternata, asarum sieboldii, pheretima aspergillum, aster tataricus, fritillaria cirrhosa, lepidii semen, pericarpium citri reticulatae, cortex mori, ephedrae herba, zingiberis rhizoma recens, tussilago farfara, platycodonis radix, fritillaria thunbergii, paeonia lactiflflora, magnolia officinalis, bupleurum chinense scutellaria baicalensis, anemarrhena rhizoma, gypsum fibrosum, eriobotryae folium, cinnamomi ramulus, zingiberis rhizoma, schisandra fructus, and perillae fructus. Another review article added atractylodes macrocephala, astragalus membranaceus, and codonopsis pilosula in Yi Fei Ping Chuan (益肺平喘) decoction (33).

13.4.2 The Mechanisms of Action of Active Ingredients in Asthma Treatment

Based on the known main mechanisms of action of β2 adrenoceptor agonists and corticosteroids in Western medicine for the treatment of asthma, the associated Chinese medical herbs were simply divided into β-adrenergic agonist, steroid-like, anticholinergics, phosphodiesterase (PDE) antagonist, leukotriene antagonist, and herbs with monoclonal effects or those affecting signaling pathways (34). For example, pericarpium citri reticulatae, ephedrae herba, tussilago farfara, and scutellaria baicalensis have β-adrenergic agonist effect; magnolia officinalis has steroidal effects; glycyrrhiza uralensis has both β-adrenergic and steroidal effects; fritillaria thunbergii has PDE inhibitor effect; cortex mori has anticholinergic effects; pheretima aspergillum has leukotriene antagonist like effect; lepidii semen, perillae fructus, and eriobotryae folium can regulate T helper cells; anemarrhena rhizoma, platycodonis radix, and paeonia lactiflora have multiple monoclonal effects for asthma.

The pathophysiology of asthma is extremely complicated, and the mechanisms of action of herbs may go far beyond the above-mentioned mechanisms. In general, TCM mainly covers four aspects of pathophysiological changes in asthma, including abnormal immune regulation and autoimmunity, abnormal effect of airway smooth muscle, abnormal effect of mucus, and airway remodeling. As we know, when the respiratory tract is exposed to allergens, the inhaled allergens are endocytosed and then presented to CD4+ T cells by dendritic cells. The allergens can also attack epithelial cells to produce cytokines (IL-33, TSLP), which facilitate dendritic cells to promote the polarization of CD4+ T cell to Th2 cells, making the Thl/Th2 balance shift to Th2 phenotype for triggering lots of harmful responses. Th2 releases many inflammatory cytokines and adhesion molecules, such as IL-4, IL-5, IL-13, GM-CSF, as well as TNF-α, TGF-β1, and VCAM-1 or ICAM-1 (35). Relying on these cytokines, Th2 attacks mast cells, eosinophils, plasma cells, B cells, epithelial cells, epithelial goblet cells, and smooth muscle cells, causing airway eosinophilia, goblet cell and epithelial cell hyperplasia, and airway hyperresponsiveness (AHR). The activated plasma cells can produce IgE that acts on basophils and mast cells to further increase the production of cytokines from the two types of cells. The stimulated basophils attack smooth muscle cells making them hypertrophy/hyperplasia. The basophils also stimulate B cells. The activated B cells can also produce IgE and are required for eosinophilic access to the lung tissue to increase vascular permeability and chemotaxis. The activated mast cells then activate innate lymphoid cells (ILC2s) and eosinophils by releasing PDG2 and IL-5, respectively. The activated ILC2s, in turn, express GATA3, ROR-α, and epithelial cytokine receptors, releasing cytokines to act on eosinophils, epithelial cells, goblet cells, mast cells, and airway smooth muscle. The ILC2s also increase the production of IgE by activating the plasma cells. The activated and recruited eosinophils can further stimulate plasma cells and dendritic cells. The epithelial cells that endocytosed the allergens also produce numerous cytokines to attack several inflammation related cells, such as Th2, eosinophils, mast cells, and ILC2s, as well as dendritic cells (31, 36).

Another lymphocyte is Th17, which can result in inflammation and autoimmune tissue damage by production of the cytokines TNF-α, IL-6, and IL-17. IL-17 can induce the recruitment of polymorphonuclear neutrophils. Although regulatory T (Treg) cell is an inflammatory suppressor, it can downregulate the production of proinflammatory cytokines and chemokines in many proinflammatory cells by secreting IL-10, an immunosuppressive factor. In addition, Treg also expresses FOXP3, a significant transcription factor playing an important role in differentiation and development. The imbalance of Treg/Th17 cells is connected to the severity of asthma (31, 35). Besides the inflammatory cells, some structural cells such as airway smooth muscle cells, epithelial cells, and fibroblasts may be involved in exacerbating asthmatic progression via increasing inflammation and airway remodeling. The release of cytokines, chemokines, and extracellular matrix (ECM) proteins, including TGF-β, IL-13, and VEGF have important roles (31). The MMP-9, MMP-12, or Rho-kinase (ROCK) proteins also contribute to the remodeling process by degrading the components of ECM (37). The intricate network of relationships between the inflammatory cells and related cytokines associated with asthma is shown in Figure 13.5.

The activation of inflammatory cells and the production of cytokines and chemokines rely on the related signaling pathways. Therefore, the active ingredients of herbs exert their anti-inflammatory, antioxidant, and other biological effects by modulating these molecular targets and related signaling pathways, which include activation of allergen-inhibited CD4+Th1 cell differentiation, inhibition of Th2 cell–related signaling pathways such as PI3K, reduction of IgE levels, improvement of the balance of IL4/INFγ, and action on the signaling pathway toll-like receptors (TLRs) of dendritic cells and the Rho-ROCK, JAK-STAT, MAPK, Wnt, VEGF, TGF-β, and Nrf-2 pathways (33, 34, 36). In addition, as multiple signal pathways in inflammation converge on the NF-κB pathway, various strategies that target NF-κB signaling have been considered for asthma treatment (33, 38). Indeed, the therapeutic efficacy of glucocorticoids is largely contributed by inhibiting NF-κB (39). The effects and possible molecular mechanisms of some TCM ingredients for treating asthma have been studied in depth; these will certainly provide more scientific basis for development of new drugs and effective treatment of asthma.

Currently, many ingredients from Chinese herbal medicines have been confirmed to be effective against asthma, including icarlin, apigenin, baicalein, vitexin, catechin, gallocatechin, hesperidin, naringin, naringenin, kaempferol, quercetin, rutin, apigenidin, cyanidin, schizandrin, glycyrrhizin, emodin, mangiferin, curcumin, cordycepin, erysotrine, matrine, mitraphylline, piperlongumine

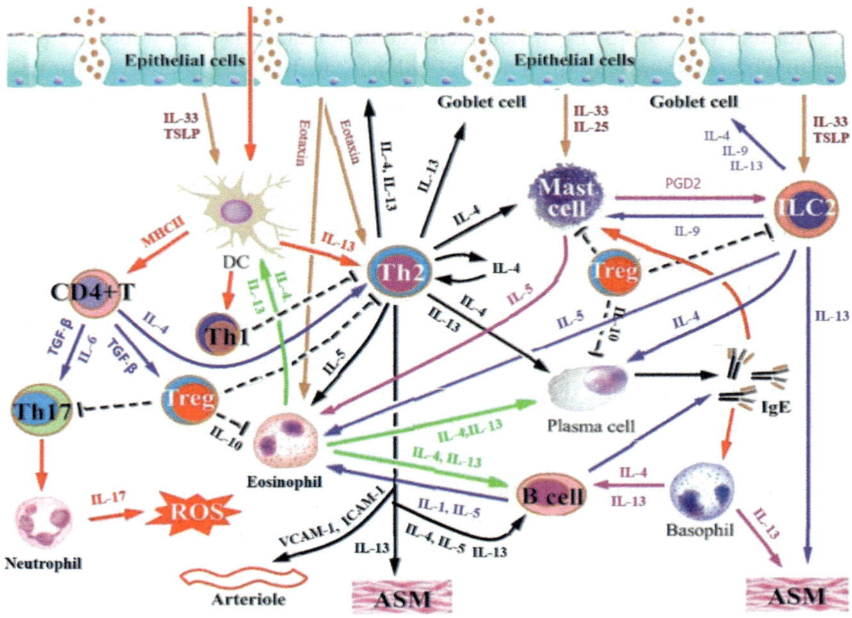

FIGURE 13.5 The role of different cells and produced cytokines, and the effect of herbal ingredients in asthma.

resveratrol, saikosaponin A and B, ginsenoside Rg, baicalin, chitin, chrysin, scholaricine, camphor, gallic acid, wedelolactone, ephedrine, ellagic acid, coumarin, luteolin, oroxylin A, galangin, frankincense, boswellic acids and so on. Most of them are categorized as flavonoids (31, 40, 41).

13.4.3 Examples of the Mechanisms of Action of Herbal Ingredients against Asthma

Quercetin, widely distributed in fruits and vegetables, is one of the main flavonoids in the human diet, and the daily requirement is 5–40 mg. Quercetin has multiple functions in treating asthma. In vitro studies with human airway epithelial cells, quercetin could inhibit neutrophil elastase induced mucin 5AC (MUC5AC) expression to decrease the hypersecretion of mucus. In studies using animal models, the anti-asthma activity of quercetin was similar to that of cromolyn sodium and dexamethasone, as it suppresses GATA-3 gene expression and increases T-box protein expression in T cells (T-bet), altering Th1/Th2 differentiation and decreasing allergic airway inflammation and AHR. Quercetin decreases IL-4, IL-5, IL-25, IL-33, IgE, thymic stromal lymphopoietin (TSLP), caspase-3, and terminal deoxynucleotidyl transferase-mediated dUTP nick end labeling (TUNEL). Conversely, quercetin increases IFN-γ. The signaling pathways of quercetin action include attenuating the PI3K/Akt/NF-κB and the PKC/EGFR/ERK signaling pathways. Quercetin also lowers epithelial and subepithelial smooth muscle thickness, inhibits eosinophil growth, and decreases goblet and mast cell numbers and activation (36). The target points of quercetin action are summarized in Figure 13.6.

Curcumin is a core ingredient of rhizoma curcumae longae, often used to treat a variety of diseases, especially chronic inflammation, such as asthma. Curcumin can block NF-κB, p38 MAPK, ERK 42/44, and JNK54/56 related signaling pathways, and the PPARγ-dependent NF-κB signaling pathway, thereby reducing AHR, airway infiltration of inflammatory cells, and production of inflammatory cytokines, as well as phospholipase A2 (PLA2), PGD2, and COX2 levels. Moreover, curcumin inhibits upregulation of MCP-1 and MUC5AC, and mucus hypersecretion. Curcumin can also activate Nrf2/HO-1 signaling, reduces TNF-α, IL-1β, IL-6, eosinophils, and exerts antioxidative effects (42).

Curcumin improves the imbalance of Treg/Th17 for activating Treg cells (43). The inhibitory effect of curcumin on iNOS and NO can suppress IL-2, IL-5, GM-CSF, and IL-4 production,

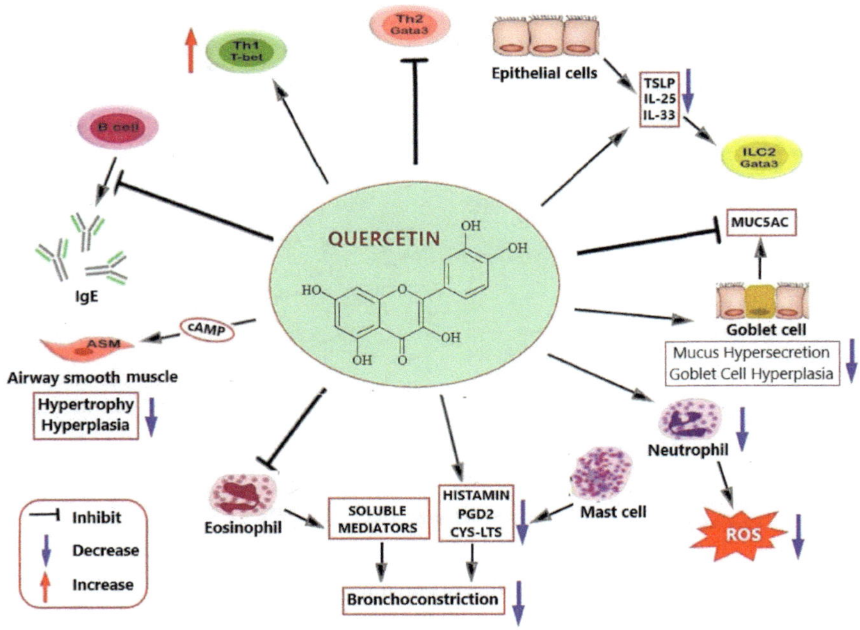

FIGURE 13.6 The possible mechanisms of quercetin in asthma.

lymphocyte proliferation, eosinophil recruitment and Th2 cell differentiation. Curcumin can also downregulate the Notch signaling pathway-associated T cell activation and proliferation. Curcumin suppresses proliferation of airway smooth muscular cells by inhibiting ERK signaling. Curcumin also reduces the activity of MMP-9 and expression of α-SMA and eotaxin for alleviating pulmonary fibrosis. Curcumin activates Wnt/β-catenin signaling pathway (possibly in dendritic cells) to reduce lung inflammation, and increase IFNγ (44, 45).

Curcumin prevents and protects from structural alterations and airway remodeling by modulating TLR-4/MAPK signaling pathway (46, 47). Curcumin also ameliorates TLR4/NF-κB/NLRP3 inflammasome signaling and pyroptosis (48). Curcumin is a histone deacetylase 1(HDAC1) inhibitor for inhibiting NF-κB expression (49).

In conclusion, the pharmacological effect of curcumin in preventing and treating asthma is achieved through its anti-inflammatory, antioxidation, and relief of trachea remodeling functions. The possible mechanisms of curcumin action is summarized in Figure 13.7.

Resveratrol possesses many means to inhibit asthma, including inhibition of inflammatory cells and production of inflammatory cytokines, AHR, oxidative stress, mucus hypersecretion, and airway remodeling. Resveratrol inhibits the proliferation and activation of Th2, Th17, mast cells and eosinophils, and inhibits DNA damage and apoptosis of bronchial epithelial cells. Resveratrol alleviates oxidative stress and repairs the mitochondrial function of lung tissue by modulating several signaling pathways, including activation of AMPK expression and inhibition of expression of iNOS, TNF-α and p47phox; reduction of ROS production; and promotion of the Nrf2 pathway to increase the production of catalase, GSH, GPx, and SOD (31, 49). Resveratrol has potential anti-inflammatory and anti-airway remodeling effects by downregulating AMPK, PI3K/Akt, TGF-β1/Smad, and HMGB1/TLR4/NF-κB signaling pathways. Resveratrol also activates SIRT1 for upregulating PTEN to inhibit PIP3 pathways. Resveratrol upregulates the expression of FOXP3a through downregulating miR34a expression. The enhanced FOXP3a subsequently upregulates ROS scavenging enzyme SOD2 and catalase expressions, and as a master regulator promotes Treg development and functions (31, 35, 50). Resveratrol inhibits proliferation of eosinophils, arrest cell cycle progression in G1/S phase by increasing p53 and p21, and concurrently reducing the expression of cyclin-dependent kinase 2 (CDK2), cyclin A and

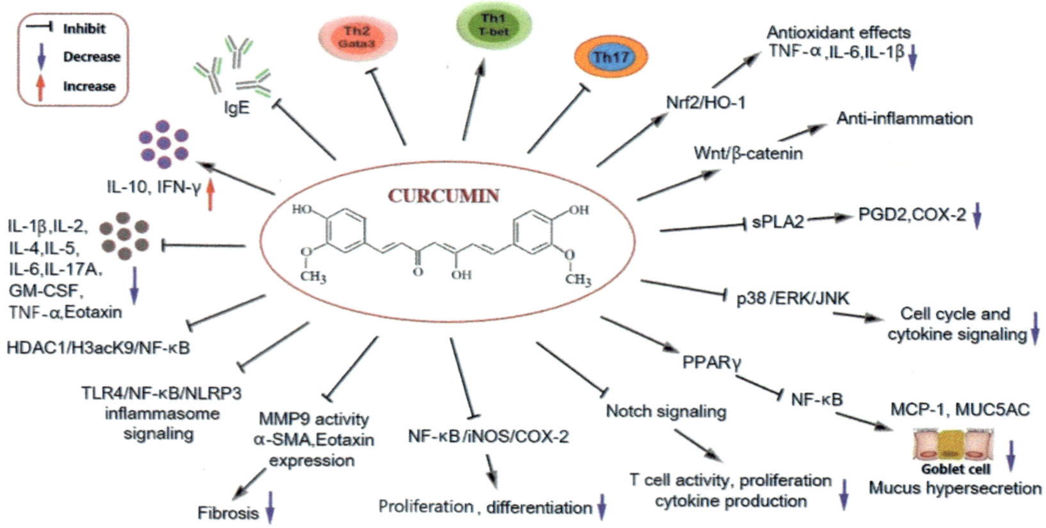

FIGURE 13.7 The possible mechanism of action of curcumin for treating asthma.

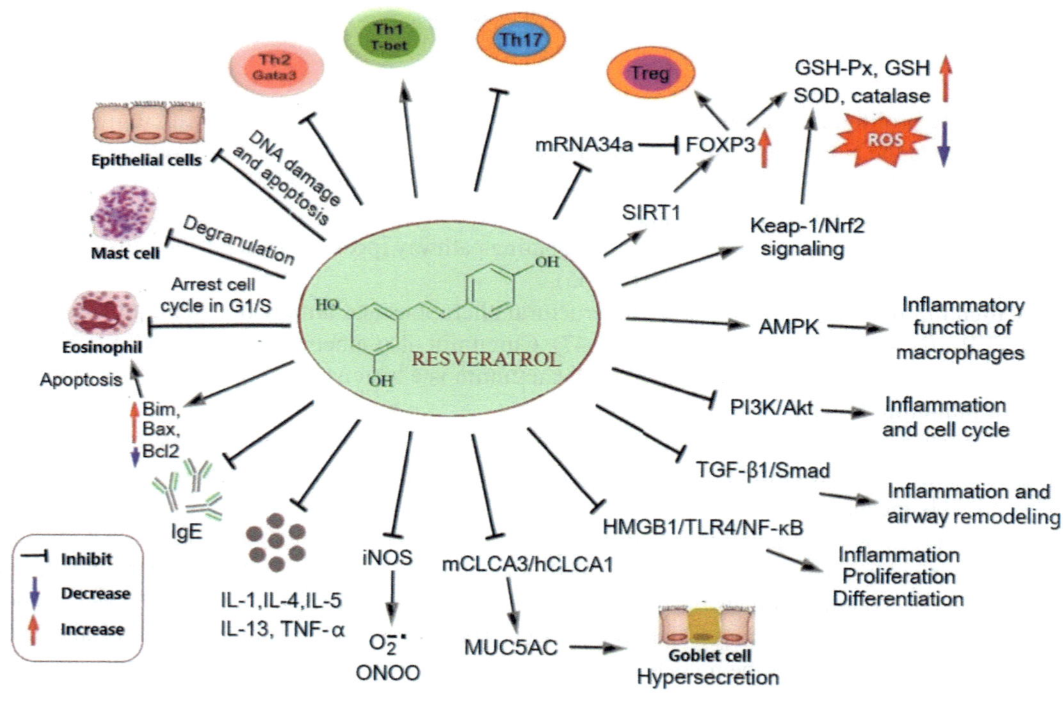

FIGURE 13.8 The possible mechanism of action of resveratrol in asthma.

cyclin E. In addition, resveratrol induces eosinophil apoptosis via upregulating Bim and Bax expressions, and downregulating Bcl-2 expression (31, 51). Resveratrol suppresses calcium-activated chloride channel (CLCA)1 signaling pathway to prevent mucus overproduction by inhibiting overexpression of MUC5AC (31, 52). Resveratrol can also reduce the expression of TGF-β1/Smad2/3 pathway in airway epithelial cells to inhibit pulmonary fibrosis (31, 53). The possible mechanisms of action of resveratrol in asthma are illustrated in Figure 13.8.

13.5 CHRONIC OBSTRUCTIVE PULMONARY DISEASE

Chronic obstructive pulmonary disease (COPD) is a progressive airflow restrictive and not completely reversible disease that is a leading global cause of morbidity and mortality. Exposure to gases from cigarette smoking and inhaled particles such as $PM_{2.5}$ are two archetypical inducing factors for causing COPD. The pathogenesis of COPD is characterized by inexorable deterioration of small airways obstruction with chronic inflammation, oxidative stress, cellular senescence, corticosteroid resistance, cell apoptosis, and airway remodeling. The changes in pulmonary histology and functions include inflammatory cell infiltration, epithelial-mesenchymal transition (EMT), ECM deposition, and smooth muscle cell (SMC) proliferation. The COPD may be the underlying cause of emphysema and fibrosis (54). These pathological changes make the airway wall thick with fixed obstruction, persistent airflow limitation, irreversible loss of lung function, and thus poor disease prognosis. The acute exacerbation caused by COPD is often associated with respiratory failure and a high mortality rate.

In Western medicine, the treatments for COPD include infection control, airway relief, and water, electrolyte, and acid-base imbalance corrections, as well as oxygenation improvement and respiratory failure combat. Although long-term oxygen supply, corticosteroid inhalation, and bronchial pipette can effectively reduce hypoxemia and airway obstruction to some extent, respiratory muscle fatigue, poor nighttime respiratory center response, and carbon dioxide retention may contribute to further disease progression for patients with respiratory failure.

According to TCM theory, the vital Qi refers to the body's ability to self-regulate, its resistance to pathogens, and self-recovery. Lungs are in charge of the vital Qi and respiratory function, whereas the kidney governs reception of Qi. COPD belongs to the category of "syndrome characterized by dyspnea" and "lung distention," and is characterized by the deficiency of vital Qi in lung, spleen, and kidney. The Chinese herbal medicines in treating COPD have the Qi-tonifying character involved in improving the defense capacity of immune system and enhancement of ATP generation capacity. Many studies have confirmed that the nourishment of the lung and kidney has dramatic and long-term effects on boosting the weak immune system, decreasing inflammatory responses, reducing pulmonary pathological impairment and airway remodeling (55).

13.5.1 The Recommended TCM Prescriptions and Frequently Used Herbs Involved

The *International Clinical Practice Guideline of Chinese Medicine Chronic Obstructive Pulmonary Disease* compiled by the World Federation of Chinese Medicine Societies recommends corresponding TCM according to the different types of COPD and the different development periods of the disease (56). In the cases with acute exacerbation of COPD, modified San ao decoction plus Shike (止嗽, relieve a cough) powder, Tongxuan Lifei (通宣理肺) pill and Xing Su Zhike (杏苏止咳, armeniacae semen-perilla frutescens in relieving cough) granule, modified Qingqi Huatan (清气化痰) pill plus Beimu Gualou (贝母瓜蒌, fritillaria thunbergii-fructus trichosanthis) powder, Xiaoqinglong decoction or granule, Tanreqing (痰热清) injection, Ting Bei (葶贝, lepidium apetalum-fritillaria cirrhosa) capsule, Xuebijing (血必净) injection, modified Banxia Houpu (半夏厚朴, pinellia ternata-magnoliae officinalis) decoction plus Sanzi Yangqin (三子养亲) decoction, Suzi Jiangqi (苏子降气) pill, Ling Gui Kechuanning (苓桂咳喘宁) capsule, modified Ditan (涤痰) decoction, Xinnaojing (醒脑静) injection, and Qingkailing (清开灵) injection are recommended for selecting use based on the specific symptoms. In the stable COPD cases, modified Renshen Hutao (人参胡桃, panax ginseng-semen juglandis) decoction plus Renshen Yangfei (人参养肺) pill, Yupingfeng (玉屏风) granule, modified Liujunzi (六君子) decoction plus Huangqi Buzhong (黄芪补中) decoction, modified Bufei Yishen (补肺益肾) formula, Bufei Huoxue (补肺活血) capsules, modified Baoyuan (保元) decoction plus Renshen Bufei (人参补肺) decoction, or Yiqi Zishen (益气滋肾) formula, and Bufei (补肺) Granules, Shengmaiyin (生脉饮) oral liquid, Yangyin Qingfei (养阴清肺) pill, Baihe Gujin (百合固金) pill, and GeJie Dingchuan (蛤蚧定喘) pill are recommended to be selected based on the individual symptoms. In addition, other techniques of treatment for COPD, such as Taiji box, acupuncture, breathing guidance, acupoint application (such as Shufei [舒肺] plaster, Xiaochuan [消喘] plaster, etc.), and Yifei (益肺) moxibustion are also effective in relieving clinical symptoms, improving exercise tolerance, delaying the decline of lung function, and improving the quality of life of COPD patients.

The 33 Chinese patented medicines and formulas contain a total of 150 herbs. Among those, the frequency of application of eight herbs is more than ten times, including pericarpium citri reticulatae, glycyrrhizae radix, armeniacae semen, poria, perilla frutescens, pinellia ternata, scutellaria baicalensis, and atractylodis macrocephalae. Fifteen herbs and gypsum fibrosum are used five to nine times: magnoliae officinalis, ophiopogon japonicus, ephedrae herba, platycodonis radix, pheretima aspergillum, astragari radix, fritillaria cirrhosa, fritillaria thunbergii, paeoniae radix rubra, panax ginseng C.A.Mey., schisandra fructus, gardenia jasminoides, coconopsis radix, aster tataricus, and angelica sinensis.

13.5.2 The Bioactive Ingredients of TCM and Their Mechanisms of Action

Based on the theories of TCM related to lung diseases, several review articles searched electronic databases and tried to determine the active ingredients of these commonly used herbs in clinical application for treating COPD. They found the following compounds to have high-frequency in application: are andrographolide, artemisinin, cimigenoside, astragaloside, astragaloside, crocin, celastrol. apigenin, hesperidin, kaempferol, naringenin, quercetin, naringenin, baicalin, curcumin, resveratrol, epigallocatechin, gallatic acid, procyanidins, acetylshikonin chrysophanol, tanshinone IIA sulfonate, salidroside, stemonine, matrine, tussilagone, epigallocatechin-3-gallate (EGCG), amygdalin, ursolic acid, liquiritin, and resveratrol (38, 57).

Because chronic inflammation and oxidative stress are the two crucial controlling mechanisms of COPD pathogenesis and disease progression, many studies have focused on the role of TCM in the two aspects of mechanisms. The mechanisms of action of some major herbal ingredients for treatment of COPD are illustrated in Figure 13.9. The involved signaling pathways include NF-κB

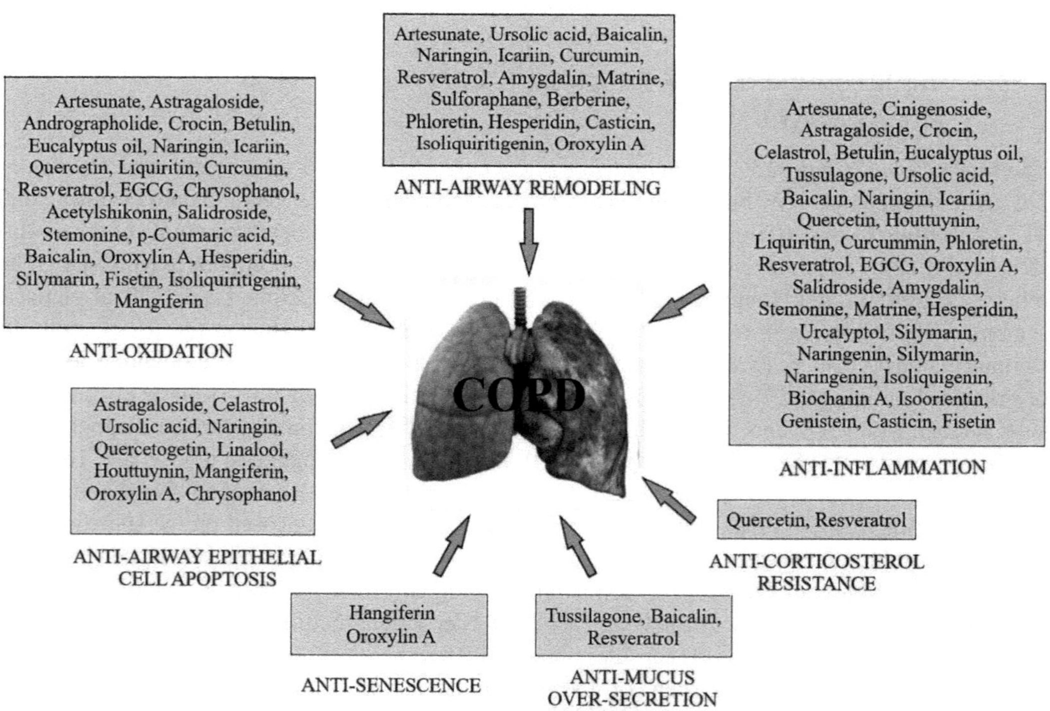

FIGURE 13.9 The diverse mechanisms of herbal ingredients targeting COPD.

(58), TLR4/MyD88/NF-κB, MAPKs (ERK, p38) (17, 46, 58, 59, 60), JNK/p38 (61), JAK3/STAT3/NF-κB (59), TGF-β1/Smad2/3 (62), AMPK/Nrf2 (58, 63), Nrf2/HO-1 (17, 58, 64), AMPK/PGC-1α (60, 65), PI3K/Akt (66), PPARγ/NF-κB (67), glucocorticoid receptor (60), and several miRs (61).

Baicalin (Figure 13.10) is an important active ingredient in dried roots of scutellaria baicalensis. In the cigarette smoke (CS)-induced COPD rat model, the baicalin can significantly improve lung function by anti-inflammatory, antioxidant, anti–airway remodeling as evidenced by reduction of pro-inflammatory cytokines such as TNF-α, IL-6, and IL-1β, and inflammatory cell infiltration, as well as an increase in IL-10 level, reduction of MDA, and increase in SOD and HO-1 levels. Baicalin improves the lung function also by reducing MUC5AC expression. In addition, baicalin also reduces MMP-2, MMP-9, and VEGF levels, suggesting the function of inhibiting airway remodeling. Baicalin can upregulate the expression of HSP72, resulting in inhibition of JNK signaling activation. Baicalin possesses the function of elevating cell viability by inhibiting apoptosis (57, 68). A literature reported that baicalin could improve hypothalamic-pituitary-adrenal axis (HPA axis) by increasing ACTH, cortisol, and leptin levels (69). Another research group indicated that the amelioration of baicalin on CS-induced airway inflammation in rats was partially attributed to the modulation of histone deacetylase 2 (HDAC2)/NF-κB/plasminogen activator inhibitor type-1 (PAI-1) signaling pathway though enhanced HDAC2 protein expression and inhibited the expression of NF-κB and its downstream target PAI-1 (70).

Liquiritin (Figure 13.11), mainly from glycyrrhizae radix, alleviates pulmonary inflammation by targeting TGF-β signaling pathway (38). Liquiritin also possesses antioxidant effect, increases SOD level, and decreases the MDA level and goblet cell hyperplasia to protect from the CS-induced lung injury (60, 71). Liquiritin activates the β2-adrenoceptor and effectively relaxes tracheal smooth

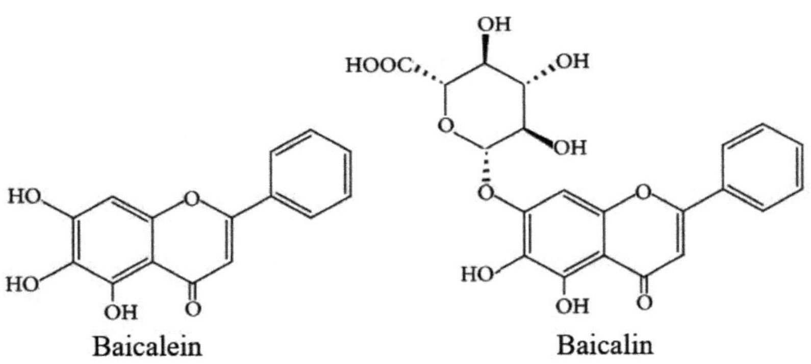

FIGURE 13.10 The structure of baicalein and baicalin.

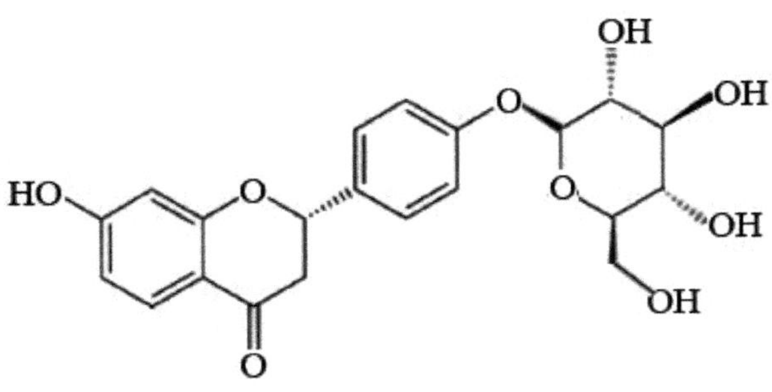

FIGURE 13.11 The structure of liquiritin.

muscle (72). Liquiritin can inhibit the abnormal upregulation of phosphorylation of nuclear receptor 4A1(Nur77), as well as the increased levels of inflammatory factors TNF-α and IL-6. Liquiritin also alleviates the secretion of eosinophil chemokine-1. As we know, the transient receptor potential cation channel subfamily V member 1 (TRPV1) and TRPA1 are expressed in subpopulations of sensory neurons and in non-nerve tissues throughout the bronchioles and alveoli. Activation of TRPV1 and TRPA1 can cause cough, bronchoconstriction, microvascular leakage, hyperresponsiveness, and excessive secretion, as well as epithelial cell death in the respiratory tract. Liquiritin inhibits TRPV1 and TRPA1 to reduce inflammatory response including decrease in expression of TGF-β, TNF-α, IL-6 IL-1β, IgE, and the counts of inflammatory cells (72, 73).

In addition, amygdalin, from armeniacae semen, ameliorates EMT via the TGF-β/Smad pathway (74). Tussilagone and EGCG from *Tussilago farfara* L., *Eriobotrya japonica* Lindl., and *Morus alba* L. can enhance anti-proliferation activity by inhibiting the NF-κB signaling pathway (38).

13.5.3 Examples of TCM Prescriptions for Treatment of COPD Studied in Depth

13.5.3.1 Bufei Yishen Formula

Bufei Yishen formula (BYF, 补肺益肾方, invigorating the lung and the kidney) is a traditional Chinese herbal formula widely used in COPD treatment. It has been confirmed that BYF can improve lung function, alleviate airway inflammation, oxidative stress, and airway remodeling. BYF is efficacious in the treatment of COPD by relieving symptoms, reducing the risk of exacerbation, and enhancing exercise endurance and life quality. BYF is constituted by 12 Chinese herbal medicines, including panax ginseng, red ginseng, astragali radix, epimedii herba, corni fructus, lycii fructus, schisandrae chinensis, fritillariae thunbergii bulbus, perillae fructus, pericarpium citri reticulatae, paeniae rubra radix, pheretima, and ardisiae japonicae herba (75).

A research group systematically studied the mechanism of action of BYF in treating COPD (76). They found 182 candidate ingredients had interactions with 195 potential targets. Most of the targets of BYF are related to activation of AP-1 family of transcription factors and MAPK. They suggested that IL-1β, IL-6, and TNF-α expressed in lung tissue of COPD patients are regulated by the MAPK/AP-1 pathway.

Indeed, BYF exerts its effects on COPD possibly by modulating lipid metabolism; reducing inflammation, mucus secretion, and oxidative stress; improving MMPs/anti-MMPs balance and collagen deposition; downregulating the EGFR/PI3K/mTOR, TLR-4/NF-κB and JNK, p38, and NF-κB pathways; and modulating cellular connectivity pathways at the systemic level (57, 77). BYF III combines with electroacupuncture can effectively suppress inflammatory response in COPD rats via regulating SIRT1/NF-κB signaling (78).

Moreover, BYF significantly inhibits Th17 cell differentiation as marked by suppressing the expressions of RORγt and IL-17. Concurrently, BYF activates Treg cell differentiation as verified by markedly increasing the expression of Foxp3 and IL-10. Finally, BYF can restore the Treg/Th17 balance via reduction of STAT3 phosphorylation in Th17 cells, and enhancement of STAT5 phosphorylation in Treg cells, and activation of adenosine receptor 2a for attenuating COPD (75).

The $PM_{2.5}$ and CS-induced COPD animals exhibit high levels of miR155 and lower levels of FOXO3a in bronchial epithelial cells. FOXO3a has a critical role in controlling the expression of genes involved in oxidative stress response, DNA damage repair, inflammation, and cellular metabolism. BYF upregulates FOXO3a expression via modulating the miR155/FOXO3a axis, thereby suppressing miR155 expression (79).

It is known that there is an accumulation of senescence cells in CS-induced COPD lung tissue, including aging alveolar epithelial cells and aging endothelial cells. The COPD cellular senescence is a state of irreversible permanent cell cycle arrest. Cyclin-dependent kinases inhibitory factors (p16, p21 and p27) accumulate in senescent cells. The senescent cells can increase secretion of senescence-related secretory phenotype (SASP) factors (IL-6, IL-8, and CCL3). BYF treatment can inhibit CS extract (CSE)-induced cellular senescence and reduce CSE-suppressed expression and secretion of klotho in lung epithelial cells, thereby reducing inflammation, oxidative stress, and cellular senescence. In addition, BYF also reverses CSE-downregulated expression of ZNH263 in BEAS-2B cells that in turn upregulate the expression and secretion of klotho (80).

13.5.3.2 Guben Zhike Decoction

Another most commonly used TCM formula is Guben Zhike decoction (固本止咳汤, GBZKD, firm the root and relieve a cough), which is composed of seven herbs: astragari radix, epimedii herba, radix stemonae, scutellaria baicalensis, atractylodis macrocephalae, paeoniae radix rubra, and saposhnikoviae divaricata. GBZKD can improve lung function and associated pathology in COPD patients. The lung bullae and infiltration of inflammatory cells can be relatively averted, and the airway cilia can be rearranged regularly in COPD animals. A research group (81) identified 287 and 184 proteins with marked regulatory roles in lung tissue of mice belonging to COPD and GBZKD experimental group, respectively, by label-free proteomic analysis. GBZKD may improve lung injury by downregulating the complement and coagulation cascades, metabolic pathways, nitrogen metabolism, tight junction, gene and protein expression, and downregulating the multiple inflammatory pathways, including JAK/STAT signal transducer and activator of transcription signaling pathway. In addition, GBZKD also improves oxidative stress by regulatory factors, such as carbonic anhydrase 1 and 2.

GBZKD can reduce IL-6 and IL-13 levels, increase the proportion of γδT cells in lung tissue, promote the secretion of keratinocyte growth factor (KGF) and KGFR. Therefore, GBZKD can effectively reduce airway inflammation and improve the structure of the airway from damage and airway epithelial integrity, and enhance the respiratory mucosal immune function (82).

13.6 PULMONARY FIBROSIS

Pulmonary fibrosis (PF) is a serious lung disorder caused by viral infections, radiation exposure, and airborne toxins. Idiopathic PF (IPF) is related to pneumonia, and it often affects the elderly and is characterized by recurring scar formation in the lung including the irregular aggregation of ECM constituents (e.g., mesenchymal cells, fibroblasts and myofibroblasts, interstitial collagens, hyaluronic acid, proteoglycans, fibronectin, laminin, tenascin-C). The pathogenesis of PF is associated with adaptive and innate immune activation, inflammation, oxidative stress, epithelial/endothelial damage, EMT, and cell apoptosis. Numerous studies have clarified the activation of TGF-β or NF-κB pathways involved in the expression of ECM genes. Oxidant stress caused by the imbalance between oxidant/antioxidant in lung of sufferers is one of the major mechanisms presents in the pathogenesis of PF (2, 38).

13.6.1 A Brief Introduction of the Pathology of Pulmonary Fibrosis

The pathogenesis of PF is briefly illustrated in Figure 13.12. The inflammatory mediators unleashed by the injured lung epithelial cells stimulate platelets for blood clot formation and stimulate neutrophils and macrophages to secrete profibrotic cytokines such as TNF-α, IL-1β, IL-13, and TGF-β1. Subsequently, bone marrow fibrocytes and resident fibroblasts propagate and develop into myofibroblasts. Besides producing IL-13, IL-25, and TSLP to stimulate profibrotic Th2 responses, the injured epithelial cells also undergo EMT to produce fibroblasts and myofibroblasts, which secrete ECM elements for fibrosis formation. TGF-β1 can also promote EMT and induce Th17 cell differentiation. T cells facilitate Th2 differentiation by producing IL-25 and IL-21. The Th2 cells produce IL-13, enabling profibrotic macrophages to release TGF-β1 and other mediators. The Th2 cells also activate the activity of collagen-secreting fibrocytes from bone marrow. As a result, myofibroblasts form to unlock ECM elements and, therefore, lead to PF development (83–85).

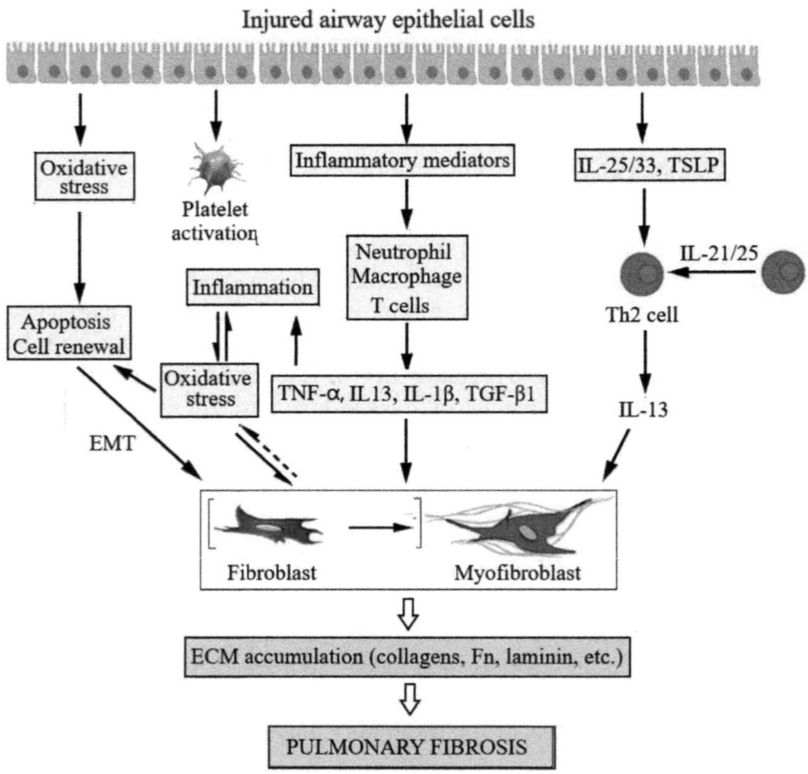

FIGURE 13.12 The pathogenesis of PF.

13.6.2 The Mechanism of Action of Some Ingredients of TCM in Treating PF

According to the pathology of PF, the therapeutic strategy includes antioxidative stress and anti-inflammation, inhibiting macrophages activation, inhibiting EMT, modulating MMP/inhibitor of TIMP balance, improving the ECM deposition, stimulating autophagy, modulating immune balance, inhibiting angiogenesis, and regulating related signaling pathways. These therapeutic functions are achieved through various active herbal ingredients in TCM. Several ingredients have been confirmed to have efficacy in treating PF in animal models in preclinical or clinical practice (83, 84). Some commonly encountered compounds in literature include terpenoids (triptolide, celastrol, tanshinone IIA, glycyrrhizic acid, andrographolide, glaucocalyxin A, and saikosaponin), flavonoids (baicalein and baicalin, quercetin, hydroxysafflor yellow A, puerarin, icariin, isoliquiritigenin, scutellarin, hesperidin, Kaempferol, and calycosin), phenolics (resveratrol, curcumin, EGCG, gallic acid, salvianolic acid B, juglanin, ellagic acid, and zingerone), glycosides (salidroside and astragalin), alkaloids (neferine, tetrahydropalmatine, oxymatrine, matrine, berberine, stemona alkaloid, and ligustrazine), saponins (astragaloside IV, dioscin, and ginsenside), quinonoids (emodin and polydatin), and others (osthole and schisandrin B).

For example, baicalein can attenuate TGF-β1/Smad signaling to improve antioxidant activity and alleviate inflammation by repressing miR-21 expression (86). Baicalein represses TGF-β1-induced fibroblast differentiation through inhibition of miR21 expression, thereby inhibiting α-SMA filament formation in fibroblasts (87). Baicalein also alleviates TGF-β1-induced Col-I production in fibroblasts via downregulation of CTGF (88). Baicalin can diminish the levels of IL-1β, IL-6, TNF-α, and TGF-β1 in lung tissues, inhibit the expression of α-SMA and vimentin, and increase the expression of E-cadherin. These results could be attributed to inhibition of the TLR4/NF-κB signaling pathway (89). Evidence proved that the effective inhibition of PF by baicalin is partially attributed to enhancing adenosine A2a receptor and then downregulating the TGF-β1-induced ERK1/2 signaling pathway (90). The major functions of the natural ingredients in anti–pulmonary fibrosis and the related targets involved are illustrated in Figure 13.13 (83, 85).

The potential therapeutic targets of PF may include antioxidative pathways to alleviate oxidative stress (e.g., inhibition of ROS, NOX, iNOS and NO); cell signaling pathways (e.g., the PI3K/Akt/mTOR pathway, G-protein coupled receptors, Rho-associated coiled-coil-forming protein kinase [RhoA/ROCK] pathway, and MAPK/JNK pathway); cytokines and chemokines (e.g., CCL2, TGF-β, and IL-13); EMT pathways (Wnt signaling pathway); growth factors (e.g., CTGF, PDGF, VEGF, and FGF, and TGF-β1); transcription factors (e.g., FoxOs, HSP 90); and others (e.g., Pentraxin 2 (PTX-2), galectin-3, and β-galactoside-binding lectin, integrin αvβ6) (84).

13.6.3 Examples of Commonly Used TCM Formulas in the Treatment of PF

According to the dialectical treatment theory of TCM, the physicians always select different formulas and add or subtract parts of components of the formula for specific individual treatment. For example, modified Maimendong (ophiopogon japonicus) decoction for patients with deficiency of both Qi and Yin (气阴两虚), Qingzao Jiufei (清燥救肺) decoction for patients with lung injury caused by pathogenic dryness (燥邪伤肺), modified Bufei (补肺) decoction for patients with deficiency of lung Qi (肺气亏虚), modified Buyang Huanwu (补阳还五) decoction for patients with Qi deficiency and blood stasis (气虚血瘀), and so on.

Here we introduce Buyang Huanwu decoction as an example to explain the principle of treatment of PF. Buyang Huanwu decoction, as a classic prescription in the *Correction on the Errors of Medical Works* (医林改错) by Wang Qingren, a doctor in the Qing Dynasty, is widely used in the treatment of IPF based on its unique curative effect of invigorating Qi and promoting blood circulation to remove meridian obstruction. Buyang Huanwu decoction is composed of astragari radix, ligusticum wallichii, pheretima aspergillum, persicae semen, angelica sinensis, paeoniae radix rubra, and safflower (91).

There is increasing literature reporting that both Buyang Huanwu decoction and its components have certain anti-fibrotic activities, and its mechanism of action is involved in improvement of lung pathology, anti-inflammation, antioxidant, regulation of ECM metabolism, anti-thrombotic effect,

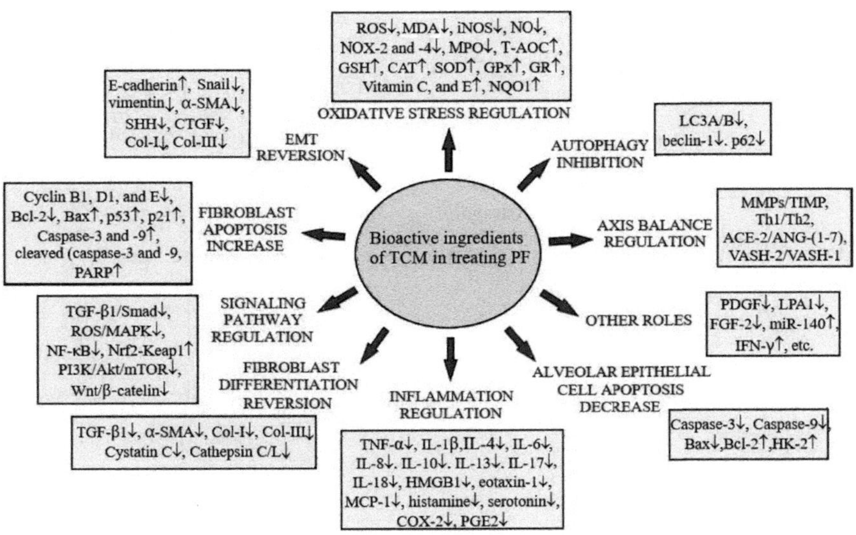

FIGURE 13.13 Emerging studies have shown the anti-fibrotic effects of the active ingredients from TCM.

regulation of angiogenesis, intervention of autophagy and apoptosis via regulation of many signaling pathways (91, 92).

Astragari radix as a monarch drug in Buyang Huanwu decoction, contains many groups of active ingredients, such as flavonoid, saponin, flavonoid polysaccharides and alkaloids. Astragari radix can regulate the metabolism of ECM, inhibiting the expression of α-SMA, Col-I and Col III. Astragaloside IV (Figure 13.14) can inhibit Smad signaling by activating miR-21 to decrease the production of EMT. Astragaloside IV also inhibit PI3K/Akt/mTOR signaling to increase autophagy via decreasing caspase-1, IL-1β, and IL-18 (92, 93). Astragalin (Figure 13.14) is effective in ameliorating oxidative stress-associated PF through disturbing airway EMT and epithelial autophagic stress (94).

Angelica sinensis, as a ministerial drug in the decoction, can protect pulmonary vascular endothelial cells from the synthesis and release of vasoconstrictor and pro-inflammatory factors, and hypercoagulable state of blood. Angelica sinensis exerts its function by decreasing the expression of CTGF and the downregulation of TGF-β1/NF-κB (86, 88). Ligustrazine (Figure 13.14), one of the major active ingredients in ligusticum wallichii, has the function of blocking PI3K/Akt/mTOR and hedgehog signaling pathways to reduce oxidative stress via increasing miR-193 expression and autophagy. Ligustrazine also activates Nrf2/HO-1 signaling. Paeoniae radix rubra contains total glycosides of peony, as well as tannin, flavonoid and volatile oils. The major content paeoniflorin (Figure 13.14) attenuates the infiltration of inflammatory cells and the deposition of ECM, and decreases the contents of hydroxyproline, Col-I, and α-SMA in myofibroblasts and ECM. Paeoniflorin could contribute to the downregulation of TGF-β1/Smad pathway and increase in the expression of IFN-γ (95).

Safflower (Figure 13.14) contains components such as safflower yellow pigment, hydroxysafflor yellow A, and total flavonoid of safflower. The target of safflower may be TGF-β1 signaling. It is proved that safflower yellow pigment can alleviate the increase of hydroxyproline content in the lung tissue of PF, inhibit the expression of TGF-β1, and increase α-SMA in stimulated fibroblasts (96). Amygdalin (Figure 13.14) is the main active ingredient of persicae semen. Pheretima aspergillum upregulates Nrf2 signaling pathway and inhibits TGF-β1 and the overexpression of HMGB1.

In the final analysis, Buyang Huanwu decoction effectively attenuates PF via processes such as regulating several signaling pathways (e.g., ERK, TGF-β1/Smad, HMGF/RAGE, PI3K/Akt,

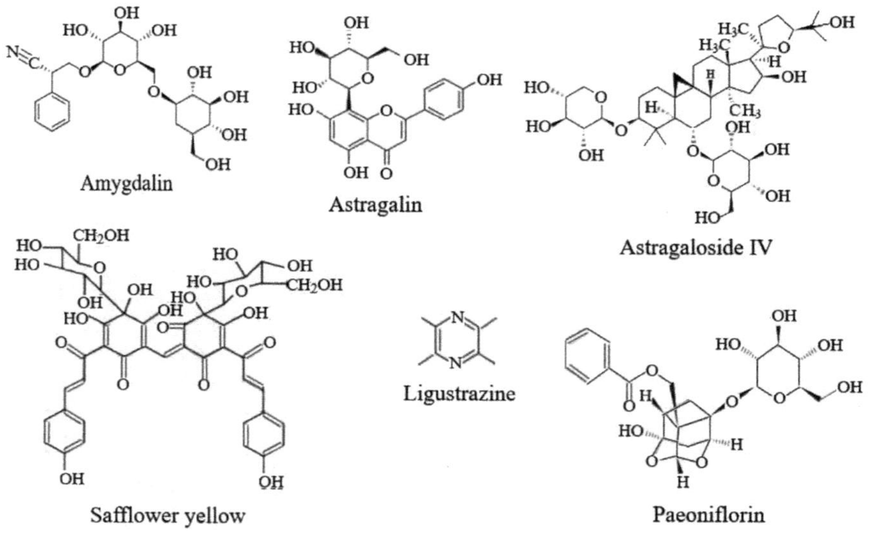

FIGURE 13.14 Some active ingredients in the Buyang Huanwu decoction

mTOR, DLL4/Notch4, and Nrf2/HO-2 signaling pathways); regulating ECM metabolism; inhibiting expression of α-SMA; differentiation of fibroblasts; and inflammation and oxidation (92, 97).

13.7 CONCLUSIONS

In conclusion, almost all TCM formulas are composed of multiple herbs, each with multiple chemical ingredients. One herb can appear in multiple formulas, and one chemical ingredient can also appear in multiple herbs. One compound can have multiple targets of action, and multiple compounds can also act on the same target concurrently. Therefore, TCM prescriptions have the characteristics of multiple composition, targets, pathways, and functions. In the prescriptions, herbs have both their own role and synergistic and mutually restrictive effects between each other for the maximum therapeutic effect and minimum toxicity. Moreover, physicians can select specific prescriptions based on the patient's specific symptoms and adjust the composition of the prescriptions to achieve personalized treatment for the goal of treating both symptoms and root causes.

The active chemical ingredients in herbs cover many classes of compounds, mainly flavonoids, terpenes, saponins, alkaloids, and phenols. Modern scientific research has found that these compounds act on various inflammatory cells, immune cells, and even structural cells. They can directly or indirectly act on biomolecules in the body, such as enzymes and mediators of signal pathways, playing roles of anti-inflammation, antioxidation, immunomodulation, anti-airway remodeling, anti–pulmonary fibrosis, regulation of water, electrolytes, acid-base balance, affecting proliferation and differentiation, and so on. They can simulate antagonists and agonists of some hormones. As part of the treatment strategy, combining with Western medicine may achieve more perfect results. Therefore, using modern scientific theories and methods to deeply explore, explain, and develop traditional Chinese medicine, and integrating TCM and Western medicine will bring greater benefits to humanity.

ABBREVIATIONS

ACE-2	angiotensin-converting enzyme 2
ACTH	adrenocorticotropic hormone
AHR	airway hyperresponsiveness
ALI	acute lung injury
AMPK	5′-AMP activated protein kinase
AP-1	activator protein-1
ARDS	acute respiratory distress syndrome
Arg-1	arginase-1
BALF	bronchoalveolar lavage fluid
3CLpro	3-chymotripsin-like protease
CDK2	cyclin-dependent kinase 2
CHOP	enhancer binding protein (c/ebp) homologous protein
CLCA1	calcium-activated chloride channel
CMI	Chinese medicine injection
COL1A1	collagen type I alpha 1 chain
Col-I	collagen-I
Col-III	collagen-III
COPD	chronic obstructive pulmonary disease
COVID-19	disease caused by SARS-CoV-2 infection
COX-2	cyclooxygenase-2
CS	cigarette smoke
CSE	cigarette smoke extract
CTGF	connective tissue growth factor
DLL4	delta-like 4

ECM	extracellular matrix
EGCG	epigallocatechin-3-gallate
EGFR	epithelial growth factor receptor
EMT	epithelial-mesenchymal transition
ERK	extracellular signal-regulated kinase
FAK	focal adhesion kinase
FDP	fibrinogen degradation product
FGF	fibroblast growth factor
FiO_2	fraction inspiration oxygen
Fn	fibronectin
FOXO3a	forkhead box O3a
FOXP3	forkhead box protein 3
GBZKD	Guben Zhike decoction
GM-CSF	granulocyte-macrophage colony–stimulating factor
GPx	glutathione peroxidase
GSH	reduced glutathione
HDAC1	histone deacetylase 1
HIF-1α	hypoxia inducible factor
HMGB	high-mobility group box 1
HO-1	heme oxygenase 1
HPA	axis hypothalamic-pituitary-adrenal axis
HSP72	heat shock protein 72
Hyp	hydroxyproline
ICAM-1	intercellular adhesion molecule-1
IgE	immunoglobulin E
IL-6	interleukin-6
ILC2s	innate lymphoid cells 2
INFγ	interferon-γ
iNOS	inducible nitric oxide synthase
IPF	idiopathic pulmonary fibrosis
JAK	Janus kinase
JNK	c-Jun *N*-terminal kinase
KGF	keratinocyte growth factor
KGFR	keratinocyte growth factor receptor
LPS	lipopolysaccharide
MAPK	mitogen-activated protein kinase
MCP-1	monocyte chemokine protein–1
MDA	malondialdehyde
MIP-1α	macrophage inflammatory protein–1α
miR34a	micro-RNA34a
MMP	matrix metalloproteinase
M^{pro}	Main protease
mTOR	mammalian target of rapamycin
MUC5AC	mucin 5AC
MyD88	myeloid differentiation factor 88
NF-κB	nuclear factor kappa B
NLRP	nucleotide binding oligomerization domain-like receptor
NO	nitric oxide
NOX	NADPH oxidase
NQO1	NAD(P)H:quinone oxidoreductase 1
Nrf2	nuclear factor erythroid-2-related factor 2

Nsp14	nonstructural protein 14, N-terminal 3′-5′ exonuclease
Nur77	phosphorylation of nuclear receptor 4A1
PAI-1	plasminogen activator inhibitor type-1
PAK1	serine/threonine p21 protein activated kinase-1
PaO_2	partial pressure of oxygen
PDE	phosphodiesterase
PDGF	platelet-derived growth factor
PF	pulmonary fibrosis
PGC-1α	peroxisome proliferators-activated receptor γ coactivator 1α
PGD2	prostaglandin D2
PI3K	phosphatidylinositol 3-kinase
PLA2	phospholipase A2
PLpro	papain-like protease
$PM_{2.5}$	particulate matter with a diameter of 2.5 μm or less
PPARγ	peroxisome proliferative activated receptor γ
PTEN	phosphatase and tensin homolog
QFLT	Qingfei Litan decoction
RAGE	receptor of advanced glycation end-product
RdRp	RNA-dependent RNA polymerase
ROCK	rho-associated coiled-coil-forming protein kinase
ROR-alpha	retinoic acid receptor-related orphan receptor-alpha
ROS	reactive oxygen species
SARS-CoV	severe acute respiratory syndrome coronavirus
SASP	senescence-related secretory phenotype
SGp	spike glycoprotein
SHH	sonic hedgehog
SIRT1	sirtuin 1
α-SMA	α-smooth muscle actin
SMC	smooth muscle cell
SOD	superoxide dismutase
STAT	signal transducer and activator transcription
TAK1	transforming growth factor–β activated kinase–1
TCM	traditional Chinese medicine
TGF-β1	transforming growth factor–β1
Th17	T helper 17 cell
TIMP	tissue inhibitor of metalloproteinase
TLR4	toll-like receptor–4
TMPRSS2	serine protease transmembrane protease serine 2
TNF-α	tumor necrosis factor–α
TPCs	two-pore channels
Treg cell	regulatory T cell
TRPA1	transient receptor potential cation channel subfamily A member 1
TRPV1	transient receptor potential cation channel subfamily V member 1
TSLP	thymic stromal lymphopoietin
TUNEL	terminal deoxynucleotidyl transferase-mediated dUTP nick end labeling
VCAM-1	vascular cell adhesion molecule-1
VEGF	vascular endothelial growth factor
Wnt	wingless-type MMTV integration site family
XBCQ	Xuanbai Chengqi decoction
ZEB2	zinc finger E-box-binding homeobox 2
ZNH263	zinc finger protein 263

REFERENCES

1. Khan, T., Khan, M. A., Mashwani, Z. U. R., Ullah, N., and Nadhman, A. (2021) Therapeutic potential of medicinal plants against COVID19: The role of antiviral medicinal metabolites. *Biocatal Agric Biotechnol.* **31**, 101890
2. Chu, H. Y., Shi, Y., Jiang, S., Zhong, Q. C., Zhao, Y. Q., Liu, Q. M., *et al.* (2017) Treatment effects of the traditional Chinese medicine Shenks in bleomycin-induced lung fibrosis through regulation of TGF-beta/Smad3 signaling and oxidative stress. *Sci Rep.* **7**, 2252
3. Zhang, Z. J. (2022) Diagnosis and treatment protocol for COVID-19 patients (Trial Version 9). *Health Care Sci.* **1**, 14–28
4. Lyu, M., Fan, G. W., Xiao, G. X., Wang, T. Y., Xu, D., Gao, J., *et al.* (2021) Traditional Chinese medicine in COVID-19. *Acta Pharm Sin B.* **11**(11), 3337–3363
5. Wang, Z. Y., Xia, Q., Liu, X., Liu, W. X., Huang, W. Z., Mei, X., *et al.* (2018) Phytochemistry, pharmacology, quality control and future research of *Forsythia suspensa* (Thunb.) Vahl: A review. *J Ethnopharmacol.* **210**(10), 318–339
6. Kang, X. M., Jin, D., Jiang, L. L., Zhang, Y. Q., Zhang, Y. H., An, X. D., *et al.* (2022) Efcacy and mechanisms of traditional Chinese medicine for COVID-19: A systematic review. *Chin Med.* **17**, 30
7. Jin, G. Y., Jin, L. L., and Jin, B. X. (2021) The rationale behind the four major anti-COVID-19 principles of Chinese herbal medicine based on systems medicine. *Acupunct Herbal Med.* **1**(2), 90–98
8. Rahman, Md. M., Bibi, S., Rahaman, Md. S., Rahman, F., Islam, F., Khan, M. S., *et al.* (2022) Natural therapeutics and nutraceuticals for lung diseases: Traditional significance, phytochemistry, and pharmacology. *Biomed Pharmacother.* **150**, 113041
9. Huang, F. F., Li, Y., Leung, E. L. H., Liu, X. H., Liu, K. F., Wang, Q., *et al.* (2020) A review of therapeutic agents and Chinese herbal medicines against SARS COV-2 (COVID-19). *Pharm Res.* **158**, 104929
10. Wang, J. Q., Zhao, H. Y., and An, Y. Z. (2022) ACE2 shedding and the role in COVID-19. *Front Cell Infect Microbiol.* **11**, 789180
11. Mandal, A., Jha, A. K., and Hazra, B. (2021) Plant products as inhibitors of coronavirus 3CL protease. *Front Pharm.* **12**, 583387
12. Berretta, A. A., Silveira, M. A. D., Capcha, J. M. C., and Jong, D. D. (2020) Propolis and its potential against SARS-CoV-2 infection mechanisms and COVID-19 disease running title: Propolis against SARS-CoV-2 infection and COVID-19. *Biomed Pharmacother.* **131**, 110622
13. Agrawal, P. K., Agrawal, C., and Blunden, G. (2021) Naringenin as a possible candidate against SARS-CoV-2 infection and in the pathogenesis of COVID-19. *Nat Prod Commun.* **16**(12), 1–16
14. Liu, Z. Y., Li, S. Y., Zhao, W. J., and Zhou, G. (2021) A systematic review and meta-analysis of effect evaluation of traditional Chinese medicine in treating acute respiratory distress syndrome. *Ann Palliat Med.* **10**(5), 5520–5532
15. Chen, Y. B., Liu Q., Han, X., Yin S. M., Wu, L., Yu, X. H., *et al.* (2020) Is Chinese medicine injection applicable for treating acute lung injury and acute respiratory distress syndrome? A systematic review and meta-analysis of randomized controlled trials. *Chin J Integr Med.* **26**(11), 857–866
16. Zhang, Y. T., Gao, Z., Jiang, F., Yan, H., Yang, B., He, Q., Luo, P., Xu, Z., and Yang, X. (2023) JAK-STAT signaling as an ARDS therapeutic target: Status and future trends. *Biochem Pharmacol.* **208**, 115382
17. Luo, Z. Y., Dong, J. C., and Wu, J. F. (2022) Impact of Icariin and its derivatives on inflammatory diseases and relevant signaling pathways. *Int Immunopharmacol.* **108**, 10886
18. Li, W. L., Li, D., Chen, Y. S., Abudou, H., Wang, H. W., Cai, J. X., *et al.* (2022) Classic signaling pathways in alveolar injury and repair involved in sepsis-induced ALI/ARDS: New research progress and prospect. *Dis Markers.* 6362344
19. Kumari, S., and Singh, R. (2022) Protective effects of intranasal curcumin on silica-induced lung damage. *Cytokine.* **157**, 155949
20. Liu, C., Xiao, K., and Xie, L. X. (2022) Advances in the regulation of macrophage polarization by mesenchymal stem cells and implications for ALI/ARDS treatment. *Front Immunol.* **13**, 928134
21. Li, Q., Hu, X. P., Sun, R. H., Tu, Y. X., Gong, F. X., and Ni, Y. (2016) Resolution acute respiratory distress syndrome through reversing the imbalance of Treg/Th17 by targeting the cAMP signaling pathway. *Mol Med Rep.* **14**, 343–348
22. Liu, B. H., He, R. Y., Zhang, L., Hao, B., Jiang, W. Y., Wang, W., and Geng, Q. (2021) Inflflammatory caspases drive pyroptosis in acute lung injury. *Front Pharmacol.* **12**, 631256

23. Wang, G. Z., Han, D., Zhang, Y. H., Xie, X. M., Wu, Y. Y., Li, S. J., *et al.* (2013) A novel hypothesis: Up-regulation of HO-1 by activation of PPARγ inhibits HMGB1-RAGE signaling pathway and ameliorates the development of ALI/ARDS. *J Thorac Dis.* **5**(5), 706–710
24. Zhu, H. H., Wang, S., Shan, C., Li, X. Q., Tan, B., Chen, Q. L., *et al.* (2021) Mechanism of protective effect of xuan-bai-cheng-qi decoction on LPS-induced acute lung injury based on an integrated network pharmacology and RNA-sequencing approach. *Respir Res.* **22**, 188
25. Diao, Y. R., Ding, Q., Xu, G. H., Li, Y. D., Li, Z. Q., Zhu, H. P., *et al.* (2022) Qingfei Litan decoction against acute lung injury/acute respiratory distress syndrome: The potential roles of anti-inflammatory and anti-oxidative effects. *Front Pharmacol.* **13**, 857502
26. Sun, L. C., Zhang, H. B., Gu, C. D., GuoS. D., Li, G., Lian, R., *et al.* (2018) Protective effect of acace, tin on sepsis-induced acute lung injury via its anti-inflflammatory and antioxidative activity. *Arch Pharm Res.* **41**, 199–210
27. Luo, L., Jiang, J. W., Wang, C., Fitzgerald, M., Hu, W. F., Zhou, Y. M., *et al.* (2020) Analysis on herbal medicines utilized for treatment of COVID-19. *Acta Pharm Sin B.* **10**(7), 1192–1204
28. Deng, Y. L., Ye, X. W., Chen, Y. F., Ren. H. M., Xia, L. T., Liu, Y., *et al.* (2021) Chemical characteristics of platycodon grandiflflorum and its mechanism in lung cancer treatment. *Front Pharmacol.* **11**, 609825
29. Pu, W. L., Bai, R. Y., Zhou, K., Peng, Y. F., Zhang, M. Y., Hottiger, M. O., *et al.* (2019) Baicalein attenuates pancreatic inflammatory injury through regulating MAPK, STAT 3 and NF-κB activation. *Int Immunopharmacol.* **72**, 204–210
30. Kim, Y. W., Zhao, R. J., Park, S. J., Lee, J. R., Cho, I. J., Yang, C. H., *et al.* (2008) Anti-inflammatory effects of liquiritigenin as a consequence of the inhibition of NF-κB-dependent iNOS and proinflammatory cytokines production. *Br J Pharmacol.* **154**, 165–173
31. Wang, W. Q., Yao, Q., Teng, F. Z., Cui, J., Dong, J. C., and Wei, Y. (2021) Active ingredients from Chinese medicine plants as therapeutic strategies for asthma: Overview and challenges. *Biomed Pharmacother.* **137**, 111383
32. Zhong, Y. Y., Hu, L. L., Chen, W. J., Wang, B., Sun, J., and Dong, J. C. (2022) Exploring the comorbidity mechanisms between asthma and idiopathic pulmonary fibrosis and the pharmacological mechanisms of Bu-Shen-Yi-Qi decoction therapy via network pharmacology. *BMC Complement Med Ther.* **22**, 151
33. Liu, J. X., Zhang, Y., Yuan, H. Y., and Liang, J. (2021) The treatment of asthma using the Chinese Materia Medica. *J Ethnopharmacol.* **269**, 113558
34. Wong, L. H., Tay, L., Robby Goh, M. W. J., Tan, T. J., Zhou, R. S., Ho, A. K. H, *et al.* (2021) Systematic review: Guideline-based approach for the management of asthma and subtypes via Chinese medicine. *Evid Based Complementary Altern Med.* 4319657
35. Alharris, E., Alghetaa, H., Seth, R., Chatterjee, S., Singh, N. P., Nagarkatti, M., and Nagarkatti, P. (2018) Resveratrol attenuates allergic asthma and associated inflammation in the lungs through regulation of miRNA-34a that targets FoxP3 in mice. *Front Immunol.* **9**, 2992
36. Jafarinia, M., Hosseini, M. S., Kasiri, N., Fazel, N., Fathi, F., Hakemi, M. G., *et al.* (2020) Quercetin with the potential effect on allergic diseases. *Allergy Asthma Clin Immunol.* **16**, 36
37. Zhou, B. W., Liu, H. M., and Jia, X. H. (2022) The role and mechanisms of traditional Chinese medicine for airway inflflammation and remodeling in asthma: Overview and progress. *Front Pharmacol.* **13**, 917256
38. Wang, J., Wu, Q. B., Ding, L., Song, S. Y., Li, Y. X., Shi, L., *et al.* (2021) Therapeutic effects and molecular mechanisms of bioactive compounds against respiratory diseases: Traditional Chinese medicine theory and high-frequency use. *Front Pharmacol.* **12**, 734450
39. Schuliga, M. (2015) NF-kappaB signaling in chronic inflammatory airway disease. *Biomolecules.* **5**, 1266–1283
40. Borghi, S. M., Zaninelli, T. H., Carra, J. B., Heintz, O. K., Baracat, M. M., Georgetti, S. R., *et al.* (2023) Therapeutic potential of controlled delivery systems in asthma: Preclinical development of flavonoid-based treatments. *Pharmaceutics.* **15**, 1
41. Amaral-Machado, L., Oliveira, W. N., Moreira-Oliveira, S. S., Pereira, D. T., Alencar, É. N., Nicolas, T., *et al.* (2020) Use of natural products in asthma treatment. *Evid Based Complementary Altern Med.* 1021258
42. Zhu, T., Chen, Z. H., Chen, G. H., Wang, D. X., Tang, S., Deng, H. J., *et al.* (2019) Curcumin attenuates asthmatic airway inflammation and mucus hypersecretion involving a PPARγ-dependent NF-κB signaling pathway in vivo and in vitro. *Mediators Inflamm.* 4927430

43. Ma, C. H., Ma, Z. Q., Fu, Q., and Ma, S. P. (2013) Curcumin attenuates allergic airway inflammation by regulation of CD4+CD25+ regulatory T cells (Tregs)/Th17 balance in ovalbumin-sensitized mice. *Fitoterapia.* **87**, 57–64
44. Yang, X., J. N., Li, H., Jiao, B., Zhang, Q. H., Zhang, Y., *et al.* (2017) Curcumin reduces lung inflammation via Wnt/β-catenin signaling in mouse model of asthma. *J Asthma.* **54**(4), 335–340
45. Jia, X. X., Zhu, T. T., Huang, Y., Zeng, X. X., Zhang, H., and Zhang, W. Xi. (2019) Wnt/β-catenin signaling pathway regulates asthma airway remodeling by influencing the expression of c-Myc and cyclin D1 via the p38 MAPK-dependent pathway. *Exp Ther Med.* **18**, 3431–3438
46. Chauhan, P. S., Dash, S. D., and Singh, R. (2014) Intranasal curcumin attenuates airway remodeling in murine model of chronic asthma. *Int Immunopharmacol.* **21**, 63–75
47. Kumari, A., Singh, D. K., Dash, D., and Singh, R. (2019) Intranasal curcumin protects against LPS-induced airway remodeling by modulating toll-like receptor-4 (TLR-4) and matrix metalloproteinase-9 (MMP-9) expression via affecting MAP kinases in mouse model. *Inflammopharmacology.* **27**(4), 731–748
48. Jaiswal, A., Dash, D., and Singh, R. (2022) Intranasal curcumin and dexamethasone combination ameliorates inflammasome (NLRP3) activation in lipopolysaccharide exposed asthma exacerbations. *Toxicol Appl Pharmacol.* **436**, 115861
49. Li, X. N., Ma, L. Y., Ji, H., Qin, Y. H., Jin, S. S., and Xu, L. X. (2018) Resveratrol protects against oxidative stress by activating the Keap-1/Nrf2 antioxidant defense system in obese-asthmatic rats. *Exp Ther Med.* **16**, 4339–4348
50. Yun, J. M., Chien, A., Jialal, I., and Devaraj, S. (2012) Resveratrol upregulates SIRT1 and inhibits cellular oxidative stress in the diabetic milieu: Mechanistic insights. *J Nutr Biochem.* **23**(7), 699–705
51. Hu, X., Wang, J., Xia, Y., Simayi, M., Ikramullah, S., He, Y. B., *et al.* (2016) Resveratrol induces cell cycle arrest and apoptosis in human eosinophils from asthmatic individuals. *Mol Med Rep.* **14**, 5231–5236
52. Ni, Z. H., Tang, J. H., Chen, G., Lai, Y. M., Chen, Q. G., Li, Z., *et al.* (2016) Resveratrol inhibits mucus overproduction and MUC5AC expression in a murine model of asthma. *Mol Med Rep.* **13**, 287–294
53. Kim, H. Y. (2017) Resveratrol in asthma: A French paradox? *Allergy Asthma Immunol Res.* **9**(1), 1–2
54. Chung, K. F., and Adcock, I. M. (2008) Multifaceted mechanisms in COPD: Inflammation, immunity, and tissue repair and destruction. *Eur Respir J.* **31**, 1334–1356
55. Ma, Q., Zhang, A. N., and Zhang, C X. (2022) Exploration of the pharmacological mechanism of Bufei Nashen pill in treating chronic obstructive pulmonary disease using network pharmacology integrated molecular docking. *Nat Prod Commun.* **17**(11), 1–17
56. Li, J. S. (2020) International clinical practice guideline of Chinese medicine: Chronic obstructive pulmonary disease. *World J Tradit Chin Med.* **6**(1), 39–50
57. Cao, X., Wang, Y., Chen, Y., Zhao, M. T., Liang, L. Y., Yang, M. R., *et al.* (2023) Advances in traditional Chinese medicine for the treatment of chronic obstructive pulmonary disease. *J Ethnopharmacol.* **307**, 116229
58. Guan, S. P., Tee, W., Ng, D. S. W., Chan, T. K., Peh, H. Y., Ho, W. E., *et al.* (2013) Andrographolide protects against cigarette smoke-induced oxidative lung injury via augmentation of Nrf2 activity. *Br J Pharmacol.* **168**, 1707–1718
59. Zhou, M. Q., Zhuo, L. Y., and Cai, C. (2018) Astragaloside IV inhibits cigarette smoke-induced pulmonary inflammation in mice. *Inflammation.* **41**, 1671–1680
60. Yang, Y., Jin, X., Jiao, X-Y, Li, J-J., Liang, L-Y., Ma, Y-Y., *et al.* (2020) Advances in pharmacological actions and mechanisms of flavonoids from traditional Chinese medicine in treating chronic obstructive pulmonary disease. *Evid Based Complementary Altern Med.* 8871105
61. Chen, Z., Chen, P, Wu, H., Shi, R., Su, W. W., Wang, Y. G., and Li, P. B. (2020) Evaluation of naringenin as a promising treatment option for COPD based on literature review and network pharmacology. *Biomolecules.* **10**, 1644
62. Lin, L., Hou, G., Han, D., Kang, J., and Wang, Q. Y. (2019) Ursolic acid protected lung of rats from damage induced by cigarette smoke extract. *Front Pharmacol.* **10**, 700
63. Mitani, A., Azam, A., Vuppusetty, C., Ito, K., Mercado, N., and Barnes, P. J. (2017) Quercetin restores corticosteroid sensitivity in cells from patients with chronic obstructive pulmonary disease. *Exp Lung Res.* **43**(9–10), 417–425
64. Li, L. Y., Zhang, C. T., Zhu, F. Y., Zheng, G., Liu, Y.-F., and Liu, K. (2022) Potential natural small molecular compounds for the treatment of chronic obstructive pulmonary disease: An overview. *Front Pharmacol.* **13**, 821941

65. Du, L. M., Huo, X. L., and Liu, Y. (2022) Regulatory effect of resveratrol on the expressions of factors and surface markers in alveolar macrophages of *Aspergillus fumigatus*-infected COPD rat model, and the associated mechanism. *Trop J Pharm Res.* **21**(10), 2139–2146
66. Wang, W., Wu, W., Wang, B., and Gao, F. (2021) Effect of houttuynia on improving lung injury in chronic obstructive pulmonary disease by regulating the TLR4 signaling pathway. *Food Sci Nutr.* **9**, 3389–3396
67. Li, Q. P., Zhang, H. Y., Yan, X. P., Zhao, Z. X., Qiu, J., Hu, L. L., Jiang, S., *et al.* (2023) Up-regulation of PPAR-γ involved in the therapeutic effect of icariin on cigarette smoke-induced inflammation. *Pulm Pharmacol Ther.* **79**, 102197
68. Hao, D. X., Li, Y. S., Shi, J., and Jiang, J. G. (2021) Baicalin alleviates chronic obstructive pulmonary disease through regulation of HSP72-mediated JNK pathway. *Mol Med.* **27**, 53
69. Wang, G. F., Mohammadtursun, N., Lv, Y. B., Zhang, H. Y., Sun, J., and Dong, J.-C. (2018) Baicalin exerts anti-airway inflammation and anti-remodelling effects in severe stage rat model of chronic obstructive pulmonary disease. *Evid Based Complementary Altern Med.* 7591348
70. Zhang, H., Liu, B. J., Jiang, S., Wu, J. F., Qi, C. H., Mohammadtursun, N., *et al.* (2021) Baicalin ameliorates cigarette smoke-induced airway inflammation in rats by modulating HDAC2/NF-κB/PAI-1 signalling. *Pharmacol Ther.* **70**, 102061
71. Guan, Y., Li, F. F., Hong, L., Yan, X. F., Tan, G. L., He, J. S., *et al.* (2012) Protective effects of liquiritin apioside on cigarette smoke-induced lung epithelial cell injury. *Fundam Clin Pharmacol.* **26**(4), 473–483
72. Qin, J. Y., Chen, J. R., Peng, F., Sun, C, Lei Yu, Chen, G-R., *et al.* (2022) Pharmacological activities and pharmacokinetics of liquiritin: A review. *J Ethnopharmacol.* **293**, 115257
73. Liu, Z. H., Wang, P. W., Lu, S. S., Guo, R., Gao, W., Tong, H. Y., *et al.* (2020) Liquiritin, a novel inhibitor of TRPV1 and TRPA1, protects against LPS induced acute lung injury. *Cell Calcium.* **88**, 102198
74. Wang, Z. Y., Fang, K. Y., Wang, G. Q., Guan, X. W., Pang, Z. Q., Guo, Y. Q., *et al.* (2019) Protective effect of amygdalin on epithelial—mesenchymal transformation in experimental chronic obstructive pulmonary disease mice. *Phytother Res.* **33**, 808–817
75. Zhao, P., Liu, X. F., Dong, H. R., Tian, Y. Ga., Feng, S. X., Zhao, D., *et al.* (2020) Bufei Yishen formula restores Th17/Treg balance and attenuates chronic obstructive pulmonary disease via activation of the adenosine 2a receptor. *Front Pharmacol.* **11**, 1212
76. Li, J. S., Zhao, P., Li, Y., Tian, Y. G., and Wang, Y. H. (2015) Systems pharmacology-based dissection of mechanisms of Chinese medicinal formula Bufei Yishen as an effective treatment for chronic obstructive pulmonary disease. *Sci Rep.* **5**, 15290
77. Li, J. S., Xie, Y., Zhao, P., Qin, Y. Q., Oliver, B. G., Tian, Y. G., *et al.* (2021) A chinese herbal formula ameliorates COPD by inhibiting the inflammatory response via downregulation of p65, JNK, and p38. *Phytomedicine.* **83**, 153475
78. Jin, F. L., Zhang, L. X., Chen, K., Miao, Y. F., Liu, Y., Tian, Y. G., and Li, J. S. (2022) Effective-component compatibility of Bufei Yishen formula III combined with electroacupuncture suppresses inflammatory response in rats with chronic obstructive pulmonary disease via regulating SIRT1/NF-κB signaling. *BioMed Res Int.* 3360771
79. Li, J. S., Wang, J., Li, Y., Zhao, P., Tian, Y. G., Liu, X. F., *et al.* (2021) Effective-component compatibility of Bufei Yishen formula protects COPD rats against PM2.5-induced oxidative stress via miR-155/FOXO3a pathway. *Ecotoxicol Environ Saf.* **228**, 112918
80. Wang, W. M., Zhang, S. H., Cui, L., Chen, Y., Xu, X. X., and Wu, L. C. (2023) Bufei Yishen formula inhibits the cell senescence in COPD by up-regulating the ZNF263 and Klotho expression. *Int J Chron Obstruct Pulmon Dis.* **18**, 533–539
81. Wang, M. Z., Liu, G. X., Xiao, Y., Cai, Z., Liu, C., Pan, L., *et al.* (2021) Proteomic analysis of a chronic obstructive pulmonary disease mouse model to determine the efficacy of treatment using Guben Zhike Decoction. *J Tradit Chin Med Sci.* **8**, 34–42
82. Wang, Y. Q., Liao, Q., Tang, S. H., Cai, Z., and Zhang, H. C. (2020) Gubenzhike recipe ameliorates respiratory mucosal immunity in mice with chronic obstructive pulmonary disease through upregulation of the γδT lymphocytes and KGF levels. *Evid Based Complementary Altern Med.* 3056797
83. Li, L. C., and Kan, L. D. (2017) Traditional Chinese medicine for pulmonary fibrosis therapy: Progress and future prospects. *J Ethnopharmacol.* **198**, 45–63
84. Hasan, M., Paul, N. C., Paul, S. K., Saikat, A. S. M., Akter, H., Mandal, M., and Lee, S.-S. (2022) Natural product-based potential therapeutic interventions of pulmonary fibrosis. *Molecules.* **27**, 1481

85. Wang, Q., Li, W. J., Hu, H. B., Lu, X. C., and Qin, S. (2023) Monomeric compounds from traditional Chinese medicine: New hopes for drug discovery in pulmonary fibrosis. *Biomed Pharmacother.* **159**, 114226
86. Gao, Y., Lu, J., Zhang, Y., Chen, Y. F., Gu, Z. L., and Jiang, X. G. (2013) Baicalein attenuates bleomycin-induced pulmonary fibrosis in rats through inhibition of miR-21. *Pulm Pharmacol Ther.* **26**(6), 649–654
87. Cui, X. J., Sun, X. H., Lu, F. Q., and Jiang, X. G. (2018) Baicalein represses TGF-β1-induced fibroblast differentiation through the inhibition of miR-21. *Toxicol Appl Pharmacol.* **358**, 35–42
88. Sun, X. G., Cui, X. J., Chen, X. H., and Jiang, X. G. (2020) Baicalein alleviated TGF β1-induced type I collagen production in lung fibroblasts via downregulation of connective tissue growth factor. *Biomed Pharmacother.* **131**, 10744
89. Zhang, Y., Liu, F., Jia, Q., Zheng, L., Tang, Q., Sai, L., *et al.* (2023) Baicalin allevia tes silica-induced lung inflammation and fibrosis by inhibiting TLR4/NF-κB pathway in rats. *Physiol Res.* **72**(2), 221–233
90. Huang, X. Y., He, Y. C., Chen, Y. F., Wu, P. L., Gui, D., Cai, H., *et al.* (2016) Baicalin attenuates bleomycin-induced pulmonary fibrosis via adenosine A2a receptor related TGF-β1-induced ERK1/2 signaling pathway. *BMC Pulm Med.* **16**, 132
91. Quan, L. X. (2023) Research progress of Buyang Huanwu decoction in the treatment of pulmonary fibrosis. *Adv Clin Med.* **13**(2), 1957–1961
92. Yang, K., Gong, X. Y., and Wang, F. (2020) Research progress on the mechanism of Buyang Huanwu decoction in the treatment of pulmonary fibrosis. *J Basic Chin Med.* **26**(7), 1034–1040 (Abstract in English)
93. Zhao, H. L., and Qu, J. L. (2021) Discussion on application of Buyang Huanwu decoction (补阳还五汤) in idiopathic pulmonary fibrosis. *Liaoning J Tradit Chin Med.* **48**(10), 53–59 (Abstract in English)
94. Cho, I. H., Choi, Y. J., Gong, J. H., Shin, D., Kang, M. K., and Kang, Y. H. (2015) Astragalin inhibits autophagy-associated airway epithelial fibrosis. *Respir Res.* **16**, 51
95. Ji, Y., Wang, T., Wei, Z. F., Lu, G. X., Jiang, S. D., Xia, Y. F., and Dai, Y. (2013) Paeoniflorin, the main active constituent of *Paeonia lactiflora* roots, attenuates bleomycin-induced pulmonary fibrosis in mice by suppressing the synthesis of type I collagen. *J Ethnopharmacol.* **149**(3), 825–832
96. Wang, L., Jin, M., Zang, B. X., and Wu, Y. (2011) Inhibitory effect of safflower yellow on pulmonary fibrosis. *Biol Pharm Bull.* **34**(4), 511–516
97. Ding, D. L., Shen, X. B., Yao, J. H., Yu, L. Z., Zhou, M. J., and Nian, S. H. (2021) Research progress on traditional Chinese medicine against treating pulmonary fibrosis. *Chin Tradit Herbal Drugs.* **52**(22), 7006–7024 (Abstract in English)

Index

C

F

G

H

I

N

O

P

Q

R

S

9781032409023